Graphene

Graphene
Fundamentals and Emergent Applications

Jamie H. Warner
Department of Materials
University of Oxford
Oxford, UK

Franziska Schäffel
Department of Materials
University of Oxford
Oxford, UK

Alicja Bachmatiuk
IFW Dresden
Helmholtzstraße 20
Dresden, Germany

Mark H. Rümmeli
IFW Dresden
Helmholtzstraße 20
Dresden, Germany

AMSTERDAM • WALTHAM • HEIDELBERG • LONDON
NEW YORK • OXFORD • PARIS • SAN DIEGO
SAN FRANCISCO • SYDNEY • TOKYO

ELSEVIER

Elsevier
225 Wyman Street, Waltham, MA 02451, USA
The Boulevard, Langford Lane, Kidlington, Oxford OX5 1GB, UK

First edition 2013

Library of Congress Cataloging-in-Publication Data
Application submitted

Warner, Jamie H.
    Graphene : fundamentals and emergent applications / Jamie H. Warner, Franziska Schäffel,
Mark Hermann Rümmeli, Alicja Bachmatiuk. – First edition.
        pages cm
    Includes bibliographical references and index.
    ISBN 978-0-12-394593-8
1. Graphene. 2. Graphene–Industrial applications. I. Schäffel, Fransizka. II. Rummeli, Mark.
III. Bachmatiuk, Alicja. IV. Title.
    QD341.H9W284 2013
    546'.681–dc23

                                2012038909

British Library Cataloguing in Publication Data
A catalogue record for this book is available from the British Library

ISBN: 978-0-12-394593-8

For information on all Elsevier publications
visit our website at www.store.elsevier.com

Printed and bound by CPI Group (UK) Ltd, Croydon, CR0 4YY

Contents

Chapter 1

Introduction

Jamie H. Warner

The Nobel prize for physics in 2010 was awarded to Sir Professor Andrei Geim and Sir Professor Kostya Novoselov, from the Manchester University, for their *'ground-breaking experiments regarding the two-dimensional material graphene'*. Reading these words carefully, there is no mention of 'discovery' and this is a debatable topic, and a similar discussion clouds the discovery of carbon nanotubes and its relation to the landmark paper of Sumio Iijima that stimulated the field of nanotubes (Iijima, 1991). In contrast, the Nobel prize in chemistry (1996) was awarded to Robert F. Curl Jr, Sir Harold W. Kroto, and Richard E. Smalley for *'their discovery of fullerenes'*, leaving no ambiguity.

Whilst it was pointed out by Professor Walt de Heer from the Georgia Tech University that numerous factual errors were made by the Nobel Committee in their scientific background document on the Nobel prize for graphene, there is no doubt that the papers in 2004 and 2005 by Novoselov et al. were instrumental in igniting the field of graphene (Novoselov et al., 2004, 2005). The Manchester group is seen as developing the 'scotch-tape' mechanical exfoliation technique for graphene production that was simple, effect, cheap, and therefore could be taken up rapidly by research groups all across the world. It is this simplicity that helped the graphene research develop at a remarkable pace and generate momentum. Although this technique had been applied for cleaving graphite for scanning tunnelling microscopy studies, there was no further development in demonstrating how it could be used to discover the superb electronic properties of graphene.

The pioneering work of Professor de Heer should be recognised, as his group developed synthetic graphene from silicon carbide precursors and undertook electronic measurements of monolayer graphene independently of the Manchester group (Berger et al., 2004). He was already aware of the wonders that graphene could offer before the 2004 report of Novoselov et al. (2004). His group published a report showing 2D electron gas properties in ultrathin epitaxial graphite films and opened a route towards scalable graphene-based nanoelectronics (Berger et al., 2004). In 2005, Professor Philip Kim's group from the Colombia University reported the observation of the quantum Hall effect and Berry's phase in graphene and extended this further with many important contributions to discovering the amazing electronic properties of

Graphene. http://dx.doi.org/10.1016/B978-0-12-394593-8.00001-1

graphene (Zhang et al., 2005). Their method for obtaining graphene was similar to that reported in Novoselov's 2004 report as is cited as such. Professor Rodney Ruoff, from the University of Texas at Austin, has also been instrumental in advancing the chemical vapour deposition growth of graphene using metal catalysts that is critical to graphene having a commercial impact. There have been numerous other leaders in the field of graphene research and far too many to mention without offending someone by unintentionally leaving them out. Instead, we celebrate all contributions to graphene research that has led to so much high impact science and helped forge/reinvigorate/establish many careers for both the young and experienced.

The boom in graphene was helped by many people who were already studying carbon nanotubes and fullerenes, simply translating their activities into this new area. The apparatus for characterising graphene is often similar to those used for nanotubes, such as transmission electron microscopy (TEM), scanning electron microscopy (SEM), electronic device fabrication, diffraction and Raman spectroscopy. This also bodes well for rapid advances in understanding the properties of new 2D crystals that may exhibit new behaviour absent in graphene.

Graphene is the building block for understanding the structure of fullerenes, nanotubes and graphite, and given the simplicity of its isolation, it is surprising it was studied last. Many people who do not work on graphene question its hype, which is fair considering the hype that surrounded the discovery of fullerenes (1985) and nanotubes (1991) and the lack of real world implementation and life-changing technology that has resulted. Perhaps the biggest difference between fullerenes and nanotubes with graphene is the manufacturing aspect. It is challenging to produce vast quantities of fullerenes that are pure, apart from C_{60} and C_{70}. Some of the most interesting and useful properties of fullerenes come by adding dopants or molecular functional groups, but this makes their separation using high-performance liquid chromatography time consuming and thus the end product expensive. Carbon nanotubes are facing criticism as being asbestos-like when inhaled. More work is needed to be sure of this as residual metal catalysts that are known to be toxic are often retained in the material. Carbon nanotubes also suffer from the problem of mixed chiralities that lead to both semiconducting and metallic electronic transport behaviour. This has limited their application in electronics, despite demonstrating their outstanding properties at the single-device level. Carbon nanotubes will continue to be intensively studied as they encompass the ideology of a 1D nanowire at its best. Graphene, on the other hand, looks to be solving these manufacturing challenges with SiC and chemical vapour depositions proving fruitful for electronic-grade graphene, whilst chemical exfoliation is excellent for solution-based processing for spray-casting and polymer blending. In order for graphene to be useful in applications, there will be a strong need to interface it with other materials, in particular semiconductor nanomaterials. The pathway for cheap, high-quality graphene suited for a variety of applications is achievable. Perhaps, it is this reason that graphene leaped ahead of carbon nanotubes in being awarded a Nobel prize, since carbon nanotubes also display amazing electronic and mechanical properties that captured the imagination and interest of the world for more than 10 years.

1.1. ABOUT THE BOOK

This book is aimed at undergraduate students towards the end of the degrees and PhD students starting out, plus anyone new entering into the field of graphene. The objective of the book is to provide the necessary basic information about graphene in a broad variety of topics. Each chapter is designed to be relatively self-contained, and as a consequence, there may be some occasions of slight repetition. The field has grown extensively over the past 9 years, and it is not possible to cover every published piece of work, and therefore, we have restricted discussions to the key findings in the field. The book starts with a description of the atomic structure of graphene, as this essentially dictates the properties of graphene. Monolayer, bilayer, trilayer and few-layer graphene are all discussed and how they stack in both AB Bernal stacking as well as Rhombohedral stacking. The relationship of graphene to nanotubes is explored, namely rolling up graphene to form a cylinder. Once an understanding of the atomic structure is obtained, we move towards describing graphene's properties: electrical, chemical, spin, mechanical and thermal. It is these amazing properties that have generated the exuberant fascination in graphene.

Any experimental researcher will need to know how to obtain graphene, and for theoretical scientists, it is important to have a real-world understanding of how graphene can be made, what can be expected in terms of material structure and what the limiting factors are for each approach. We have included Chapter 4 as summarising methods for getting graphene in your hands. It covers the Manchester 'Scotch-tape' mechanical exfoliation, solution-phase chemical exfoliation, bottom-up chemical methods using molecular precursors, chemical-vapour deposition and silicon carbide. A section on how to transfer graphene to arbitrary substrates is included as this is an essential part of getting graphene where you want it, which is often on an insulating substrate or partially suspended like a drum-skin.

Too often, when reviewing papers for journals, poor characterisation has limited the chance of the work being accepted for publication. Effective characterisation is the key for proving you have what you claim, and then being able to draw the right conclusions about from the results. We encourage the use of multiple techniques, with each one providing a piece of the jigsaw that you can fit together for a solid and robust conclusion. We could not cover all characterisation techniques, and recommend further reading on angular-resolved photoemission spectroscopy and X-ray photoemission spectroscopy for graphene characterisation, which are not covered in this book.

The final chapter presents an overview of seven key application areas that graphene has shown promise in: electronic devices, spintronics, transparent conducting electrodes (TCE), Nano-Electro-Mechanical Systems (NEMS), free-standing membranes, energy and super-strong composites. Whilst it was the outstanding properties of graphene in electronic devices that received the most attention, this area is seen as one of the most challenging in terms of developing a commercial product that will replace silicon electronics due to the absence of an appreciable band-gap. It seems that the niche area of high-frequency electronics may be more suitable for graphene than logic-based transistors. Graphene TCEs have already been used to make touch-screens and outperform Indium Tin Oxide (ITO) on flexible substrates in terms of durability. The major limiting factor here is that the sheet resistance of graphene is still too high. Further advancements in materials design through substitutional doping, intercalation and multilayering will solve this problem in the near future.

We hope this book serves well to get you started and leads to more activity in graphene research. Time will tell whether graphene can live up to its promise, but right now, graphene research is at its pinnacle, and it is never too late to join in the fun.

REFERENCES

Berger, C., Song, Z., Li, T., Li, X., Ogbazghi, A.Y., Feng, R., Dai, Z., Marchenkov, A.N., Conrad, E.H., First, P.N., de Heer, W.A., 2004. Ultrathin epitaxial graphite: 2D electron gas properties and a route toward graphene-based nanoelectronics. J. Phys. Chem. B 108, 19912–19916.

Iijima, S., 1991. Helical microtubules of graphitic carbon. Nature 354, 56–58.

Novoselov, K.S., Geim, A.K., Morozov, S.V., Jiang, D., Zhang, Y., Dubonos, S.V., Grigorieva, I.V., Firsov, A.A., 2004. Electric field effect in atomically thin carbon films. Science 306, 666–669.

Novoselov, K.S., Jiang, D., Schedin, F., Booth, T.J., Khotkevich, V.V., Morozov, S.V., Geim, A.K., 2005. Two-dimensional atomic crystals. Proc. Natl. Acad. Sci. 102, 10451–10453.

Zhang, Y., Tan, Y.-W., Stormer, H.L., Kim, P., 2005. Experimental observation of the quantum hall effect and Berry's phase in graphene. Nature 438, 201–204.

Chapter 2

The Atomic Structure of Graphene and Its Few-layer Counterparts

Franziska Schäffel

University of Oxford, Oxford, UK

2.1. GRAPHENE

In order to understand the atomic structure of graphene, it is helpful to first gain an understanding of the peculiarities of elemental carbon as well as its three-dimensional (3D) allotropes. The general interest in carbon arises from the variety of structural forms in which this element is available. This variety results from a special electron configuration of carbon that provides the ability to form different types of valence bonds to various elements, including other carbon atoms, through atomic orbital hybridisation. Carbon has the atomic number 6 and therefore, electrons occupy the $1s^2$, $2s^2$, $2p_x^1$ and $2p_y^1$ atomic orbitals as illustrated in Fig. 2.1a (ground state). It is a tetravalent element, i.e. only the four exterior electrons participate in the formation of covalent chemical bonds.

When forming bonds with other atoms, carbon promotes one of the 2s electrons into the empty $2p_z$ orbital, resulting in the formation of hybrid orbitals. In diamond the 2s-energy level hybridises with the three 2p levels to form four energetically equivalent sp^3-orbitals that are occupied with one electron each (Fig. 2.1b). The four sp^3-orbitals are oriented with largest possible distance from each other; they therefore point towards the corners of an imaginary tetrahedron. The sp^3-orbitals of one carbon atom overlap with the sp^3-orbitals of other carbon atoms, forming the 3D diamond structure. The high hardness of diamond results from the strong binding energy of the C–C bonds.

In graphite only two of the three 2p-orbitals partake in the hybridisation, forming three sp^2-orbitals (Fig. 2.1c). The sp^2-orbitals are oriented perpendicular to the remaining 2p-orbital, therefore lying symmetrically in the X–Y plane at 120° angles. Thus, sp^2-carbon atoms form covalent in-plane bonds

Graphene. http://dx.doi.org/10.1016/B978-0-12-394593-8.00002-3

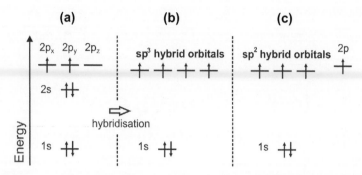

FIGURE 2.1 Atomic orbital diagram of a carbon atom. The four electrons in the doubly occupied spherical 2s orbital and the half occupied dumbbell-shaped 2p-orbitals participate in the chemical bonding of carbon. (a) Ground state, (b) sp³-hybridised as in diamond and (c) sp²-hybridised as in graphite and graphene.

affecting the planar hexagonal "honeycomb" structure of graphite. While the in-plane σ-bonds within the graphene layers (615 kJ/mol) are even stronger than the C–C bonds in sp³-hybridised diamond (345 kJ/mol), the interplane π-bonds, formed by the remaining 2p-orbitals, have a significantly lower binding energy, leading to an easy shearing of graphite along the layer plane. A single layer of graphite (the so-called graphene layer) has a lattice constant $a = \sqrt{3}a_0$ where $a_0 = 1.42$ Å is the nearest neighbour interatomic distance (Haering, 1958). The interplane distance between two adjacent graphene layers in AB stacked graphite is 3.35 Å (Haering, 1958).

The term 'graphene' is often incorrectly used for ultrathin graphite layers. Strictly it only refers to a quasi-two-dimensional isolated monolayer of carbon atoms that are arranged in a hexagonal lattice (Novoselov et al., 2005a). As will be discussed in detail in chapter 3, the electronic properties of graphene depend strongly on the number of graphene layers (Geim and Novoselov, 2007). Only single-layer graphene (SLG) and bilayer graphene (BLG) are zero-gap semi-conductors with only a single type of electrons and holes, respectively. In the case of the so-called few-layer graphene (FLG, 3 to <10 layers), the conduction and valence bands start to overlap, and several charge carriers appear (cf. Section 3.1) (Morozov et al., 2005; Partoens and Peeters, 2006). Thicker graphene structures are considered as thin films of graphite.

The hexagonal lattice of graphene is shown in Fig. 2.2a with an armchair and a zigzag edge highlighted in grey. The unit cell of graphene is a rhombus (grey) with a basis of two nonequivalent carbon atoms (A and B). The black and white circles represent sites of the corresponding A and B triangular sublattices. In cartesian coordinates the real space basis vectors of the unit cell a_1 and a_2 are written as

$$a_1 = \begin{pmatrix} \sqrt{3}a/2 \\ a/2 \end{pmatrix} \quad \text{and} \quad a_2 = \begin{pmatrix} \sqrt{3}a/2 \\ -a/2 \end{pmatrix} \tag{2.1}$$

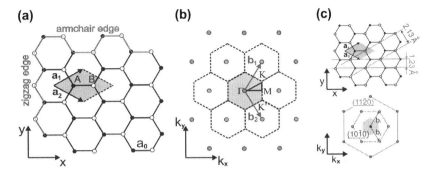

FIGURE 2.2 Crystal structure of graphene: (a) 2D hexagonal lattice of graphene in real space with basis vectors a_1 and a_2. The unit cell is highlighted in grey. It contains two nonequivalent carbon atoms A and B, each of which span a triangular sublattice as indicated with black and white atoms, respectively. An armchair and a zigzag edge are highlighted in grey. (b) Reciprocal lattice (dashed) with reciprocal lattice vectors b_1 and b_2. The first Brillouin zone is marked grey and the high symmetry points Γ, M, K and K' are indicated. (c) Demagnified views of the real (upper panel) and reciprocal lattice (lower panel), respectively. Two sets of lattice planes with $d = 2.13$ Å and $d = 1.23$ Å are highlighted with dotted and full lines in the real space lattice. In the reciprocal lattice the corresponding diffraction spots are marked with a dotted and full hexagon, accordingly.

with $a = \sqrt{3}a_0$ (Dresselhaus et al., 1995). A section of the corresponding reciprocal lattice is depicted in Fig. 2.2b together with the first Brillouin zone (grey hexagon). The reciprocal basis vectors b_1 and b_2 can be expressed as

$$b_1 = \begin{pmatrix} 2\pi/\sqrt{3}a \\ 2\pi/a \end{pmatrix} \quad \text{and} \quad b_2 = \begin{pmatrix} 2\pi/\sqrt{3}a \\ -2\pi/a \end{pmatrix} \tag{2.2}$$

and thus, the reciprocal lattice constant is $4\pi/\sqrt{3}a$. The high symmetry points (Γ, M, K and K') are also indicated in Fig. 2.2b.

The reciprocal lattice is a geometrical construction that is very useful in describing diffraction data (cp. Section 5.4). It is an array of points where each point corresponds to a specific set of lattice planes of the crystal in real space, as illustrated in Fig. 2.2c. The lower panel of Fig. 2.2c shows the schematic of the graphene diffraction pattern. The diffraction peaks in the corners of the dotted hexagon correspond to the $\{10\bar{1}0\}$ planes of graphene with a lattice spacing of $d = 2.13$ Å as shown in the real space lattice in the upper panel of Fig. 2.2c. In the same way, the $\{11\bar{2}0\}$ planes with a shorter lattice spacing of $d = 1.23$ Å give rise to the six diffraction spots of the larger hexagon (full lines) in reciprocal space. Further, the reciprocal lattice is generally used to depict a material's electronic band structure, plotting the bands along specific reciprocal directions within the Brillouin zone (e.g. from Γ to M or K in graphene). In this context the two points K and K' (also known as Dirac points) are of particular importance. Their coordinates in reciprocal space can be expressed as

$$K = \begin{pmatrix} 2\pi/\sqrt{3}a \\ 2\pi/3a \end{pmatrix} \quad \text{and} \quad K' = \begin{pmatrix} 2\pi/\sqrt{3}a \\ -2\pi/3a \end{pmatrix} \tag{2.3}$$

(Castro Neto et al., 2009). Graphene is a zero-gap semiconductor since the conduction and valence bands touch at the Dirac points and exhibit a linear dispersion (Novoselov et al., 2005a). The electronic density of states is zero at the Dirac points (cf. Section 3.1). This topology of the bands gives rise to unique and exotic electronic transport properties. The charge carriers are massless, which affects an extreme intrinsic carrier mobility (Morozov et al., 2008; Novoselov et al., 2005a). This makes graphene a promising candidate for applications in new generation molecular electronics, e.g. for graphene-based interconnects and field-effect transistors (cf. Chapter 6).

The stability of isolated two-dimensional (2D) crystals has long been studied theoretically (Mermin and Wagner, 1966; Mermin, 1968; Peierls, 1934, 1935) leading to the conclusion that long-range order cannot exist in two dimensions due to thermal fluctuations resulting in melting of a 2D crystal at a finite temperature. This finding was supported by experiments looking into the growth of thin films (Venables et al., 1984; Zinke-Allmang et al., 1992). Below a thin film thickness of the order of a couple of atomic layers, the films became unstable, i.e. they segregated or decomposed. Therefore, the discovery of graphene came as a huge surprise. Graphene was first isolated from graphite via mechanical exfoliation and was found to be stable on a supporting substrate under ambient conditions (Novoselov et al., 2004, 2005a; Zhang et al., 2005). Later, experimental (Bangert et al., 2009; Bao et al., 2009; Geringer et al., 2009; Ishigami et al., 2007; Lui et al., 2009; Meyer et al., 2007a,b; Wilson et al., 2010) as well as theoretical (Fasolino et al., 2007; Katsnelson and Geim, 2008; Thompson-Flagg et al., 2009) studies helped in understanding the phenomenon, showing in-plane and out-of-plane distortions of the graphene lattice, which seem to be essential for the structural stability of graphene. These corrugations can, however, effectively limit the electronic properties of graphene (Katsnelson and Geim, 2008).

Scanning probe microscopy (SPM) has been used to explore the topography of graphene and chemically modified graphene (CMG), i.e. graphene oxide (GO) or reduced graphene oxide (rGO), on a variety of substrates (Geringer et al., 2009; Ishigami et al., 2007; Lui et al., 2009; Wilson et al., 2010). When supported on SiO_2 substrates, graphene and CMG primarily follow the underlying topography of the SiO_2 substrate (Ishigami et al., 2007; Wilson et al., 2010). Depositing graphene on atomically smooth substrates like mica provides "ultraflat" graphene samples where intrinsic rippling is suppressed by interfacial van der Waals interactions (Lui et al., 2009).

Meyer et al. were the first to study the topography of free-standing graphene membranes (Meyer et al., 2007a,b). They suspended mechanically exfoliated graphene across a metal support grid and investigated these membranes via nanoarea electron diffraction (ED). They observed a broadening of the diffraction peaks with respect to the tilt angle, which cannot be explained by standard diffraction behaviour of 3D crystals (Spence, 2003; Williams and Carter, 2009). Figure 2.3a shows an ED pattern of monolayer graphene at zero tilt angle with the inner hexagon of diffraction peaks obtained from the $\{10\bar{1}0\}$ planes and the outer second hexagon from $\{11\bar{2}0\}$ planes, as schematically introduced in Fig. 2.2c. In Fig. 2.3b and c ED patterns of tilted monolayer graphene are depicted. Clearly a peak broadening can be made out with increasing tilt angle. The full width at half maximum (FWHM) of the peak width is plotted as a function of the tilt angle in Fig. 2.3d. Further details on the ED analysis will be discussed in the characterisation chapter of this book, together with an explanation as to how the peak broadening arises (cf. Section 5.5). From their data Meyer et al. concluded that the graphene membranes are not perfectly flat. For an SLG membrane the peak broadening could be related to microscopic corrugations with out-of-plane deformations of ≈ 1 nm and a variation of the surface normal of about $5°$. The lateral size of the ripples was estimated to be between 5 nm and 20 nm (Meyer et al., 2007a). The suggested atomic model is shown in Fig. 2.3e (perspective view). The microscopic roughness becomes notably smaller for BLG and disappears for thicker graphitic

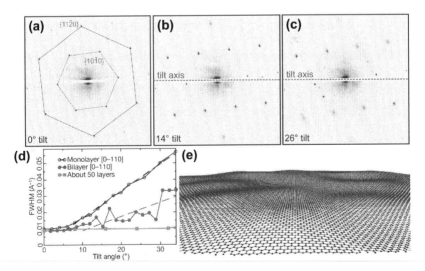

FIGURE 2.3 Analysis of the roughness of graphene membranes via nanoarea ED: ED patterns of (a) an untilted, (b) $14°$-tilted, (c) $26°$-tilted monolayer of graphene. A clear broadening of the $\{10\bar{1}0\}$ and $\{11\bar{2}0\}$ peaks can be observed upon tilting. (d) FWHM of the peak width as a function of tilt angle for monolayer graphene, BLG and thin graphite with about 50 layers. (e) Schematic atomic model of a corrugated graphene monolayer (perspective view). *Adapted by permission from Macmillan Publishers Ltd: Nature (Meyer et al., 2007a), Copyright (2007).*

membranes (Fig. 2.3d), indicating that the corrugations are intrinsic to graphene membranes.

Graphene oxide (GO) and reduced graphene oxide (rGO) have been studied intensively as alternative materials to graphene. Details on the atomic structure of GO and rGO, their chemical composition and preparation techniques will be given in Sections 2.4 and 4.3. Wilson et al. presented results from an ED study on suspended GO and rGO membranes Wilson et al. (2009),(2010). While the positions of the diffraction peaks are indistinguishable from those of graphene (Wilson et al., 2009), the broadening behaviour with tilt angle differs. For graphene, the peaks broaden linearly with tilt angle, which could be related to 'long-wavelength' ripples that undulate with a wavelength larger than the coherence length of the electron beam (i.e. approx. 1–10 nm) (Wilson et al., 2010). In the case of GO a clear nonlinear dependence was observed. The latter has been attributed to distortions occurring on the length scale of a few nanometres or even less, i.e. "short-wavelength" ripples (Wilson et al., 2010). Distortions of this magnitude correspond to atomic displacements of the order of 10% of the carbon–carbon distance. This points to a large strain in the lattice that may be induced by the functional hydroxyl and epoxide groups present on both sides of the GO layer (Lerf et al., 1998; Thompson-Flagg et al., 2009; Wilson et al., 2010). This suggests that there are fundamental structural and topographical differences between chemically modified and mechanically cleaved graphene.

While ED allows one to identify rippling, it is not sufficient to determine if the observed corrugations are of a static or a dynamic nature. However, direct visualisation of ripples on suspended graphene has also been achieved. Using atomic force microscopy (AFM), Wilson et al. found that GO is significantly rougher when free-standing as compared to GO deposited on atomically smooth substrates. They observed distortions with length scales as small as 10 nm that are comparable to the size of the AFM tip applied (Wilson et al., 2010). Direct visualisation of the ripples in suspended mechanically exfoliated graphene was achieved using aberration-corrected scanning transmission electron microscopy (STEM), revealing ripples with amplitudes of ≈ 0.5 nm and widths of ≈ 5 nm (Bangert et al., 2009). Here, changes in the corrugation pattern were observed to occur within a time period of several seconds.

2.2. BILAYER, TRILAYER AND FEW-LAYER GRAPHENE

As pointed out in Section 2.1, the electronic properties of graphene depend very drastically on the number of layers (Geim and Novoselov, 2007). Generally, the graphene community distinguishes between single-layer, bilayer and few-layer graphene, the latter of which refers to graphene with a layer number of less than 10. As mentioned above, a structure consisting of more than 10 graphene layers can be considered as a graphite thin film since it essentially exhibits the electronic properties of graphite.

A second important aspect influencing the electronic band structure of graphene is the way the graphene layers are stacked (Aoki and Amawashi, 2007; Latil and Henrard, 2006; Varchon et al., 2008). Three different stacking sequences of graphene sheets can occur: simple hexagonal (AAA..., space group: *P6/mmm*), hexagonal, so-called Bernal stacking (ABAB..., space group: *P6₃/mmc*) and rhombohedral (ABC..., space group: $R\,\overline{3}\,m$) (Charlier et al., 1994). AB-Bernal stacking and ABC-stacking are schematically illustrated in Fig. 2.4. For AB-Bernal stacking, alternate layers have the same projections on the basal plane (X–Y plane). The interlayer spacing is 3.35 Å. In ABC stacked rhombohedral graphite the third layer is shifted with respect to the first and the second layers. The fourth layer then has the same projection on the basal plane as the first layer. In this case the interlayer separation is 3.37 Å (Haering, 1958). Disordered graphite, also termed turbostratic graphite, does not exhibit a preferred stacking order. Although the adjacent graphene layers are parallel, they contain so-called rotational stacking faults, i.e. they are rotated relative to each other with no preferred orientation.

It is widely known that AB-Bernal stacking is most commonly observed in graphite and FLG. In bulk graphite, the volume fraction of AB:ABC: turbostratic is reported to be about 80:14:6 (Haering, 1958). However, AA stacking, where the carbon atoms of the second layer come to a rest directly above the first layer, has repeatedly been reported to occur at folds of SLG and BLG (Liu et al., 2009; Roy et al., 1998). Further, AA-stacked nanofilms of graphite oxide have been observed (Horiuchi et al., 2003). ABC stacking has been observed in multilayer graphene epitaxially grown on silicon-terminated SiC (Norimatsu and Kusunoki, 2010). Further, FLG grown via chemical vapour deposition (CVD) can also adopt ABC rhombohedral stacking (Warner et al., 2012). Multilayer graphene grown on the carbon-terminated ($000\overline{1}$) face of SiC obtains a high density of rotational disorder (Hass et al., 2007, 2008; Varchon et al., 2008). These multilayered structures are of interest, since the

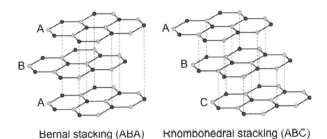

Bernal stacking (ABA)   Rhombohedral stacking (ABC)

FIGURE 2.4   AB-Bernal stacking (left) and ABC Rhombohedral stacking (right): in AB-Bernal stacking the third layer is located above the first layer; right: in ABC stacking the third layer is shifted with respect to the first and the second layers. The fourth layer is then located above the first layer. *Reprinted by permission from Macmillan Publishers Ltd: Nat. Phys. (Yacoby, 2011), Copyright (2011).*

introduction of rotational stacking faults into AB-Bernal stacked graphene causes decoupling of adjacent graphene layers. As a result the dispersion relation close to the K-point is altered from a parabolic (AB) to a linear band behaviour (rotational stacking fault). Thus, electronic properties typical for isolated SLG can be observed in bilayer and multilayer graphene structures (Hass et al., 2008; Latil et al., 2007). Rotational stacking faults have also been found in FLG prepared by liquid phase exfoliation (Warner et al., 2009). Graphene sheets with a relative rotation between them give rise to Moiré patterns in scanning tunnelling microscopy (STM) or transmission electron microscopy (TEM) images (Varchon et al., 2008; Warner et al., 2009).

In the previous section, it was pointed out that SLG exhibits intrinsic microscopic corrugations apparently essential for its structural stability and that rippling has also been observed in BLG. In their ED studies Meyer et al. observed that rippling is less pronounced in BLG as compared to SLG (cp. Fig. 2.3d) (Meyer et al., 2007a,b). While for SLG a deviation of the surface normal from the mean surface of $5°$ is observed, it is only $2°$ for BLG. Meyer et al. also directly visualised static ripples in BLG using convergent beam electron diffraction with the sample offset from the beam focus (Meyer et al., 2007b). Local variations of the orientation of the graphene membrane translate into intensity variations within the diffraction spots. They found these intensity variations to be in agreement with the $\pm 2°$ deviation of the mean surface normal derived from the ED-peak broadening.

2.3. RELATIONSHIP OF GRAPHENE TO CARBON NANOTUBES

One can image the atomic structure of a carbon nanotube (CNT) as a strip of graphene rolled up into a tubular structure. In general, two different types of CNTs are distinguished: single-wall CNTs (SWCNTs), which can be envisaged as cylinders obtained by rolling up planar sheets or strips of SLG, and multiwall CNTs (MWCNTs) consisting of multiple, concentrically arranged graphene cylinders. The spacing between two adjacent nanotube walls exceeds that of single crystal graphite by approximately 3–5% (Ajayan and Ebbesen, 1997; Iijima, 1991).

MWCNTs were first identified by S. Iijima in 1991, who examined carbon soot obtained from arc discharge of graphite electrodes via TEM (Iijima, 1991). He observed coaxial tubes of graphene sheets as presented in Fig. 2.5a. Two years later, SWCNTs formed by only a single rolled-up layer of graphene (cf. Fig. 2.5b) were successfully synthesised by adding transition-metal catalysts to one of the graphite electrodes (Bethune et al., 1993; Iijima and Ichihashi, 1993). These tubes are characterised by a small and uniform diameter of the order of 1 nm, with lengths up to several mm. With such extreme aspect ratios this 1D system quickly sparked interest in theoretical as well as experimental research (Hamada et al., 1992; Odom et al., 1998; Wildöer et al., 1998).

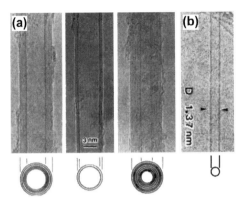

FIGURE 2.5   Transmission electron micrographs of (a) MWCNTs with five, two and seven concentric graphene sheets (from left to right) and (b) a SWCNT with a diameter of 1.37 nm. *Reprinted by permission from Macmillan Publishers Ltd: Nature (Iijima, 1991) and (Iijima and Ichihashi, 1993), Copyright (1991 and 1993).*

A large variety of methods has been developed over the years to synthesise CNTs. The most established approaches include the above-mentioned arc discharge technique (Ebbesen et al., 1996; Iijima, 1991), laser ablation (Guo et al., 1995) and CVD (José-Yacamán et al., 1993; Ren et al., 1998). These methods yield CNT material with different characteristics that may be applied in different fields. For example, laser ablation is advantageous in that it produces CNT material that can easily be purified by acid treatment and that the thus obtained CNTs in general contain very few structural defects. In comparison to laser ablation and arc discharge, CVD is promising due to its upward scalability, low cost, rather low production temperatures and the possibility to grow vertically or horizontally aligned CNTs. CVD thus holds significant potential for the integration of CNTs into nanoelectronic devices (Huang et al., 2003; José-Yacamán et al., 1993; Lee et al., 2000; Li et al., 1996; Ren et al., 1998).

As discussed above, a SWCNT can be imagined as a single honeycomb graphene sheet that is rolled into a hollow cylinder. This virtual process of rolling-up can be carried out in different directions, e.g. armchair, zigzag, or any direction in between. In this way an infinite number of different SWCNT structures with varying diameters and different helical geometries can be generated. The CNT structure is schematically illustrated in Fig. 2.6a. A CNT can be specified by its diameter d and the chiral angle θ. Starting from a graphene sheet – a planar honeycomb lattice – a strip of graphene (marked white) is rolled up in such a way that the lattice points O and A, and B and B', come into coincidence (Dresselhaus et al., 1995). The vector **OA** is the so-called chiral or Hamada vector C_h. It represents an unambiguous notation of a SWCNT structure and can be expressed as

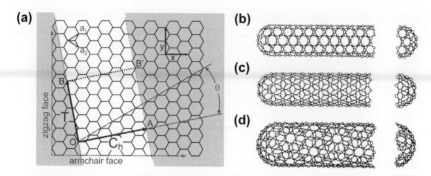

FIGURE 2.6 Atomic structure of CNTs: (a) hexagonal graphene lattice. A strip of graphene (white) is rolled up along the chiral vector $C_h = 4 \cdot a_1 + 2 \cdot a_2$ into a (4,2) chiral SWCNT. The translation vector $T = 4 \cdot a_1 - 5 \cdot a_2$ runs along the SWCNT axis. (b–d) SWCNTs can be classified into three types, i.e. (b) armchair (n,n), (c) zigzag $(n,0)$ and (d) chiral (n,m) nanotubes. The specific chiral vectors for the SWCNTs displayed are (b) (4,2), (c) (5,5), and (d) (9,0), respectively. *Reprinted from Dresselhaus et al. (1995). Copyright (1995), with permission from Elsevier.*

a linear combination of the real space basis vectors a_1 and a_2 (cp. Eqn. (2.1)) of the graphene lattice

$$C_h = n \cdot a_1 + m \cdot a_2 \equiv (n, m) \qquad (2.4)$$

where n and m are integers and $0 \leq |m| \leq |n|$ (Hamada et al., 1992). The chiral vector is generally used for SWCNT classification. In "achiral" arrangements, i.e. armchair (n,n) and zigzag $(n,0)$ nanotubes, the honeycomb lattice is oriented parallel to the nanotube axis as presented in Fig. 2.6b,c, respectively. The chiral vector of armchair nanotubes is aligned with the armchair face (Fig. 2.6a) so that they exhibit an armchair pattern along the circumference. In these nanotubes the two carbon–carbon bonds on opposite sides of each hexagon are arranged perpendicular to the nanotube axis. In contrast, in a zigzag nanotube the opposing carbon–carbon bonds are oriented parallel to the nanotube axis and the chiral vector is aligned with the zigzag face of the graphene sheet. All other nanotubes with a chiral vector (n,m) – an example is presented in Fig. 2.6d – are termed "chiral" or "helical" nanotubes. The specific chiral vectors for the SWCNTs displayed in Fig. 2.6 are (4,2) (Fig. 2.6a), (5,5), (9,0), and (10,5) (Fig 2.6b–d) (Dresselhaus et al., 1995).

The circumference of a SWCNT is determined by the length of the chiral vector. Thus the CNT diameter d directly relates to the chiral vector and is defined as

$$d = \frac{C_h}{\pi} = \frac{a \cdot \sqrt{m^2 + mn + n^2}}{\pi}. \qquad (2.5)$$

The chiral angle θ is defined as the angle between the zigzag direction a_1 and the chiral vector C_h as marked in Fig. 2.6a. It is thus equal to the tilt angle of the hexagons with respect to the SWCNT axis and can be directly deduced from the chiral angle as follows:

$$\cos \theta = \frac{C_h \cdot a_1}{|C_h| \cdot |a_1|} = \frac{2n + m}{2 \cdot \sqrt{m^2 + mn + n^2}}. \tag{2.6}$$

The value of θ is in the range $0° \leq |\theta| \leq 30°$ due to the hexagonal symmetry of the graphene lattice. The achiral zigzag and armchair nanotubes possess chiral angles of $\theta = 0°$ and $\theta = 30°$, respectively. All other nanotubes with chiral angles of $-30° < \theta < 30°$ are chiral, with right-handed helices for $\theta > 0$ and left-handed helices for $\theta < 0$ (Saito et al., 1998).

While the vectors a_1 and a_2 define the area of the unit cell of the graphene lattice, the unit cell of a SWCNT is given by the rectangle bounded by the chiral vector C_h and the vector T, which is the 1D translation vector of the nanotube. The translation vector is oriented perpendicular to C_h and is defined by the first lattice point reached by a vector that runs parallel to the tube axis and originates in O. The translation vector can be expressed as

$$T = t_1 \cdot a_1 + t_2 \cdot a_2 \quad \text{with} \quad t_1 = \frac{2m + n}{d_R}, \quad t_2 = -\frac{2n + m}{d_R}, \tag{2.7}$$

where d_R is the greatest common divisor of $(2m + n)$ and $(2n + m)$. The translation vector of the nanotube described in the example in Fig. 2.6a is $T = 4 \cdot a_1 - 5 \cdot a_2 \equiv (4; -5)$. The number of hexagons per nanotube unit cell, N, can be determined by dividing the surface area of the SWCNT unit cell by the area of the graphene unit cell

$$N = \frac{C_h \times T}{a_1 \times a_2} = \frac{2(m^2 + mn + n^2)}{d_R}. \tag{2.8}$$

It has been shown that CNTs are terminated by a fullerene-like hemispherical cap on each end. The effect of these caps on the mechanical and electronic properties of the nanotubes can be neglected due to the extreme CNT aspect ratio of typically 10^4 to 10^7 (Hata et al., 2004; Saito et al., 1998). However, they seem to play an important role at the nucleation stage of CNT growth (Rümmeli et al., 2007). Further, due to intertube van der Waals interactions, SWCNTs tend to agglomerate to bundles with hexagonal structure.

Theoretical studies show that the electronic properties of CNTs are strongly dependant on their geometric structure (Dresselhaus et al., 1995; Hamada et al., 1992; Mintmire et al., 1992; Saito et al., 1998). While graphene is a zero-gap semiconductor (cf. Chapter 3), CNTs can be metallic or semiconducting depending on their diameter and chirality, i.e. their chiral indices (n,m). A nanotube can be considered metallic at room temperature when $n - m = 3r$

(where r is an integer). Armchair nanotubes (n,n), sometimes also called type-I metallic nanotubes, are the only zero-gap nanotubes. The other (n,m) nanotubes that fulfil the condition for metallic nanotubes are in fact tiny-gap semiconductors and called type-II metallic nanotubes (Charlier et al., 2007). Semiconducting nanotubes are obtained when $n - m = 3r \pm 1$. It follows that two-thirds of all SWCNTs are semiconductors and one third is metallic (or semimetallic). The band gap energy is inversely proportional to the nanotube diameter. Very large diameter nanotubes are zero-gap semiconductors since the electronic properties of a graphene sheet are recovered (Charlier et al., 2007). Control of the electronic properties remains a major problem in CNT research, to date. However, attempts to disperse SWCNT bundles and separate semiconducting from metallic species are advancing (Hersam, 2008; Lemasson et al., 2011).

Since this book is primarily focused on graphene, not on CNTs, the structure and electronic properties of CNTs cannot be discussed in detail here. The reader is referred to respective books and reviews on CNTs (e.g. (Charlier et al., 2007; Reich et al., 2004; Saito et al., 1998)).

2.4. OTHER LAYERED 2D CRYSTALS

2.4.1. Introduction

Due to the fact that graphene's exceptional properties are a result of the electron confinement in two dimensions the hunt for other 2D materials with novel and exciting properties has already begun. Graphite is not the only layered material. A large variety of layered crystals with strong in-plane bonds and weak van der Waals-like inter-plane bonds exists which could potentially be exfoliated into 2D materials. The following chapter comprises a review of research on 2D materials other than graphene, including inorganic materials, such as boron nitride (BN), transition metal oxide (TMO), transition metal dichalcogenide (TMD), and silicene nanosheets, as well as derivatives of graphene such as GO, graphane and fluorographene. Besides these, other candidates for future 2D structures include 2D polymers (for which graphene is a natural example) (Bieri et al., 2009; Sakamoto et al., 2009), covalent organic frameworks (Dienstmaier et al., 2011) and metal-organic frameworks (Mas-Ballesté et al., 2009, 2011); these will, however, not be elaborated on within this short review. Looking at this long list of materials it is clear that research in 2D materials will not end with graphene.

Layered compounds do not only include metals, semiconductors and insulators (Wilson and Yoffe, 1969) but also superconductors (Novoselov et al., 2005b) and thermoelectric materials (Tang et al., 2007), thus providing a vast playground for investigating their 2D properties. While graphene is envisaged as a potential conductor in nanoelectronics, reliable insulators will also be

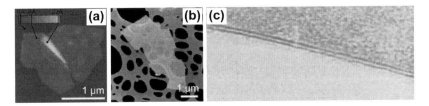

FIGURE 2.7   Alternative 2D crystals prepared by micromechanical cleavage: (a) AFM image of a 2D NbSe$_2$ layer; (b) scanning electron micrograph of a Bi$_2$Sr$_2$CaCu$_2$O$_x$ single layer; (c) TEM micrograph of a MoS$_2$ double layer. When the crystal is backfolded, the number of layers can be detected by the number of dark lines. The distance between the two dark lines in (c) corresponds to 6.5 Å, which is in agreement with the interlayer distance of bulk MoS$_2$. *Reprinted with permission from Novoselov et al. (2005b). Copyright (2005) National Academy of Sciences, U.S.A.*

required at this scale. Here, materials such as BN nanosheets (Golberg et al., 2010) or TMOs and perovskites (Osada and Sasaki, 2012) are of interest.

Extending their early search from graphene to other 2D materials, Novoselov et al. examined the suitability of micromechanical cleavage in order to prepare single sheets of other layered crystals such as TMDs (NbSe$_2$, MoS$_2$) and superconductors (Bi$_2$Sr$_2$CaCu$_2$O$_x$) (Novoselov et al., 2005b). TMDs consist of layers of hexagonally arranged metal atoms that are sandwiched between two layers of chalcogen atoms (i.e. S, Se or Te) (Coleman et al., 2011). Bi$_2$Sr$_2$CaCu$_2$O$_x$ is a high-temperature superconductor with a perovskite structure where the supercurrent flows within the weakly coupled 2D copper-oxide layers. Novoselov et al. could demonstrate that these 2D systems are stable under ambient conditions and exhibit high crystal quality, despite the old assumption that thin films become thermodynamically unstable below a certain thickness (Novoselov et al., 2005b). Preparation of these 2D crystals was carried out by micromechanical cleavage, i.e. rubbing the respective bulk crystal against another surface. Among the resulting flakes, the authors always observed single layers, which has been confirmed by AFM analysis. Figure 2.7 shows examples of mechanically exfoliated NbSe$_2$ (Fig. 2.7a), Bi$_2$Sr$_2$CaCu$_2$O$_x$ (Fig. 2.7b) and MoS$_2$ (Fig. 2.7c).

2.4.2. Boron Nitride Nanosheets

Novoselov et al. also commented on the possibility of obtaining monolayers of hexagonal boron nitride (h-BN) (Novoselov et al., 2005b). h-BN is the structural analogue of graphite with alternating B and N atoms, substituting the C atoms within the 2D honeycomb lattice. Similar to graphite, the in-plane bonds are covalent and the bonds between the layers are weak and slightly ionic in h-BN (Pacilé et al., 2008). However, the stacking sequence of the atomic planes differs. While graphite predominantly shows AB Bernal stacking (cp. Section 2.2), the layers in h-BN are stacked with boron atoms on top of nitrogen atoms and vice versa (Pacilé et al., 2008). Not considering the discrepancy between B and N atoms, h-BN, therefore, exhibits AAA stacking order (Zhi et al., 2009). A schematic of the h-BN structure is presented in Fig. 2.8a.

Although h-BN and graphite have a similar atomic structure, their electronic properties are distinctly different. While graphite is a semimetal with anisotropic resistivity, i.e. $(0.4-5) \cdot 10^{-4}$ Ωcm along the basal plane and 0.2–1 Ωcm perpendicular to the plane, h-BN is an insulator (or wide-band semiconductor) with a band gap of about 5.2 eV (Paine and Narula, 1990).

As a consequence of the structural similarity, many research groups started looking for ways to synthesise boron nitride nanotubes (BNNTs) after CNTs where first identified in 1991. It has been shown that BNNTs can be fabricated by arc discharge (Loiseau et al., 1996), laser heating (Golberg et al., 1996) and CVD (Lourie et al., 2000), i.e. using very similar synthesis methods as for CNTs. However, an effective method for the large-scale synthesis of high-purity small-diameter BNNTs has not been found to date (Golberg et al., 2010). BNNTs can be envisaged as graphene-like h-BN sheets rolled up into a hollow cylinder. Similar to their carbon counterparts one can differentiate between single- and multiwall BNNTs. Preparation of BN thin films by CVD methods

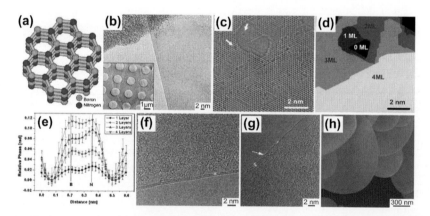

FIGURE 2.8 (a) Schematic representation of the atomic structure of BN. (b) TEM micrograph of the backfolded edge of a mechanically cleaved seven-layer h-BN sheet. The dark lines along the fold provide a signature of the layer number. The inset shows a larger section of the h-BN nanosheet (BNNS) spanned across a Quantifoil™ gold grid with 1.3-μm sized holes in the amorphous carbon film. *Reprinted with permission from Pacilé et al. (2008). Copyright (2008), American Institute of Physics.* (c) Atomic-resolution TEM micrograph of a one- to four-layer h-BN sheet, together with (d) its thickness map indicating the number of layers. (e) Intensity line profile along the BN unit cell, with relative phase plotted vs. the distance. *(a, c–e) Reprinted with permission from Alem et al. (2009). Copyright (2009) by the American Physical Society.* (f, g) TEM micrographs of chemically exfoliated h-BN sheets. *Reprinted with permission from Han et al. (2008). Copyright (2008), American Institute of Physics.* (h) SEM image of BN nanosheets as produced by CVD. *Reprinted with permission from Gao et al. (2009a). Copyright (2009) American Chemical Society.*

has also been studied for some time (Andújar et al., 1996). A notable example was the formation of a regular double-layer BN nanomesh on a Rh(111) surface by exposure to borazine $(HBNH)_3$ (Corso et al., 2004). When graphene was discovered later in 2004, research into BN thin films revived, with the focus now on extracting monolayers of BN.

Pacilé et al. prepared ultrathin h-BN attached to SiO_2 substrates as well as freely suspended by repeatedly pealing h-BN powders using adhesive tape and examined them using TEM (Pacilé et al., 2008). They obtained h-BN flakes consisting of just a few layers with continuity over several microns, thus realising the first isolation of few-layer h-BN by micromechanical cleavage. As an example, Fig. 2.8b shows a TEM micrograph of a seven-layer h-BN flake. Alem et al. prepared suspended h-BN single- and multilayers in a similar way and examined them with atomic resolution using an ultrahigh-resolution TEM (Alem et al., 2009). Figure 2.8c shows a high-resolution TEM (HRTEM) micrograph of a h-BN flake with layer numbers varying from one to four as indicated in the corresponding thickness map in Fig. 2.8d. Atoms are imaged with white contrast. The hexagonal honeycomb structure shows clearly in the image. Although boron and nitrogen have similar atomic numbers, their scattering power is different, with nitrogen exhibiting slightly stronger scattering than boron: Fig. 2.8e shows intensity line profiles through the BN unit cell, revealing that the intensity profiles of adjacent columns of atoms in even-layer h-BN are symmetric, while they are asymmetric in odd-layer h-BN. Alem et al. demonstrate that the number of h-BN layers can be determined using such intensity profiles. Further, they investigate the formation of defects in h-BN resulting from the so-called knock-on damage, which will be discussed in more detail in Section 5.4. In Fig. 2.8c missing atoms and larger triangular holes can be made out in the h-BN sheet. In h-BN the energy thresholds for knock-on damage of boron and nitrogen atoms are 74 keV and 84 keV, respectively (Zobelli et al., 2007). As a result of this, monovacancies appear to be predominantly missing boron atoms, and the 'internal edges' of the larger triangular defects exhibit a nitrogen-terminated zigzag configuration (Alem et al., 2009). Armchair edges terminated with alternating boron and nitrogen atoms are also observed, however less frequently as compared to the nitrogen-terminated zigzag edges.

Han et al. were the first to chemically exfoliate h-BN (Han et al., 2008). h-BN crystals were dispersed in a 1,2-dichloroethane solution of poly-(m-phenylenevinylene-co-2,5-dictoxy-p-phenylenevinylene) (PmPV) and sonicated for 1 h to break up the crystals into few-layer h-BN. A similar approach has previously been applied to exfoliate graphite and prepare ultra-smooth nanoribbons (Li et al., 2008b). Sonication effectively breaks up the h-BN crystals along the weak interlayer bonds into monolayer and few-layer sheets. To some extent, sonication also breaks in-plane bonds, resulting in a reduction of flake size. TEM inspection of the as-prepared h-BN sheets revealed sheet dimensions on the order of several microns. Few-layer h-BN flakes as well as monolayer flakes have been observed. Figure 2.8 shows examples of a h-BN double-layer (Fig. 2.8f) and a h-BN sheet with a mixed number of layers, i.e. a triple layer at the bottom left and a monolayer at the top right side (Fig. 2.8g) (Han et al., 2008).

Zhi et al. reported on large-scale fabrication of 2D BNNSs and their application as components in composite materials (Zhi et al., 2009). They prepared milligramme quantities of BNNSs by sonicating h-BN microparticles in *N,N*-dimethylformamide, a strong polar solvent, for 10 h and centrifuging the solutions in order to remove larger BN particles. Further, they fabricated polymethylmetracrylate (PMMA)/BNNSs composites and were able to show a strong reduction of the coefficient of thermal expansion for very-low BNNSs fractions in the PMMA, thus demonstrating that BNNSs effectively restrict the mobility of polymer chains. In addition, they demonstrated that the embedment of BNNSs in PMMA effected mechanical reinforcement, i.e. an enhanced elastic modulus and increased strength, indicating that the mechanical load can be transferred to the BNNSs through interfacial interactions (Zhi et al., 2009).

A last example of the variety of ways in which BNNSs can be obtained is the catalyst-free CVD synthesis of BNNSs as reported by Gao et al., (2009a) In this process, B_2O_3 and melamine powders are mechanically mixed. After evacuation of the CVD oven, N_2 is introduced as reaction gas. Gao et al. carried out BNNS synthesis at reaction temperatures of 1000–1350 °C for 1 h, yielding large quantities of a white powder. They found that with increasing synthesis temperature the nanosheet thickness decreased, with the lowest thickness of 20 nm observed at 1300 °C. Figure 2.8h shows an scanning electron microscopy (SEM) image of BN nanosheets observed after growth at a temperature of 1300 °C. Gao et al. also showed that these BNNSs have an onset temperature of oxidation of 850 °C, which is very high as compared to graphene sheets that are stable only up to approximately 400 °C (Gao et al., 2009a). This highlights that BN nanostructures may potentially replace carbon nanostructures in high-temperature applications.

2.4.3. Transition Metal Dichalcogenides

Apart from BN, other inorganic 2D materials are also increasingly investigated. Notable examples are the TMDs (Coleman et al., 2011; Smith et al., 2011). TMDs have a stoichiometry of MX_2, where M refers to metal atoms arranged in a hexagonal layer and sandwiched between two layers of chalcogen atoms X, typically S, Se or Te (Coleman et al., 2011). TMDs can exhibit a variety of different properties depending on the combination of transition metal and chalcogen and on the coordination and oxidation state of the metal atoms. With regard to electronic properties, this variety includes metals, semiconductors and insulators (Wilson and Yoffe, 1969). Even superconducting properties have been reported for some TMDs (Gamble and Silbernagel, 1975). Thus, exfoliation of these materials could potentially result in interesting new 2D phenomena.

TMDs have previously been exfoliated in liquids. For example, exfoliation of metallic TaS_2 and NbS_2 by intercalation of hydrogen and water has already been demonstrated in the 1970s and 1980s (Liu et al., 1984; Murphy and Hull, 1975). In 1986, Joensen et al. reported exfoliation of semiconducting MoS_2 in monolayers by lithium intercalation and following reaction with water (Joensen et al., 1986). Moreover, the exfoliation of WS_2 by lithium intercalation has been successfully accomplished (Miremadi and Morrison, 1988; Yang and Frindt, 1996). With the discovery of graphene in 2004, the search for graphene-like analogues of TMDs, e.g. MoS_2 and WS_2, started anew, now equipped with more advanced characterisation tools for unambiguous identification of monolayers (Matte et al., 2010). The intercalation methods used are, however, fairly time-consuming and very sensitive to the reaction environment.

Recently, Coleman et al. demonstrated that bulk TMD crystals can be exfoliated into mono- and few-layer nanosheets using a variety of organic solvents (Coleman et al., 2011). In their studies on liquid-phase exfoliation of graphene, they previously suggested that the energy cost of exfoliation is minimal for solvents whose surface energy matches that of graphene (Hernandez et al., 2008). Testing more than 25 solvents, they found that this concept can also be employed to exfoliate other 2D materials such as MoS_2, WS_2 and h-BN with promising solvents being N-methyl-pyrrolidone (NMP) and isopropanol (IPA) (Coleman et al., 2011). The method involves sonicating the powders in the respective solvent in a low-power sonic bath for approximately 1 h. The resulting dispersions are then centrifuged to remove large aggregates, and the supernatant is decanted. Coleman et al. report that for MoS_2 and WS_2, dark-green dispersions were obtained and the BN dispersion appeared milky white. HRTEM revealed very thin sheets of all three materials. In contrast to observations from MoS_2 and WS_2 exfoliated by a lithium intercalation that results in a strong deviation from the hexagonal lattice structure, Coleman et al. could demonstrate that the hexagonal symmetry of their materials is preserved after liquid-phase exfoliation (Coleman et al., 2011; Gordon et al., 2002). Optimised dispersions are stable over periods of hundreds of hours, thus providing a reasonably large window for further processing. Coleman et al. were therefore also able to demonstrate the preparation of hybrid and composite films and their deposition, using vacuum filtration or spray coating. Mixing their dispersions with CNT or graphene dispersions resulted in an increase in the dc conductivity σ_{dc}. WS_2/SWCNT hybrid films revealed an increased conductivity σ_{dc} without a degrading Seebeck coefficient, S, resulting in a high power factor ($S^2 \sigma_{dc}$) which is critical in the search for efficient thermoelectric materials (Coleman et al., 2011). Further, Coleman et al. reported exfoliation of other inorganic layered compounds, including $MoSe_2$, $MoTe_2$, $TaSe_2$, $NbSe_2$, Bi_2Te_3 and $NiTe_2$. They found that all these materials can be dispersed in cyclohexylpyrrolidone

(CHP) by a point-probe sonication. In STEM analysis all samples were found to contain thin layer flakes (Coleman et al., 2011). For example, Bi_2Te_3 is a very promising low temperature thermoelectric material. Therefore, the achievement to exfoliate this material and the possibility to simply mix it with other dispersions (e.g. SWCNTs) in order to obtain a hybrid material with improved thermoelectric properties (as demonstrated for the WS_2/SWCNT hybrid) illustrates the versatility and the potential of liquid-phase exfoliation for 2D material research. This method is not only insensitive to air and water but can also potentially be scaled up in order to obtain large quantities of exfoliated material and allows for the formation of hybrid films with enhanced properties (Coleman et al., 2011).

While Coleman et al. reported on exfoliation of 2D materials in organic solvents, exfoliation in an aqueous environment would be highly appealing for large-scale applications. Smith et al. demonstrated that a number of layered

FIGURE 2.9 (a) Photograph of a dispersion of MoS_2 in water, stabilised by sodium cholate. (b-d) TEM and HRTEM micrographs of a MoS_2 flake, revealing that the hexagonal lattice has been preserved throughout the exfoliation process. The inset in b) shows a typical ED pattern. (e) Photograph of dispersions of WS_2, $MoTe_2$, $MoSe_2$, $NbSe_2$, $TaSe_2$, and BN stabilised in water by sodium cholate. *Reprinted with permission from Smith et al. (2011). Copyright (2011) John Wiley and Sons.*

crystals can be exfoliated in water as long as a stabiliser is present to prevent rapid reaggregation that would otherwise occur driven by the large surface energy of TMDs (Smith et al., 2011). They primarily tested ionic surfactants as stabilisers due to their van der Waals binding to the exfoliated nanosheets and subsequent electrostatic stabilisation. The dispersion process involves probe sonication of the mixture of powder, at first MoS_2, and an aqueous solution of the surfactant sodium cholate for 30 min. As in the case of organic solvents the resultant dispersions are then centrifuged to remove unexfoliated powder. After decantation, dark dispersions, as shown in Fig. 2.9a, are obtained. The stability of these surfactant-coated nanosheets has been characterised by the zeta potential since their stabilisation is based on electrostatic repulsion (Lotya et al., 2009; Smith et al., 2011). The zeta potential of the sodium cholate coated MoS_2 flakes was found to be stable over weeks at -40 mV and robust against pH variation. TEM analysis revealed a large number of extremely thin 2D flakes consisting of approximately 2 to 9 stacked MoS_2 monolayers. In Fig. 2.9b a bright field TEM micrograph of a typical MoS_2 flake is presented. The inset shows a typical ED pattern revealing that the hexagonal lattice structure has been preserved throughout this exfoliation process. This has been confirmed by HRTEM investigation of the flakes. In Fig. 2.9c and d the respective HRTEM micrographs are presented, clearly showing the hexagonal symmetry. Smith et al.'s approach is not limited to MoS_2. They demonstrated the preparation of stable dispersions of WS_2, $MoTe_2$, $MoSe_2$, $NbSe_2$, $TaSe_2$ and BN exfoliated into thin flakes in aqueous solutions of sodium cholate. A photograph of the dispersions of these inorganic layered compounds is presented in Fig. 2.9e. They also processed a TMO, namely MnO_2, without expecting to find a 2D phase after the dispersion process. However, the resultant dispersion did contain thin layer flakes of MnO_2, a finding that illustrates the generality of this approach. As indicated above, the method is fairly easy to implement, versatile and very robust (Smith et al., 2011). Further, it can be carried out in ambient conditions, is scalable, and allows for the processing of films, hybrids and composites in a similar fashion as discussed for organic solvent exfoliation. Therefore, it holds great potential for the production of hybrids with tunable conductivity and improved thermoelectric properties.

2.4.4. Transition Metal Oxides

As mentioned above, a strong interest in 2D materials arises from the need for 2D insulators that are essential for many nano- and microelectronic devices such as memories, capacitors and gate dielectrics. There are various TMOs that have been exfoliated into 2D sheets by protonation, followed by intercalation with bulky organic ions, resulting in the electrostatic repulsion of the layers. Most commonly, tetrabutylammonium (TBA^+) is used as intercalator, but also tetramethylammonium and ethylammonium have successfully been utilised. These procedures have been recently reviewed by Osada and Sasaki, (2012). The versatility of the approach is illustrated by the fact that the oxides that have been processed into nanosheets include titanium, manganese, cobalt, tantalum, molybdenum, ruthenium and tungsten oxides as well as nanosheets of several perovskites [(Osada and Sasaki, 2012) and references therein]. Remarkably, many of these oxide-exfoliation studies have been carried out, long before the discovery of graphene.

A notable example is the synthesis of titania nanosheets with a lateral size of several tens of micrometres (Osada et al., 2006; Tanaka et al., 2003). Single crystals of potassium lithium titanite with a lateral size of a few millimetres were converted into an acid-exchanged form of $H_{1.07}Ti_{1.73}O_4 \cdot H_2O$ and then reacted with aqueous TBA^+, resulting in strong swelling and exfoliation of nanosheets. Figure 2.10a shows a TEM micrograph of a very large titania nanosheet obtained in this way. These nanosheets had lateral sizes up to 100 μm with an average size of about 30 μm. Selected area electron diffraction (SAED) confirmed the single-crystalline nature of the nanosheets. The SAED-spot pattern is shown as an inset in Fig. 2.10a. The faint TEM contrast of the titania sheet indicates that its thickness is very low. This has been confirmed by AFM analysis. Figure 2.10b shows an AFM image together with the corresponding

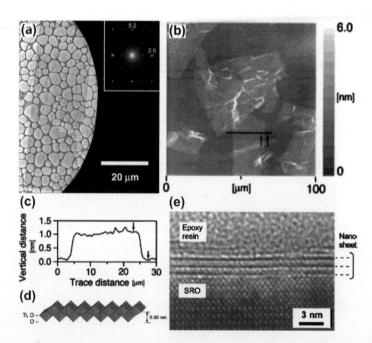

FIGURE 2.10 (a) TEM micrograph of a very large titania nanosheet. The inset shows the respective SAED pattern. (b) AFM image of titania nanosheets. (c) AFM height profile measured along the black line in (b). *Reprinted with permission from Tanaka et al. (2003). Copyright (2003) American Chemical Society.* (d) Structural model of a 2D titania nanosheet. (e) Cross-sectional HRTEM micrograph of a multilayer titania nanosheet (N = 5) as deposited onto a SrRuO₃ epitaxial film. *Reprinted with permission from Osada et al. (2006). Copyright (2006) John Wiley and Sons.*

height profile in Fig. 2.10c, measured along the black line in Fig. 2.10b, revealing nanosheets with a thickness as low as ≈ 1.0 nm and a low thickness distribution between different sheets (Tanaka et al., 2003). In Fig. 2.10d, a structural model of a titania nanosheet ($Ti_{0.87}O_2$) is depicted, showing that a Ti atom is coordinated with six oxygen atoms forming TiO_6 octahedra that are edge-linked to produce a 2D lattice (Osada et al., 2006). The layer height measured in AFM is slightly above the value predicted by the atomic model (cf. Fig. 2.10d). This slight increase, of sometimes up to 1 nm, is generally attributed to a surface-sheet interface due to the presence of intercalator ions or hydration layers (Mas-Ballesté et al., 2011).

Titania ($Ti_{0.87}O_2$) nanosheets have attracted broad attention due to the dielectric properties of the bulk TiO_2 system. While anatase TiO_2 has a dielectric constant in the range of $\varepsilon_r = 30\text{--}45$ (Kim et al., 2005; 2006), rutile TiO_2 exhibits a very high dielectric constant of $\varepsilon_r > 100$ (Diebold, 2003), making it highly appealing for application in high-κ devices. Osada et al. demonstrated that titania ($Ti_{0.87}O_2$) nanosheets can effectively function as a high-κ material in dielectric films (Osada et al., 2006). These sheets are 2D components that can be self-assembled into multilayers in a layer-by-layer approach via a solution-based process. Figure 2.10e shows a cross-sectional HRTEM micrograph of the multilayer film formed of $Ti_{0.87}O_2$ nanosheets as deposited onto atomically flat (001) $SrRuO_3$ epitaxial films. A well-ordered stacked structure is revealed, corresponding to layer-by-layer assembly of titania nanosheets. These ultrathin films exhibit a very high relative dielectric constant of $\varepsilon_r \approx 125$ at thicknesses down to 10 nm and a leakage current density of 10^{-9} to 10^{-7} Acm^{-2}, which is three to five orders of magnitude lower than in rutile TiO_2 films (Osada et al., 2006).

2.4.5. Silicene

Recently, 2D hexagonal silicon, also known as silicene, has started to raise interest among researchers. Silicene exhibits a honeycomb structure similar to that of graphene and has received strong theoretical attention. From density functional theory (DFT) calculations, it is known that silicon and germanium can exhibit stable, 2D, low-buckled, honeycomb structures (Cahangirov et al., 2009). While the 2D germanium is metallic, silicene has an electronic structure very similar to that of graphene with a linear dispersion around the K point (Lebègue and Eriksson, 2009). On the one hand, silicon appears to have advantages over carbon in that it might be more easily interfaced with existing electronic devices and technologies. On the other hand, synthesis of silicon in a graphene-like structure has proven to be very challenging (Lebègue and Eriksson, 2009).

First experiments involved the room temperature deposition of silicon atoms onto an Ag(110) surface from a direct-current-heated piece of silicon wafer, resulting in the formation of straight, high aspect-ratio Si nanowires with a characteristic width of 16 Å and a metallic character (Aufray et al., 2010; De Padova et al., 2008; Léandri et al., 2005). STM revealed Si nanowires with lengths of up to 30 nm and a density of $\approx 1.4 \cdot 10^{12}/cm^2$ (Fig. 2.11a). The nanowires are perfectly aligned along the $[\bar{1}10]$ direction of the Ag(110) surface. High-resolution STM studies revealed that the Si atoms are arranged in a graphene-like honeycomb structure that possibly indicates the synthesis of silicene (Fig. 2.11b) (Aufray et al., 2010). According to theoretical calculations, such silicene nanowires exhibit electronic and magnetic properties depending strongly on their size and geometry (Cahangirov et al., 2009). De Padova et al. studied the electronic structure of silicene nanoribbons as-grown on Ag(110) by angle-resolved photoelectron spectroscopy (ARPES)

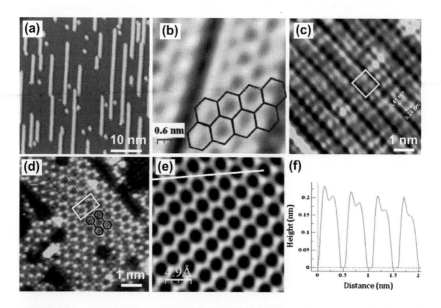

FIGURE 2.11 (a) Topographic STM image of silicon nanowires and nanodots deposited on Ag(110) at room temperature. *Reprinted from Léandri et al. (2005). Copyright (2005), with permission from Elsevier.* (b) Atomic resolution filled-state STM image of a silicon nanowire with the silicon hexagons in a honeycomb arrangement. *Reprinted with permission from Aufray et al. (2010), Copyright (2010), American Institute of Physics.* (c) Atomic resolution filled-state STM image of the p(3 × 3) superstructure formed during deposition of Si onto Ag(001) at 230 °C. The p(3 × 3) unit cell is indicated with a white rectangle. (d) Room temperature filled-state STM image of the more complex p(7 × 4) superstructure formed during deposition of Si onto Ag(001) beyond one monolayer. The local (7 × 4) unit cell is represented by the white rectangle. *Reprinted from Léandri et al. (2007). Copyright (2007), with permission from Elsevier.* (e) Atomic resolution filled-state STM image of the silicene sheet with graphene-like structure deposited on Ag(111) at 250 °C. (f) Line-intensity profile acquired along the direction indicated in (e), revealing two sublattices in the silicene sheet. *Reprinted with permission from Lalmi et al. (2010). Copyright (2010), American Institute of Physics.*

measurements (De Padova et al., 2010). They observed the Si bands to be separated by a gap of about 0.5 eV and centred 0.6 eV below the Fermi energy. This was attributed to the interaction with the Ag(110) substrate (De Padova et al., 2010). A similar behaviour has been observed for the band dispersion near the K point of metal-supported graphene (Wintterlin and Bocquet, 2009).

In order to investigate the role of surface orientation, Léandri et al. also carried out deposition studies of Si atoms onto Ag(001) substrates (Léandri et al., 2007). When deposited at 230 °C, the Si atoms first arranged into a p(3 × 3) superstructure (Fig. 2.11c) where the silicon tetramers with a quasi-rectangular shape self-organised on the surface. With further Si deposition beyond one monolayer, a local p(7 × 4) superstructure developed from the p(3 × 3) structure (Fig. 2.11d). Here, the unit cell is composed of two joint hexagons (white reactangle in Fig. 2.11d). These observations led to the proposal of a tentative atomic model that consists of a modified Si(111) bilayer on top of the Ag(001) surface with the $[\bar{1}10]$ direction of the Si(111) plane aligned with the [110] direction of the Ag surface (Léandri et al., 2007).

Epitaxial growth of a silicene sheet with graphene-like structure was for the first time achieved via condensation of Si on an Ag(111) surface held at 220–250 °C at a Si-deposition rate below 0.1 monolayer per minute (Lalmi et al., 2010). Highly ordered Si monolayers were observed that were arranged in a honeycomb lattice covering the substrate. Figure 2.11e shows a filled-state atomically resolved STM image of the silicene sheet, revealing the honeycomb structure. The line profile shown in Fig. 2.11f, which is obtained along the white line in Fig. 2.11e, reveals that the structure actually consists of two silicon sublattices occupying positions at different heights with a height difference of ≈ 0.02 nm, indicating – in line with theoretical predictions – possible sp^2–sp^3 hybridisations (Cahangirov et al., 2009; Lalmi et al., 2010). The authors further determine a Si–Si nearest neighbour distance of 0.19 ± 0.01 nm, which is, however, significantly shorter than that reported from DFT calculations (≈0.225 nm) (Cahangirov et al., 2009). Later, Vogt et al. suggested that the results presented by Lalmi et al. may refer to a clean Ag(111) surface only, since the pure Ag(111)–(1 × 1) surface can mimic a honeycomb-like appearance in STM caused by a tip-induced contrast reversal (Vogt et al., 2012). Vogt et al. studied one-atom-thick Si sheets formed on Ag(111) at temperatures between 220 and 260 °C and provided evidence, from STM and ARPES, for the synthesis of epitaxial silicene sheets. A silicon adlayer with honeycomb-like structure is formed with the Si atoms located either on top or between Ag atoms. Vogt et al. find the average Si–Si distance to be 0.22 ± 0.01 nm, a finding that is also in excellent agreement with the theoretical predictions for silicene (Vogt et al., 2012; Cahangirov et al., 2009).

TMOs can be epitaxially grown on Ag(111) as well. As an example, alternating adsorption of Mg and O_2 at a substrate temperature of 300 K led to the formation of a flat unreconstructed polar MgO(111) surface on Ag(111)

(Kiguchi et al., 2003). Another group reported on the formation of a nonpolar 2-monolayer ZnO(0001) film on Ag(111) (Tusche et al., 2007). ZnO was grown by pulsed laser deposition from a stoichiometric target under an oxygen background pressure at 300 K. In these ZnO sheets, the Zn atoms almost completely relax into the plane of O atoms, adopting a planar threefold coordination in a h-BN-like structure.

It should be pointed out that epitaxially grown thin metal oxide films, as well as epitaxially grown silicene films discussed above, are always grown on a substrate and are thus surface dependent. Therefore, strictly, they are not 2D nanosheets. In order to test their stability in two dimensions, one would have to bring them into a suspended, free-standing form by, for example, transferring them onto a TEM grid or across a trench.

Efforts have been made to achieve a 'true' 2D material in the case of silicon nanosheets, however, by means of a different method: chemical exfoliation. Nakano et al. reported on the chemical exfoliation of calcium disilicide ($CaSi_2$) (Nakano et al., 2006). The electrostatic interaction between the Ca^{2+} and the Si^- layers is strong, which requires an adjustment of the charge on the Si layer prior to exfoliation. This was achieved by doping $CaSi_2$ with Mg. $CaSi_{1.85}Mg_{0.15}$ was then immersed in propylamine hydrochloride, which resulted in the deintercalation of the Ca ions, leaving a mixture of silicon sheets and insoluble black metallic solid (Nakano et al., 2006). After sedimentation, the suspension appeared light brown. The monolayer flakes were single crystalline and exhibited a hexagonal lattice with a lattice constant of 0.82 nm, roughly twice that of the (111) plane structure of bulk silicon. X-ray photoelectron spectroscopy revealed a composition of $Si:Mg:O = 7.0:1.3:7.5$. The larger lattice constant could be attributed to a superlattice structure comprising Si, Mg and O atoms (Nakano et al., 2006). Therefore, these chemically prepared sheets are not silicene, i.e. pure silicon sheets, but much rather oxygen-capped and Mg-doped 2D silicon nanosheets that may be applicable for future optoelectronic devices.

2.4.6. Graphene Oxide and Reduced Graphene Oxide

GO and its reduced form rGO have been studied intensively as alternative materials to graphene due to the ease with which their production can be scaled up. The solution-based preparation method that is used to produce GO and rGO will briefly be covered here in order to elucidate the structural differences as compared to mechanically exfoliated graphene. Further preparational details will be provided in Section 4.3.

The technique involves the oxidation of graphite powder in the presence of strong acids and oxidants (Eda et al., 2008; Hummers and Offeman, 1958; Park and Ruoff, 2009). The resulting graphite oxide product has a C:O atomic ratio of 2.0 to 2.9 and a layered structure that is similar to graphite but with a much larger and irregular layer spacing ranging from 0.6 to 1.1 nm depending on the preparation procedure, the remaining interlamellar water content and the graphite precursor (Dreyer et al., 2010; He et al., 1998; Hirata et al., 2004; Horiuchi et al., 2003; Hummers and Offeman, 1958; Wilson et al., 2009). The chemical composition of fully oxidised graphite oxide has been estimated to be approximately $C_4O(OH)$ (de la Cruz and Cowley, 1963). As opposed to hydrophobic graphite, graphite oxide is a highly hydrophilic layered material and can easily be exfoliated by sonication in water, yielding stable dispersions that consist mostly of single-layer sheets, the so-called GO (Gómez-Navarro et al., 2007; Li et al., 2008a; Stankovich et al., 2007). AFM analysis revealed a layer height of approximately 1 nm (Gómez-Navarro et al., 2007; Schniepp et al., 2006; Stankovich et al., 2006a). Theory has predicted a height of 0.78 nm (Schniepp et al., 2006), but the additional height can be attributed to oxygen-containing groups on both sides of the GO layer (Fig. 2.12a) (Gómez-Navarro et al., 2007). Reduction of GO can be accomplished by using a strong chemical reducing agent, e.g. hydrazine, that leads to a significant removal of oxygen (Stankovich et al., 2006a, 2007).

While GO is insulating, rGO is conductive but exhibits strongly reduced conductivity as compared to mechanically exfoliated graphene (Park and Ruoff, 2009). It has been suggested that electronic transport in rGO occurs via the electron hopping over varying distances between the nonoxidised graphene islands that have a size on the order of several nanometres (Gómez-Navarro et al., 2007).

GO can further be covalently functionalised. For example, treatment with organic isocyanates yielded GO with reduced hydrophilic properties that can be exfoliated in polar aprotic solvents, enabling the preparation of graphene-based composite materials (Stankovich et al., 2006b,c). Moreover, the exfoliation of the as-prepared GO in various organic solvents has been demonstrated (Paredes et al., 2008).

The atomic structures of GO and rGO are still a matter of debate due to a rather random functionalisation of each layer and compositional variations depending on the preparation method used. Figure 2.12a shows the chemical structure of a graphite oxide layer as derived from solid-state nuclear magnetic resonance spectroscopy by He et al. (1998). It includes the main features on which a consensus has been reached in the graphene community: graphitic regions of a few nanometres in size appear to be randomly inter-mixed with islands of oxygen-functionalised atoms. The most prevalent functional groups, decorating either side of the basal plane, are hydroxyl and epoxide groups (Cai et al., 2008; Gómez-Navarro et al., 2007; He et al., 1996, 1998; Lerf et al., 1998). It should be noted that the structural model in Fig. 2.12a only shows the chemical connectivity of the functional groups – not their spatial orientation with respect to the basal plane. He et al. concluded that aromatic regions, double-bonds and epoxide groups effect the formation of an almost-flat carbon grid, with only the carbon atoms that are attached to hydroxyl groups causing a slight distortion of the layers (He et al., 1998).

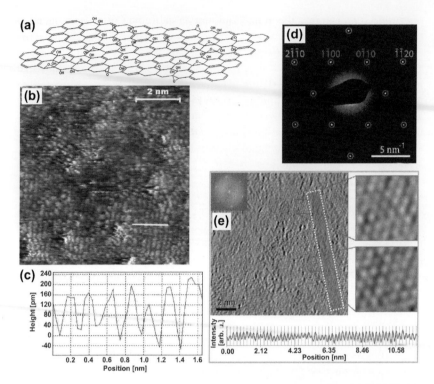

FIGURE 2.12 (a) Proposed atomic model of (one layer of) graphite oxide. Graphitic regions are interrupted by strongly oxidised regions where functional hydroxyl and epoxide groups are located. For clarity, carbonyl and carboxylic edge groups are not shown in the model. *Reprinted from He et al. (1998). Copyright (1998), with permission from Elsevier.* (b) Atomic-scale STM image of the basal plane of rGO nanosheets. (c) Line profile taken along the white line marked in b). *Reprinted with permission from Paredes et al. (2009). Copyright (2009) American Chemical Society.* (d) SAED pattern of a single GO sheet characteristic for a hexagonal lattice. (e) HRTEM micrograph of a single GO sheet with a FFT of the image (inset top left), magnified views as marked with grey boxes and a contrast profile measured along the length of the white dotted box, revealing periodicity over a distance of more than 10 nm. *Reprinted with permission from Wilson et al. (2009). Copyright (2009) American Chemical Society.*

At the sheet edges, carbonyl and carboxylic groups are located, which are not included in Fig. 2.12a for clarity. It is due to the interplane functionalisation that the graphite oxides layers are less strongly bound as compared to graphite and can readily be exfoliated in aqueous solutions (Becerril et al., 2008). If the functional groups are removed, the GO sheets flocculate and precipitate (Stankovich et al., 2006a).

A useful method to determine the degree of disorder in a graphitic material is Raman spectroscopy – a method of analysis that will be explained in detail in Section 5.2. While in graphite or graphene, a sharp G band is obtained that is characteristic for sp^2-hybridised carbon, this G band is broadened and blue-shifted in graphite oxide and GO and an additional disorder-induced D band appears (Kudin et al., 2008; Wilson et al., 2009).

Although it has been known for decades from ED studies that graphite oxide retains the strong crystalline order of graphite (de la Cruz and Cowley, 1962, 1963; Scholz and Boehm, 1969), recent work suggests that GO may predominantly be amorphous or semiamorphous (Mkhoyan et al., 2009; Paredes et al., 2009; Wang et al., 2008b). Using STM, Paredes et al. were able to show that the atomic structure of rGO nanosheets significantly differs from that of pristine graphene prepared by mechanical exfoliation (Paredes et al., 2009). The rGO sheets displayed an undulated globular topology that can be attributed to distortions in the basal plane, induced by the strong oxidation during their preparation. From their atomic-scale STM analysis, the authors could identify ordered domains of a few nanometres in size, intermixed with other domains that showed no ordered patterns. In Fig. 2.12b, an STM image is presented that shows a larger ordered domain, in the bottom right corner, together with a few smaller ones. Figure 2.12c shows a line profile taken from the large domain along the white line in Fig. 2.12b, revealing a peak-to-peak distance of ≈ 0.22 to 0.24 nm, close to the ≈ 0.25 nm periodicity, typically observed for defect-free highly oriented pyrolytic graphite (HOPG). However, long-range atomic-scale order across a rGO sheet has not been observed by Paredes et al. (2009).

Wilson et al. conducted a detailed structural analysis of GO, using a combined HRTEM and SAED approach (Wilson et al., 2009). Figure 2.12d shows a SAED pattern from monolayer GO. The sharp diffraction spots are arranged in a sixfold pattern characteristic of a hexagonal crystal structure similar to that of graphite oxide; sharp spots further indicate short-range order over a length scale of the coherence length of the electron beam (a few nm). Further, SAED patterns from regions several micrometres in size were found to be similar, which is indicative of a long-range orientational order within these areas and suggests that the GO sheets are single crystals. The authors did not find other diffraction spots, which shows that the functional groups present on GO do not form an ordered array. Wilson et al. also extracted the carbon–carbon spacing from the SAED pattern and found it to be indistinguishable from that of graphene (0.1421 ± 0.0007 nm), a finding that strengthens the assumption of structural similarity of GO and graphene, as it implies that the carbon atoms are sp^2-hybridised and form a planar arrangement (cf. Section 2.1). This does not exclude the presence of sp^3-hybridised carbon atoms (which could be distributed randomly across GO); it merely shows that the crystalline carbon component is predominantly and structurally equivalent to that of graphene. Figure 2.12e shows a HRTEM image of monolayer GO together with an FFT (inset) and magnified views that clearly reveal crystalline order (Wilson et al., 2009). A contrast profile, measured along the length of the white dotted box in Fig. 2.12e, reveals periodicity over a distance of more than 10 nm (Wilson et al., 2009), which is clearly larger than the large ordered domain, observed by Paredes et al. (cf. Fig. 2.12b) (Paredes et al., 2009). Other regions appear to be obscured by disordered material (Fig. 2.12e). In HRTEM studies of graphene, often a significant amount of amorphous adsorbates is observed on the surface [e.g. (Meyer et al., 2008)]. Therefore, it is hard to determine whether the disordered regions are oxidised carbons formed and attached, as a result of the GO preparation or spurious surface contaminants adsorbed onto the GO surface (Wilson et al., 2009).

2.4.7. Graphane and Fluorographene

Graphene can be modified by subjecting it to other elements such as hydrogen or fluorine, resulting in the formation of so-called graphane and fluorographene, respectively. Figure 2.13a shows a schematic of the atomic structures of graphane and fluorographene, which are fairly similar. They shall be described in more detail in the following.

Graphane is a stoichiometric derivative of graphene that can be derived from a graphene monolayer by attaching a hydrogen atom to each carbon atom, thus forming a fully saturated hydrocarbon with the formula $(CH)_n$. Its existence has been predicted theoretically (Sofo et al., 2007). In this material, the sp^2-carbon atoms are altered to sp^3, resulting in a modified structure, the removal of the conducting π-bands and the opening of a band gap (Boukhvalov et al., 2008; Sofo et al., 2007). By exposing mechanically exfoliated graphene to a cold hydrogen plasma for typically two hours, Elias et al. demonstrated that graphene can indeed react with atomic hydrogen to form graphane (Elias et al., 2009). They were able to show that the highly conductive zero-overlap semi-metal could be hydrogenated and thus converted into an insulator. It was possible to restore the original conductive state by annealing the sample at 450 °C in Ar atmosphere for 24 h. This reversibility makes this material interesting for use in hydrogen storage.

ED studies revealed that even with prolonged exposure of graphene to the hydrogen plasma, the hexagonal symmetry – and hence the crystallinity – is retained (Elias et al., 2009). However, the lattice constant decreased drastically from $a = 2.46$ Å in graphene to $a \approx 2.42$ Å in graphane (Elias et al., 2009). In graphane hydrogen atoms are attached to the graphene A and B

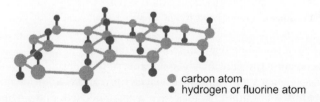

    ● carbon atom
    ● hydrogen or fluorine atom

FIGURE 2.13 Schematic representation of the atomic structures of graphane and fluorographene.

sublattices from opposite sides, which results in 'buckling' of the in-plane carbon atoms and thus a compressed in-plane periodicity. The change from planar sp^2 to tetrahedrally coordinated sp^3 carbon increases the carbon–carbon bond length from $a_0 = 1.42$ Å in graphene to $a_0 \approx 1.53$ Å in graphane and also changes the bond angles (Elias et al., 2009).

While full hydrogenation of a graphene sheet from both sides results in the formation of graphane, single-sided hydrogenation leads to so-called hydrogenated graphene, which is thermodynamically unstable (Elias et al., 2009). One-sided hydrogenation of graphene supported on a substrate becomes energetically favourable due to the existence of microscopic rippling (Meyer et al., 2007a). Since these corrugations are randomly distributed, single-sided hydrogenated graphene is expected to be a disordered material as opposed to the two-sided graphane (Elias et al., 2009).

Fluorographene is another stoichiometric derivative of graphene in which a fluorine atom is attached to each carbon atom, as illustrated in Fig. 2.13. Fluorinating graphene is of interest since stable graphene derivatives may be formed due to the strong bond between carbon and fluorine. Nair et al. succeeded in the formation of fluorographene by fluorinating mechanically cleaved graphene through exposure to atomic fluorine formed by decomposing xenon difluoride at 70 °C (Nair et al., 2010). The obtained partially and fully fluorinated graphene samples were found to be excellent insulators. Fully fluorinated membranes exhibit a high mechanical strength and stiffness, are chemically inert and are thermally stable up to 400 °C even in air. This thermal stability is even better as compared to Teflon®. Fluorographene membranes retain perfect hexagonal symmetry with a unit cell slightly expanded ($a \approx 2.48$ Å) as compared to graphene (Nair et al., 2010).

As pointed out above, this short review on 2D materials other than graphene cannot be exhaustive. However, graphene appears to be only the tip of the iceberg of various 2D materials that have yet to be fully explored. In the attempt to grasp the vast diversity of composition, structure, and functionality available, 2D research will continue to attract and challenge researchers. The theoretical and experimental tools developed for graphene research can, to some degree, also be applied to other 2D materials and, therefore, may accelerate 2D research as a whole.

2.5. NANOSTRUCTURED GRAPHENE

2.5.1. Introduction

Due to graphene's exceptional electronic properties, strong efforts currently go into pushing forward nanoelectronic applications. However, the gapless band structure of truly 2D graphene makes it unsuitable for the direct use in graphene-based field-effect transistors, which is one of the most widely discussed graphene application in electronics. Therefore, in order to accomplish graphene's integration in semiconducting nanoelectronics, it will be necessary to produce graphene nanoribbons (GNRs) with widths below 10 nm, thus introducing further confinement and opening a reasonably large band gap (Barone et al., 2006). The band gap has been reported to strongly depend on the nanoribbon width and the atomic edge structure (Cresti and Roche, 2009; Han et al., 2007; Son et al., 2006). For example, armchair-terminated GNRs can be metallic or semiconducting depending on the ribbon's width (Brey and Fertig, 2006). Besides, it is also the chemical constitution of the ribbon's edge that has a strong influence on the electronic properties of GNRs (Lee and Cho, 2009; Wassmann et al., 2008). Thus, it is highly desirable to find preparation methods that provide GNRs with crystallographically and chemically well-defined edges and atomically smooth-edge termination. In this way, the negative impact of electron and phonon scattering at the edges of the ribbon on its electrical and thermal conduction properties, respectively, can be reduced (Balandin et al., 2008; Cresti and Roche, 2009; Zhao et al., 2008).

Various experimental methods have been employed to structure graphene. They can be roughly divided into two main categories: lithographical approaches, such as electron-beam lithography (Chen et al., 2007; Han et al., 2007), nanowire lithography (Bai et al., 2009), helium-ion lithography (Bell et al., 2009; Lemme et al., 2009), and SPM-based lithography (Giesbers et al., 2008; Tapasztó et al., 2008), and chemical approaches, such as sonochemical cutting (Li et al., 2008b; Wu et al., 2010), catalytic hydrogenation (Ci et al., 2008; Datta et al., 2008) and carbothermal reduction of SiO_2 (Nemes-Incze et al., 2010). GNR fabrication has also been carried out by 'unzipping' CNTs (Elías et al., 2010; Jiao et al., 2009, 2010a; Kosynkin et al., 2009). In order to prepare GNRs in large-scale quantities, CVD approaches were tested; however, the product of these attempts would better be described as graphitic ribbons since the number of graphene layers is large (predominantly larger than 10) and the resulting ribbon width is of the order of 100 nm (Campos-Delgado et al., 2008; Mahanandia et al., 2008). In addition to these approaches, one impressive experiment of bottom-up growth of GNRs has been reported (Cai et al., 2010). In the following, these methods will be discussed and the degree of (edge) control that can be achieved at the atomic level will be evaluated.

2.5.2. Patterning Graphene via Lithography

Fabrication of GNRs with a sufficiently small ribbon width for opening a band gap was first achieved by means of electron beam lithography (Chen et al., 2007; Han et al., 2007). Mechanically exfoliated graphene was patterned by electron beam lithography and then exposed to oxygen plasma. The electron-beam resist served as an etching mask. In Fig. 2.14a an SEM image of parallel GNRs with varying width oriented along the same crystallographic direction of the graphene layer is shown. GNRs with widths ranging from 10 to 100 nm have been fabricated (Han et al., 2007). A second SEM image in Fig. 2.14b depicts GNRs with different orientations on the same graphene sheet (Han et al., 2007). In Fig. 2.14c, the energy gap E_{gap} is plotted as a function of the GNR width W. This plot shows clearly that the energy gap scales inversely with the ribbon width. Nanoribbons with a width of ≈ 15 nm exhibit energy gaps of up to ≈ 0.2 eV. In the inset, the energy gap is plotted as a function of the relative angle θ for devices fabricated along different orientations on the graphene sheet (cf. Fig. 2.14b). These values appear to be randomly scattered, not showing any sign of crystallographic directional dependence. This finding suggests that the edge structure is not well-defined and that edge effects mask the influence of crystallographic orientation on the electronic properties of the GNRs that has been theoretically predicted (Barone et al., 2006; Son et al., 2006).

Although it has been demonstrated that electron beam lithography can generally be used to pattern GNRs and study the opening of a band gap due to transverse electron confinement, for practical applications a larger band gap and smoother edges would be desirable. It was found, however, that GNRs with widths down to 16 nm prepared by electron beam lithography exhibit an impressive breakdown-current density of more than 10^8 A/cm^2, which is comparable with the current-carrying capacity of single-walled CNTs (Murali et al., 2009). In terms of control at the atomic level, state-of-the-art electron beam lithography and subsequent oxygen plasma etching do not allow atomic-level resolution; the resulting GNRs exhibit an edge roughness of approximately 1–3 nm (Bai et al., 2009; Chen et al., 2007). Further, the oxygen plasma

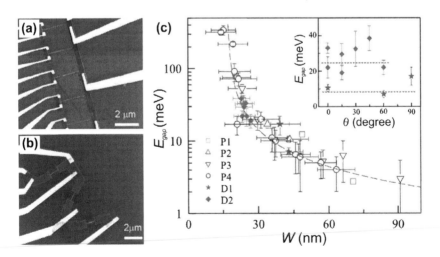

FIGURE 2.14 SEM images of GNRs prepared by electron beam lithography (a) with varying diameter in parallel arrangement and (b) arranged along different angles on the same graphene sheet. (c) Energy gap E_{gap} plotted as a function of the nanoribbon width W. The inset shows E_{gap} as a function of the relative orientation θ of the nanoribbons. *Reprinted with permission from Han et al. (2007). Copyright (2007) by the American Physical Society.*

can affect the properties of these GNRs due to residual oxygen remaining at the GNR edges (Lee and Cho, 2009).

The width of graphene nanostructures can be further reduced by means of high-temperature gas-phase oxidation (Wang and Dai 2010). GNRs with widths ranging from 20 to 30 nm were prepared via electron beam lithography and subsequently, exposed to a mixed atmosphere of oxygen and ammonia at high temperature. Careful tuning of the etching conditions, i.e. gas mixture and temperature, led to graphene being slowly and isotropically etched from the edges inwards due to the higher chemical reactivity of the edge carbon atoms. In this way, the GNRs could be uniformly narrowed from ≈ 20 to 8 nm; further narrowing, however, often resulted in breaks due to edge roughness and variations in the GNR widths after lithographic patterning (Wang and Dai, 2010). A field-effect transistor prepared from a GNR with a width of only ≈ 4 nm exhibited a very high room temperature on–off ratio of $\approx 1 \times 10^4$ and the band gap was estimated to be ≈ 0.4 eV (Wang and Dai, 2010). From Raman spectroscopic measurements, it was concluded that the edge roughness of the GNRs increased during the etching process. Thus, the controlled gas-phase etching represents a tool to manipulate the size of various graphene nanostructures beyond the capability of lithography. However, its success is currently limited by the edge roughness introduced in the patterning process and further increased in the etching process (Wang and Dai, 2010).

GNRs with sub-10 nm width can also be prepared by means of nanowire lithography (Bai et al., 2009; Fasoli et al., 2009). In this process chemically synthesised silicon nanowires are deposited onto graphene and used as the physical protection mask during an oxygen-plasma treatment. The GNR width can be controlled by varying nanowire diameter and etching time. GNRs with widths down to 6 nm have been fabricated in this way (Bai et al., 2009). Branched and crossed GNRs can be fabricated as well if the nanowires are deposited in crossed arrangements. Using this approach, it has been possible to produce room-temperature GNR field-effect transistors from GNRs with a width below 10 nm (Bai et al., 2009). It is suggested that the obtained etching edges of the GNRs are smoother in comparison to edges of GNRs that have been prepared by means of electron beam lithography (Bai et al., 2009). Therefore, nanowire lithography shows the potential to create interesting graphene nanostructures with sub-10 nm measures. However, a challenge lies in controlling the deposition of the nanowire masks on the graphene layer. For large-scale parallel-device integration, a deposition method with positional and directional control will be required. Positional control could be achieved by using metal masks fabricated by means of electron beam lithography instead of randomly depositing nanowires (Lian et al., 2010). This approach could potentially be scaled to produce sub-10 nm GNRs; so far, however, a significant edge roughness is observed.

A similar approach for structuring graphene is nanosphere lithography, where a well-ordered planar array of spheres serves as a mask during an

oxygen-plasma treatment (Cong et al., 2009). A monolayer of colloidal particles, e.g. polystyrene spheres, self assembles on a water surface and is then
transferred onto a graphene layer. During the oxygen-plasma treatment, areas
that are not protected by the spheres are etched, resulting in the formation of
a 2D array of graphene nanodiscs. Varying the sphere diameter results in the
fabrication of graphene nanodiscs with different dimensions and periodicities.
The as-produced graphene nanodisc arrays feature round edges, i.e. edges not
following specific crystallographic directions, and a high density of edges (per
unit area), implicating a high edge reactivity which may offer advantages in
chemical-edge functionalisation (Cong et al., 2009).

An inverse structure, i.e. a graphene nanomesh, can be generated by means
of block copolymer lithography (Bai et al., 2010). The nanomesh is created
using a self-assembled block copolymer with a periodic cylindrical domain
arrangement as etching mask. Neck widths as small as ≈ 7 nm have been
achieved (Bai et al., 2010). A different method that has been employed to
fabricate graphene nanomeshes with nanoribbon widths below 10 nm is
nanoimprint lithography (Liang et al., 2010).

Another area of research aims to develop 'direct-writing' processes for
nanostructuring graphene (Bell et al., 2009; Fischbein and Drndić, 2008; Zhou
et al., 2010). In contrast to electron beam lithography, these methods do not
require the use of photoresists or masks during structuring and thus may
establish the possibility to prepare clean graphene nanostructures. For example,
it has been demonstrated that graphene can be structured using a focused
(≈ 1 nm) electron beam inside a TEM (Fischbein and Drndić, 2008). This
approach allows for structuring with few-nanometre precision. However,
it requires graphene to be suspended on a TEM grid. Preparational transfer
steps that are necessary to obtain suspended graphene sheets often involve
polymer scaffolds or solvents that leave residual adsorbates on the transferred
graphene, to a certain degree attenuating the advantage of mask-free
structuring.

A focused laser beam has been used to structure GO deposited on quartz
substrates (Zhou et al., 2010). Sample areas that are hit by the laser are
locally heated up to 500 °C, resulting in localised oxidative burning of the
GO. In this way, periodic microchannels have been prepared that were only
a few micrometres wide (Zhou et al., 2010). Yet, the resolution of laser
structuring is inherently limited by the wavelength of the laser. In order to
fabricate sub-10 nm nanoribbons techniques with higher resolution have to
be applied.

A helium ion microscope promises high resolution imaging as well as
high precision structuring through ion beam milling. Both the resolution of
a helium ion microscope and the minimum width of a line etched by it are
determined by the size of the helium ion probe (≈ 0.5 nm) and the interaction of the helium ion beam with the sample. The small atomic mass of
the helium ions and a high accelerating voltage result in a relatively small

interaction volume with the surface (Bell et al., 2009; Scipioni et al., 2007). Using this technique, graphene devices with feature sizes of about 10 nm have been successfully demonstrated (Lemme et al., 2009). GNRs with widths as small as 5 nm and an aspect ratio of 60:1 have also been produced (Pickard and Scipioni, 2009). Even GNRs with a stepped width, i.e. 20 nm width at the ends and 5 nm width in the middle of the GNR, have been fabricated, highlighting the versatility of this structuring tool (Pickard and Scipioni, 2009). If the lattice orientation of the graphene layer is known it should, in principle, be possible to structure GNRs with armchair or zigzag edges. Due to the small probe size, the edges should be fairly smooth; however, it remains an open question if atomic resolution can be achieved.

Scanning probe microscopy techniques, i.e. AFM and STM, achieve atomic resolution imaging of flat surfaces and allow for patterning of material structures at the nanoscale (Tseng et al., 2005). In SPM-based lithographic processes, the crystallographic lattice orientation can be determined with high precision before structuring. Therefore, these processes should, in principle, be feasible for the GNR device fabrication with edge control. The attempts that have been made to achieve this goal will briefly be described in the following.

Graphene can be structured by means of local anodic oxidation in an AFM (so called AFM lithography) (Giesbers et al., 2008). In a humid environment that allows for the formation of a water meniscus between the sample and the AFM tip, a positive bias voltage is applied to the graphene sheet and the AFM tip is grounded. In this way, graphene located underneath the AFM tip can be oxidised via an electrochemical oxidation reaction and graphene-oxide tracks can be formed by moving the AFM tip across the graphene sample. The width of the oxidised area depends on the apex of the AFM tip as well as on the bias voltage (Giesbers et al., 2008; Masubuchi et al., 2011). The GNRs can be fabricated by writing two parallel oxidised lines into the graphene. The smallest structures/GNRs that have been patterned in this way were about 25 nm wide (Weng et al., 2008). By varying the bias voltage, it is also possible to tune the degree of oxidation (Masubuchi et al., 2011). Local anodic oxidation, however, works only if the oxidation reaction is started at a graphene edge, possibly due to the higher reactivity of the edges as compared to the basal graphene plane; even at very high voltages (≈ 40 V), it has not been possible to initiate the oxidation reaction within the area of the graphene sheet itself (Giesbers et al., 2008). Compared to electron beam lithography, AFM lithography does not provide major advantages since the crystallographic orientation control has not been demonstrated to date and the resolution, which is limited by the size of the apex of the AFM tip, is not likely to improve (Biró and Lambin, 2010). Further, multitip systems that are essential in order to achieve simultaneous production of multiple devices at industrially relevant patterning times still have to be refined (Koelmans et al., 2012).

In contrast, STM lithography is a nanostructuring technique that enables simultaneous nanoscale control over both GNR width and crystallographic orientation (Tapasztó et al., 2008). Once the lattice orientation is determined by high-resolution imaging patterning is achieved by applying a bias potential, which is significantly higher than the bias used for imaging, and moving the STM tip across the sample. The reaction that effects the patterning underneath the STM tip is based on the local dissociation of water adsorbed on the graphitic substrate under the electron flux emitted from the tip (McCarley et al., 1992). In this way, the preparation of GNRs with widths down to ≈2.5 nm has been demonstrated. These GNRs can selectively be designed to follow specific crystallographic directions; the fabrication of zigzag and armchair GNRs as well as the realisation of a GNR with a 30° kink connecting an armchair and a zigzag ribbon was shown (Tapasztó et al., 2008). Figure 2.15 shows examples of GNRs prepared by means of STM lithography.

An advantage of using the STM for nanostructuring is that the electronic band structure of the as-prepared GNRs can be studied by scanning tunnelling spectroscopy. Armchair GNRs with widths of 10 and 2.5 nm and suitably regular edges exhibited energy gaps of 0.18 and 0.5 eV, respectively (Tapasztó et al., 2008). The latter is comparable with the energy gap of Germanium (0.67 eV) and allows for their integration in room temperature transistors. Although, as described above, orientation control can be achieved using STM lithography, some edge disorder of less than ≈0.3 nm still occurs (Fig. 2.15b,c). Recent investigations on the possibility of parallel STM lithography indicate that structuring of complex nanoscale patterns is feasible using multiple-tip systems (Biró et al., 2011). However, similar to AFM lithography, these techniques still need to be refined.

In addition to the methods described above, a wide variety of soft lithographical methods, transfer printing, microcontact printing and solvent-assisted

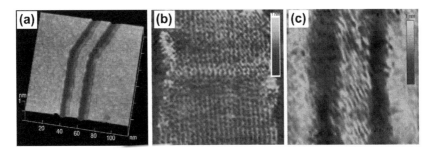

FIGURE 2.15   GNRs prepared by means of STM lithography: (a) GNR with 8 nm width and a 30° kink connecting an armchair and a zigzag ribbon; (b) STM image (12×12 nm^2) of an armchair GNR with 10 nm width; (c) STM image (15×15 nm^2) of an armchair GNR with 2.5 nm width. *Reprinted by permission from Macmillan Publishers Ltd: Nat. Nanotechnol. Tapasztó et al. (2008). Copyright (2008).*

micromolding, have been introduced to structure graphene or to print graphene or graphene-oxide inks (Hong and Jang, 2012). These techniques offer many advantages, including experimental simplicity, high throughput and low cost. Soft stamps, e.g. poly(dimethylsiloxane) stamps, are flexible, easy to fabricate and offer the possibility to transfer material onto flexible substrates, e.g. in a roll-to-roll process. However, they wear out quickly and are only capable of feature resolution of the order of a few micrometres (Allen et al., 2009; Hendricks et al., 2008). Having said that, much smaller structures can be transferred if the graphene is prepatterned using another technique. As an example, nanoimprint lithography has been used to pattern 18 nm wide GNRs on HOPG over an area of more than 5000 μm^2. Subsequently, these GNR arrays were transfer printed onto SiO_2/Si substrates by means of electrostatic force-assisted printing (Wang et al., 2011). Soft lithographical techniques are clearly advancing as they hold great potential for cost-effective large-scale production of graphene nanostructures.

2.5.3. Sonochemical Cutting of Graphene

An impressive method for the fabrication of GNRs with sub-10 nm width and, at the same time, smooth edges has been reported by Li et al. (2008b). They produced GNRs by means of a chemical route that involved dispersing exfoliated graphite in a 1,2-dichloroethane (DCE) solution of PmPV by sonication for 30 min. PmPV is a conjugated polymer that is known to adsorb onto the walls of SWCNTs via π-stacking; here, it covalently functionalises the exfoliated graphite during sonication, and a homogeneous suspension is formed. During sonication, the suspended graphene sheets are chemo-mechanically broken up into various smaller structures, including an appreciable amount of GNRs. After sonication, the suspension contains GNRs with widths ranging from sub-10 to 60 nm, a topographic height between 1 and 1.8 nm, which suggest single-, double- and trilayer GNRs, and an average length of ≈ 1 μm. Figure 2.16a shows an AFM image acquired from a typical sub-10 nm GNR (highlighted with a white arrow). This GNR exhibits a width of $W \sim 2 \pm 0.5$ nm and has been assembled into a field-effect transistor device. In Fig. 2.16b an AFM image of a GNR field-effect transistor with $W \sim 60 \pm 0.5$ nm is presented. The GNR edges are smooth with an edge roughness well below the ribbon width even for sub-10 nm ribbons (Li et al., 2008b; Wang et al., 2008a). Further, bent GNRs with sharp kinks that include an angle of 120° are observed suggesting that sono-chemical cutting occurs along well-defined lattice directions (i.e. the zigzag or armchair direction).

This assumption is further supported by Raman spectroscopy studies by Gupta et al., who did not detect a measurable D peak on some of the examined 2–3 nm wide GNRs, although both edges of the ribbon were simultaneously exposed to the laser excitation (Gupta et al., 2009). This

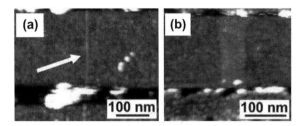

FIGURE 2.16   GNRs prepared by sonochemical cutting: (a) AFM images of a GNR with a width of $W \sim 2 \pm 0.5$ nm that has been assembled into a field-effect transistor device; (b) AFM image of a GNR field-effect transistor with $W \sim 60 \pm 0.5$ nm. *Reprinted with permission from Wang et al. (2008a). Copyright (2007) by the American Physical Society.*

suggests that these chemically produced GNRs exhibit almost perfect zigzag edges. The D peak is a feature in a Raman spectrum that is related to the amount of disorder in a sp^2-carbon sample (cf. Section 5.2). While a perfect armchair edge does give rise to a D peak, this peak cannot be produced by a perfect zigzag edge due to momentum conservation (Cançado et al., 2004; Casiraghi et al., 2009). However, as pointed out by Gupta et al., high resolution imaging techniques should be employed to confirm this finding (Gupta et al., 2009).

Field-effect transistors based on these high quality sub-10 nm GNRs have been found to exhibit similar electronic properties as compared to small-diameter CNT devices (Li et al., 2008b; Wang et al., 2008a). However, in contrast to typical samples of SWCNTs that contain one-third metallic species, the sub-10 nm GNRs were all semiconducting. The field-effect transistors were characterised by on–off ratios of more than 10^5 at room temperature and an energy gap of up to ≈ 0.4 eV (Li et al., 2008b; Wang et al., 2008a).

2.5.4. Crystallographically Selective Structuring of Graphene Through Anisotropic Etching

Other possible ways to achieve graphene patterning with crystallographic orientation control are anisotropic etching techniques such as catalytic hydrogenation (Ci et al., 2008; Datta et al., 2008), catalytic oxidation (Severin et al., 2009) and carbothermal reduction of SiO_2 (Nemes-Incze et al., 2010). These techniques rely on the different reactivity of the graphene/graphite edges, which result in different reaction rates for carbon removal from specific sites/edges.

Graphene can be patterned by means of catalytic hydrogenation and catalytic oxidation reactions. This approach involves the dispersion of catalytic nanoparticles on a graphite or graphene sheet and the subsequent exposure of the as-prepared sample to hydrogen or oxygen at elevated temperatures. The thermally activated nanoparticles then start to act as 'knives', cutting trenches along specific crystallographic directions of the graphene or graphite support. The predominant alignment of their movement with a specific crystallographic direction is intrinsic to the process; further the etching is dependant on the interaction of the specific catalyst with the graphene edges (Ci et al., 2008; Datta et al., 2008; Schäffel et al., 2009; Severin et al., 2009).

As a first step of the catalytic hydrogenation reaction, the catalyst particle, typically Fe, Ni or Co, dissociates molecular hydrogen. Atomic hydrogen then diffuses to the graphene edge, where it reacts with carbon to form methane. In this way, all catalyst particles that lie at a graphene edge "eat up" graphitic carbon starting from that edge and produce etch channels with widths equal to the particle size.

Figure 2.17a shows an SEM image of a HOPG surface where nanosized Ni nanoparticles have etched channels in a heat treatment at 1000 °C under Ar/H_2 gas flow. Several straight channels have formed. Each of them starts from a graphite-step edge, demonstrating that an edge is necessary to initiate the etching process. In Fig. 2.17b, an STM image is shown where the etching directions are indicated, as determined through comparison with the crystallographic orientation of the graphene flake in the high resolution STM image in Fig. 2.17c. Interestingly, channels with widths of more than 10 nm appear to be parallel to a $[11\bar{2}0]$ (zigzag) direction, whereas narrower channels are predominantly aligned with an $[1\bar{1}00]$ (armchair) direction (Ci et al., 2008). This observation suggests that patterning of graphene with structurally defined edge termination can be achieved via catalytic hydrogenation.

The formation of sub-10 nm GNRs by means of this approach has been reported by Campos et al. (2009). The authors carried our catalytic hydrogenation experiments on monolayer graphene using Ni nanoparticles as catalysts. They found that when an etching nanoparticle gets into close vicinity (about 10 nm) of a previously etched trench, it is deflected before reaching the trench. Moreover, they observed trenches where the catalytic particle first approaches another trench and is then deflected to continue etching in a direction parallel to the other trench that results in the formation of GNRs with widths of 10 nm and below (Campos et al., 2009). Catalytic hydrogenation has also been reported using nonmetal SiO_2 nanoparticles as cutting tools (Gao et al., 2009b). This approach opens up a possibility of orientation-controlled graphene patterning without metal contamination.

Catalytic oxidation has been reported using Ag nanoparticles as catalysts (Severin et al., 2009). The process is based on oxygen adsorption and dissociation on the surface of the Ag nanoparticle and subsequent diffusion of atomic oxygen towards the graphene edge, where CO and CO_2 formation takes place. Annealing at 650 °C in air led to the formation of straight and spiralling trenches at extremely high speeds of up to 250 nm/s. The authors attributed the formation of spiralling trenches to inhomogeneous catalytic activity of the

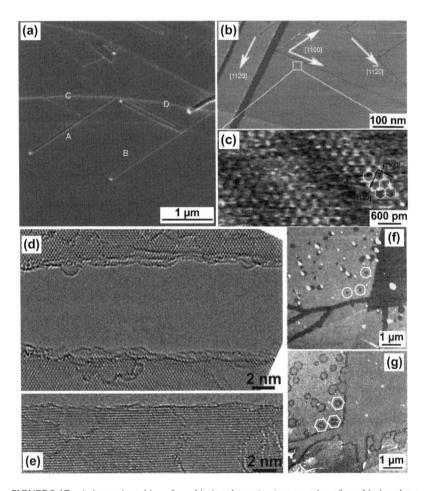

FIGURE 2.17   Anisotropic etching of graphite/graphene: (a–e) nanocutting of graphite/graphene by catalytic hydrogenation: (a) SEM image of channels cut by Ni nanoparticles through a graphite surface. Etching starts from graphite steps and straight channels are formed; (b) STM image of nanocut graphite: typical etch orientations are along the zigzag and armchair lattice directions, as derived from the high-resolution STM image in (c). *Reprinted with permission from Ci et al. (2008). Copyright (2008) Springer.* (d, e) HRTEM micrographs of typical edge structures obtained from catalytically etching suspended FLG using Co nanoparticles. *Reprinted with permission from Schäffel et al. (2011). Copyright (2011) American Chemical Society.* (f, g) Carbothermal reduction of graphene: (f) AFM image of etch pits formed during graphene oxidation in N_2/O_2 atmosphere at 500 °C. Three pits are highlighted with circles; (g) the respective etch pits (marked with hexagons) have grown into hexagonal holes during a subsequent heat treatment at 700 °C in Ar. *Reprinted with permission from Nemes-Incze et al. (2010). Copyright (2010) Springer.*

particles due to contaminations. In the case of the straight trenches, high resolution STM studies reveal edges with a peak-to-peak roughness below 2 nm (Severin et al., 2009).

Catalytic etching of the suspended FLG has also been demonstrated (Schäffel et al., 2011). Co nanoparticles were deposited on the suspended FLG flakes and exposed to an Ar/H_2 gas mixture containing 25 % H_2 at 800 °C. Interestingly, nanoparticles that were located close to the flake edge etched channels into the flake, at the same time fully cutting through all the graphene layers, having no support beneath. This made it possible to study the edge structure of these cuts in greater detail, using HRTEM. Figure 2.17d shows an HRTEM micrograph of an etch channel along a zigzag direction, revealing a typical edge roughness of 1–2 nm, that is comparable with what was reported in the context of catalytic oxidation reactions (Severin et al., 2009). In the best case scenario, sub-nanometre edge smoothness of ≈ 0.6 nm has been observed over a significant edge length (cf. Fig. 2.17e) (Schäffel et al., 2011). These results demonstrate the potential of orientational structuring of graphene via catalytic etching. However, atomically smooth edges have not yet been reported. Further, in the experiments conducted, to date, catalyst particles were always placed in a random manner. In order to achieve the controlled fabrication of GNRs, the deposition of the catalysts will have to be carried out with positional control at nanometre scale instead, which remains a challenge.

The third crystallographically selective etching technique mentioned above, the carbothermal reduction of SiO_2, does not involve the use of catalytic nanoparticles. Here, graphene is prepared by, for example, mechanical exfoliation and is then placed onto a thermally oxidised Si wafer (Nemes-Incze et al., 2010). During an exposure to a N_2/O_2 mixture at 500 °C, circular etch pits form on the graphene surface. Figure. 2.17f shows an AFM image where three etch pits have been marked with circles. In a second heat treatment at 700 °C in Ar, these etch pits grow in size at a rate of about 0.2 nm/min, forming hexagonal etch patterns as can be seen in the AFM image depicted in Fig. 2.17g. These hexagonal holes develop through carbothermal reduction of the SiO_2 substrate in the presence of a graphene edge via the reaction $SiO_2 + C \xrightarrow{700\ °C} SiO(g) + CO(g)$ (Nemes-Incze et al., 2010). The hexagonal patterns are all oriented in the same direction relative to each other, indicating that carbon removal occurs at different rates from different edges. High-resolution STM images revealed the crystallographic orientation of the graphene flake. From this, the edge type could be derived. All edges were found to be of the zigzag type, suggesting that carbon atoms at an armchair edge have a higher reaction rate under these conditions.

This technique also allows for positional control of the etched structures. For example, using the AFM tip as an indentation tool allows for prepatterning of defect sites. The use of multiple tip systems permits parallel production of localised defect structures (Biró et al., 2011). The size of the hexagonal holes can be adjusted by varying the annealing time; in this way, the widths of the remaining graphene strips, i.e. GNRs with well-defined zigzag edge structure, can be tuned as well. Even more complex architectures, as e.g. Y-junctions, can easily be obtained (Nemes-Incze et al., 2010). Confocal Raman spectroscopy has confirmed the formation of well-defined zigzag edges (Krauss et al., 2010). Hardly, any signature of the D peak could be detected along the edges, which indicates that these edges were almost purely of the zigzag type.

2.5.5. Graphene Nanoribbon Formation by 'Unzipping' Carbon Nanotubes

Various techniques have been reported to fabricate GNRs, by 'unzipping' CNTs. These approaches involve plasma etching of partly embedded CNTs (Jiao et al., 2009), solution-phase oxidative processes (Jiao et al., 2010a; Kosynkin et al., 2009) as well as catalytic cutting of CNTs (Elías et al., 2010).

Jiao et al. report on the production of GNRs by unzipping multiwalled carbon nanotubes that were partly embedded in a polymer film by means of Ar plasma etching (Jiao et al., 2009). In a first step, MWCNTs were deposited on a Si substrate and a PMMA layer was spin-coated on top. After curing, the PMMA–MWCNT film was peeled off, leaving MWCNTs with a narrow strip of side wall not covered by PMMA. During plasma treatment, the unprotected MWCNT side walls were removed leaving single-, bi- and multilayer GNRs, depending on the number of walls of the original MWCNTs and the etching time. The starting MWCNTs had an average diameter of ≈ 8 nm. Figure 2.18a shows an AMF image of raw MWCNTs as dispersed on a Si substrate. After plasma etching, single- and few-layer GNRs with smooth edges and widths between 10 and 50 nm have been obtained (cf. Fig. 2.18b-j). However, as opposed to sonochemically cut GNRs, a small D peak has been detected from these unzipped CNTs, indicating that the edges are not pure zigzag edges but contain some disorder (Gupta et al., 2009; Jiao et al., 2009).

Further, aligned GNR arrays can be obtained by means of unzipping partially embedded and horizontally aligned CNTs through Ar-plasma etching (Jiao et al., 2010b). Here, CNTs with a mean diameter of ≈ 1.4 nm were employed, and thus, the width distribution of the GNRs fell into a smaller range of 2–8 nm. In these studies approximately 80% of the CNTs have been transformed into narrow GNRs. Also, the fabrication of crossbar structures by transferring an aligned GNR array onto another GNR array with a 90° rotation has been demonstrated (Jiao et al., 2010b).

Longitudinal unzipping of MWCNTs has also been achieved in a solution-based oxidative process yielding almost 100% nanoribbon structures (Kosynkin et al., 2009). MWCNTs with a starting diameter of 40–80 nm were oxidised by suspending them in concentrated H_2SO_4 followed by the addition of $KMnO_4$ and stirring for 1 h at 55–70 °C. During this treatment, the MWCNT side walls are attacked and gradually unzipped to form nanoribbons. Due to the relatively large outer CNT diameters, the nanoribbon width was found to be larger than

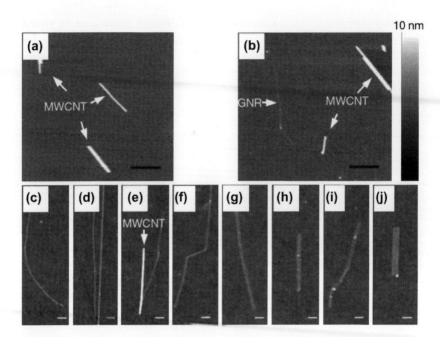

FIGURE 2.18   GNR formation through unzipping of partially embedded CNTs through Ar plasma etching: (a) AFM image of the raw MWCNTs; (b) AFM of the sample after Ar plasma treatment: thin GNRs and thicker MWCNTs are apparent. (black scale bars: 1 μm); (c-j) Single- or few-layer GNRs of different widths W and heights H: (c) $W \sim 7$ nm, $H \sim 1.8$ nm, (d) $W \sim 8$ nm, $H \sim 1.8$ nm (left) and $W \sim 13$ nm, $H \sim 2.0$ nm (right), (e) $W \sim 15$ nm, $H \sim 0.9$ nm, (f) $W \sim 17$ nm, $H \sim 1.0$ nm, (g) $W \sim 25$ nm, $H \sim 1.1$ nm, (h) $W \sim 33$ nm, $H \sim 1.4$ nm, (i) $W \sim 45$ nm, $H \sim 0.8$ nm, (j) $W \sim 51$ nm, $H \sim 1.9$ nm. (white scale bars: 100 nm). *Reprinted by permission from Macmillan Publishers Ltd: Nature Jiao et al. (2009). Copyright (2009).*

100 nm. Narrower GNRs have been obtained from unzipping SWCNTs via $KMnO_4$ oxidation; however, their subsequent disentanglement and dispersion appeared to be challenging (Kosynkin et al., 2009). The as-prepared nano-ribbons possessed oxygen-containing groups, such as epoxy and hydroxyl functionalities on the basal planes and carbonyl and carboxylic groups at their edges, similar to the atomic structure of GO (cf. Section 2.4) and have been found to be poor conductors. An additional reduction in aqueous hydrazine (N_2H_4) or by annealing in H_2 restored the conductivity; however, the electronic characteristics remained inferior to those of the mechanically exfoliated graphene and GNRs derived from other techniques (Kosynkin et al., 2009).

GNRs can also be produced by sonochemical unzipping of mildly gas-phase oxidised MWCNTs in an organic solvent (Jiao et al., 2010a). In a first step, MWCNTs were calcined in air at 500 °C. This treatment is known to etch MWCNTs at defect sites and ends without affecting the pristine CNT side walls. Subsequently, the CNTs were dispersed in a DCE organic solution of

PmPV by sonication. After ultracentrifugation, a high percentage of GNRs (>60%) was found in the supernatant. The GNRs were found to be single-, bi- or trilayered with widths between 10 and 30 nm. A small D peak has been detected from these GNRs suggesting that oxidation and sonication did hardly affect the defect density (Jiao et al., 2010a). HRTEM showed GNRs with Moiré patterns, which indicates that these few-layer GNRs do not exhibit AA or AB stacking order as a result from the random stacking of the layers within the MWCNT (Xie et al., 2011). While most of the GNRs exhibited a random chiral angle distribution (the chiral angle being defined as the angle between the GNR axis and the zigzag direction), a small fraction of bilayer GNRs was observed, where either both layers were oriented close to the armchair (30°) or the zigzag (0°) direction. Many GNRs showed straight edges with an edge roughness below 1 nm (Xie et al., 2011). The high yield and good quality of GNRs obtained through mild oxidation and subsequent sonochemical cutting of MWCNT make this approach highly appealing for future large-scale integration.

Nonoxidised GNRs with width >100 nm have been produced by splitting MWCNTs with potassium vapour (Kosynkin et al., 2011). Here, potassium and MWCNTs with diameters in the range of 40–80 nm were placed into a glass tube, evacuated and sealed. During a heat treatment at 250 °C for 14 h, a potassium intercalation compound was formed. Subsequent quenching with ethanol affected the longitudinal splitting of the CNTs. Exfoliation of the split CNTs has then been achieved via sonication in chlorosulfonic acid. The as-received GNRs were free from oxidative damage and exhibited conductivities parallelling the electronic properties of mechanically exfoliated graphene (Kosynkin et al., 2011).

Unzipping of MWCNTs was also demonstrated by catalytic etching using Co and Ni nanoparticles (Elías et al., 2010). Similar to the studies that report on the structuring of graphite or graphene through catalytic hydrogenation (cp. Fig. 2.17), Co and Ni catalyst particles can also cut trenches into MWCNTs when exposed to hydrogen at elevated temperatures. Some MWCNTs were found to be partially or fully cut along the tube axis; others were cut in two (Elías et al., 2010). Thus, control of the cutting direction is still lacking due to the mixed chiralities of the MWCNTs. An alternative may be to employ chirality-selected CNT; yet, the preparation of chirality-selected CNTs is costly. Also the edge structure of the longitudinally cut ribbons appeared to be rather rough.

2.5.6. Bottom-up Fabrication of Graphene Nanostructures

As compared to the top-down approaches introduced above, bottom-up growth protocols for GNR fabrication promise atomically precise control with regard to the ribbon-edge structure (Cai et al., 2010; Yang et al., 2008). Cai et al. demonstrated the surface-supported formation of well-defined GNRs with widths of ≈ 1 nm and lengths of ≈ 30 nm through surface-assisted coupling of molecular precursors into linear polyphenylenes and their subsequent cyclodehydrogenation (Cai et al., 2010). The schematic in Fig. 2.19a sketches the basic fabrication steps for an armchair GNR of width $N = 7$. The 10,10'-dibromo-9,9'-bianthryl precursor monomers (left panel) were thermally evaporated and adsorbed on a Au(111) surface that was kept at 200 °C to induce dehalogenation and monomer diffusion on the substrate and hence, the formation of polymer chains through covalent interlinking of the dehalogenated intermediates (middle panel). A second heat treatment at 400 °C induced intramolecular cyclodehydrogenation of the polymer chain and hence the fully aromatic armchair GNR has been obtained (right panel). A high-resolution STM image of the as-prepared GNRs is shown in Fig. 2.19b. Comparison with STM simulation revealed that atomically precise and fully hydrogenated armchair GNRs with width $N = 7$ are obtained. A molecular model of the armchair GNR is overlaid in the bottom right of the image. The topology, width and edge periphery of the GNR products can be controlled by the structure of the precursor monomers. For example, Cai et al. also demonstrated the synthesis of chevron-type nanoribbons using 6,11-dibromo-1,2,3,4-tetraphenyltriphenylene precursor monomers that have a different functionality pattern (Fig. 2.19c). Further, chemically doped GNRs may be synthesised from precursor monomers with phenyl rings substituted by

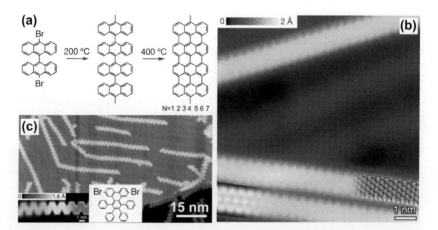

FIGURE 2.19 Bottom-up growth of GNRs: (a) reaction steps for an armchair GNR of width $N = 7$ as derived from 10,10'-dibromo-9,9'-bianthryl precursor monomers; (b) high-resolution STM image of the $N = 7$-armchair GNRs. A molecular model of the GNR is overlaid in the bottom right of the image; (c) STM image of chevron-type GNRs. The insets show a high-resolution STM image together with an overlaid atomic model as well as the 6,11-dibromo-1,2,3,4-tetraphenyl-triphenylene precursor monomer. *Reprinted by permission from Macmillan Publishers Ltd: Nature Cai et al. (2010). Copyright (2010).*

heterocycles forming GNRs with heteroatoms at specific positions along the GNR edge (Cai et al., 2010).

To summarise, this chapter shows that a vast variety of techniques is available to structure graphene. However, the reported top-down approaches have not yet demonstrated control at the atomic level that is essential for many applications. Possible exceptions are the sonochemical cutting of graphene (Li et al., 2008b) and the carbothermal reduction of SiO_2 (Nemes-Incze et al., 2010). Atomically precise bottom-up fabrication of GNRs is another promising approach and can be carried out at moderate temperatures that are compatible with current semiconductor processing (Cai et al., 2010). Many efforts have been made to characterise the graphene edge structure at the atomic level; here, a thorough study comparing the different structuring techniques and making use of different characterisation tools, including HRTEM and Raman spectroscopy, would be desirable. To date, large-scale high yield fabrication of GNRs with narrow widths and smooth edges remains a great challenge.

REFERENCES

Ajayan, P.M., Ebbesen, T.W., 1997. Nanometre-size tubes of carbon. Rep. Prog. Phys. 60, 1025–1062.

Alem, N., Erni, R., Kisielowski, C., Rossell, M.D., Gannett, W., Zettl, A., 2009. Atomically thin hexagonal boron nitride probed by ultrahigh-resolution transmission electron microscopy. Phys. Rev. B 80, 155425.

Allen, M.J., Tung, V.C., Gomez, L., Xu, Z., Chen, L.-M., Nelson, K.S., Zhou, C., Kaner, R.B., Yang, Y., 2009. Soft transfer printing of chemically converted graphene. Adv. Mater. 21, 2098–2102.

Andújar, J.L., Bertran, E., Maniette, Y., 1996. Microstructure of highly oriented, hexagonal, boron nitride thin films grown on crystalline silicon by radio frequency plasma-assisted chemical vapor deposition. J. Appl. Phys. 80, 6553–6555.

Aoki, M., Amawashi, H., 2007. Dependence of band structures on stacking and field in layered graphene. Solid State Commun. 142, 123–127.

Aufray, B., Kara, A., Vizzini, S., Oughaddou, H., Léandri, C., Ealet, B., Le Lay, G., 2010. Graphene-like silicon nanoribbons on Ag(110): a possible formation of silicene. Appl. Phys. Lett. 96, 183102.

Bai, J., Duan, X., Huang, Y., 2009. Rational fabrication of graphene nanoribbons using a nanowire etch mask. Nano Lett. 9, 2083–2087.

Bai, J., Zhong, X., Jiang, S., Huang, Y., Duan, X., 2010. Graphene nanomesh. Nature Nanotechnol. 5, 190–194.

Balandin, A.A., Ghosh, S., Bao, W., Calizo, I., Teweldebrhan, D., Miao, F., Lau, C.N., 2008. Superior thermal conductivity of single-layer graphene. Nano Lett, 8, 902–907.

Bangert, U., Gass, M.H., Bleloch, A.L., Nair, R.R., Geim, A.K., 2009. Manifestation of ripples in free-standing graphene in lattice images obtained in an aberration-corrected scanning transmission electron microscope. Phys. Stat. Sol. A 206, 1117–1122.

Bao, W., Miao, F., Chen, Z., Zhang, H., Jang, W., Dames, C., Lau, C.N., 2009. Controlled ripple texturing of suspended graphene and ultrathin graphite membranes. Nature Nanotechnol. 4, 562–566.

Barone, V., Hod, O., Scuseria, G.E., 2006. Electronic structure and stability of semiconducting graphene nanoribbons. Nano Lett. 6, 2748–2754.

Becerril, H.A., Mao, J., Liu, Z., Stoltenberg, R.M., Bao, Z., Chen, Y., 2008. Evaluation of solution-processed reduced graphene oxide films as transparent conductors. ACS Nano 2, 463–470.

Bell, D.C., Lemme, M.C., Stern, L.A., Williams, J.R., Marcus, C.M., 2009. Precision cutting and patterning of graphene with helium ions. Nanotechnol. 20, 455301.

Bethune, D.S., Klang, C.H., De Vries, M.S., Gorman, G., Savoy, R., Vazquez, J., Beyers, R., 1993. Cobalt-catalysed growth of carbon nanotubes with single-atomic-layer walls. Nature 363, 605–607.

Bieri, M., Treier, M., Cai, J., Aït-Mansour, K., Ruffieux, P., Gröning, O., Gröning, P., Kastler, M., Rieger, R., Feng, X., Müllen, K., Fasel, R., 2009. Porous graphenes: two-dimensional polymer synthesis with atomic precision. Chem. Commun. 45, 6919–6921.

Biró, L.P., Lambin, P., 2010. Nanopatterning of graphene with crystallographic orientation control. Carbon 48, 2677–2689.

Biró, L.P., Nemes-Incze, P., Dobrik, G., Hwang, C., Tapasztó, L., 2011. Graphene nanopatterns with crystallographic orientation control for nanoelectronic applications. Diam. Relat. Mater. 20, 1212–1217.

Boukhvalov, D.W., Katsnelson, M.I., Lichtenstein, A.I., 2008. Hydrogen on graphene: electronic structure, total energy, structural distortions and magnetism from first-principles calculations. Phys. Rev. B 77, 035427.

Brey, L., Fertig, H A, 2006. Electronic states of graphene nanoribbons studied with the Dirac equation. Phys. Rev. B 73, 235411.

Cahangirov, S., Topsakal, M., Aktürk, E., Sahin, H., Ciraci, S., 2009. Two- and one-dimensional honeycomb structures of silicon and germanium. Phys. Rev. Lett. 102, 236804.

Cai, W., Piner, R.D., Stadermann, F.J., Park, S., Shaibat, M.A., Ishii, Y., Yang, D., Velamakanni, A., An, S.J., Stoller, M., An, J., Chen, D., Ruoff, R.S., 2008. Synthesis and solid-state NMR structural characterization of ^{13}C-labeled graphite oxide. Science 321, 1815–1817.

Cai, J., Ruffieux, P., Jaafar, R., Bieri, M., Braun, T., Blankenburg, S., Muoth, M., Seitsonen, A.P., Saleh, M., Feng, X., Müllen, K., Fasel, R., 2010. Atomically precise bottom-up fabrication of graphene nanoribbons. Nature 466, 470–473.

Campos, L.C., Manfrinato, V.R., Sanchez-Yamagishi, J.D., Kong, J., Jarillo-Herrero, P., 2009. Anisotropic etching and nanoribbon formation in single-layer graphene. Nano Lett. 9, 2600–2604.

Campos-Delgado, J., Romo-Herrera, J.M., Jia, X., Cullen, D.A., Muramatsu, H., Kim, Y.A., Hayashi, T., Ren, Z., Smith, D.J., Okuno, Y., Ohba, T., Kanoh, H., Kaneko, K., Endo, M., Terrones, H., Dresselhaus, M.S., Terrones, M., 2008. Bulk production of a new form of sp^2 carbon: crystalline graphene nanoribbons. Nano Lett. 8, 2773–2778.

Cançado, L.G., Pimenta, M.A., Neves, B.R.A., Dantas, M.S.S., Jorio, A., 2004. Influence of the atomic structure on the Raman spectra of graphite edges. Phys. Rev. Lett. 93, 247401.

Casiraghi, C., Hartschuh, A., Qian, H., Piscanec, S., Georgi, C., Fasoli, A., Novoselov, K.S., Basko, D.M., Ferrari, A.C., 2009. Raman spectroscopy of graphene edges. Nano Lett. 9, 1433–1441.

Castro Neto, A.H., Guinea, F., Peres, N.M., 2006. Drawing conclusions from graphene. Physics World 19, 33–37.

Castro Neto, A.H., Guinea, F., Peres, N.M.R., Novoselov, K.S., Geim, A.K., 2009. The electronic properties of graphene. Rev. Mod. Phys. 81, 109–162.

Charlier, J.-C., Gonze, X., Michenaud, J.-P., 1994. First-principle study of the stacking effect on the electronic properties of graphite(s). Carbon 32, 289–299.

Charlier, J.-C., Blase, X., Roche, S., 2007. Electronic and transport properties of nanotubes. Rev. Mod. Phys. 79, 677–732.

Chen, Z., Lin, Y.-M., Rooks, M.J., Avouris, P., 2007. Graphene nano-ribbon electronics. Physica E 40, 228–232.

Ci, L., Xu, Z., Wang, L., Gao, W., Ding, F., Kelly, K.F., Yakobson, B.I., Ajayan, P.M., 2008. Controlled nanocutting of graphene. Nano Res. 1, 116–122.

Coleman, J.N., Lotya, M., O'Neill, A., Bergin, S.D., King, P.J., Khan, U., Young, K., Gaucher, A., De, S., Smith, R.J., Shvets, I.V., Arora, S.K., Stanton, G., Kim, H.-Y., Lee, K., Kim, G.T., Duesberg, G.S., Hallam, T., Boland, J.J., Wang, J.J., Donegan, J.F., Grunlan, J.C., Moriarty, G., Shmeliov, A., Nicholls, R.J., Perkins, J.M., Grieveson, E.M., Theuwissen, K., McComb, D.W., Nellist, P.D., Nicolosi, V., 2011. Two-dimensional nanosheets produced by liquid exfoliation of layered materials. Science 331, 568–571.

Cong, C.X., Yu, T., Ni, Z.H., Liu, L., Shen, Z.X., Huang, W., 2009. Fabrication of graphene nanodisk arrays using nanosphere lithography. J. Phys. Chem. C 113, 6529–6532.

Corso, M., Auwärter, W., Muntwiler, M., Tamai, A., Greber, T., Osterwalder, J., 2004. Boron nitride nanomesh. Science 303, 217–220.

Cresti, A., Roche, S., 2009. Range and correlation effects in edge disordered graphene nano-ribbons. New J. Phys. 11, 095004.

Datta, S.S., Strachan, D.R., Khamis, S.M., Charlie Johnson, A.T., 2008. Crystallographic etching of few-layer graphene. Nano Lett. 8, 1912–1915.

de la Cruz, F.A., Cowley, J.M., 1962. Structure of graphitic oxide. Nature 196, 468–469.

de la Cruz, F.A., Cowley, J.M., 1963. An electron diffraction study of graphitic oxide. Acta Crystallogr. 16, 531–534.

De Padova, P., Quaresima, C., Perfetti, P., Olivieri, B., Léandri, C., Aufray, B., Vizzini, S., Le Lay, G., 2008. Growth of straight, atomically perfect, highly metallic silicon nanowires with chiral asymmetry. Nano Lett. 8, 271–275.

De Padova, P., Quaresima, C., Ottaviani, C., Sheverdyaeva, P.M., Moras, P., Carbone, C., Topwal, D., Olivieri, B., Kara, A., Oughaddou, H., Aufray, B., Le Lay, G., 2010. Evidence of graphene-like electronic signature in silicene nanoribbons. Appl. Phys. Lett. 96, 261905.

Diebold, U., 2003. The surface science of titanium dioxide. Surf. Sci. Rep. 48, 53–229.

Dienstmaier, J.F., Gigler, A.M., Goetz, A.J., Knochel, P., Bein, T., Lyapin, A., Reichlmaier, S., Heckl, W.M., Lackinger, M., 2011. Synthesis of well-ordered COF monolayers: surface growth of nanocrystalline precursors versus direct on-surface polycondensation. ACS Nano 5, 9737–9745.

Dresselhaus, M.S., Dresselhaus, G., Saito, R., 1995. Physics of carbon nanotubes. Carbon 33, 883–891.

Dreyer, D.R., Park, S., Bielawski, C.W., Ruoff, R.,S., 2010. The chemistry of graphene oxide. Chem. Soc. Rev. 39, 228–240.

Ebbesen, T.W., Lezec, H.J., Hiura, H., Bennett, J.W., Ghaemi, H.F., Thio, T., 1996. Electrical conductivity of individual carbon nanotubes. Nature 382, 54–56.

Eda, G., Fanchini, G., Chhowalla, M., 2008. Large-area ultrathin films of reduced graphene oxide as a transparent and flexible electronic material. Nature Nanotechnol. 3, 270–274.

Elias, D.C., Nair, R.R., Mohiuddin, T.M.G., Morozov, S.V., Blake, P., Halsall, M.P., Ferrari, A.C., Boukhvalov, D.W., Katsnelson, M.I., Geim, A.K., Novoselov, K.S., 2009. Control of graphene's properties by reversible hydrogenation: evidence for graphane. Science 323, 610–613.

Elías, A.L., Botello-Méndez, A.R., Meneses-Rodríguez, D., Jehová González, V., Ramírez-González, D., Ci, L., Muñoz-Sandoval, E., Ajayan, P.M., Terrones, H., Terrones, M., 2010. Longitudinal cutting of pure and doped carbon nanotubes to form graphitic nanoribbons using metal clusters as nanoscalpels. Nano Lett. 10, 366–372.

Fasoli, A., Colli, A., Lombardo, A., Ferrari, A.C., 2009. Fabrication of graphene nanoribbons via nanowire lithography. Phys. Stat. Sol. B 246, 2514–2517.

Fasolino, A., Los, J.H., Katsnelson, M.I., 2007. Intrinsic ripples in graphene. Nature Mater. 6, 858–861.

Fischbein, M.D., Drndić, M., 2008. Electron beam nanosculpting of suspended graphene sheets. Appl. Phys. Lett. 93, 113107.

Gamble, F.R., Silbernagel, B.G., 1975. Anisotropy of the proton spin–lattice relaxation time in the superconducting intercalation complex $TaS_2(NH_3)$: structural and bonding implications. J. Chem. Phys. 63, 2544–2552.

Gao, R., Yin, L., Wang, C., Qi, Y., Lun, N., Zhang, L., Liu, Y.-X., Kang, L., Wang, X., 2009. High-yield synthesis of boron nitride nanosheets with strong ultraviolet cathodoluminescence emission. J. Phys. Chem. C 113, 15160–15165.

Gao, L., Ren, W., Liu, B., Wu, Z.-S., Jiang, C., Cheng, H.-M., 2009b. Crystallographic tailoring of graphene by nonmetal SiO_x nanoparticles. J. Am. Chem. Soc. 131, 13934–13936.

Geim, A.K., Novoselov, K.S., 2007. The rise of graphene. Nature Mater. 6, 183–191.

Geringer, V., Liebmann, M., Echtermeyer, T., Runte, S., Schmidt, M., Rückamp, R., Lemme, M.C., Morgenstern, M., 2009. Intrinsic and extrinsic corrugation of monolayer graphene deposited on SiO_2. Phys. Rev. Lett. 102, 076102.

Giesbers, A.J.M., Zeitler, U., Neubeck, S., Freitag, F., Novoselov, K.S., Maan, J.C., 2008. Nanolithography and manipulation of graphene using an atomic force microscope. Solid State Commun. 147, 366–369.

Golberg, D., Bando, Y., Eremets, M., Takemura, K., Kurashima, K., Yusa, H., 1996. Nanotubes in boron nitride laser heated at high pressure. Appl. Phys. Lett. 69, 2045–2047.

Golberg, D., Bando, Y., Huang, Y., Terao, T., Mitome, M., Tang, C., Zhi, C., 2010. Boron nitride nanotubes and nanosheets. ACS Nano. 4, 2979–2993.

Gómez-Navarro, C., Weitz, R.T., Bittner, A.M., Scolari, M., Mews, A., Burghard, M., Kern, K., 2007. Electronic transport properties of individual chemically reduced graphene oxide sheets. Nano Lett. 7, 3499–3503.

Gordon, R.A., Yang, D., Crozier, E.D., Jiang, D.T., Frindt, R.F., 2002. Structures of exfoliated single layers of WS_2, MoS_2 and $MoSe_2$ in aqueous suspension. Phys. Rev. B 65, 125407.

Guo, T., Nikolaev, P., Thess, A., Colbert, D.T., Smalley, R.E., 1995. Catalytic growth of single-walled nanotubes by laser vaporization. Chem. Phys. Lett. 243, 49–54.

Gupta, A.K., Russin, T.J., Gutiérrez, H.R., Eklund, P.C., 2009. Probing graphene edges via Raman scattering. ACS Nano. 3, 45–52.

Haering, R.R., 1958. Band structure of rhombohedral graphite. Can. J. Phys. 36, 352–362.

Hamada, N., Sawada, S., Oshiyama, A., 1992. New one-dimensional conductors: graphitic microtubules. Phys. Rev. Lett. 68, 1579–1581.

Han, M.Y., Özyilmaz, B., Zhang, Y., Kim, P., 2007. Energy band-gap engineering of graphene nanoribbons. Phys. Rev. Lett. 98, 206805.

Han, W.-Q., Wu, L., Zhu, Y., Watanabe, K., Taniguchi, T., 2008. Structure of chemically derived mono- and few-atomic-layer boron nitride sheets. Appl. Phys. Lett. 93, 223103.

Hass, J., Feng, R., Millán-Otoya, J.E., Li, X., Sprinkle, M., First, P.N., de Heer, W.A., Conrad, E.H., Berger, C., 2007. Structural properties of the multilayer graphene/$4H$-SiC-$(000\bar{1})$ system as determined by surface x-ray diffraction. Phys. Rev. B 75, 214109.

Hass, J., Varchon, F., Millán-Otoya, J.E., Sprinkle, M., Sharma, N., de Heer, W.A., Berger, C., First, P.N., Magaud, L., Conrad, E.H., 2008. Why multilayer graphene on $4H$-SiC-$(000\bar{1})$ behaves like a single sheet of graphene. Phys. Rev. Lett. 100, 125504.

Hata, K., Futaba, D.N., Mizuno, K., Namai, T., Yumura, M., Iijima, S., 2004. Waterassisted highly efficient synthesis of impurity-free single-walled carbon nanotubes. Science 306, 1362–1364.

He, H., Riedl, T., Lerf, A., Klinowski, J., 1996. Solid-state NMR studies of the structure of graphite oxide. J. Phys. Chem. 100, 19954–19958.

He, H., Klinowski, J., Forster, M., Lerf, A., 1998. A new structural model for graphite oxide. Chem. Phys. Lett. 287, 53–56.

Hendricks, T.R., Lu, J., Drzal, L.T., Lee, I., 2008. Intact pattern transfer of conductive exfoliated graphite nanoplatelet composite films to polyelectrolyte multilayer platforms. Adv. Mater. 20, 2008–2012.

Hernandez, Y., Nicolosi, V., Lotya, M., Blighe, F.M., Sun, Z., De, S., Mc-Govern, I.T., Holland, B., Byrne, M., Gun'Ko, Y.K., Boland, J.J., Niraj, P., Duesberg, G., Krishnamurthy, S., Goodhue, R., Hutchison, J., Scardaci, V., Ferrari, A.C., Coleman, J.N., 2008. High-yield production of graphene by liquid-phase exfoliation of graphite. Nature Nanotechnol. 3, 563–568.

Hersam, M.C., 2008. Progress towards monodisperse single-walled carbon nanotubes. Nature Nanotechnol. 3, 387–394.

Hirata, M., Gotou, T., Horiuchi, S., Fujiwara, M., Ohba, M., 2004. Thin-film particles of graphite oxide 1: high-yield synthesis and flexibility of the particles. Carbopn 42, 2929–2937.

Hong, J.-Y., Jang, J., 2012. Micropatterning of graphene sheets: recent advances in techniques and applications. J. Mater. Chem. 22, 8179–8191.

Horiuchi, S., Gotou, T., Fujiwara, M., Sotoaka, R., Hirata, M., Kimoto, K., Asaka, T., Yokosawa, T., Matsui, Y., Watanabe, K., Sekita, M., 2003. Carbon nanofilm with a new structure and property. Jpn. J. Appl. Phys. 42, L1073–L1076.

Huang, S., Cai, X., Liu, J., 2003. Growth of millimeter-long and horizontally aligned single-walled carbon nanotubes on flat substrates. J. Am. Chem. Soc. 125, 5636–5637.

Hummers, W.S., Offeman, R.E., 1958. Preparation of graphitic oxide. J. Am. Chem. Soc. 80, 1339.

Iijima, S., Ichihashi, T., 1993. Single-shell carbon nanotubes of 1-nm diameter. Nature 363, 603–605.

Iijima, S., 1991. Helical microtubules of graphitic carbon. Nature 354, 56–58.

Ishigami, M., Chen, J.H., Cullen, W.G., Fuhrer, M.S., Williams, E.D., 2007. Atomic structure of graphene on SiO_2. Nano Lett. 7, 1643–1648.

Jiao, L., Zhang, L., Wang, X., Diankov, G., Dai, H., 2009. Narrow graphene nanoribbons from carbon nanotubes. Nature 458, 877–880.

Jiao, L., Wang, X., Diankov, G., Wang, H., Dai, H., 2010a. Facile synthesis of high-quality graphene nanoribbons. Nature Nanotechnol. 5, 321–325.

Jiao, L., Zhang, L., Ding, L., Liu, J., Dai, H., 2010b. Aligned graphene nanoribbons and crossbars from unzipped carbon nanotubes. Nano Res. 3, 387–394.

Joensen, P., Frindt, R.F., Morrison, S.R., 1986. Single-layer MoS_2. Mater. Res. Bull. 21, 457–461.

José-Yacamán, M., Miki-Yoshida, M., Rendón, L., Santiesteban, J.G., 1993. Catalytic growth of carbon microtubules with fullerene structure. Appl. Phys. Lett. 62, 657–659.

Katsnelson, M.I., Geim, A.K., 2008. Electron scattering on microscopic corrugations in graphene. Phil. Trans. R. Soc. A 366, 195–204.

Kiguchi, M., Entani, S., Saiki, K., Goto, T., Koma, A., 2003. Atomic and electronic structure of an unreconstructed polar MgO(111) thin film on Ag(111). Phys. Rev. B 68, 115402.

Kim, W.D., Hwang, G.W., Kwon, O.S., Kim, S.K., Cho, M., Jeong, D.S., Lee, S.W., Seo, M.H., Hwang, C.S., Min, Y.-S., Chob, Y.J., 2005. Growth characteristics of atomic layer deposited TiO_2 thin films on Ru and Si electrodes for memory capacitor applications. J. Electrochem. Soc. 152, C552–C559.

Kim, J.Y., Jung, H.S., No, J.H., Kim, J.-R., Hong, K.S., 2006. Influence of anatase-rutile phase transformation on dielectric properties of sol-gel derived TiO_2 thin films. J. Electroceram 16, 447–451.

Koelmans, W.W., Peters, T., Berenschot, E., de Boer, M.J., Siekman, M.H., Abelmann, L., 2012. Cantilever arrays with self-aligned nanotips of uniform height. Nanotechnol. 23, 135301.

Kosynkin, D.V., Higginbotham, A.L., Sinitskii, A., Lomeda, J.R., Dimiev, A., Price, B.K., Tour, J.M., 2009. Longitudinal unzipping of carbon nanotubes to form graphene nanoribbons. Nature 458, 872–876.

Kosynkin, D.V., Lu, W., Sinitskii, A., Pera, G., Sun, Z., Tour, J.M., 2011. Highly conductive graphene nanoribbons by longitudinal splitting of carbon nanotubes using potassium vapor. ACS Nano 5, 968–974.

Krauss, B., Nemes-Incze, P., Skakalova, V., Biró, L.P., Klitzing, K. v., Smet, J.H., 2010. Raman scattering at pure graphene zigzag edges. Nano Lett. 10, 4544–4548.

Kudin, K.N., Ozbas, B., Schniepp, H.C., Prud'homme, R.K., Aksay, I.A., Car, R., 2008. Raman spectra of graphite oxide and functionalized graphene sheets. Nano Lett. 8, 36–41.

Lalmi, B., Oughaddou, H., Enriquez, H., Kara, A., Vizzini, S., Ealet, B., Aufray, B., 2010. Epitaxial growth of a silicene sheet. Appl. Phys. Lett. 97, 223109.

Latil, S., Henrard, L., 2006. Charge carriers in few-layer graphene films. Phys. Rev. Lett. 97, 036803.

Latil, S., Meunier, V., Henrard, L., 2007. Massless fermions in multilayer graphitic systems with misoriented layers: ab initio calculations and experimental fingerprints. Phys. Rev. B 76, 201402.

Léandri, C., Le Lay, G., Aufray, B., Girardeaux, C., Avila, J., Dávila, M.E., Asensio, M.C., Ottaviani, C., Cricenti, A., 2005. Self-aligned silicon quantum wires on Ag(110). Surf. Sci. 574, L9–L15.

Léandri, C., Oughaddou, H., Aufray, B., Gay, J.M., Le Lay, G., Ranguis, A., Garreau, Y., 2007. Growth of Si nanostructures on Ag(001). Surf. Sci. 601, 262–267.

Lebègue, S., Eriksson, O., 2009. Electronic structure of two-dimensional crystals from ab initio theory. Phys. Rev. 79, 115409.

Lee, G., Cho, K., 2009. Electronic structures of zigzag graphene nanoribbons with edge hydrogenation and oxidation. Phys. Rev. B 79, 165440.

Lee, C.J., Park, J., Kim, J.M., Huh, Y., Lee, J.Y., No, K.S., 2000. Low-temperature growth of carbon nanotubes by thermal chemical vapor deposition using Pd, Cr, and Pt as co-catalyst. Chem. Phys. Lett. 327, 277–283.

Lemasson, F., Tittmann, J., Hennrich, F., Stürzl, N., Malik, S., Kappes, M.M., Mayor, M., 2011. Debundling, selection and release of SWNTs using fluorene-based photocleavable polymers. Chem. Commun. 47, 7428–7430.

Lemme, M.C., Bell, D.C., Williams, J.R., Stern, L.A., Baugher, B.W.H., Jarillo-Herrero, P., Marcus, C.M., 2009. Etching of graphene devices with a helium ion beam. ACS Nano 3, 2674–2676.

Lerf, A., He, H., Forster, M., Klinowski, J., 1998. Structure of graphite oxide revisited. J. Phys. Chem. B 102, 4477–4482.

Li, W.Z., Xie, S.S., Qian, L.X., Chang, B.H., Zou, B.S., Zhou, W.Y., Zhao, R.A., Wang, G., 1996. Large-scale synthesis of aligned carbon nanotubes. Science 274, 1701–1703.

Li, D., Müller, M.B., Gilje, S., Kaner, R.B., Wallace, G.G., 2008a. Processable aqueous dispersions of graphene nanosheets. Nature Nanotechnol. 3, 101–105.

Li, X., Wang, X., Zhang, L., Lee, S., Dai, H., 2008b. Chemically derived, ultrasmooth graphene nanoribbon semiconductors. Science 319, 1229–1232.

Lian, C., Tahy, K., Fang, T., Li, G., Xing, H.G., Jena, D., 2010. Quantum transport in graphene nanoribbons patterned by metal masks. Appl. Phys. Lett. 96, 103109.

Liang, X., Jung, Y.-S., Wu, S., Ismach, A., Olynick, D.L., Cabrini, S., Bokor, J., 2010. Formation of bandgap and subbands in graphene nanomeshes with sub-10 nm ribbon width fabricated via nanoimprint lithography. Nano Lett. 10, 2454–2460.

Liu, C., Singh, O., Joensen, P., Curzon, A.E., Frindt, R.F., 1984. X-ray and electron microscopy studies of single-layer TaS_2 and NbS_2. Thin Solid Films 113, 165–172.

Liu, Z., Suenaga, K., Harris, P.J.F., Iijima, S., 2009. Open and closed edges of graphene layers. Phys. Rev. Lett. 102, 015501.

Loiseau, A., Willaime, F., Demoncy, N., Hug, G., Pascard, H., 1996. Boron nitride nanotubes with reduced numbers of layers synthesized by arc discharge. Phys. Rev. Lett. 76, 4737–4740.

Lotya, M., Hernandez, Y., King, P.J., Smith, R.J., Nicolosi, V., Karlsson, L.S., Blighe, F.M., De, S., Wang, Z., McGovern, I.T., Duesberg, G.S., Coleman, J.N., 2009. Liquid phase production of graphene by exfoliation of graphite in surfactant/water solutions. J. Am. Chem. Soc. 131, 3611–3620.

Lourie, O.R., Jones, C.R., Bartlett, B.M., Gibbons, P.C., Ruoff, R.S., Buhro, W.E., 2000. CVD growth of boron nitride nanotubes. Chem. Mater. 12, 1808–1810.

Lui, C.H., Liu, L., Mak, K.F., Flynn, G.W., Heinz, T.F., 2009. Ultraflat graphene. Nature 462, 339–441.

Mahanandia, P., Nanda, K.K., Prasad, V., Subramanyam, S.V., 2008. Synthesis and characterization of carbon nanoribbons and single crystal iron filled carbon nanotubes. Mater. Res. Bull. 43, 3252–3262.

Mas-Ballesté, R., Castillo, O., Sanz Miguel, P.J., Olea, D., Gómez-Herrero, J., Zamora, F., 2009. Towards molecular wires based on metal-organic frameworks. Eur. J. Inorg. Chem. 20, 2885–2896.

Mas-Ballesté, R., Gómez-Navarro, C., Gómez-Herrero, J., Zamora, F., 2011. 2D materials: to graphene and beyond. Nanoscale 3, 20–30.

Masubuchi, S., Arai, M., Machida, T., 2011. Atomic force microscopy based tunable local anodic oxidation of graphene. Nano Lett. 11, 4542–4546.

Matte, H. S. S. Ramakrishna, Gomathi, A., Manna, A.K., Late, D.J., Datta, R., Pati, S.K., Rao, C.N.R., 2010. MoS_2 and WS_2 analogues of graphene. Angew. Chem. Int. Ed. 49, 4059–4062.

McCarley, R.L., Hendricks, S.A., Bard, A.J., 1992. Controlled nanofabrication of highly oriented pyrolytic graphite with the scanning tunneling microscope. J. Phys. Chem. 96, 10089–10092.

Mermin, N.D., Wagner, H., 1966. Absence of ferromagnetism or antiferromagnetism in one- or two-dimensional isotropic Heisenberg models. Phys. Rev. Lett. 17, 1133–1136.

Mermin, N.D., 1968. Crystalline order in two dimensions. Phys. Rev. 176, 250–254.

Meyer, J.C., Geim, A.K., Katsnelson, M.I., Novoselov, K.S., Booth, T.J., Roth, S., 2007a. The structure of suspended graphene sheets. Nature 446, 60–63.

Meyer, J.C., Geim, A.K., Katsnelson, M.I., Novoselov, K.S., Obergfell, D., Roth, S., Girit, C., Zettl, A., 2007b. On the roughness of single- and bi-layer graphene membranes. Solid State Commun. 143, 101–109.

Meyer, J.C., Kisielowski, C., Erni, R., Rossell, M.D., Crommie, M.F., Zettl, A., 2008. Direct imaging of lattice atoms and topological defects in graphene membranes. Nano Lett. 8, 3582–3586.

Mintmire, J.W., Dunlap, B.I., White, C.T., 1992. Are fullerene tubules metallic? Phys. Rev. Lett. 68, 631–634.

Miremadi, B.K., Morrison, S.R., 1988. The intercalation and exfoliation of tungsten disulfide. J. Appl. Phys. 63, 4970–4974.

Mkhoyan, K.A., Contryman, A.W., Silcox, J., Stewart, D.A., Eda, G., Mattevi, C., Miller, S., Chhowalla, M., 2009. Atomic and electronic structure of graphene-oxide. Nano Lett. 9, 1058–1063.

Morozov, S.V., Novoselov, K.S., Schedin, F., Jiang, D., Firsov, A.A., Geim, A.K., 2005. Two-dimensional electron and hole gases at the surface of graphite. Phys. Rev. B 72, 201401.

Morozov, S.V., Novoselov, K.S., Katsnelson, M.I., Schedin, F., Elias, D.C., Jaszczak, J.A., Geim, A.K., 2008. Giant intrinsic carrier mobilities in graphene and its bilayer. Phys. Rev. Lett. 100, 016602.

Murali, R., Yang, Y., Brenner, K., Beck, T., Meindl, J.D., 2009. Breakdown current density of graphene nanoribbons. Appl. Phys. Lett. 94, 243114.

Murphy, D.W., Hull, G.W., 1975. Monodispersed tantalum disulfide and adsorption complexes with cations. J. Chem. Phys. 62, 973–978.

Nair, R.R., Ren, W., Jalil, R., Riaz, I., Kravets, V.G., Britnell, L., Blake, P., Schedin, F., Mayorov, A.S., Yuan, S., Katsnelson, M.I., Cheng, H.-M., Strupinski, W., Bulusheva, L.G., Okotrub, A.V., Grigorieva, I.V., Grigorenko, A.N., Novoselov, K.S., Geim, A.K., 2010. Fluorographene: a two-dimensional counterpart of Teflon. Small 6, 2877–2884.

Nakano, H., Mitsuoka, T., Harada, M., Horibuchi, K., Nozaki, H., Takahashi, N., Nonaka, T., Seno, Y., Nakamura, H., 2006. Soft synthesis of single-crystal silicon monolayer sheets. Angew. Chem. 118, 6451–6454.

Nemes-Incze, P., Magda, G., Kamarás, K., Biró, L.P., 2010. Crystallographically selective nano-patterning of graphene on SiO_2. Nano Res. 3, 110–116.

Norimatsu, W., Kusunoki, M., 2010. Selective formation of ABC-stacked graphene layers on SiC(0001). Phys. Rev. B 81, 161410.

Novoselov, K.S., Geim, A.K., Morozov, S.V., Jiang, D., Zhang, Y., Dubonos, S.V., Grigorieva, I.V., Firsov, A.A., 2004. Electric field effect in atomically thin carbon films. Science 306, 666–669.

Novoselov, K.S., Geim, A.K., Morozov, S.V., Jiang, D., Katsnelson, M.I., Grigorieva1, I.V., Dubonos, S.V., Firsov, A.A., 2005a. Two-dimensional gas of massless Dirac fermions in graphene. Nature 438, 197–200.

Novoselov, K.S., Jiang, D., Schedin, F., Booth, T.J., Khotkevich, V.V., Morozov, S.V., Geim, A.K., 2005b. Two-dimensional atomic crystals. Proc. Natl. Acad. Sci. USA 102, 10451–10453.

Odom, T.W., Huang, J.-L., Kim, P., Lieber, C.M., 1998. Atomic structure and electronic properties of single-walled carbon nanotubes. Nature 391, 62–64.

Osada, M., Sasaki, T., 2012. Two-dimensional dielectric nanosheets: novel nanoelectronics from nanocrystal building blocks. Adv. Mater. 24, 210–228.

Osada, M., Ebina, Y., Funakubo, H., Yokoyama, S., Kiguchi, T., Takada, K., Sasaki, T., 2006. High-κ dielectric nanofilms fabricated from titania nanosheets. Adv. Mater. 18, 1023–1027.

Pacilé, D., Meyer, J.C., Girit, Ç.Ö., Zettl, A., 2008. The two-dimensional phase of boron nitride: few-atomic-layer sheets and suspended membranes. Appl. Phys. Lett. 92, 133107.

Paine, R.T., Narula, C.K., 1990. Synthetic routes to boron nitride. Chem. Rev. 90, 73–91.

Paredes, J.I., Villar-Rodil, S., Martínez-Alonso, A., Tascón, J.M.D., 2008. Graphene oxide dispersions in organic solvents. Langmuir 24, 10560–10564.

Paredes, J.I., Villar-Rodil, S., Solis-Fernandez, P., Martinez-Alonso, A., Tascon, J.M.D., 2009. Atomic force and scanning tunneling microscopy imaging of graphene nanosheets derived from graphite oxide. Langmuir 25, 5957–5968.

Park, S., Ruoff, R.S., 2009. Chemical methods for the production of graphenes. Nature Nanotechnol. 4, 217–224.

Partoens, B., Peeters, F.M., 2006. From graphene to graphite: electronic structure around the K point. Phys. Rev. B 74, 075404.

Peierls, R., 1934. Bemerkungen über Umwandlungstemperaturen. Helv. Phys. Acta 7, 81–83.

Peierls, R., 1935. Quelques propriétés typiques des corps solides. Ann. Inst. Henri Pointcare 5, 177–222.

Pickard, D., Scipioni, L., 2009. Graphene Nano-Ribbon Patterning in the ORION® PLUS. http://www.microscopy.info/Content/pdf/ORION_PLUS_Graphene.pdf.

Reich, S., Thomsen, C., Maultzsch, J., 2004. Carbon Nanotubes. Wiley-VCH, Berlin.

Ren, Z.F., Huang, Z.P., Xu, J.W., Wang, J.H., Bush, P., Siegal, M.P., Provencio, P.N., 1998. Synthesis of large arrays of well-aligned carbon nanotubes on glass. Science 282, 1105–1107.

Roy, H.V., Kallinger, C., Sattler, K., 1998. Study of single and multiple foldings of graphitic sheets. Surf. Sci. 407, 1–6.

Rümmeli, M.H., Schäffel, F., Kramberger, C., Gemming, T., Bachmatiuk, A., Kalenczuk, R.J., Rellinghaus, B., Büchner, B., Pichler, T., 2007. Oxide-driven carbon nanotube growth in supported catalyst CVD. J. Am. Chem. Soc. 129, 15772–15773.

Saito, R., Dresselhaus, G., Dresselhaus, M.S., 1998. Physical Properties of Carbon Nanotubes. Imperial College Press, London.

Sakamoto, J., van Heijst, J., Lukin, O., Schlüter, A.D., 2009. Two-dimensional polymers: just a dream of synthetic chemists? Angew. Chem. Int. Ed. 48, 1030–1069.

Schäffel, F., Warner, J.H., Bachmatiuk, A., Rellinghaus, B., Büchner, B., Schultz, L., Rümmeli, M.H., 2009. Shedding light on the crystallographic etching of multi-layer graphene at the atomic scale. Nano Res. 2, 695–705.

Schäffel, F., Wilson, M., Bachmatiuk, A., Rümmeli, M.H., Queitsch, U., Rellinghaus, B., Briggs, G.A.D., Warner, J.H., 2011. Atomic resolution imaging of the edges of catalytically etched suspended few layer graphene. ACS Nano 5, 1975–1983.

Schniepp, H.C., Li, J.-L., McAllister, M.J., Sai, H., Herrera-Alonso, M., Adamson, D.H., Prud'homme, R.K., Car, R., Saville, D.A., Aksay, I.A., 2006. Functionalized single graphene sheets from splitting graphite oxide. J. Phys. Chem. B 110, 8535–8539.

Scholz, W., Boehm, H.P., 1969. Betrachtungen zur Struktur des Graphitoxids. Z. Anorg. Allg. Chem. 369, 327–340.

Scipioni, L., Stern, L., Notte, J., 2007. Applications of the helium ion microscope. Micros. Today 15, 12–15.

Severin, N., Kirstein, S., Sokolov, I.M., Rabe, J.P., 2009. Rapid trench channeling of graphenes with catalytic silver nanoparticles. Nano Lett. 9, 457–461.

Smith, R.J., King, P.J., Lotya, M., Wirtz, C., Khan, U., De, S., O'Neill, A., Duesberg, G.S., Grunlan, J.C., Moriarty, G., Chen, J., Wang, J., Minett, A.I., Nicolosi, V., Coleman, J.N., 2011. Large-scale exfoliation of inorganic layered compounds in aqueous surfactant solutions. Adv. Mater. 23, 3944–3948.

Sofo, J.O., Chaudhari, A.S., Barber, G.D., 2007. Graphane: a two-dimensional hydrocarbon. Phys. Rev. B 75, 153401.

Son, Y.-W., Cohen, M.L., Louie, S.G., 2006. Energy gaps in graphene nanoribbons. Phys. Rev. Lett. 97, 216803.

Spence, J.C.H., 2003. High-resolution Electron Microscopy. Oxford University Press.

Stankovich, S., Piner, R.D., Chen, X., Wu, N., Nguyen, S.T., Ruoff, R.S., 2006a. Stable aqueous dispersions of graphitic nanoplatelets via the reduction of exfoliated graphite oxide in the presence of poly(sodium 4-styrenesulfonate). J. Mater. Chem. 16, 155–158.

Stankovich, S., Piner, R.D., Nguyen, S.T., Ruoff, R.S., 2006b. Synthesis and exfoliation of isocyanate-treated graphene oxide nanoplatelets. Carbon 44, 3342–3347.

Stankovich, S., Dikin, D.A., Dommett, G.H.B., Kohlhaas, K.M., Zimney, E.J., Stach, E.A., Piner, R.D., Nguyen, S.T., Ruoff, R.S., 2006c. Graphene-based composite materials. Nature 442, 282–286.

Stankovich, S., Dikin, D.A., Piner, R.D., Kohlhaas, K.A., Kleinhammes, A., Jia, Y., Wu, Y., Nguyen, S.T., Ruoff, R.S., 2007. Synthesis of graphene-based nanosheets via chemical reduction of exfoliated graphite oxide. Carbon 45, 1558–1565.

Tanaka, T., Ebina, Y., Takada, K., Kurashima, K., Sasaki, T., 2003. Oversized titania nanosheet crystallites derived from flux-grown layered titanate single crystals. Chem. Mater. 15, 3564–3568.

Tang, X., Xie, W., Li, H., Zhao, W., Zhang, Q., Niino, M., 2007. Preparation and thermoelectric transport properties of high-performance p-type Bi_2Te_3 with layered nanostructure. Appl. Phys. Lett. 90, 012102.

Tapasztó, L., Dobrik, G., Lambin, P., Biró, L.P., 2008. Tailoring the atomic structure of graphene nanoribbons by scanning tunnelling microscope lithography. Nature Nanotechnol. 3, 397–401.

Thompson-Flagg, R.C., Moura, M.J.B., Marder, M., 2009. Rippling of graphene. EPL 85, 46002.

Tseng, A.A., Notargiacomo, A., Chen, T.P., 2005. Nanofabrication by scanning probe microscope lithography: a review. J. Vac. Sci. Technol. B 23, 877–894.

Tusche, C., Meyerheim, H.L., Kirschner, J., 2007. Observation of depolarized $ZnO(0001)$ monolayers: formation of unreconstructed planar sheets. Phys. Rev. Lett. 99, 026102.

Varchon, F., Mallet, P., Magaud, L., Veuillen, J.-Y., 2008. Rotational disorder in few-layer graphene films on $6H$-SiC-$(000\overline{1})$: a scanning tunneling microscopy study. Phys. Rev. B 77, 165415.

Venables, J.A., Spiller, G.D.T., Hanbücken, M., 1984. Nucleation and growth of thin films. Rep. Prog. Phys. 47, 399–459.

Vogt, P., De Padova, P., Quaresima, C., Avila, J., Frantzeskakis, E., Carmen Asensio, M., Resta, A., Ealet, B., Le Lay, G., 2012. Silicene: compelling experimental evidence for graphenelike two-dimensional silicon. Phys. Rev. Lett. 108, 155501.

Wang, X., Dai, H., 2010. Etching and narrowing of graphene from the edges. Nature Chem 2, 661–665.

Wang, X., Ouyang, Y., Li, X., Wang, H., Guo, J., Dai, H., 2008a. Room-temperature all-semiconducting sub-10-nm graphene nanoribbon field-effect transistors. Phys. Rev. Lett. 100, 206803.

Wang, G., Yang, J., Park, J., Gou, X., Wang, B., Liu, H., Yao, J., 2008b. Facile synthesis and characterization of graphene nanosheets. J. Phys. Chem. C 112, 8192–8195.

Wang, C., Morton, K.J., Fu, Z., Li, W.-D., Chou, S.Y., 2011. Printing of sub-20 nm wide graphene ribbon arrays using nanoimprinted graphite stamps and electrostatic force assisted bonding. Nanotechnol. 22, 445301.

Warner, J.H., Rümmeli, M.H., Gemming, T., Büchner, B., Briggs, G.A.D., 2009. Direct imaging of rotational stacking faults in few layer graphene. Nano Lett. 9, 102–106.

Warner, J.H., Mukai, M., Kirkland, A.I., 2012. Atomic structure of ABC rhombohedral stacked trilayer graphene. ACS Nano 6, 5680–5686.

Wassmann, T., Seitsonen, A.P., Saitta, A.M., Lazzeri, M., Mauri, F., 2008. Structure, stability, edge states, and aromaticity of graphene ribbons. Phys. Rev. Lett. 101, 096402.

Weng, L., Zhang, L., Chen, Y.P., Rokhinson, L.P., 2008. Atomic force microscope local oxidation nanolithography of graphene. Appl. Phys. Lett. 93, 093107.

Wildöer, J.W.G., Venema, L.C., Rinzler, A.G., Smalley, R.E., Dekker, C., 1998. Electronic structure of atomically resolved carbon nanotubes. Nature 391, 59–62.

Williams, D.B., Carter, C.B., 2009. Transmission Electron Microscopy – A Textbook for Materials Science. Springer.

Wilson, J.A., Yoffe, A.D., 1969. The transition metal dichalcogenides discussion and interpretation of the observed optical, electrical and structural properties. Adv. Phys. 18, 193–335.

Wilson, N.R., Pandey, P.A., Beanland, R., Young, R.J., Kinloch, I.A., Gong, L., Liu, Z., Suenaga, K., Rourke, J.P., York, S.J., Sloan, J., 2009. Graphene oxide: structural analysis and application as a highly transparent support for electron microscopy. ACS Nano 3, 2547–2556.

Wilson, N.R., Pandey, P.A., Beanland, R., Rourke, J.P., Lupo, U., Rowlands, G., Römer, R.A., 2010. On the structure and topography of free-standing chemically modified graphene. New J. Phys. 12, 125010.

Wintterlin, J., Bocquet, M.-L., 2009. Graphene on metal surfaces. Surf. Sci. 603, 1841–1852.

Wu, Z.-S., Ren, W., Gao, L., Liu, B., Zhao, J., Cheng, H.-M., 2010. Efficient synthesis of graphene nanoribbons sonochemically cut from graphene sheets. Nano Res. 3, 16–22.

Xie, L., Wang, H., Jin, C., Wang, X., Jiao, L., Suenaga, K., Dai, H., 2011. Graphene nanoribbons from unzipped carbon nanotubes: atomic structures, Raman spectroscopy, and electrical properties. J. Am. Chem. Soc. 133, 10394–10397.

Yacoby, A., 2011. Tri and tri again. Nat. Phys. 7, 925–926.

Yang, D., Frindt, R.F., 1996. Li-intercalation and exfoliation of WS_2. J. Phys. Chem. Solids 57, 1113–1116.

Yang, X., Dou, X., Rouhanipour, A., Zhi, L., Joachim Räder, H., Müllen, K., 2008. Two-dimensional graphene nanoribbons. J. Am. Chem. Soc. 130, 4216–4217.

Zhang, Y., Tan, Y.-W., Stormer, H.L., Kim, P., 2005. Experimental observation of the quantum Hall effect and Berry's phase in graphene. Nature 438, 201–204.

Zhao, P., Choudhury, M., Mohanram, K., Guo, J., 2008. Computational model of edge effects in graphene nanoribbon transistors. Nano Res. 1, 395–402.

Zhi, C., Bando, Y., Tang, C., Kuwahara, H., Golberg, D., 2009. Large-scale fabrication of boron nitride nanosheets and their utilization in polymeric composites with improved thermal and mechanical properties. Adv. Mater. 21, 2889–2893.

Zhou, Y., Bao, Q., Varghese, B., Tang, L.A.L., Tan, C.K., Sow, C.-H., Loh, K.P., 2010. Micro-structuring of graphene oxide nanosheets using direct laser writing. Adv. Mater. 22, 67–71.

Zinke-Allmang, M., Feldman, L.C., Grabow, M.H., 1992. Clustering on surfaces. Surf. Sci. Rep. 16, 377–463.

Zobelli, A., Gloter, A., Ewels, C.P., Seifert, G., Colliex, C., 2007. Electron knock-on cross section of carbon and boron nitride nanotubes. Phys. Rev. B 75, 245402.

Properties of Graphene

Electronic Properties

Christopher S. Allen and Jamie H. Warner

University of Oxford, Oxford, UK

3.1.1. INTRODUCTION

The huge scientific and technological interest in graphene has largely been driven by its electronic properties. A good approximation to the band structure of mono-layer graphene can be obtained from a simple nearest-neighbour tight binding calculation. Inspection of this band structure immediately reveals three electronic properties of mono-layer graphene which have excited such interest in this material: the vanishing carrier density at the Dirac points, the existence of pseudo-spin and the relativistic nature of carriers.

In this section we aim to give an introduction to the electronic transport properties of graphene in order to highlight why it has generated so much interest. We begin by examining the band structure of graphene and discussing its implications on electron transport. We then go on to describe how to extract important material quantities such as mobility from transport measurements and proceed to introduce the more advanced topics of the quantum Hall effect (QHE), Klein tunnelling and graphene nanoribbons (GNRs). There are many fascinating transport properties, such as the fractional QHE, which we do not cover here for the sake of brevity and simplicity. The interested reader is directed towards review articles in the literature, particularly those by Castro Neto et al. (2009) and Das Sarma et al. (2011), for in-depth discussions of the transport properties of graphene.

Graphene. http://dx.doi.org/10.1016/B978-0-12-394593-8.00003-5

3.1.2. THE BAND STRUCTURE OF GRAPHENE

Each carbon atom in the graphene lattice is connected to its three nearest neighbours by strong in-plane covalent bonds. These are known as σ bonds and are formed from electrons in the 2s, $2p_x$ and $2p_y$ valence orbitals. The fourth valence electron occupies the $2p_z$ orbital which is oriented perpendicular to the plane of the graphene sheet, and as a consequence, does not interact with the in-plane σ electrons. The $2p_z$ orbitals from neighbouring atoms overlap resulting in delocalised π (occupied or valence) and π^* (unoccupied or conduction) bands. Most of the electronic properties of graphene can be understood in terms of these π bands.

The unit cell of the hexagonal graphene lattice consists of two atoms separated by $a_{C-C} = 1.42$ Å. It is, as we shall see, more instructive to describe the graphene lattice as two interspersed triangular sub-lattices, commonly denoted A and B (Fig. 3.1.1). This sub-lattice description was used by Wallace (1947) in the first calculation of the band structure of mono-layer graphene some 57 years before the publication of Geim and Novosolov's seminal paper.

The band structure of mono-layer graphene can be adequately described using a simple nearest neighbour tight-binding approach considering a single π electron per atom (Castro Neto, 2009; Charlier, 2007; Reich 2002; Wallace, 1947). The resultant dispersion relation can be written

$$E^{\pm}(k_x, k_y) = \pm \gamma_0 \sqrt{1 + 4\cos\frac{\sqrt{3}k_x a}{2}\cos\frac{k_y a}{2} + 4\cos^2\frac{k_y a}{2}} \qquad 3.1.1$$

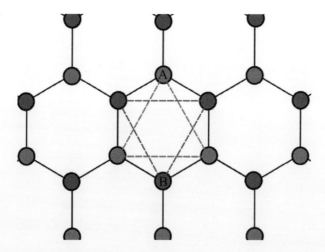

FIGURE 3.1.1 Sub-lattice description of graphene. Each atom on the A sub-lattice is surrounded by three B sub-lattice atoms and vice-versa.

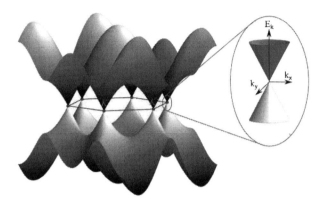

FIGURE 3.1.2   Band structure of graphene showing the conductance and valence bands meeting at the Dirac points (marked with blue dots). Inset is a closeup of one of the Dirac points showing the linear dispersion relation at small values of k.

with $a = \sqrt{3}a_{C-C}$, and γ_0 the nearest neighbour overlap integral which takes a value between 2.5 and 3 eV (Reich, 2002).

The band structure of graphene calculated using Eqn 3.1.1 is shown in Fig. 3.1.2. The valence and conduction bands meet at the high symmetry K and K' points (marked with blue dots in Fig. 3.1.2). In intrinsic (un-doped) graphene each carbon atom contributes one electron completely filling the valence band and leaving the conduction band empty. As such the Fermi level, E_F is situated precisely at the energy where the conduction and valence bands meet. These are known as the Dirac or charge neutrality points. Consideration of the dispersion relation can be limited to just two of the Dirac points ($K = [2\sqrt{3}\pi/3a, 2\pi/3a]$ and $K' = [2\sqrt{3}\pi/3a, -2\pi/3a]$), the others being equivalent through translation by a reciprocal lattice vector. These two Dirac points in reciprocal space can be directly related to the two real space graphene sub-lattices, K being due to electrons on sub-lattice A and K' due to electrons on sub-lattice B.

Expanding Eqn 3.1.1 close to the Dirac point (K or K') results in the famous dispersion relation showing the linear relationship between energy, $E(k)$ and momentum, k:

$$E^{\pm}(k) = \hbar v_F |k - K| \qquad\qquad 3.1.2$$

where $k = (k_x, k_y)$ and $v_F = (\sqrt{3}\gamma_0 a/2\hbar) \approx 1 \times 10^6$ ms^{-1}. The region of the dispersion relation close to $K(K')$ is plotted in the inset to Fig. 3.1.2, showing the linear nature of the Dirac cones.

In summary, by employing a simple nearest neighbour tight binding description the band structure of graphene can be calculated. This reveals three important features which to a large extent define the nature of electron transport through this material:

1. The occupied valence and empty conduction bands meet at Dirac point at which the density of states (DOS) is zero. Graphene is therefore best described as a zero-gap semiconductor, with vanishing DOS at the Dirac point but no energy gap between the valence and conduction bands.
2. Close to the Fermi energy the band structure of graphene can be described in terms of two inequivalent Dirac cones situated at K and K'. In order for an electron to scatter from K to K' requires a large momentum change. Electron transport in graphene can therefore be thought of as occurring in parallel through the K and K' Dirac cones (corresponding to the two graphene sublattices A and B). As such charge carriers in graphene have, in addition to orbital and spin quantum numbers, a valley or pseudo-spin quantum number with a degeneracy of 2. The term pseudospin is used due to the analogy with real spin, the two are however, completely independent of one and other.
3. Close to the Dirac point the graphene dispersion relation is linear in nature. This linear-dispersion relation is well described by the relativistic Dirac equation (Castro Neto et al., 2009). In this description the charge carriers (electrons or holes) are massless Dirac Fermions travelling with a group velocity of $v_F \approx 1 \times 10^6$ ms^{-1}.

3.1.3. TRANSPORT EXPERIMENTS IN GRAPHENE

The isolation of single layer graphene (SLG) by Novoselov and Geim (Novoselov et al., 2004) began a frenzy of experimental activity investigating this novel material. In this section we will not attempt to provide an exhaustive overview of transport experiments on graphene (the speed at which the field is moving renders this a somewhat fruitless task). Instead we aim to give the reader an introduction to experimental realisations of some of the transport properties which we predicted in the previous section and introduce some other important and interesting topics.

Most of the early experiments which we describe in this section were performed on exfoliated graphene transferred to Si/SiO$_2$ substrates. Electronic devices are routinely fabricated from graphene produced from numerous other methods and on a variety of substrates, details of which can be found in the literature.

3.1.3.1. Modulation of Carrier Density with Gate Voltage

The standard (but by no means only) device geometry for performing electrical measurements on graphene samples is the Hall bar as shown in Fig. 3.1.3 (a). Graphene is transferred to a heavily doped silicon substrate capped with an insulating silicon oxide layer of several hundreds of nanometres.

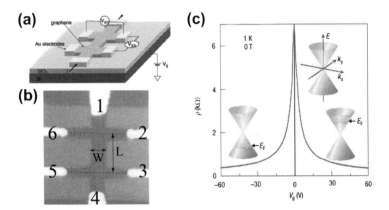

FIGURE 3.1.3 (a) Schematic and (b) optical microscope image of a typical graphene Hall bar device. *Modified from Jiang et al. (2007). Copyright (2007) Elsevier.* (c) Modulation of the resistivity of a graphene sample by the application of a gate voltage. *Modified from Geim et al. (2007). Copyright (2007) Nature Publishing Group.*

Figure 3.1.3(b) shows a graphene sample that has been etched into a Hall bar geometry by means of a lithographically patterned resist mask followed by plasma etching (Jiang, 2007). A four point measurement can be performed by driving a current between electrodes 1 and 4, I_{14}, and measuring the voltage drop between electrodes 2 and 3, V_{23}, (or 5 and 6, V_{56}). The resistivity of the graphene sample is then defined as

$$\rho_{xx} = \left(\frac{W}{L}\right)\left(\frac{V_{23}}{I_{14}}\right) \qquad 3.1.3$$

and has units of Ohms. W is the width and L the length of the graphene channel between the voltage probes.

By electrically contacting the heavily doped Si, an electric field can be applied across the graphene sample by means of application of a gate voltage (V_g). This acts to tune the Fermi level, E_F (and thus the carrier density) of the graphene (see inset to Fig. 3.1.3 (c)). As E_F approaches the Dirac point the number of available carriers decreases and there is a resultant increase in resistivity reaching a maximum at the Dirac point ($V_g = 0$ V in Fig. 3.1.3(c)). The ambipolar nature of the field effect in graphene allows for the study of electron transport when E_F is above the Dirac point (in the conduction band) and hole transport when E_F is below the Dirac point (in the valence band)

In Fig. 3.1.3 (c), the $\rho(V_g)$ curve is symmetric about the Dirac point which sits at $V_g = 0$ V as is expected for intrinsic graphene. In practice it is often the case that the Dirac point is only reached by application of a gate voltage of several tens of volts due to doping of the graphene sample. The Dirac point can be brought back closer to $V_g = 0$ V by removal of surface contaminants via post fabrication annealing in ultrahigh vacuum or a H_2/Ar atmosphere (Ishigami, 2007) or by the application of a high current density through the sample (Moser, 2007).

Although theoretically the density of charge carriers should go to zero at the Dirac point, experimentally there remains a finite conductivity of the order of $4e^2/h$ (Novoselov, 2005; Zhang, 2005). This is due to various contributions including thermally excitation (Dorgan, 2010), the presence of charged impurities (Chen et al., 2008) and ripples in the graphene layer (Katsnelson, 2008). Whether there is an intrinsic explanation for a minimum conductivity in graphene remains an open question (Geim, 2007).

3.1.3.2. Mobility and Density of Carriers

The Drude model defines conductivity σ (the inverse of resistivity ρ) in terms of two important material properties, carrier density n and mobility μ:

$$\rho^{-1} = \sigma = ne\mu \qquad 3.1.4$$

In experiments on graphene n and μ have generally been determined from either field effect or Hall effect measurements.

Using field effect measurements, the density of carriers in a graphene sample, n can be estimated from the surface charge density induced by application of a gate voltage (Novoselov, 2004)

$$n = \varepsilon_0\varepsilon_r V_g/te \qquad 3.1.5$$

ε_0 is the permittivity of free space, ε_r the relative permittivity of the dielectric (generally SiO_2), t the dielectric thickness and e the electron charge. As $n = 0$ at the Dirac point, any doping of the sample is compensated for by replacing V_g in Eqn 3.5 by $(V_g - V_{gD})$ with V_{gD} the gate voltage at the Dirac point.

The field effect mobility, μ_{FE} can be simply extracted from the gate voltage dependence of conductivity:

$$\mu_{FE} = \frac{d\sigma}{dV_g}\frac{1}{c_g} \qquad 3.1.6$$

where σ is the conductivity of the sample and C_g the gate capacitance which can be calculated from $C_g = ne/(V_g - V_{gD})$ (Bolotin et al., 2008). Values for μ_{FE} in excess of $100,000 \text{ cm}^2\text{V}^{-1}\text{s}^{-1}$ have been reported for graphene on boron nitride (Dean et al., 2010).

An alternate technique for the determination of the carrier density and mobility is to perform a Hall effect measurement in which the transverse resistivity $\rho_{xy} = (W/L)(V_{26}/I_{14})$ [see Fig. 3.1.3(b)] of a sample is measured in the presence of an out of plane magnetic field (a detailed description of performing Hall measurements is given in chapter 5.8). The carrier density is then given by

$$n = B/e\rho_{xy} \qquad 3.1.7$$

and the Hall co-efficient defined as

$$R_H = \frac{1}{ne} \qquad\qquad 3.1.8$$

Once the carrier density is known the Hall mobility can be found.

$$\mu_H = \frac{1}{ne\rho_{xx}} \qquad\qquad 3.1.9$$

Hall mobilities in excess of $200,000 \ \mathrm{cm^2 V^{-1} s^{-1}}$ have been reported for suspended graphene devices with carrier densities as low as $5 \times 10^9 \ \mathrm{cm^{-2}}$ (Bolotin et al., 2008; Du et al., 2008).

3.1.3.3. Quantum Hall Effect

The presence of a uniform magnetic field perpendicular to the plane of a two dimensional (2D) conductor causes the carriers to move in circular cyclotron orbitals with a characteristic cyclotron frequency, ω_C. Treating quantum mechanically these orbitals are quantised leading to the formation of discrete Landau levels (LL) in the energy spectrum. If a Hall effect measurement is performed on a high quality sample in a sufficiently large magnetic field at an appropriate temperature, the LLs manifest themselves as oscillations in ρ_{xx} (known as Shubnikov-de Haas oscillations) and an exact quantisation of σ_{xy} (see Fig. 3.1.4). In the regions where ρ_{xx} goes to zero there is a corresponding plateaux in σ_{xy}.

There are three important differences between the QHE in graphene and that of conventional 2D systems. Due to the massless Dirac like nature of the charge carriers in graphene the eigenenergies of the LLs are given by (Gusynin and Sharapov, 2005; Zheng and Ando, 2002)

$$E_n = \sqrt{2eB\hbar v_F^2 |l|}, \quad l = 0, \pm 1, \pm 2, \pm 3, ..., \qquad 3.1.10$$

This is in contrast to conventional 2D systems which follow

$$E_n = \hbar\omega_c (l + 1/2) \qquad\qquad 3.1.11$$

The most obvious difference between the QHE in graphene and that in conventional 2D systems is the existence of a LL at zero energy (seen as the peak in ρ_{xx} at $n = 0$ in Fig. 3.1.4). The QHE, and the existence of a zero-energy LL were reported concurrently by two groups and provided direct and compelling experimental evidence for the massless Dirac like nature of charge carriers in graphene (Novosolov et al., 2005; Zhang et al., 2005).

Furthermore in graphene the rule for quantisation of σ_{xy} takes the form (Gusynin and Sharapov, 2005; Zheng and Ando, 2002).

$$\sigma_{xy} = g\left(l + \frac{1}{2}\right)\frac{e^2}{h}$$

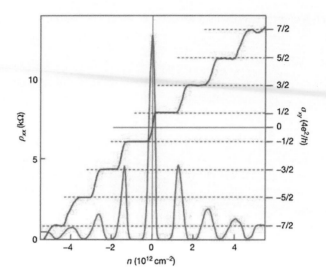

FIGURE 3.1.4 The quantum Hall effect in graphene. *Modified from Novoselov et al. (2005). Copyright (2005) Nature Publishing Group.*

Compared with in conventional 2D systems

$$\sigma_{xy} = gl\frac{e^2}{h}$$

where g is the total degeneracy spin and valley degeneracy. The existence of conductance plateaus at half integer values of the quantum of conductance (e^2/h) in graphene is a consequence of the linear dispersion of the band structure for massless Dirac Fermions.

Lastly as the LL separation scales as $\sqrt{l}$ in graphene as opposed to l in standard 2D systems (see Eqns 3.1.10 and 3.1.11) the energy level spacing, $\Delta_E = E_{l+1} - E_l$ can be large for low energies. This has enabled the QHE effect to be observed at room temperature in graphene, the only material in which this has been achieved (Novoselov et al., 2007).

For the case of bilayer graphene, the positions of the LL are the same as for monolayer graphene. Due to an additional 'layer-'degree of freedom, the jump in σ_{xy} for the 0th (zeroth) LL is $8e^2/h$ in bilayer graphene as opposed to $4e^2/h$ in monolayer (Novoselov et al., 2006).

3.1.3.4. Klein Tunnelling

When an electron is incident on a potential barrier, quantum mechanics (in the form of the Schrodinger equation) tells us that there is a finite probability that the electron will tunnel through the barrier, with tunnelling probability decreasing exponentially with both barrier height and width. For the case of Dirac particles, the situation is quite different.

When a Dirac electron is incident on a potential barrier it will traverse the potential barrier as a hole, emerging on the other side as an electron once again. This is known as Klein tunnelling. Due to the need to mode-match electron states outside the barrier with hole states inside the barrier the transmission probability increases with increased barrier height, reaching unity for an infinite barrier (Klein, 1929).

In graphene the transmission of electrons at normal incidence to a potential barrier has been found to always be unity (due to the massless nature of the charge carriers and the existence of pseudospin). At angles away from normal incidence the transmission varies as $\cos^2\theta$ for a perfectly sharp barrier (Cheianov and Falko, 2006; Katsnelson et al., 2006).

Klein tunnelling in graphene and the related difficulty in confining electrons represents a real barrier to the development of graphene based transistors. It does, however open up interesting lines of investigation into the collimation of electrons in graphene (Allain and Fuchs, 2011; Katsnelson et al., 2006).

Experimentally signatures of Klein tunnelling have been observed in conductivity measurements of top-gated p-n-p or n-p-n graphene devices (Gorbachev et al., 2008; Huard et al., 2007; Stander et al., 2009; Young and Kim 2009). These have shown good agreement with theoretical predictions based on Klein tunnelling (Rossi et al., 2010).

3.1.3.5. Graphene Nano-ribbons

The lack of band gap and the inability to confine electrons (due to Klein tunnelling) is a major hurdle to the development of graphene electronics. One approach to engineering a band gap is to pattern graphene into thin nano-ribbons.

When the lateral dimensions of a graphene device are reduced a band gap can be opened. The size of the band gap is determined by the state of the edge and the width of the nanoribbon (Nakada et al., 1996). Tight binding calculations predict that GNRs terminated with 'armchair' edges (Fig. 3.1.5(a)) can be either metallic or semi-conducting depending on their width. Nanoribbons terminated with 'zigzag' edges (Fig. 3.1.5(b)) are always metallic. In order for the GNRs to have band gaps of the same order as the commonly used semi-conducting materials (e.g. Si(1.14 eV), GaAs(1.43 eV)) widths of less than 2 nm are likely to be necessary (Barone et al., 2006).

GNRs have been fabricated via chemical methods (Li et al., 2008), unzipping of carbon nanotubes (Jiao et al., 2009) and by plasma etching of masked graphene sheets (Han et al., 2007; Wang and Dai 2010). It has been shown that the size of the band gap is inversely proportional to the width of the ribbon (see Fig. 3.1.6). GNR FETs down to ~10 nm have been fabricated using hydrogen-silsesquioxane as an etch mask (Wang and Dai, 2010;

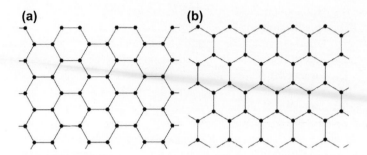

(a)　　　　　　　　　　　　**(b)**

FIGURE 3.1.5　(a) Graphene armchair and (b) zigzag edges.

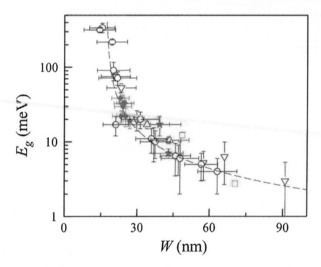

FIGURE 3.1.6　Experimentally determined band gap E_g for GNRs of varying widths W fabricated by plasma etching. *Modified from Han et al. (2007). Copyright (2007) by the American Physical Society.*

Wang et al., 2012), and sub-10 nm GNR FETs have been achieved using chemical fabrication techniques (Wang et al., 2008).

Beyond engineering of the band gap, there have been theoretical predictions of interesting magnetic edge states in zig-zag GNRs (Abanin et al., 2006; Son et al., 2006; Wakabayashi et al., 1999) and some experimental evidence suggesting their existence (Joly et al., 2010; Kobayashi, 2006). However, in order to fully understand the impact of edge structure on the electronic properties of GNRs it is necessary to be able to control the nature of the edges. This remains a challenge.

REFERENCES

Abanin, D.A., Lee, P.A., Levitov, L.S., 2006. Spin-filtered edge states and quantum Hall effect in graphene. Phys. Rev. Lett. 96, 176803.

Allain, P.E., Fuchs, J.N., 2011. Klein-tunneling in graphene: optics with massless electrons. Eur. Phys. J. B 83, 301–337.

Barone, V., Hod, O., Scuseria, G.E., 2006. Electronic structure and stability of semiconducting graphene nanoribbons. Nano Lett. 6 (12), 2748–2754.

Bolotin, K.I., Sikes, K.J., Jiang, Z., Klima, M., Fudenberg, G., Hone, J., Kim, P., Stormer, H.L., 2008. Ultrahigh electron mobility in suspended graphene. Solid State Commun. 146, 351–355.

Castro Neto, A.H., Guinea, F., Peres, N.M.R., Novoselov, K.S., Geim, A.K., 2009. The electronic properties of graphene. Rev. Mod. Phys. 81, 109–162.

Charlier, J.-C., Blase, X., Roche, S., 2007. Electronic and transport properties of carbon nanotubes. Rev. Mod. Phys. 79, 677–732.

Cheianov, V.V., Falko, V.I., 2006. Selective transmission of Dirac electrons and ballistic magnetoresistance of n-p junctions in graphene. Phys. Rev. B 74, 041403(R).

Chen, J.-H., Jang, C., Adam, S., Fuhrer, M.S., Williams, E.D., Ishigami, M., 2008. Charged-impurity scattering in graphene. Nat. Phys. 4, 377–381.

Das Sarma, S., Adam, S., Hwang, E.H., Rossi, E., 2011. Electronic transport in two-dimensional graphene. Rev. Mod. Phys. 83, 407–470.

Dean, C.R., Young, A.F., Meric, I., Lee, C., Wang, L., Sorgenfrei, S., Watanabe, K., Taniguchi, T., Kim, P., Shepard, K.L., Hone, J., 2010. Boron nitride substrates for high-quality graphene electronics. Nat. Nano 5 (10), 722–726.

Dorgan, V.E., Myung-Ho, B., Pop, E., 2010. Mobility and saturation velocity in graphene on SiO_2. Appl. Phys. Lett 97, 082112.

Du, X., Skachko, I., Barker, A., Andrei, E.Y., 2008. Approaching ballistic transport in suspended graphene. Nat. Nano. 3 (8), 491.

Geim, A.K., Novoselov, K.S., 2007. The rise of graphene. Nat. Mat. 6, 183–191.

Gorbachev, R.V., Mayorov, A.S., Savchenko, A.K., Horsell, D.W., Guinea, F., 2008. Conductance of p-n-p graphene structures with "air-bridge" top gates. Nano Lett. 8 (7), 1995–1999.

Gusynin, V.P., Sharapov, S.G., 2005. Magnetic oscillations in planar systems with the Dirac-like spectrum of quasiparticle excitations. II Transport properties. Phys. Rev. B 71, 125124.

Han, M.Y., Özyilmaz, B., Zhang, Y., Kim, P., 2007. Energy band-gap engineering of graphene nanoribbons. Phys. Rev. Lett. 98, 206805.

Huard, B., Sulpizo, J.A., Stander, N., Todd, K., Yang, B., Goldhaber-Gordon, D., 2007. Transport measurements across a tunable potential barrier in graphene. Phys. Rev. Lett. 98, 236803.

Ishigami, M., Chen, J.H., Cullen, W.G., Fuhrer, M.S., Williams, E.D., 2007. Atomic structure of graphene on SiO_2. Nano Lett. 7 (6), 1643–1648.

Jiang, Z., Zhang, Y., Tan, Y.-W., Stormer, H.L., Kim, P., 2007. Quantum Hall effect in graphene, Solid State Communications 143, 14–19.

Jiao, L., Zhang, L., Wang, X., Diankov, G., Dai, H., 2009. Narrow graphene nanoribbons from carbon nanotubes. Nature 458, 877.

Joly, V.L.J., Kiguchi, M., Hao, S.-J., Takai, K., Enoki, T., Sumii, R., Amemiya, K., Muramatsu, H., Hayashi, T., Kim, Y.A., Endo, E., Campos-Delgado, J., López-Urías, F., Botello-Méndez, A., Terrones, H., Terrones, M., Dresselhaus, M.S., 2010. Observation of magnetic edge state in graphene nanoribbons. Phys. Rev. B 81, 245248.

Katsnelson, M.I., Geim, A.K., 2008. Electron scattering on microscopic corrugations in graphene. Phil. Trans. R. Soc. A 366, 195–204.

Katsnelson, M.I., Novoselov, K.S., Geim, A.K., 2006. Chiral tunnelling and the Klein paradox in graphene. Nature Phys. 2, 620.

Klein, O., 1929. Die Reflexion von Elektronen an einem Potentialsprung nach der relativistischen Dynamik von Dirac. Z. Phys. 53, 157–165.

Kobayashi, Y., Fukui, K., Enoki, T., Kusakabe, K., Kaburagi, Y., 2005. Observation of zigzag and armchair edges of graphite using scanning tunnelling microscopy and spectroscopy. Phys. Rev. B 71, 193406.

Li, X., Wang, X., Zhang, L., Lee, S., Dai, H., 2008. Chemically derived, ultrasmooth graphene nanoribbon semiconductors. Science 319, 1229–1232.

Moser, J., Barreiro, A., Bachtold, A., 2007. Current-induced cleaning of graphene. Appl. Phys. Lett. 91, 163513.

Nakada, K., Fujita, M., Dresselhaus, G., Dresselhaus, M.S., 1996. Edge state in graphene ribbons: nanometer size effect and edge shape dependence. Phys. Rev. B 54 (24), 17954–17961.

Novoselov, K.S., Geim, A.K., Morozov, S.V., Jiang, D., Zhang, Y., Dubonos, S.V., Grigorieva, I.V., Firsov, A.A., 2004. Electric field effect in atomically thin carbon films. Science 306, 666–669.

Novoselov, K.S., Geim, A.K., Morozov, S.V., Jiang, D., Katsnelson, I.V., Grigorieva, I.V., Dubonos, S.V., Firsov, A.A., 2005. Two-dimensional gas of massless Dirac fermions in graphene. Nature 438, 197–200.

Novoselov, K.S., McCann, E., Morozov, S.V., Falko, V.I., Katsnelson, M.I., Zeitler, U., Jiang, D., Schedin, F., Geim, A.K., 2006. Unconventional quantum Hall effect and Berry's phase of 2π in bilayer graphene. Nature Physics 2, 177.

Novoselov, K.S., Jiang, Z., Zhang, Y., Morozov, S.V., Stormer, H.L., Zeitler, U., Maan, J.C., Boebinger, G.S., Kim, P., Geim, A.K., 2007. Room-temperature quantum Hall effect in graphene. Science 315, 1379.

Reich, S., Maultzsch, J., Thomsen, C., Ordejón, P., 2002. Tight-binding description of graphene. Phys. Rev. B 66, 035412.

Rossi, E., Bardarson, J.H., Brouwer, P.W., Das Sarma, S., 2010. Signatures of Klein tunnelling in disordered graphene p-n-p junctions. Phys. Rev. B 81, 121408.

Son, Y.-W., Cohen, M.L., Louis, S.G., 2006. Half-metallic graphene nanoribbons. Nature 444, 347.

Stander, N., Huard, B., Goldhaber-Gorder, D., 2009. Evidence for Klein tunnelling in graphene p-n junctions. Phys. Rev. Lett. 102, 026807.

Wakabayashi, K., Fujita, M., Ajiki, H., Sigrist, M., 1999. Electronic and magnetic properties of nanographite ribbons. Phys. Rev. B 59, 8271–8282.

Wallace, P.R., 1947. The band theory of graphite. Phys. Rev. 71 (9), 622–634.

Wang, X., Dai, H., 2010. Etching and narrowing of graphene from the edges. Nat. Chem. 2, 661–665.

Wang, X., Ouyang, Y., Li, X., Wang, H., Guo, J., Dai, H., 2008. Room temperature all-semiconducting sub-10-nm graphene nanoribbon field-effect transistors. Phys. Rev. Lett. 100, 206803.

Wang, S.H., Tahy, C., Nyakiti, L.O., Wheeler, V.D., Myers-Ward, R.L., Eddy Jr., C.R., Gaskill, D.K., Xing, H., Seabaugh, A., Jeena, D., 2012. Fabrication of top-gated epitaxial graphene nanoribbon FETs using hydrogen-silsequioxane. J. Vac. Sci. Technol. B 30 (3), 03D104.

Young, A., Kim, P., 2009. Quantum interference and Klein tunnelling in graphene heterojunctions. Nature Phys. 5, 222.

Zhang, Y., Tan, Y.-W., Stormer, H.L., Kim, P., 2005. Experimental observation of the quantum Hall effect and Berry's phase in graphene. Nature 438, 201.

Zheng, Y., Ando, T., 2002. Hall conductivity of a two-dimensional graphite system. Phys. Rev. B 65, 245420.

Chapter 3.2

Chemical Properties of Graphene

Jamie H. Warner

University of Oxford, Oxford, UK

3.2.1. INTRODUCTION

Chemical reactions to modify the structure of graphite have been undertaken for more than 150 years. In graphite there are two main structural changes of interest brought about by chemical processes, exfoliating graphene layers from the bulk graphite, or intercalating material within the layers. Chemical exfoliation of graphite to obtain graphene is discussed in Chapters 4.2 and 4.3. Graphite intercalation compounds are formed by inserting either atomic or molecular species between each graphene layer in graphite (Dresselhaus and Dresselhaus, 1981). Metals such as potassium, caesium, rubidium, and lithium are often used. Graphite with intercalated lithium is used in commercial lithium ion batteries for reversible charge storage. When moving from the 3D material of graphite to the 2D material of graphene, new avenues for chemistry emerge. Chemical reactions are focussed either at the edge of graphene sheets or to the bulk lattice. The atoms at the edge of a graphene sheet are different to those within the lattice due to the fact that edge atoms are missing neighbours. Atoms within the main lattice of graphene are identical in their chemical nature and represent a pure aromatic system. When the size of a graphene sheet shrinks to the nanoscale/molecular level it becomes a polycyclic aromatic hydrocarbon (Wu et al., 2007).

Graphene is generally explained as having a hexagonal 2D array of carbon atoms with sp^2 bonding, whereas diamond and amorphous carbon materials as having sp^3 bonding. Sp^2 and sp^3 are descriptions of carbon–carbon (C–C) bonds based on hybridisation and the concept of mixing atomic orbitals. The s and p describe the atomic orbitals involved in the bond. Hybridisation theory was developed by Pauling (1931) to explain simple molecules, such as methane (CH_4). It works well for organic compounds but has problems when d orbitals are involved in bonding. A single carbon atom has six electrons that are distributed across the s and p shells as $1s^2 2s^2 2p_x^1 2p_y^1 2p_z^0$. In a C–C bond a linear combination of s and p wavefunctions can result in the hybridisation. For sp^3 bonding, the $2s$ orbital mixes with the three $2p$ orbitals to form four sp^3 orbitals. Sp^2 hybridisation involves molecules with a double bond between carbons atoms. A π bond is required for the double bond and only three σ bonds are formed per carbon atom. Therefore the $2s$ orbital is mixed with only two of the three $2p$ orbitals forming

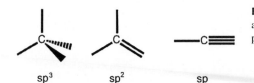

FIGURE 3.2.1 Schematics of sp^3, sp^2 and sp bonding, with tetrahedral, trigonal planar and linear geometries respectively.

three sp^2 orbitals with one p orbital remaining. Figure 3.2.1 shows schematics of sp^3, sp^2 and sp bonding, with tetrahedral, trigonal planar and linear geometries respectively.

Chemical functionalisation of the main graphene sheet (not the edges) is achieved by either covalent or non-covalent methods. Covalent functionalisation requires the breaking of sp^2 bonds and can be achieved using a wide range of reactions. Non-covalent functionalisation relies on van der Waals forces often due to π–π stacking between aromatic molecules and the graphene lattice, much in the same way that graphene sheets are held together in graphite.

3.2.2. COVALENT FUNCTIONALISATION OF GRAPHENE

Graphene oxide and reduced graphene oxide lend themselves to covalent functionalisation due to the presence of defects in the graphene lattice that act as sites for reactivity. Chapter 4.3 covers this area- and therefore- we shall concentrate on covalent functionalisation of pristine graphene based around the disruption of sp^2 bonds.

Graphene is relatively chemically inert and has been used to provide resistance against hydrogen peroxide and protect metal surfaces from oxidation even after heating to 200 °C for 4 hours (Chen et al., 2011). We remove polymers such as polymethylmethacrylate (PMMA) and hydrocarbon surface contamination from graphene by simply heating suspended graphene in air for 2 days at 350 °C. Unlike fullerenes and carbon nanotubes which have curved surfaces that induce strain in sp^2 bonds that help facilitate chemical functionalisation, graphene is flat (Niyogi et al., 2002). One of the first chemical modifications to pristine graphene was to react it with atomic hydrogen to form graphane (Elias et al., 2009). Elias et al. exposed their mechanically exfoliated graphene to a cold hydrogen plasma for 2 hours. Changes in the electron diffraction pattern of graphene were observed after exposure to hydrogen to form graphane, with a lattice contraction seen by a shift in the position of diffraction spots (Elias et al., 2009). The change in the atomic structure of graphene to graphane is illustrated in the atomic models of Figures 3.2.2(a)-(c). The electron diffraction reveals changes in the structural properties that lead to changes in the electronic properties (Elias et al., 2009). Graphane showed insulating behaviour and decreased carrier mobilities. Upon further annealing of graphane at 450 °C for 24 hours in Ar atmosphere, graphene was restored along with all its characteristic properties (Elias et al., 2009). Balog et al. (2010) showed bandgap opening in graphene by patterned

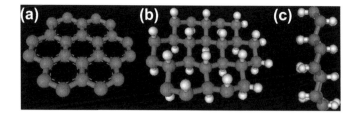

FIGURE 3.2.2 Atomic models for perspective view of (a) graphene and (b) graphane. White atoms are hydrogen. (c) Side view graphane.

adsorption of atomic hydrogen on a Moire pattern produced from graphene grown on an Ir(111) substrate. Figure 3.2.3(a)–(e) show scanning tunnelling microscope (STM) images of the Moire pattern of graphene on Ir(111) for (a) no hydrogen, (b) a very low dose of hydrogen, and (c)–(e) 15 s, 30 s and 50 s exposure to hydrogen (Balog et al., 2010).

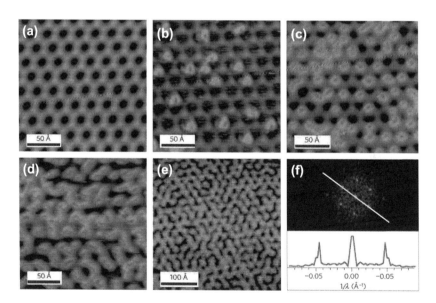

FIGURE 3.2.3 STM images of hydrogen adsorbate structures following and preserving the Moiré pattern of graphene on Ir(111).(a) Moiré pattern of clean graphene on Ir(111) with the superlattice cell indicated. (b)–(e) Graphene exposed to atomic hydrogen for very low dose, 15 s, 30 s and 50 s, respectively. The data show the evolution of hydrogen structures along the bright parts of the Moiré pattern with increasing hydrogen dose. (f) Fourier transform of the image in (e), illustrating that hydrogen adsorbate structures preserve the Moiré periodicity. The inset in (f) shows a line profile through the Fourier transform along the line indicated. The separation of the peaks corresponds to a real-space distance of 21.5 Å, which is equal to 25 Å $\times \cos(30°)$, confirming the Moiré superlattice periodicity. *Reproduced from Balog (2010). Copyright (2010) Nature Publishing Group.*

Other atomic species such as fluorine can also be added to graphene in a reversible manner (Cheng et al., 2010; Nair et al., 2010). Fluorographene is insulating with a band gap of 3 eV and is inert and stable in air up to 400 °C (Nair et al., 2010). Whilst atomic species, such as hydrogen and fluorine can chemically react with graphene and modify the structure, they do not necessarily offer a means for further chemical functionalisation where molecules are covalent attached to the surface of graphene. Oxidisation of graphene to form graphene oxide is one way of introducing reaction sites for molecular attachment. Electrochemical oxidation of graphene is a way in which more control can be achieved over the oxidisation process. When the surface charge becomes sufficient, oxygen atoms from an electrolyte form covalent bonds with a graphitic surface by converting conjugated carbon atoms from sp^2 to sp^3 (Bekyarova et al., 2012). Ramesh et al. used nitric acid to electrochemically oxidised epitaxial graphene grown from SiC substrates. A new reaction channel that was not present in graphite was identified. They studied the field effect transistor performance and found the electro-oxidised channel had higher on-off ratio and mobility, which was explained as being due to the electro-oxidisation removal of the defective top layers and the exposure of high quality internal graphene layers (Bekyarova et al., 2012).

Diazonium reagents are one of the most effective chemical systems for functionalising the basal plane of graphite and thus the main lattice of graphene. They consist of a general form $R\text{-}N_2^+X^-$ where R is an organic residue such as aryl group and X is an inorganic or organic anion. Diazonium compounds have been used to graft aryl groups to carbon nanotubes (Bahr et al., 2001), diamond (Kuo et al., 1999), HOPG (Liu and McCreery, 1995) and glassy carbon (Delamar et al., 1992). Bekyarova et al. (2009) spontaneously grafted aryl groups to epitaxial graphene grown by annealing SiC by reducing 4-nitrophenyl diazonium tetrafluoroborate (NPD), shown schematically in Fig. 3.2.4. The chemical reaction resulted in a surface coverage of 1×10^{15} nitrophenyl groups per cm^2 of graphene (Bekyarova et al., 2009). Carbon centres were transformed from sp^2 to sp^3 and resulted in a barrier to electron flow by introducing a band gap. This showed that covalent functionalisation of few-layer graphene (FLG) could change the electronic properties from near-metallic to semiconducting

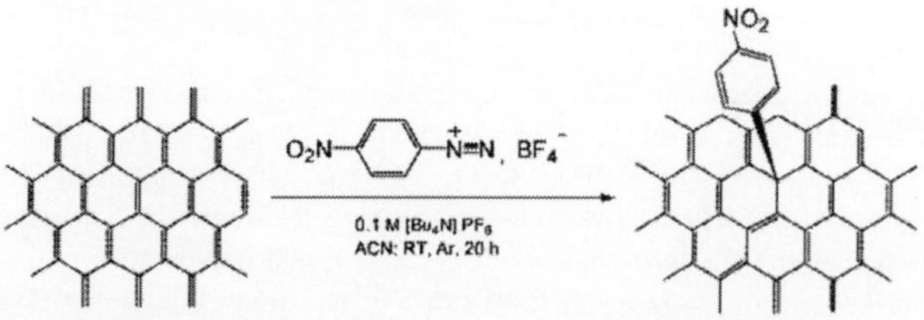

FIGURE 3.2.4 Schematic illustration of the spontaneous grafting of aryl groups to epitaxial graphene via reduction of 4-nitrophenyl diazonium (NPD) tetrafluoroborate. *Reprinted with permission from Bekyarova (2009). Copyright (2009) American Chemical Society.*

(Bekyarova et al., 2009). Further work has shown that this reaction depends upon how the graphene was obtained and whether it is single or multilayered (Koehler et al., 2010; Niyogi et al., 2010). Niyogi et al. (2011) used STM to image the surface of epixatial graphene before and after nitrophenyl functionalisation, shown in Fig. 3.2.5 and observed threefold symmetric patterns associated with the redistribution of the local DOS due to defect sites. The STM images also showed that the functionalisation occurred on the same graphene sub-lattice (Niyogi et al., 2011).

Koehler et al. (2009) produced patterned unmasked areas on the surface of graphite using photoresist lithography and then exposed those open regions to highly diluted diazonium reagents. Figure 3.2.6 shows the outline of the procedure; first the surface of graphite is patterned using photoresist lithography to provide windows to the surface for reaction, next, the chemical derivatisation is undertaken and the exposed regions of graphite are functionalised, finally the

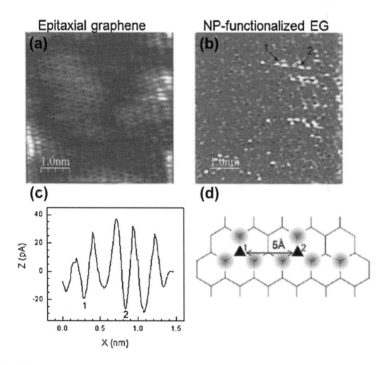

FIGURE 3.2.5 Superimposed STM images and FFT-filtered STM images. (a) 2D-FFT filtered STM image of pristine epitaxial graphene superimposed on an STM image of pristine expitaxial graphene after subtracting noise. (b) 2D-FFT-filtered STM image of nitrophenyl functionalised epitaxial graphene superimposed on an STM image of nitrophenyl functionalised epitaxial graphene. (c) Line scan of panel b, where the local density of states (LDOS) is dominated by superstructures. (d) Schematic representation of atoms labelled 1 and 2 in panel b. The distance is measured from the STM line scan shown in panel c. *Reprinted with permission from Niyogi (2011). Copyright (2011) American Chemical Society.*

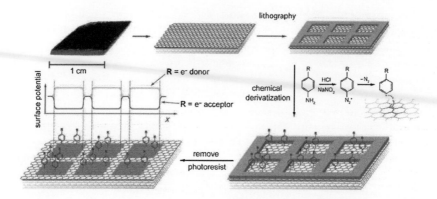

FIGURE 3.2.6 Patterned functionalisation of an HOPG surface: first, a patterned mask is created by photolithography on the top graphene layer of a graphene stack. The unmasked regions are then exposed to a diazonium reagent. The photoresist is removed prior to analysis. *Reprinted from Koehler (2009). Copyright (2009) WILEY-VCH Verlag GmbH & Co.*

photoresist is removed leaving patterned graphite surface. Not only did the reaction with the diazonium reagent covalently attach groups to the surface of graphene, it also changed the surface potential. Simply changing the functional group, R, indicated in the chemical derivatisation, from an electron donor to acceptor allowed control over the resulting surface potential (Koehler et al., 2009).

Choi et al. (2009) used azidotrimethylsilane to chemically modify epitaxial graphene and used high resolution photoemission spectroscopy to confirm that the bonding between the nitrene radicals and graphene was covalent. By changing the amount of nitrene dosing, they were able to control the band gap of their graphene.

Diels–Alder (DA) chemistry is another approach for covalent functionalisation of graphene. Figure 3.2.6 shows a schematic representation from Bekyarova et al. (2012) of the DA reaction (cyclo-addition and cyclo-reversion) in its simplest form. It has been used to functionalise fullerenes and carbon nanotubes and has naturally been extended to graphene. Sarkar et al. (2011) showed that graphene can act as either the diene or dienophile due to its zero band gap. They performed cycloaddition and also reversion using solution processed graphene, epitaxial graphene and highly ordered pyrolitic graphite (HOPG). Tetracyanoethyelene in dichloromethane was used as the dienophile and the Diels-Alder reaction with graphene occurred at room temperature within 3 hours. Figure 3.2.8(a) shows the schematic representation of the reaction. Raman spectroscopy was used to measure the disruption to the pristine graphene lattice (Sarkar et al., 2011). When 2,3-dimethoxy-1,3-butadiene was used as a diene, graphene and graphite acted as a dienophile (Sarkar et al., 2011) (Figs 3.2.7 and 3.2.8).

Light can also be used to promote chemical reactivity between graphene and molecules. Liu et al. (2009) showed that laser irradiation of graphene in a benzoyl

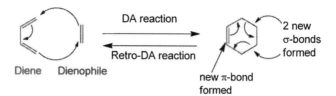

FIGURE 3.2.7 Schematic representation of the Diels–Alder (DA) reaction between a diene (1,3-butadiene) and dienophile (ethylene), illustrating the Diels–Alder (DA) cycloaddition and cyclo-reversion reactions in their simplest form; the forward reaction leads to the formation of a new six-membered ring via simultaneous creation of two new σ-bonds and one new π-bond. *Reprinted from Bekyarova (2012). Copyright (2012) Institute of Physics.*

peroxide solution resulted in holes in the basal plane due to photo-oxidisation of a defect rich area, shown in Fig. 3.2.9. Raman spectroscopy taken before and after the laser excitation showed the emergence of a strong D peak at 1343 cm^{-1}. The introduction of sp^3 defects in the graphene resulted in a decrease in the conductivity and charge mobility. They found that monolayer graphene was nearly 14 times more reactive than bilayer graphene (Liu et al., 2009).

When a graphene sheet is bent the sp^2 bonds deviate from a planar geometry, and this introduces strain, which was utilised in carbon nanotube chemistry (Niyogi

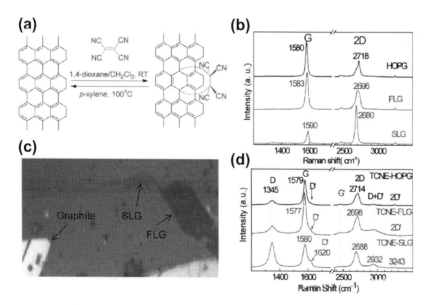

FIGURE 3.2.8 Room temperature Diels–Alder (DA) reaction between graphene (diene) and tetracyanoethylene (TCNE) dienophile. (a) Schematic representation of the reaction. (b) Micrograph showing a large piece of HOPG, SLG, and FLG on a Si substrate. (c, d) Raman spectra of HOPG, FLG and SLG (c) before and (d) after DA reaction with TCNE. *Reprinted with permission from Sarkar (2011). Copyright (2011) American Chemical Society.*

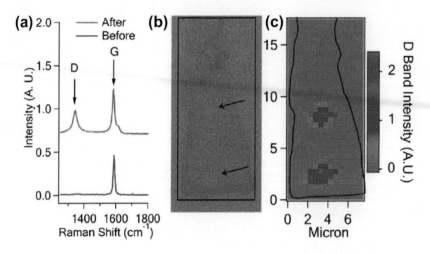

FIGURE 3.2.9 (a) Raman spectra ($\lambda_{ex} = 514.5$ nm, 0.4 mW) of the same SLG before and after the photochemical reaction. (b) Optical image of a SLG after the reaction. The contrast was enhanced to highlight the graphene (shown in red). The arrows indicate the holes resulted from the prolonged laser exposure. (c) Intensity map of the D band for the boxed. *Reprinted with permission from Liu (2009). Copyright (2009) American Chemical Society.*

et al., 2002). The variability in the diazonium reactions with graphene obtained by different forms may be related not only to the variation in layer numbers but also differences in the surface morphology. Boukhvalov and Katsnelson (2008, 2009) have shown that chemical functionalisation of graphene depends upon the corrugation of graphene as well as defects.

3.2.3. NONCOVALENT FUNCTIONALISATION OF GRAPHENE

In graphite, individual layers of graphene are held together by van der Waals forces arising from the π–π stacking of aromatic rings. For small molecules that contain aromatic rings, there may also be strong van der Waals forces that bind them to monolayer graphene and provide a means for functionalisation graphene without the need for covalent bonding. This has the advantage that it doesn't disrupt the sp^2 bonding network, but has the disadvantage of generally being weaker in strength than covalent attachment. Depositing atoms or molecules onto the surface of graphene often results in doping. In fact graphene exposed to air, moisture or hydrocarbon residues is often doped and gate-sweeps in a graphene FET show the Dirac point well away from 0 V. It then requires some form of annealing under vacuum or inert atmosphere to bring the Dirac point back to a 0 V back gate value in a FET measurement. Ohta et al. (2006) showed that simply depositing potassium atoms onto the surface of bilayer graphene grown by annealing SiC led to n-type doping due to the donation of valance electrons and the formation of dipoles. However, molecular doping of graphene rather than atomic doping offers significantly more exotic variances and has important consequences for how graphene is interfaced in thin film applications such as solar cells, photodetectors and energy storage devices. P-type doping of epitaxial graphene grown from SiC occurs when the electron accepting molecule tetrafluoro-tetracyanoquinodimethane (F4-TCNQ), shown in Fig. 3.2.10(a), is deposited onto its surface by thermal evaporation (Chen et al., 2007). Photoemission spectroscopy revealed the p-type doping was due to electron transfer from graphene to the F4-TCNQ, shown schematically in Fig. 3.2.10(b). As a comparison, C_{60} was also thermally evaporated onto graphene and results showed little charge transfer occurred (Chen et al., 2007).

Wang and Hersam (2009) also studied how epitaxial graphene from SiC responds to molecules adhered by van der Waals forces but this time using perylene-3,4,9,10-tetracarboxylic-3,4,9,10-dianhydride (PTCDA) (shown in Fig. 3.2.11(a)), which is a planar perylene based molecule with carboxylic acid anhydride side groups. STM images showed the PTCDA self assembled into monolayers on the graphene surface, shown in Fig. 3.2.11(b)–(g), and STS results indicated minimal doping of the graphene at room temperature (Wang and Hersam, 2009).

Barja et al. (2010) studied how F4-TCNQ and 7,7′,8,8′-tetracyano-p-quinodimethane (TCNQ) self organised on graphene grown on Ir(111). The major difference in this work compared to the previous mentioned work in (Chen et al., 2007) is that graphene on Ir(111) forms a strong Moire pattern compared to SiC. They showed that it is the intermolecular interaction, which is repulsive for F4-TCNQ and attractive for TCNQ, that controls the molecular ordering. Imaging of the HOMO and LUMO orbitals revealed they were nearly identical to a free molecule and that minimal charge transfer to the graphene layer occurred.

Other aromatic molecules with possible strong π–π interactions with graphene, such as 1,5-naphthalenediamine (Na–NH_2), 9,10-dimethylanthracene

(a)　　(b)

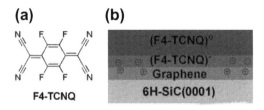

F4-TCNQ

(F4-TCNQ)⁰
(F4-TCNQ)⁻
⊕ Graphene ⊕ ⊖
6H-SiC(0001)

FIGURE 3.2.10 Schematic drawings of (a) structure of tetrafluoro-tetracyanoquinodimethane (F4-TCNQ) and (b) the charge transfer at the F4-TCNQ/graphene interface. Electron transfer from graphene to F4-TCNQ only occurred at the interface, where F4-TCNQ (in direct contact with graphene) is negatively charged and graphene is positively charged. F4-TCNQ in multilayers remains its neutral state (uncharged). *Reprinted with permission from Chen (2007). Copyright (2007) American Chemical Society.*

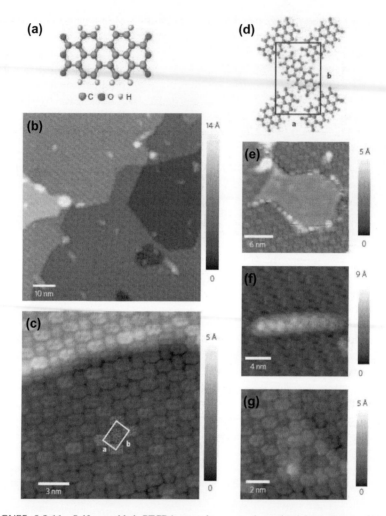

FIGURE 3.2.11 Self-assembled PTCDA monolayer on the epitaxial graphene substrate. (a) Molecular structure of PTCDA. (b) Monolayer coverage of PTCDA on epitaxial graphene. (c) Molecular-resolution STM image of the PTCDA monolayer. The PTCDA molecular structure and unit cell outline are overlaid. The monolayer continuously follows the graphene sheet over the SiC step edge. (d) PTCDA herringbone unit cell, with the lattice vectors a and b shown. (e) PTCDA surrounding a step edge where the graphene sheet is not continuous. (f) PTCDA continuously covers a graphene-subsurface nanotube defect. (g) A bright protrusion that does not disrupt the PTCDA monolayer and is potentially attributed to a sixfold scattering-centre defect. ($V_s = -2.0$ V, $I = 0.05$ nA for all five STM images.). *Reprinted from Wang (2009). Copyright (2009) Nature Publishing Group.*

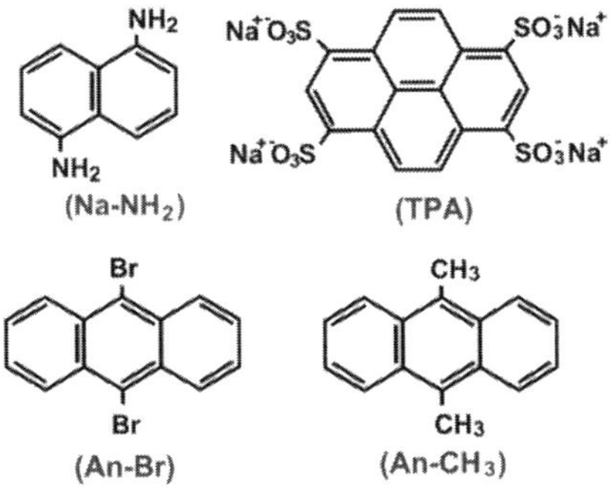

FIGURE 3.2.12 Chemical structures of the aromatic molecules used as dopants in (Dong et al., 2009). *Reproduced from Dong et al. (2009), Fig. 1. Copyright (2009) WILEY-VCH Verlag GmbH & Co.*

(An–CH₃), 9,10-dibromoanthracene (An–Br) and tetrasodium 1,3,6,8-pyrenetetrasulfonic acid, (shown in Fig. 3.2.12) were investigated by Dong et al. (2009). For this work, mechanically exfoliated graphene from HOPG was used. Raman spectroscopy showed that monolayer graphene was doped with holes when electron withdrawing aromatic molecules were deposited on the surface or doped with electrons when a electron donating molecule was used instead (Dong et al., 2009). Su et al. (2009) showed larger pyrene based molecules, such as 3,4,9,10-perylenetetracarboxylic diimide bisbenzenesulfonic acid, which has stronger π–π interactions with graphene can be used to dope it. The functionality of the perylene and pyrene molecules that bind to the graphene surface by van der Waals can be extended by adding other molecular groups to the perylene/pyrene to generate nonplanar geometry. Kozhemyakina et al. (2010) used a photoluminescent dendronised perylene bisimide to non-covalently attach to graphene in solution. Bai et al. (2009) non-covalently functionalised graphene sheets with the large-conducting polymer sulfonated polyaniline (SPANI) to produce a water soluble and electroactive composite. The SPANI exhibits strong π–π interactions with graphene and composite films showed high conductivity and good electrocatalytic activity. Water dispersed non-covalent functionalised graphene sheets have also been produced using 1-pyrenebutyrate (Xu et al., 2008).

3.2.4. SUMMARY

The basal plane of graphene (i.e. the main lattice) can be chemically functionalised by disrupting the pristine lattice and introducing covalent bonds or by relying on van der Waals π–π interactions to non-covalently adhere molecules to the surface. Both have their advantages and disadvantages, which should be considered when choosing which approach is suitable for your application. The chemical properties of graphene and how it can be interfaced with other materials will be a crucial aspect of incorporating graphene into applications. Graphene as a stand-alone material has limited value and other nanomaterials and molecular structures need to be merged. In this chapter we have seen how atoms and molecules can dope graphene to yield either p or n-type material and also open up a band gap. It is likely that the key aspect to graphene's success in many applications revolves around the ability to manipulate its chemical structure in order to tailor its electronic properties. Due to the prior work already undertaken on graphite, nanotube and fullerene chemistry, we expect that progress in this area can be rapid as it builds upon an already extensive platform of carbon chemistry.

REFERENCES

Bahr, J.L., Yang, J., Kosynkin, D.V., Bronikowski, M.J., Smalley, R.E., Tour, J.M., 2001. Functionalization of carbon nanotubes by electrochemical reduction of aryl diazonium salts: a bucky paper electrode. J. Am. Chem. Soc. 123, 6536–6542.

Bai, H., Xu, Y., Zhao, L., Li, C., Shi, G., 2009. Non-covalent functionalization of graphene sheets by sulfonated polyaniline. Chem. Commun. 1667–1669.

Balog, R., Jorgensen, B., Nilsson, L., Andersen, M., Rienks, E., Bianchi, M., Fanetti, M., Laegsgaard, E., Baraldi, A., Lizzit, S., Sljivancanin, Z., Besenbacher, F., Hammer, B., Pederson, T.G., Hofmann, P., Hornekaer, L., 2010. Bandgap opening in graphene induced by patterned hydrogen adsorption. Nat. Mater. 9, 315–319.

Barja, S., Garnica, M., Hinarejos, J.J., Vazquez de Parga, A.L., Martin, N., Miranda, R., 2010. Self-organization of electron acceptor molecules on graphene. Chem. Commun. 46, 8198–8200.

Bekyarova, E., Itkis, M.E., Ramesh, P., Berger, C., Sprinkle, M., de Heer, W.A., Haddon, R.C., 2009. Chemical modification of epitaxial graphene: spontaneous grafting of aryl groups. J. Am. Chem. Soc. 131, 1336–1337.

Bekyarova, E., Sarkar, S., Niyogi, S., Itkis, M.E., Haddon, R.C., 2012. Advances in the chemical modification of epitaxial graphene. J. Phys. D 45, 154009.

Boukhvalov, D.W., Katsnelson, M.I., 2008. Chemical functionalization of graphene with defects. Nano Lett. 8, 4373–4379.

Boukhvalov, D.W., Katsnelson, M.I., 2009. Enhancement of chemical activity in corrugated graphene. J. Phys. Chem. C 113, 14176–14178.

Chen, W., Chen, S., Qi, D.C., Gao, X.Y., Wee, A.T.S., 2007. Surface p-type doping of epitaxial graphene. J. Am. Chem. Soc. 129, 10418–10422.

Chen, S., Brown, L., Levendorf, M., Cai, W., Ju, S.-Y., Edgeworth, J., Li, X., Magnuson, C.W., Velamakanni, A., Piner, R.D., Kang, J., Park, J., Ruoff, R.S., 2011. Oxidation resistance of graphene-coated Cu and Cu/Ni Alloy. ACS Nano 5, 1321–1327.

Cheng, S.-H., Zou, K., Okino, F., Gutierrez, H.R., Gupta, A., Shen, N., Eklund, P.C., Sofo, J.O., Zhu, J., 2010. Reversible fluorination of graphene: evidence of a two-dimensional wide band gap semiconductor. Phys. Rev. B 81, 205435.

Choi, J., Kim, K.-J., Kim, B., Lee, H., Kim, S., 2009. Covalent functionalization of epitaxial graphene by azidotrimethylsilane. J. Phys. Chem. C 113, 9433–9435.

Delamar, M., Hitmi, R., Pinson, J., Saveant, J.M., 1992. Covalent modification of carbon surfaces by grafting of functionalized aryl radicals produced from electrochemical reduction of diazonium salts. J. Am. Chem. Soc. 114, 5883–5884.

Dong, X., Fu, D., Fang, W., Shi, Y., Chen, P., Li, L.-J., 2009. Doping Single-layer graphene with aromatic molecules. Small 5, 1422–1426.

Dresselhaus, M.S., Dresselhaus, G., 1981. Intercalation compounds of graphite. Adv. Phys. 30 (2), 139–326.

Elias, D.C., Nair, R.R., Mohiuddin, T.M.G., Morozov, S.V., Blake, P., Halsall, M.P., Ferrari, A.C., Boukhvalov, D.W., Katsnelson, M.I., Geim, A.K., Novoselov, K., 2009. Control of graphene's properties by reversible hydrogenation: evidence for graphane. Science 323, 610–613.

Koehler, F.M., Luechinger, N.A., Ziegler, D., Athanassiou, E.K., Grass, R.N., Rossi, A., Hierold, C., Stemmer, A., Stark, W.J., 2009. Permanent pattern-resolved adjustment of the surface potential of graphene-like carbon through chemical functionalization. Angew. Chem. Int. Ed. 48, 224–227.

Koehler, F.M., Jacobsen, A., Ensslin, K., Stampfer, C., Stark, W.J., 2010. Selective chemical modification of graphene surfaces: distinction between single- and bilayer graphene. Small 6, 1125.

Kozhemyakina, N.V., Englert, J.M., Yang, G., Spiecker, E., Schmidt, C.D., Hauke, F., Hirsch, A., 2010. Non-covalent chemistry of graphene: electronic communication with dendronized perylene bisimides. Adv. Mater. 22, 5483–5487.

Kuo, T.C., McCreery, R.L., Swain, G.M., 1999. Electrochem. Solid State Lett. 2, 288–290.

Liu, Y.C., McCreery, R.L., 1995. Reactions of organic monolayers on carbon surfaces observed with unenhanced raman spectroscopy. J. Am. Chem. Soc. 117, 11254–11259.

Liu, H., Ryu, S., Chen, Z., Steigerwald, M.L., Nuckolls, C., Brus, L.E., 2009. Photochemical reactivity of graphene. J. Am. Chem. Soc. 131, 17099–17101.

Nair, R.R., Ren, W., Jalil, R., Riaz, I., Kravets, V.G., Britnell, L., Blake, P., Schedin, F., Mayorov, A.S., Yuan, S., Katsnelson, M.I., H-M Cheng, Strupinski, W., Bulusheva, L.G., B Okotrub, A., Grigorieva, I.V., Grigorenko, A.N., Novoselov, K.S., Geim, A.K., 2010. Fluorographene: a two-dimensional counterpart of teflon. Small 6, 2877–2884.

Niyogi, S., Hamon, M.A., Hu, H., Zhao, B., Bhowmik, P., Sen, R., Itkis, M.E., Haddon, R.C., 2002. Chemistry of single-walled carbon nanotubes. Acc. Chem. Res. 35, 1105–1113.

Niyogi, S., Bekyarova, E., Itkis, M.E., Zhang, H., Sheppard, K., Hicks, J., Spinkle, M., Berger, C., Lau, C.N., de Heer, W.A., Conrad, E.H., Haddon, R.C., 2010. Spectroscopy of covalently functionalized graphene. Nano Lett. 10, 4061–4066.

Niyogi, S., Bekyarova, E., Hong, J., Khizroev, S., Berger, C., de Heer, W.A., Haddon, R.C., 2011. Covalent chemistry for graphene electronics. J. Phys. Chem. Lett. 2, 24872011.

Ohta, T., Bostwick, A., Seyller, T., Horn, K., Rotenberg, E., 2006. Controlling the electronic structure of bilayer graphene. Science 313, 951–954.

Pauling, L., 1931. The nature of the chemical bond. Application of results obtained from the quantum mechanics and from a theory of paramagnetic susceptibility to the structure of molecules. J. Am. Chem. Soc. 53, 1367–1400.

Sarkar, S., Bekyarova, E., Niyogi, S., Haddon, R.C., 2011. Diels–Alder chemistry of graphite and graphene: graphene as diene and dienophile. J. Am. Chem. Soc. 133, 3324.

Su, Q., Pang, S., Alijani, V., Li, C., Feng, X., Mullen, K., 2009. Composites of graphene with large aromatic molecules. Adv. Mater. 21, 3191–3195.

Wang, Q.H., Hersam, M.C., 2009. Room-temperature molecular-resolution characterization of self-assembled organic monolayers on epitaxial graphene. Nat. Chem. 1, 206–211.

Wu, J., Pisula, W., Mullen, K., 2007. Graphene as potential material for electronics. Chem. Rev. 107, 718–747.

Xu, Y., Bai, H., Lu, G., Li, C., Shi, G., 2008. Flexible graphene films via the filtration of water-soluble noncovalent functionalized graphene sheets. J. Am. Chem. Soc. 130, 5856–5857.

Electron Spin Properties of Graphene

Jamie H. Warner

University of Oxford, Oxford, UK

3.3.1. INTRODUCTION

Spin is a fundamental physical intrinsic property of an elementary particle and provides a degree of freedom associated with angular momentum. In quantum mechanics, angular momentum consists of orbital angular momentum and spin, which is not represented in classical mechanics. Different elementary particles possess different spin numbers and electrons have spin half. The spin angular momentum vector $\mathbf{S}$ can be expressed in Cartesian coordinates as $\mathbf{S} = S_x, S_y, S_z$, with only one component required to have a value of $\pm 1/2$. Convention sets this to be the z-component resulting in $S_z = \pm 1/2$, which are thought of as spin up and down states. A magnetic moment arises from the spin angular momentum $\mathbf{S}$ expressed by:

$$\mu_e = g\mu_B S$$

where μ_B is the Bohr magneton and g is the g-factor. The energy of a spin's magnetic moment can be expressed as:

$$E = \mu_e \cdot B$$

where B is the applied magnetic field strength. When the magnetic field is zero ($B_0 = 0$), the energy of the spin up and down states are equal. However, under an applied magnetic field a spin state aligns parallel or antiparallel with the field and results in an energy difference between them, known as the Zeeman effect, shown schematically in Fig. 3.3.1.

Electron spin resonance (ESR) is a commonly used technique that exploits the Zeeman effect to probe paramagnetic properties of materials using electromagnetic radiation and magnetic fields. Electromagnetic radiation is

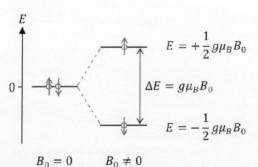

FIGURE 3.3.1 Schematic illustration of the Zeeman effect.

absorbed by a material in a magnetic field when its energy is equal to the energy difference between the two spin states, essentially flipping the spin state. In ESR the electromagnetic field is kept constant and the magnetic field is swept until resonance occurs. If both spin states are occupied then absorption of the photon cannot take place, and therefore it is often the case that unpaired electrons are probed with ESR.

A semiconducting quantum interference device (SQUID) is a highly sensitive magnetometer based on a Josephson junction and is also often used to study magnetism in materials. It is effective at distinguishing whether a material exhibits ferromagnetism, paramagnetism or diamagnetism. Ferromagnets retain magnetisation without an external magnetic field, paramagnets are attracted only in the presence of a magnetic field, whilst diamagnets form a magnetic field opposite to the applied external magnetic field. Both paramagnets and diamagnets lose their magnetisation once an external field has been removed. All materials respond either diamagnetically or paramagnetically to an applied external magnetic field. For materials that exhibit ferromagnetism or paramagnetism, the diamagnetism component becomes negligible.

3.3.2. SPIN AND MAGNETISM IN GRAPHITE

When considering electron spins in graphene and graphite, there are two types of spins we should consider, static spins associated with defects and conduction electron spins. Studies of electron spin properties of graphite have been ongoing since Castle's work in 1953, where a paramagnetic signal from graphite was attributed to conduction electrons (Castle, 1953). The graphite studied by Castle was obtained from a steelmaking process, which is known to result in iron impurities in graphite that are now thought likely to be the source of the signal. Similar results were reported a year later by Hennig et al., and they observed the signal disappearing under high-temperature vacuum annealing, then reappearing when annealing in air at high temperature. This led to the conclusion that the signals were an effect of oxygen in the sample rather than from conduction electrons (Hennig et al., 1954).

A more detailed analysis of the electron spin properties of single crystal graphite was reported by Wagoner in 1960 taking into account graphite band theory (Elliott, 1954; McClure, 1956; Slonczewski and Weiss, 1958; Wagoner, 1960). Wagoner postulated that the spin signal disappearance observed by Hennig et al. was simply due to broadening and the inability to continue to detect it. The lineshape of the ESR signal from graphite reported by Wagoner exhibited a Dysonian lineshape, that is characteristic of conduction-electron spin in metals (Dyson, 1955). In 1962, polycrystalline graphite was examined and it also showed similar results to the single crystal ESR study, but with more information regarding the effect of impurities (Singer and Wagoner, 1962). The discovery of fullerenes and then carbon nanotubes, reinvigorated the study of electron spins in carbon materials. However, it has always remained a challenge to be sure that all impurities are removed such that the true intrinsic behaviour can be studied.

There are several reports of spin resonance signals arising from carbon nanotube samples, with often contradictory results (Bandow et al., 1998; Corzilius et al., 2006; Nafradi et al., 2006; Salvetat et al., 2005; Shen et al., 2003; Petit et al., 1997). However, in most cases there was a lack of rigorous multiple characterisation techniques to fully elucidate the purity of the samples before ESR was undertaken. Impurities in carbon materials plague the understanding of intrinsic magnetism and this is evident by the report of magnetic carbon in 2001 by Makarova et al., followed by its retraction in 2006 due to contamination by iron (Makarova et al., 2001, 2006). In 2010, we showed that ESR signals in SWNTs were associated with metal impurities and once they were completely removed the ESR signal in the X-band region disappeared (Zaka et al., 2010). This is further supported when considering spin–orbit interactions in nanotubes (Huertas-Hernando et al., 2006).

Doping of fullerenes with metals, atoms or molecules can result in electron spin properties that are amongst the best for molecular materials in terms of spin coherence times (Brown et al., 2010). They have been well documented and due to the wide range of different material types, a conclusive and comprehensive understanding has been developed. Given that fullerenes are 0D structures, unpaired electrons are well localised either within the cage on encapsulated atoms or delocalised across the cage itself.

There have been several reports of ferromagnetism in graphite samples, both untreated (Esquinazi et al., 2002; Kopelevich et al., 2000) and irradiated (Barzola-Quiquia et al., 2007, 2008; Esquinazi et al., 2003; Xia et al., 2008). Cervenka et al. (2009) used both SQUID measurements and magnetic force microscopy to study ferromagnetism in HOPG. Whilst it is acknowledged that Fe impurities exist even in high quality HOPG, they excluded these as an origin of the ferromagnetism in HOPG and instead attribute it to localised electron states at grain boundaries, which consist of a 2D array of defects.

3.3.3. MAGNETISM AND SPIN IN GRAPHENE

Theoretical studies predict magnetic moments for impurities and vacancies in graphene (Faccio et al., 2008; Krasheninnikov et al., 2009; López-Sancho et al., 2009; Yazyev, 2008). Yazyev (2008) concluded that only single-atom defects could induce ferromagnetism in disordered graphene. López-Sancho et al. (2009) predicted that dislocations alter the magnetic structure of the unperturbed lattice due to the presence of odd-membered rings and expected similar results for single pentagons, heptagons or 5–7 configurations. Figure 3.3.2 shows the spin distribution for single atom defects and Fig. 3.3.3 for dislocations

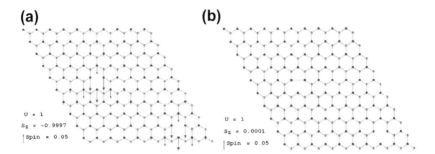

FIGURE 3.3.2 Spin distribution in a lattice with two vacancies of the same sublattice with $U = 1$. Right: same configuration in the presence of a pentagon for the same value of U. *Reproduced from López-Sancho et al. (2009), Fig. 1. Copyright (2009) American Physical Society.*

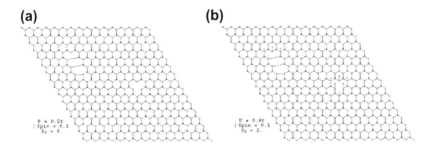

FIGURE 3.3.3 Spin structure for two different configurations of dislocations and a vacancy with $U = 0.3$ with total spin polarisations $S = 0$ (left) and $S = 1$ (right). *Reproduced from López-Sancho et al. (2009), Fig. 3. Copyright (2009) American Physical Society.*

(López-Sancho et al., 2009). The calculations of Harigaya and Enoki (2002) suggest that in order for magnetism to arise in stacked hexagonal nanographites with zig-zag edges, they must adopt near A–B stacking.

As the size of graphene sheets is reduced, edge states become important and there have been several theoretical predictions of unique spin and magnetic states for well defined graphene nano-geometries (Fernandez-Rossier and Palacios, 2007; Lee et al., 2005; Pisani et al., 2007; Son et al., 2006; Wakabayashi et al., 1999; Wang et al., 2009). Figure 3.3.4 shows (a) the spectrum of singly occupied states of a bow-tie-shaped Graphene nano-flake (GNF) populated by spin-up ($1u$ and $2u$) and spin-down ($1d$ and $2d$) electrons, (b) the isodensity surface of the total spin distribution showing opposite spins localised at opposite sides and (c) wavefunctions of the four singly occupied states (Wang et al., 2009). In 1997, Wakabayashi et al. used

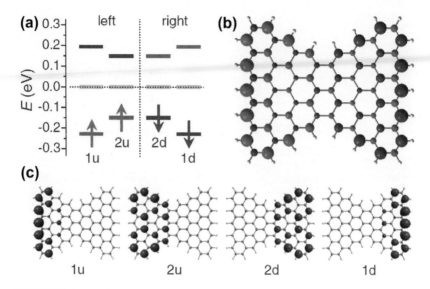

FIGURE 3.3.4 (a) The spectrum of singly occupied states of a bow-tie-shaped GNF populated by spin-up (1*u* and 2*u*) and spin down (1*d* and 2*d*) electrons. (b) Isodensity surface of the total spin distribution showing opposite spins localised at opposite sides. (c) Wavefunctions of the four singly occupied states. *Reproduced from Wang et al. (2009), Fig. 2. Copyright (2009) American Physical Society.*

a tight-binding model to predict the magnetic susceptibility of GNRs with zig-zag edges had a crossover from high-temperature diamagnetic to low-temperature paramagnetic behaviour (Wakabayashi et al., 1999). The work of Pisani et al. (2007) showed that spin polarisation is a possible stabilisation mechanism of GNRs with zig-zag edges. Pisani et al. (2007) found that by allowing the system to be spin polarised, stable magnetic states were found and the spin density of these are shown in Fig. 3.3.5 for (a) antiferromagnetic and (b) ferromagnetic cases. The challenge in realising the predictive magnetism in graphene nanostructures is the strict requirements of atomically precise edge terminations are very difficult to obtain. However if this can be mastered through either bottom-up or top-down approaches then it paves the way for interesting magnetoresistance device experiments that are important for spintronics and quantum spin logic, such as that presented in Fig. 3.3.6.

Despite the significant advances in the theoretical work on spin and magnetism of graphene and nanostructured graphene, the experimental confirmation is slow and agreement within the community is yet to be fully resolved. Wang et al. (2009) reported room temperature ferromagnetism using a SQUID in graphene that was obtained by chemical exfoliation. They used the Hummer method to obtain graphene oxide and then reduced

(a) **(b)**

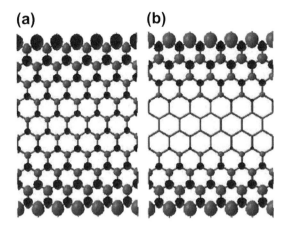

FIGURE 3.3.5 Isovalue surfaces of the spin density for the antiferromagnetic case (a) and ferromagnetic case (b). The dark grey (red online) surface represents spin up density and the black (blue online) surface spin down density. The range of isovalues is $[-0.28: 0.28]\mu_B/\text{Å}^3$ in case (a) and $[-0.09: 0.28]\mu_B/\text{Å}^3$ in case (b). *Reproduced from Pisani et al. (2007), Fig. 7. Copyright (2007) American Physical Society.*

it using hydrazine, followed by thermal annealing in Argon for 3 hours. This process is known to induce a large number of defects. Figure 3.3.7 shows their magnetisation curves for graphene that was annealed at (a) 400 °C and (b) 600 °C. In order to understand how magnetic impurities impacted on their magnetisation studies they measured graphene oxide samples before they were reduced in hydrazine and these showed no ferromagnetism. Three different SQUID systems were used and five different samples, all showing room temperature ferromagnetism they attributed as an intrinsic property of their graphene (Wang et al., 2009). The origin of the ferromagnetism was stated as 'from the long-range coupling of spin units existing as defects in graphene sheets, which are generated in the annealing process'. The graphite used in this study was flake graphite, which is known to contain more metal impurities than HOPG. Whilst this work provides very interesting results for magnetism in graphene, the lack of other detailed characterisations of the structure of the graphene after it has been annealed make it difficult to place full confidence in the conclusions drawn.

Sepioni et al. (2010) reported conflicting results with no ferromagnetism from graphene obtained by chemical exfoliation. They used high purity HOPG and performed ultrasonic chemical exfoliation in N-Methyl-2-pyrrolidone (NMP) to obtain small fragments of graphene in solution. This process yields graphene that has less defects than reduced graphene oxide and may be one explanation of the different results compared to Wang et al., 2009. The samples

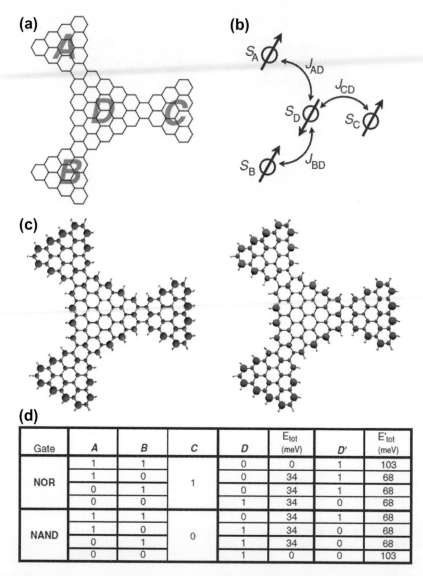

(d)

Gate	A	B	C	D	E_{tot} (meV)	D'	E'_{tot} (meV)
NOR	1	1	1	0	0	1	103
	1	0		0	34	1	68
	0	1		0	34	1	68
	0	0		1	34	0	68
NAND	1	1	0	0	34	1	68
	1	0		1	34	0	68
	0	1		1	34	0	68
	0	0		1	0	0	103

FIGURE 3.3.6 (a) Reconfigurable spin logic NOR and NAND gate based on of a tri-bow-tie GNF structure with $n_A = n_B = n_C = 4$, $n_D = 6$, $m = 1$ (A, B, and D are two inputs and one output, respectively, and C is the programming bit). (b) A scheme of the localised spins and the couplings ($2J_{XY} = 34$ meV). (c) Two distinct spin configurations corresponding to 1110 and 0110 for the *ABCD* spins, respectively. (d) The truth table of the programmable logic gate and the total energy E_{tot} of the operation configuration. D' and E'_{tot} are the error output and the corresponding energy ($E'_{tot} > E_{tot}$). *Reproduced from Wang et al. (2009), Fig. 4. Copyright (2009) American Chemical Society.*

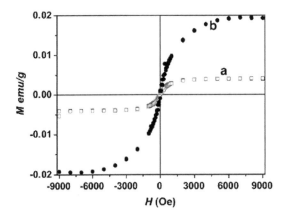

FIGURE 3.3.7 Magnetisation hysteresis loops at 300 K in the range of $-10\,\text{kOe} < H < +10\,\text{kOe}$. (a) Sample graphene annealed at 400 °C (0), $M_s = 0.004$ emu/g. (b) Sample graphene annealed at 600 °C (b), $M_s = 0.020$ emu/g. *Reproduced from Wang et al. (2009), Fig. 1. Copyright (2009) American Chemical Society.*

were strongly diamagnetic, with some low-temperature paramagnetism tentatively associated with a small number of noncompensated spins that survived after interactions between different zigzag segments in each of the sub-100 nm samples of random shape are taken into account (Wang et al., 2009). They did not rule out bi-layer and trilayer graphene as some alternative means of the paramagnetism. The authors clearly state that the main conclusion of their work is the absence of ferromagnetism in graphene down to 2 K (Sepioni et al., 2010).

Nair et al. (2012) continued on this work in investigating the magnetism in graphene by introducing point defects (vacancies) and adatoms. Fluorine adatoms were attached to the graphene sheets by the decomposition of xenon difluoride at 200 °C and resulted in strong paramagnetism that increased with fluorine concentration, shown in Fig. 3.3.8. Vacancies were introduced by high irradiation of the graphene by high-energy protons and carbon ($C4^+$) ions and also led to paramagnetism (Nair et al., 2012). The authors claim this provides the most unambiguous direct support for many theories discussing graphene's magnetic properties. They found no ferromagnetism in any of their samples (Nair et al., 2012). Importantly, they did find ferromagnetism in several HOPG samples they investigated (ZYA- ZYB- and ZYH-grade from NT-MDT), but not all (SPI-2 and SPI-3 from SPI supplies). Scanning electron microscopy combined with energy dispersive X-ray spectroscopy linked this directly to the presence of Fe impurities in the graphite. Previous reports of ferromagnetism in HOPG used ZYA- (Esquinazi et al., 2002, 2003) and ZYH-grade (Cervenka et al., 2009) HOPG that are likely to contain these magnetic inclusions.

The study of paramagnetism in graphene using ESR has been limited and again is susceptible to the same spurious effects of impurities seen in the SQUID measurements. Ciric et al. (2009) mechanically exfoliated HOPG with scotch tape and stacked several pieces in parallel for their ESR measurements at X-band. An ESR signal with a single Lorentzian lineshape

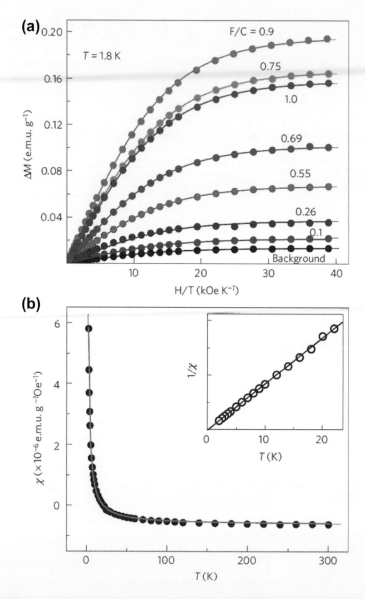

FIGURE 3.3.8 Paramagnetism due to fluorine adatoms. (a) Magnetic moment ΔM (after subtracting linear diamagnetic background) as a function of parallel field H for different F/C ratios. Symbols are the measurements and solid curves are fits to the Brillouin function with $S = 1/2$ and assuming $g = 2$ (the fits weakly depend on g). (b) Example of the dependence of susceptibility $\chi = M/H$ on T in parallel $H = 3$ kOe for CF_x with $x = 0.9$; symbols are the measurements and the solid curve is the Curie law calculated self-consistently using the M/H dependence found in (a). Inset: inverse susceptibility versus T demonstrating a linear, purely paramagnetic behaviour with no sign of magnetic ordering. *Reproduced from Nair et al. (2012), Fig. 2. Copyright (2012) Nature Publishing Group.*

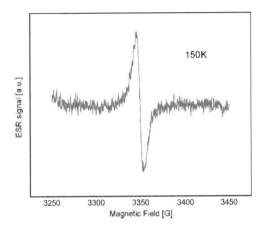

FIGURE 3.3.9 Typical ESR line recorded at 150 K for the assembly of stacked graphene flakes. The ESR line has a Lorentzian shape; it is centred on $g = 2$, having a linewidth of 6.5 G. *Reproduced from Ciric et al. (2009), Fig. 3. Copyright (2009) WILEY-VCH Verlag GmbH & Co.*

was observed, shown in Fig. 3.3.9. This is different to the Dysonian lineshape often observed for graphite. Curie-like behaviour was seen at temperatures below 70 K and the authors indicate this is due to the strong coupling between defects and conduction electrons (Ciric et al., 2009). The T_1 relaxation time was determined to be 55 ns. The challenge in interpreting this data lies in the possibility that the HOPG may contain Fe impurities and that the sample still contains some thin graphite flakes. Nonetheless, this is one of the only reports of the ESR of graphene and provides important information on the paramagnetism in graphene. Dora et al. (2010) conducted a rigorous theoretical approach for understanding the expected ESR signals from graphene. Their theory of spin relaxation includes intrinsic, Bychkov-Rashba, and ripple spin–orbit coupling- (Dora et al., 2010). By taking into account recent experimental results of Tombros et al. (2007) they find intrinsic spin–orbit coupling dominates. A plot showing the limit of detection for ESR of graphene was determined in Dora et al., 2010, as a function of spin relaxation rate Γ and chemical potential μ, shown in the top panel in Fig. 3.3.10, and the expected ESR linewidth is presented below.

3.3.4. SUMMARY

Elucidating the intrinsic spin and magnetic properties of graphite and graphene is tricky and has been complicated by impurities in samples. It is obvious that care must be taken to ensure all forms of contamination are prevented, such as from poor quality solvents, to eliminating the use of metal tweezers. Most inroads into understanding magnetism have come from SQUID measurements due to its high sensitivity, whereas ESR measurements of the paramagnetism have been limited due to the large number of spins required for detection. Careful work by Nair et al. (2012) has shown that ferromagnetism in graphite and graphene is associated with metal inclusions,

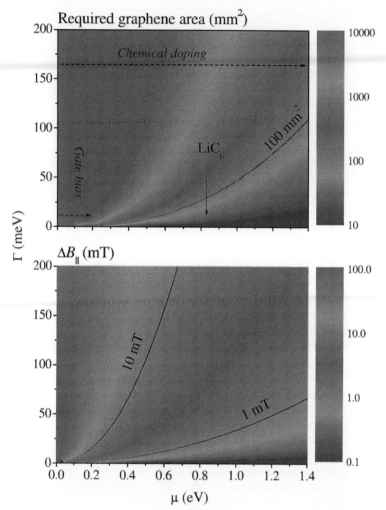

FIGURE 3.3.10 Limit of ESR detection for graphene as a function of μ and Γ in units of the graphene area (upper panel) for an in-plane magnetic field. The arrows show the maximum chemical potential by gate bias and by chemical doping, and the solid curve indicates the area border of 100 mm^2. Expected ESR linewidth, $\Delta B_\parallel$ (lower panel), solid lines show two selected linewidths, 1 and 10 mT. *Reproduced from Dora et al. (2010), Fig. 3. Copyright (2010) WILEY-VCH Verlag GmbH & Co.*

even in supposedly high purity HOPG. Weak paramagnetism at low-temperatures is reported for relatively defect free graphene, and the use of ion irradiation to add vacancies can increase the paramagnetism. Further insights into the electron spin properties of graphene can be obtained through electronic transport measurements with spin-polarised electrodes, but this will be covered in Chapter 6.2: Spintronics. If advances in the bottom-up fabrication of well ordered graphene nanostructures continues then this may provide a clean and clear route to studying ensembles of identical graphene nano-structures and provide more insights into how edge states influence para-magnetism and ferromagnetism.

REFERENCES

Bandow, S., Asaka, S., Zhao, X., Ando, Y., 1998. Purification and magnetic properties of carbon nanotubes. Appl. Phys. A 67, 23–27.

Barzola-Quiquia, J., Esquinazi, P., Rothermel, M., Spemann, D., Butz, T., Garcia, N., 2007. Phys. Rev. B 76, 161403.

Barzola-Quiquia, J., Höhne, R., Rothermel, M., Setzer, A., Esquinazi, P., Heera, V., 2008. Eur. Phys. J. B 61, 127–130.

Brown, R.M., Ito, Y., Warner, J., Ardavan, A., Shinohara, H., Briggs, G.A.D., Morton, J.J.L., 2010. Electron spin coherence in metallofullerenes: Y, Sc and La@C82. Phys. Rev. B 82.

Castle Jr., J.G., 1953. Paramagnetic resonance absorption in graphite. Phys. Rev. 92, 1063.

Cervenka, J., Katsnelson, M.I., Flipse, C.F.J., 2009. Room temperature ferromagnetism in graphite driven by two-dimensional networks of point defects. Nat. Phys. 5, 840–844.

Ciric, L., Sienkiewicz, A., Nafradi, B., Mionic, M., Magrez, A., Forro, L., 2009. Towards electron spin resonance of mechanically exfoliated graphene. Phys. Status Solidi B 246, 2558–2561.

Corzilius, B., Gembus, A., Weiden, N., Dinse, K.P., Hata, K., 2006. EPR characterization of catalyst-free SWNT and N@C60-based peapods. Phys. Stat. Sol. (b) 243, 3273–3276.

Dora, B., Muranyi, F., Simon, F., 2010. Electron spin dynamics and electron spin resonance in graphene. Eur. Phys. Lett. 92, 17002.

Dyson, F.J., 1955. Electron spin resonance absorption in metals. II. Theory of electron diffusion and the skin effect. Phys. Rev. 98, 349–359.

Elliott, R.J., 1954. Theory of the effect of spin-orbit coupling on magnetic resonance in some semiconductors. Phys. Rev. 96, 266–279.

Esquinazi, P., Setzer, A., Hohne, R., Semmelhack, C., Kopelevich, Y., Spemann, D., Butz, T., Kohlstrunk, B., Losche, M., 2002. Phys. Rev. B 66, 024429.

Esquinazi, P., Spemann, D., Höhne, R., Setzer, A., Han, K.H., Butz, T., 2003. Phys. Rev. Lett. 91, 227201.

Faccio, R., et al., 2008. Phys. Rev. B 77, 035416.

Fernandez-Rossier, J., Palacios, J.J., 2007. Magnetism in graphene nanoislands. Phys. Rev. Lett. 99, 177204.

Harigaya, K., Enoki, T., 2002. Mechanism of magnetism in stacked nanographite with open shell electrons. Chem. Phys. Lett. 351, 128–134.

Hennig, G.R., Smaller, B., Yasaitis, E.L., 1954. Paramagnetic resonance absorption in graphite. Phys. Rev. 95, 1088–1089.

Huertas-Hernando, D., Guinea, F., Brataas, A., 2006. Spin-orbit coupling in curved graphene, fullerenes, nanotubes, and nanotube caps. Phys. Rev. B 74, 155426.

Kopelevich, Y., Esquinazi, P., Torres, J.H.S., Moehlecke, S.J., 2000. Low. Temp. Phys. 119, 691–702.

Krasheninnikov, A.V., et al., 2009. Phys. Rev. Lett. 102, 126807.

Lee, H., Son, Y.-W., Park, N., Han, S., Yu, J., 2005. Magnetic ordering at the edges of graphitic fragments: magnetic tail interactions between the edge-localized states. Phys. Rev. B 72, 174431.

López-Sancho, M.P., de Juan, F., Vozmediano, M.A.H., 2009. Magnetic moments in the presence of topological defects in graphene. Phys. Rev. B 79, 075413.

Makarova, T.L., Sundqvist, B., Hohne, R., Esquinazi, P., Kopelevich, Y., Scharff, P., Davydov, V.A., Kashevarova, L.S., Rakhmanina, A.V., 2001. Magnetic carbon. Nature 413, 716–718.

Makarova, T.L., Sundqvist, B., Hohne, R., Esquinazi, P., Kopelevich, Y., Scharff, P., Davydov, V.A., Kashevarova, L.S., Rakhmanina, A.V., 2006. Retraction: Magnetic Carbon 440, 707.

McClure, J.W., 1956. Diamagnetism of graphite. Phys. Rev. 104, 666–671.

Nafradi, B., Nemes, N.M., Feher, T., Forro, L., Kim, Y., Fischer, J.E., Luzzi, D.E., Simon, F., Kuzmany, H., 2006. Electron spin resonance of single-walled carbon nanotubes and related structures. Phys. Stat. Sol. (b) 243, 3106–3110.

Nair, R.R., Sepioni, M., I-Tsai, L., Lehtinen, O., Keinonen, J., Krasheninnikov, A.V., Thomson, T., Geim, A.K., Grigorieva, I.V., 2012. Spin-half paramagnetism in graphene induced by point defects. Nat. Phys. 8, 199–202.

Petit, P., Jouguelet, E., Fischer, J.E., Rinzler, A.G., Smalley, R.E., 1997. Electron spin resonance and microwave resistivity of single-wall carbon nanotubes. Phys. Rev. B 56, 9275–9278.

Pisani, L., Chan, J.A., Montanari, B., Harrison, N.M., 2007. Electronic structure and magnetic properties of graphitic ribbons. Phys. Rev. B. 75, 064418.

Salvetat, J.P., Feher, T., L'Huillier, C., Beuneu, F., Forro, L., 2005. Anomalous electron spin resonance behaviour of single-walled carbon nanotubes. Phys. Rev. B 72, 075440.

Sepioni, M., Nair, R.R., Rablen, S., Narayanan, J., Tuna, F., Winpenny, R., Geim, A.K., Grigorieva, I.V., 2010. Limits on intrinsic magnetism in graphene. Phys. Rev. Lett. 105, 207205.

Shen, K., Tierney, D.L., Pietra, T., 2003. Electron spin resonance of carbon nanotubes under hydrogen adsorption. Phys. Rev. B 68, 165418.

Singer, L.S., Wagoner, G., 1962. Electron spin resonance in polycrystalline graphite. J. Chem. Phys. 37, 1812–1817.

Slonczewski, J.C., Weiss, P.R., 1958. Band structure of graphite. Phys. Rev. 109, 272–279.

Son, Y.-W., Cohen, M.L., Louie, S.G., 2006. Half-metallic graphene nanoribbons. Nature 444, 347.

Tombros, N., Jozsa, C., Popinciuc, M., Jonkman, H.T., van Wees, B.J., 2007. Electron spin transport and spin precession in single graphene layers at room temperature. Nature 448, 571–575.

Wagoner, G., 1960. Spin resonance of charge carriers in graphite. Phys. Rev. 118, 647–653.

Wakabayashi, K., Fujita, M., Ajiki, H., Sigrist, M., 1999. Electronic and magnetic properties of nanographite ribbons. Phys. Rev. B 59, 8271–8282.

Wang, W.L., Yazyev, O.V., Meng, S., Kaxiras, E., 2009. Topological frustration in graphene nanoflakes: magnetic order and spin logic devices. Phys. Rev. Lett. 102, 157201.

Wang, Y., et al., 2009. Room-temperature ferromagnetism of graphene. Nano Lett. 9, 220–224.

Xia, H., et al., 2008. Adv. Mater. 20, 4679–4683.

Yazyev, O.V., 2008. Phys. Rev. Lett 101, 037203.

Zaka, M., Ito, Y., Wang, H., Yan, W., Robertson, A., Wu, Y.A., Ruemmeli, M.H., Staunton, D., Hashimoto, T., Morton, J.J.L., Ardavan, A., Briggs, G.A.D., Warner, J.H., 2010. Electron paramagnetic resonance investigation of purified catalyst-free single walled carbon nanotubes. ACS Nano 4, 7708–7716.

⟨ Chapter 3.4 ⟩

The Mechanical Properties of Graphene

Mark H. Rümmeli

IFW Dresden, Germany

As already discussed in Chapter 2, in graphene the C atoms adopt a covalent trigonal bonding scheme also referred to as sp^2 hybridisation. One of the more significant aspects in determining bond strength is the degree of overlap between atomic orbitals. One of the key advantages of hybridised systems is that, according to the principle of maximum overlapping, bonding should be very strong. Indeed, it turns out that the strongest C–C chemical bond is in sp^2 carbon, viz. graphene (McWeeny and Coulson, 1980). The chemical bond strength plays a strong role in the physical and mechanical properties of a material such as melting point, activation energy of phase transition, tensile and shear strength and hardness (Kittel, 2004). Given the three-fold coordinated C–C bond is the strongest chemical bond we might anticipate graphene to exhibit exciting mechanical properties.

3.4.1. ELASTIC PROPERTIES AND INTRINSIC STRENGTH

The maximum stress that a pristine material without defects can support before failure is referred to as the intrinsic strength of the material. In sp^2 carbon materials like graphene and carbon nanotubes (rolled up graphene) its intrinsic strength is considered greater than that of any other material (Zhao et al., 2002). Having said that, experimental determination of the intrinsic strength and related mechanical characteristics of monolayer graphene was lacking. Early work confirming a high mechanical strength from sp^2 carbon was achieved using carbon nanotubes (Salvetat et al., 1999; Tombler et al., 2000; Yu et al., 2000a). However due to practical difficulties such as uncertainty in the sample geometry, unknown load distribution and defects quantification were difficult. The isolation of monolayer graphene (Wei et al., 2003; Geim and Novoselov, 2007) paved the way for mechanical studies on individual graphene sheets, in particular with Atomic Force Microscopy (AFM)-based techniques. One of the pioneering works investigated the fundamental resonant frequencies for both suspended single and multilayer graphene sheets at room temperature in vacuum ($<10^{-6}$ torr) (Bunch et al., 2007). They produced their resonator by placing exfoliated (through the peeling process) graphene sheets over predefined trenches which had been etched into a SiO_2 surface which yields a doubly clamped beam to the SiO_2 surface by van der Waals attraction. Two actuation routes were explored: namely an electrical route in which an applied Radio Frequency (RF) signal generates an electrostatic force between the graphene sheet and the substrate and an optical method in which the suspended graphene sheet is irradiated by a diode laser modulated at a frequency f. The laser in essence heats the graphene sheet thus the modulated laser beam induces a periodic contraction/expansion of the graphene membrane. Regardless of the actuation technique employed fundamental resonance frequencies in the megahertz (MHz) range were shown. In the limit of small tension, the resonance frequency, f_o scales as t/L^2 where t is the thickness of the graphene sheet and L its length. Plots comparing their experimental with theoretical fits showed scatter, most noticeably for the thinnest samples. This was attributed tension resulting from the fabrication process of the resonators. Thus to determine the effective spring constants and extract a Young's modulus the team turned to static deflection measurements (Frank et al., 2007). In this case the static force applied at the centre of the beam (graphene membrane) and under uniaxial strain can be expressed as:

$$k = 16.230Ew(t/L)^3 + 4.93T/L$$

where k is the effective spring constant, E is the Young's modulus, w the width of the beam and T the tension in the beam. The expression is valid so long as one is within the linear regime in Hooke's law.

A schematic of the setup used is given in Fig. 3.4.1.

The extraction of the effective spring constant (k) of the graphene layers is accomplished from the deflection of the AFM tip whose own spring constant is already known. In essence, a graph of the force exerted on the tip against the displacement of the graphene sheets yields a linear curve from which, according to Hooke's law, the slope corresponds to the effective spring constant of the graphene sheets. They obtained values ranging from 1 to 5 N/m for sheets with thicknesses from 2 to 8 nm. Values for monolayer graphene were obtained by another team, again using an AFM nanoindentation approach (Lee et al., 2008). In their experiment circular wells (with diameters of 1.5 μm and 1 μm) were prepared on a Si wafer (with a 300 nm SiO_2 surface) by nanoimprint

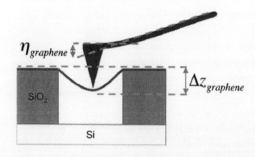

FIGURE 3.4.1 A schematic of an AFM tip that is deflected while pushing down on a suspended graphene sheet. $\eta_{graphene}$ is measured by the AFM, while $\Delta Z_{graphene}$ is calculated. *Copyright permission Frank et al. (2007).*

lithography and reactive ion etching. Graphite flakes were then mechanically deposited on the substrate surface. Raman spectroscopy and optical microscopy were employed to identify monolayer flakes. Non-contact mode AFM imaging confirmed the membranes stretched tautly over the wells, thus the membrane is clamped around the entire hole circumference (by van der Waals forces). Again, using AFM tips with known spring constants, initial force-displacement followed by load-reversal studies showed no hysteresis. This indicates that monolayer graphene demonstrates elastic behaviour. Later, the films were once again indented at the same rate, but this time to failure. They obtained a fracture strain of ca. 25%. In other words graphene can expand by nearly 25% before failing! This compares quite well with theoretical estimates which suggest fracture strain around 20% (Ansari et al., 2011; Liu et al., 2007; Ogata and Shibutani, 2003). The breaking strength was 42 Nm^{-1}, which they argue represents the intrinsic strength of a defect free sheet of graphene. Assuming an effective thickness of 0.335 nm for graphene[1], their data correspond to a Young's modulus of $E = 1.0$ terrapascals (TPa) and an intrinsic strength $\sigma_{int} = 130$ gigapascals (GPa). Maintaining the assumption of an effective thickness of 0.335 nm, the Elastic stiffness (Young's Modulus × thickness) corresponds to ca. 335 N/m^2. Their experiments confirmed graphene as the strongest material ever measured. To put this into context, the experiment is the equivalent to stretching a membrane over a cup and then pressing a pencil tip into the membrane and measuring the force required to puncture it (Bourzac, 2008). In the case of graphene, the pencil would be able to support a car on top! Even so, theory suggests graphene may be even stronger. A semi-empirical unbinding tensile force model predicts an intrinsic strength $\sigma_{int} = 162.7$ GPa for graphene, which is 20% higher than the experimental value (Xu et al., 2011).

Graphene, like most materials, when exposed to a uniaxial tension tends to contract in the direction perpendicular to the applied force. Expansion occurs in the direction of the applied force. The ratio of the two strains is known as Poisson's ratio. Molecular dynamics (MD) simulations determine a poisons ratio, ν_b, of 0.21 in good agreement with published data (Zhao et al., 2009). However, in the case of a nanoribbon, the Poisson's ratio relative to the bulk value increases as the diagonal length of the nanoribbon reduces. The Young's modulus of a nanoribbon decreases as the nanoribbon shortens. However in terms of strain (under uniaxial stress), the strain is not isotropic and depends on the applied stress direction e.g. armchair or zig-zag direction. For a given nanoribbon the Young's modulus is larger along the zig-zag direction, whereas the Poisson's ratio is larger in the armchair direction (Zhao et al., 2009). Thus, GNRs exhibit a size dependence on their

---

1. There is often a wide dispersion in theoretically derived aspects of graphene's mechanical properties which can principally be attributed to the uncertainty in the thickness of graphene (Scarpa et al., 2009).

mechanical properties. This is not so surprising in that for many materials their strength mainly depends on size, temperature and strain rate (the amount of deformation over the change in time). Indeed, temperature and strain-rate dependencies have been shown both experimentally and theoretically in single walled carbon nanotubes (Walters et al., 1999; Wei et al., 2003; Yu et al., 2000b). In the case of graphene, at the time of writing no experimental data were available. However, simulations are available and show the mechanical properties of graphene sheets are strongly dependant on temperature and strain rate (Han et al., 2009; Zhao and Aluru, 2010). Simulated stress vs. strain curves show graphene has a non-linear elastic behaviour. In addition, the fracture strength and fracture strain decrease significantly as temperature rises. The Young's modulus is not so temperature sensitive and decreases with increasing temperature, viz. graphene gets softer with increasing temperature. Nonetheless, it suggests graphene, as compared to other materials can be a strong material even at high temperatures. At low temperatures (0 K) simulations suggest graphene is intrinsically susceptible to brittle cleavage fracture (Liu et al., 2007).

Quantum mechanics and quantum molecular dynamics studies investigating the fracture process on GNRs show that the critical mechanical loads for failure and buckling of zig-zag oriented nanoribbons are larger than those for armchair graphene. The simulations indicate that for both armchair and zig-zag graphene ribbons the breakage begins at the outmost atom layer (i.e. the outside edge) when under external mechanical load (Gao and Hao, 2009).

3.4.2. ADHESION, TEARING AND CRACKING OF GRAPHENE

At the nanoscale, the influence of van der Waals forces increases. This is because even though van der Waals forces decrease with decreasing particle size, it does so to a lesser extent than inertial forces such as gravity and drag. The van der Waals forces of adhesion depend on surface topography such that surface roughness can result in a greater area of contact between two structures increasing the van der Waals force of attraction as well as the tendency for mechanical interlocking. In the case of graphene, van der Waals forces clamp the graphene to substrates and also hold together individual graphene layers to form FLG or graphite. Indeed, this force is responsible for clamping graphene over holes as used in many of the AFM-indentation studies to determine the mechanical properties of graphene and its derivatives. A clever variation of graphene membranes over cavities was used to determine the adhesion energy of graphene with a silicon oxide substrate (Koenig et al., 2011). In essence, two exfoliated flakes were deposited over an etched SiO_2 substrate with predefined microcavities. This yielded graphene membranes between 1 and 5 layers over the microcavities. To run the test, a so called pressurised blister test, the sample is first placed in a pressure chamber with nitrogen gas for several days. Nitrogen gas can then diffuse through the silicon oxide into the microcavities. Upon removing the sample, the pressure difference causes the membranes to bulge upwards (blister) increasing the volume of the microcavity. AFM is then used to determine the blisters profile. From the AFM data and mathematical solutions, one can extract the adhesion energy. For monolayer graphene the adhesion energy is $0.45\,\mathrm{Jm}^{-2}$ and $0.31\,\mathrm{Jm}^{-2}$ for FLG (2-5 layers). These adhesion energies are larger than adhesion energies generally obtained for mirco-mechanical structures. They are in fact more comparable with solid-liquid adhesion energies. The reason for this is attributed to the extreme flexibility of graphene which enables it to conform to the substrate topography, even for very smooth substrates. In essence its interaction with the substrate is more liquid-like than solid-like. Another study involving both first principles, molecular dynamics and experiments investigated the tearing of graphene sheets from adhesive substrates (Sen et al., 2010). The study showed the resulting nanoribbon geometry to be controlled by the graphene-substrate adhesion energy and by the number of torn layers. In general the width of a ribbon being torn from graphene narrows or tapers as the tear progresses, more so for high adhesion. The tear edges are composed of zig-zag and armchair edges. In addition, pentagon-heptagon defects can form at the crack tip due to high local stresses (which are adhesion dependant). The presence of these stresses allows the crack to bend sharply, thereby efficiently reducing the width of the tearing ridge.

3.4.3. THE ROLE OF DEFECTS AND STRUCTURAL MODIFICATION ON THE MECHANICAL PROPERTIES

Typically the actual tensile strength of most materials is far below their estimated or theoretical values. This discrepancy arises, in a word, because of defects. Defects are imperfections in a material that usually occur at size scales ranging from a few nanometres to a few millimetres. Defects can have a profound impact on the macroscopic properties of materials, and graphene is no different. From the stance of mechanical properties defects alter the manner in which stress is carried through the material. Various types of defects are possible in graphene and include vacancies, Stone–Wales defects, substitutional impurities (or doping), grain boundaries, adatoms or molecules and various structural modifications or variants. These are now discussed in the context of their effect on the mechanical properties of graphene.

3.4.3.1. Vacancies

By vacancies we refer to missing atoms in graphene's honeycomb lattice. Vacancies fall under the general defect category of point defects. Vacancies start upward one missing atom, the so called monoatomic vacancy. In addition, the concentration (and hence spacing) also needs to be considered when looking to understand their influence on the mechanical properties of graphene. In general experimental evidence for the effect of vacancies on the strength of graphene are limited, although there is of course experimental evidence for the presence of vacancies or holes in graphene. Molecular dynamics simulations on monoatomic vacancies show a linearly decreasing Young's modulus as the defect concentration increases. For a 4%-monoatomic vacancy concentration the Young's modulus reduction is about 12% (Hao et al., 2011). A different set of molecular dynamics simulations compared single- and double-atomic vacancy defects (Ansari et al., 2011). The double-atomic vacancy consists of two monoatomic vacancies neighbouring each other. They found that for defect spacings above 46.86 Å, there is little difference between single- or double-atomic vacancies. However, as compared to defect free graphene the critical strain for failure was less. As the defects get closer to each other (below 46.86 Å), the critical strain for failure drops markedly, especially for the case of double vacancies. The variation of the Young's modulus with separation distance was only slightly affected by the separation distance. The effect of significantly larger vacancies in the form of slits with lengths up to 10 Å has also been investigated using MD simulations and Quantised Fracture Mechanics for temperatures between 300 K and 2400 K (Zhao and Aluru, 2010). As expected the fracture strain drops with an increase in slit width. In addition, the fracture strength drops with increasing temperature. For pristine graphene at 2400 K the fracture strength is about 40% of its value at room temperature (300 K).

3.4.4. GRAPHENE DERIVATIVES

3.4.4.1. Graphene Oxide

The cost effectiveness and easy scale-up potential of the approaches used to fabricate graphene oxide make it of great interest not only as a means to yield graphene (by chemical exfoliation of graphite followed by a reduction process) but also as a material in its own right. This is because graphene oxide is easily dispersed and is a promising agent in composites. Unlike traditional composites based on carbon, composites using graphene oxide provide dramatically improved thermal, electrical and mechanical properties even at very low loadings of the nanofiller component. Hence, understanding the mechanical properties of graphene oxide are of tremendous interest. Graphene oxide differs from graphene in that it has hydroxyl, epoxy and carboxyl side groups attached to a graphitic backbone. In addition, graphene oxide often contains lattice defects due to the production process. Typically graphene oxide sheets are several hundred nanometres in diameter, have linear wrinkles and have a higher surface roughness as compared to pristine graphene.

Experimentally the mechanical properties of graphene oxide, like graphene, have been probed using AFM. Using AFM imaging in the contact mode, Rouff and colleagues (Suk et al., 2010) determined the Young's modulus and pre-stress for free-standing one-, two- and three-layer graphene oxide platelets over open holes by using a mapping method based on the finite element method. The SLG oxide membranes showed an effective Young's modulus of 156.5 ± 23.4 GPa (assuming zero bending stiffness) and an elastic stiffness of 109.6 ± 16.7 N/m using a thickness of 0.7 nm. In short the Young's modulus is about one fifth that of pristine graphene (ca. 1 TPa). The effective Young's modulus for two- and three-layer membranes were 223.9 ± 17.7 GPa and 229.5 ± 27.0 GPa, respectively. It is argued that the close similarity between the values for one, two and three layers of graphene oxide indicate that the bonding between layers (two- and three-layer systems) is sufficiently strong to avoid interlayer sliding.

The degree of functionalisation, the type of molecular structure involved in the functionalisation and the molecular weight of the functional groups can be expected to affect the mechanical properties of functionalised graphene. To this end, molecular dynamics and molecular mechanics simulations have been implemented to investigate the elastic properties of chemically functionalised graphene (Zheng et al., 2010a, b). The study revealed that the Young's modulus decreases linearly with increasing levels of –OH (random) functionalisation. A reduction of 33% was found for a functionalisation level of 15%. More precisely, the data show detrimental effects as in-plane sp^2 hybridised bonds are changed to off-plane sp^3-hybridised bonds. Two negative effects are identified: (1) destruction of the local π bond in the sheet and (2) sp^3 off-plane structures are more easily bent under tension weakening the sheet. It is then obvious, that as one increases the degree of functionalisation so the graphene oxide sheet will get weaker. To investigate the role of molecular structure, three structures with similar molecular weights were investigated. They were propyl ($-C_3H_7$ MW $= 43$), carboxyl (–COOH MW $= 45$) and methyl hydroperoxide ($-CH_2-O-OH$ MW $= 47$) groups. As found with simply –OH functionalisation, the Young's modulus decreases with increasing functionalisation degree with –COOH showing the greatest reduction and $-CH_2-O-OH$, the least. This is argued to be due to –COOH having the greatest binding energy. To investigate the role of molecular weight, six alkyl functional groups were chosen: $-CH_3$ (MW $= 15$), $-C_2H_5$ (MW $= 29$), $-C_3H_7$ (MW $= 43$), $-C_4H_9$ (MW $= 57$), $-C_5H_{11}$ (MW $= 71$) and $-C_6H_{13}$ (MW $= 85$). The Young's modulus was found to be insensitive to the molecular weight of the functional groups. This was attributed to the similarity of the binding energy irrespective of the molecular weight. The same study showed that the loading of functional groups on graphene reduce the shear modulus. In addition, the critical wrinkling strain, viz. the strain at which the wrinkling instability takes place is also reduced with functionalisation and explains why graphene oxide typically has many wrinkles.

The bending properties of monolayer graphene oxide sheets have also been investigated. In this study the investigators used the tip of an atomic force microscope to reversibly fold and unfold graphene sheets multiple times as illustrated in Fig. 3.4.2 (Schniepp et al., 2008). For the most part the graphene sheets folded and unfolded along the same bending lines. With supporting Density Functional Theory (DFT) studies they argued that folding a graphene sheet along a pre-existing kink is energetically more favourable as compared to a perfectly flat sheet. The pre-existing kinks are argued to arise from 5-8-5 defects. These defects can give rise to curvature (Schniepp et al., 2006) and so form a kink. 5-8-5 kinks are argued to occur through the reduction of epoxide lines during synthesis.

3.4.4.2. Reduced Graphene Oxide

The chemical reduction of graphene oxide is an alternative route to obtain graphene. However, the efficiency of the reduction process is limited so that residual oxygenated functional groups remain. In addition lattice defects are also incurred during the reduction process. AFM tip-induced deformation experiments show a Young's modulus of 250 GPa (similar to graphene oxide). Another study found a Young's modulus of 185 GPa (Robinson et al., 2008). The experimental variation can be attributed to differences in the experiment and also in differences in residual functional groups.

The conductivity of reduced graphene oxide was found to scale inversely with the elastic modulus (Gómez-Navarro et al., 2008). This is argued to occur due to oxygen bridges stiffening the sheet (Incze et al., 2004). Thus, a higher oxygen content leads to a higher elastic modulus but a lower conductivity. However, unreduced sheets with higher oxygen contents and conductivities of several orders of lower magnitude did not show an enhanced elastic modulus. This indicates another aspect affects the mechanical behaviour. Probably, this is due to the presence of structural defects (holes or vacancies) which limit the overall elasticity of the sheets.

3.4.4.3. Fluorographene and Graphane

DFT calculations for different stochiometric configurations of graphene and graphene fluoride show the Young's modulus for both graphene and flourographene are smaller than that of pristine graphene (Leenaerts et al., 2010). This can be attributed to the induced sp^3 hybridisation in these systems. In the case of graphene the Young's modulus is highly anisotropic such that the values roughly halve along the direction of crumpling. In the case of flourographene the Young's modulus is pretty much isotropic. This difference is argued to probably be caused by deformations in the flourographene lattice due to the charged F atoms. Theoretical estamites for the Young's modulus of

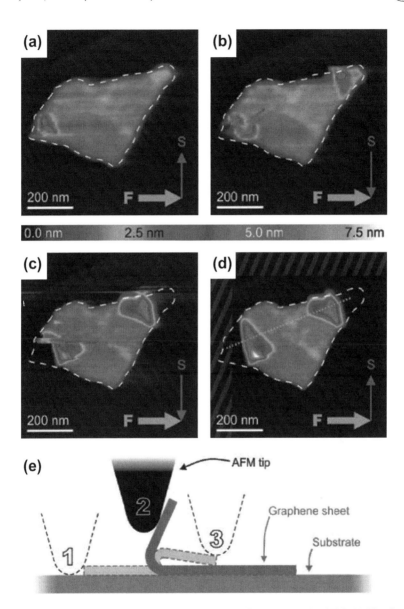

FIGURE 3.4.2 Folding a functionalised graphene sheet by contact-mode AFM. (a) The sheet is in almost completely flat conformation. (b) The top right and bottom left edges are partially folded by the AFM probe. (c) Folding progresses. (d) The folded sheet reaches a stable conformation. The hatched area was not part of the actual scan but added later to make comparison with (a)–(c) easier. The green arrows in (b) and (c) indicate lines in which folding events occurred. (e) Illustration of the folding process: (1) the tip first lifts a part of the sheet (green) off the substrate (grey). (2) The sheet is bent until it touches itself (3). *From Schniepp et al. (2008).*

flourographene lie between 215 and 253 Nm^{-1} (Leenaerts et al., 2010; Muñoz et al., 2010). This compares with a value of $100 \pm 30 \, Nm^{-1}$ for, from an experimental AFM, nano-indentation value (Nair et al., 2010). The difference between theory and experiment (some 50%) might be due to the presence of defects (e.g. holes and vacancies) in the flourographene membranes used in the experiments.

3.4.4.4. Polycrystalline Graphene

Large area graphene is best grown by CVD over metals, usually Ni and Cu. However, these substrates are polycrystalline and this tends to lead to graphene consisting of many single crystalline grains. Differences in the orientations of these individual grains leads to them stitching together with structural disorder, usually through pentagon–heptagon pairs (Huang et al., 2011; Nemes-Incze, 2011). These 5-7 pairs essentially form a grain boundary so that a large area of CVD-grown graphene (over a polycrystalline metal) form a patchwork graphene quilt. The shear modulus for CVD grown FLG (6–8 nm thick) over Ni was investigated using a double paddle oscillator technique at low temperature (0.4 K) (Liu et al., 2010). An average shear modulus of 53 GPa was obtained which is some five times larger than graphite. SLG is popularly grown over Cu by CVD. Again, using oscillator-based techniques at low temperature, the shear modulus for monolayer CVD-grown graphene over Cu was obtained. In this case the value averaged at 280 GPa (Liu et al., 2012). The authors indicate that the striking difference between single and multi-layer graphene may lie in the transition of the shear restoring force from intralayer covalent to interlayer van der Waals interactions. It should be noted that these experiments were accomplished with graphene on a support.

Huang et al. (2011) measured the failure strength of freestanding SLG using a AFM nano-indentation approach. The CVD-grown material was over Cu foils. From repeated measurements, they found an average failure load of around 100 nN. This compares with an average fracture load of 1.7 μN for single-crystal exfoliated graphene (Lee et al., 2008). This suggests grain boundaries in graphene significantly weaken it. More rigorous studies on CVD-grown graphene (again over Cu) were conducted by the same team (Ruiz-Vargas et al., 2011). The graphene membranes were transferred from the Co foil onto silcon nitride grids with arrays of prepatterned holes via a PMMA-based transfer. Force-deflection studies obtained by pushing the AFM cantilever down onto a graphene membrane were performed for 60 membranes. The average elastic modulus was 55 N/m (Full width at Half Maximum (FWHM) was 50 N/m). This is about six times lower than that for pristine graphene and strongly suggests grain boundaries weaken graphene's ultimate strength. Spatially resolved nanoindentation experiments where the vertical load attained before membrane failure is recorded as a function of tip position were performed. The study directly showed the weakening effect of the grain boundaries where tears in the membrane were seen to follow the direction and path of the grain boundary. The in-plane breaking stress for most samples was approximately 35 GPa. This low value is at odds with atomistic calculations on stitched graphene sheets, which suggest sheets with large-angle tilt boundaries are as strong as the pristine material. On the other hand, graphene sheets with low-angle tilt boundaries are much weaker. This tilt angle difference is related to the strain in the bonds at the boundaries (Grantab et al., 2010). The low experimental values as compared to the theory discussed above are argued to occur due to a shear component existing in the membranes. In effect, through external shear control, one could change the membranes out of plane stiffness.

3.4.4.5. Folded Graphene – Grafold

Geometry reconstruction can play an important role in mechanical performance. For example metal panels or sheets are often designed with folds to introduce added strength and torsional rigidity. Graphene, with a strength 200 times greater than that of steel, is already the strongest known material to exist. However, molecular dynamics simulations suggest structure reconstruction may lead to enhanced mechanical properties (Zheng et al., 2011). The simulations confirm the enormous expansion possible with graphene, but highlight that under compression out of plane pleats form, viz. graphene can only withstand elongation. The researchers then went on to investigate how the mechanical characteristics of a graphene ribbon folded to form a semi-graphene-like, semi-carbon-nanotube-like structure alter. Various ribbon widths and lengths with two- and three-folds were investigated. Configurations with different widths were found to have similar stress–strain relations. In contrast, changes in length do make a difference, longer folded ribbons or grafold seem to have a higher tensile strength. The number of folds also seems to be important. Grafold with two folds have higher strengths and fracture strains than three-fold grafold. Fracture or tearing apparently occurs at the junction of a curved and planar stripe (see Fig. 3.4.3). This is where the distance between two folding layers is at a minimum and stronger interlayer interactions may account for bond breakage initiating at this site. Unlike graphene, grafold can withstand compressive strain. Compressive stress from 10 to 25 GPa or even higher can be applied depending on the individual structure. In other words, from a compression point of view, graphene can be strengthened through folding.

3.4.4.6. Graphene Paper

Graphene paper can be formed by the layer-by-layer assembly of graphene oxide or graphene nanosheets. This is usually accomplished through flow

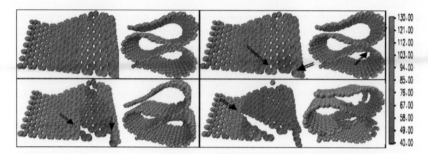

FIGURE 3.4.3 Bond breaking, crack nucleation and growth in GRA40 L20 3folds. In each subfigure the right panel demonstrates the three-dimensional visualisation and the left panel presents only one layer of the grafold. (top left panel) Per-atom stress distribution at the regime far from crack. (top right panel) Stress distribution before bond breaking. As indicated by the black arrows, the first breakage of a C–C bond would occur at the junction of a curved and a planar stripe. (bottom left panel) Crack propagation. The stress degrades at the broken region and enlarges at the crack propagation tip. (bottom right panel) Fracture of grafold. *From Zheng et al. (2011).*

induced assembly of aqueous graphene nanosheet dispersion via vacuum filtering and possibly a subsequent annealing step (Chen et al., 2008; Ranjbartoreh et al., 2011). Graphene paper has a shiny metallic luster. When examining it under a scanning electron microscope the surface is revealed as relatively smooth, while the edges show a layered structure. X-ray Diffraction (XRD) data show the spacing to range between 0.34 and 0.4 nm. Graphene papers are highly flexible and bendable. The ultimate strengths for graphene paper and graphene oxide paper have been measured as 70–78 MPa and 31–52 GPa, respectively their stiffness values were 10–16 N/mm and 21–23 N/mm respectively; that is graphene papers has a better Young's modulus than graphene oxide paper, while graphene oxide has a superior stiffness. Heat treated graphene papers has a remarkable hardness (ca. 217 Kg f/mm^2) which is nearly twice that of carbon steel and its yielding strength (ca. 6.4 Tpa) is several times higher than carbon steel. In short graphene papers have outstanding bending rigidity under bending which is about thirteen times that of carbon steel! Thus, graphene papers are of great interest as a material for engineering applications.

3.4.5. GRAPHENE-BASED COMPOSITES

The exciting mechanical properties of graphene and its derivatives along with the relative ease with which many derivatives (e.g. graphene oxide, reduced graphene oxide and exfoliated graphite) can be produced in large scale make them ideal candidates for incorporation into a variety of functional materials taking advantage not only of graphene's mechanical properties but also many other of its properties (Huang et al., 2011). In terms of graphene-based composites, using graphene solely for enhanced mechanical properties graphene has proven itself very attractive, particularly as only low levels of graphene are required to make significant improvements. For example studies on graphene/epoxy nanocomposite beam structures show a 52% increase in critical buckling load for only a 0.1% weight fraction of graphene platelets into the epoxy matrix (Rafiee et al., 2009). This important because epoxy composite materials, while extremely lightweight are prone to fracture. Functionalised graphene sheets are remarkably effective at enhancing the fracture toughness, fracture energy stiffness, strength and fatigue resistance of epoxy polymers at significantly lower nanofiller loading fractions as compared to other nanofillers (e.g. carbon nanotubes). This is attributed to the graphene sheets enhanced specific surface area, two-dimensional geometry and strong nanofiller-matrix adhesion. Nonetheless, improvements in the dispersion of the graphene nanosheets are desirable, particularly for composites with higher graphene loadings where even better performance may be expected (Rafiee et al., 2010).

REFERENCES

Ansari, R., Motevalli, B., Montazeri, A., Ajori, S., 2011. Fracture analysis of monolayer graphene sheets with double vacancy defects via MD simulation. Solid State Commun. 151 (17), 1141–1146.

Bourzac, K., July 17, 2008. Strongest material ever tested (graphene, praised for its electrical properties, has been proven the strongest known material.). Technol. Rev. http://www.impactlab.net/2008/07/18/strongest-known-material/. http://www.technologyreview.com/news/410479/strongest-material-ever-tested/ publ. by MIT.

Bunch, J.S., van der Zande, A.M., Verbridge, S.S., Frank, I.W., Tanenbaum, D.M., Parpia, J.M., Craighead, H.G., McEuen, P.L., 2007. Electromechanical resonators from graphene sheets. Science 315 (5811), 490–493 [MP9].

Chen, H., Müller, M.B., Gilmore, K.J., Wallace, G.G., Li, D., 2008. Mechanically strong, electrically conductive, and biocompatible graphene paper. Adv. Mater. 20 (18), 3557–3561.

Frank, I.W., Tanenbaum, D.M., van der Zande, A.M., McEuen, P.L., 2007. Mechanical properties of suspended graphene sheets. J. Vac. Sci. Technol. B 25 (6) 2558(4).

Gao, Y., Hao, P., 2009. Mechanical properties of monolayer graphene under tensile and compressive loading. Phys. E 41 (8), 1561–1566.

Geim, A.K., Novoselov, K.S., 2007. The rise of graphene. Nat. Mater. 6, 183–191.

Gómez-Navarro, C., Burghard, M., Kern, K., 2008. Elastic properties of chemically derived single graphene sheets. Nano Lett. 8 (7), 2045–2049.

Grantab, R., Shenoy, V.B., Ruoff, R.S., 2010. Anomalous strength characteristics of tilt grain boundaries in graphene. Science 330 (6006), 946–948.

Han, T., He, P., Zheng, B., Dependence of the Tensile Behavior of Single Graphene Sheet on Temperature and Strain Rate, 17th International Conference on Composites or Nano Engineering, ICCE-17 Honolulu, Hawaii, USA, July 26-August 1, 2009, Supplement 2009.

Hao, F., Fang, D., Xu, Z., 2011. Mechanical and thermal transport properties of graphene with defects. Appl. Phys. Lett. 99 (4) 041901(3).

Huang, P.Y., Ruiz-Vargas, C.S., van der Zande, A.M., Whitney, W.S., Levendorf, M.P., Kevek, J.W., Garg, S., Alden, J.S., Hustedt, C.J., Zhu, Y., Park, J., McEuen, P.L., Muller, D.A., 2011. Grains and grain boundaries in single-layer graphene atomic patchwork quilts. Nature 469 (7330), 389–392.

Huang, X., Yin, Z., Wu, S., Qi, X., Qi., He, Q. Zhang, Yan, Q., Boey, F., Zhang, H., 2011. Graphene-based materials: synthesis, characterization, properties, and applications. Small 7 (4), 1876–1902.

Incze, A., Pasturel, A., Peyla, P., 2004. Mechanical properties of graphite oxides: ab initio simulations and continuum theory. Phys. Rev. B 70 (21) 212103(4).

Kittel, C., 2004. Introduction to Solid State Physics, eight ed. John Wiley & Sons. ISBN-10: 0471680575, ISBN-13: 978–0471680574.

Koenig, S.P., Boddeti, N.G., Dunn, M.L., Bunch, J.S., 2011. Ultrastrong adhesion of graphene membranes. Nat. Nanotechnol. 6, 543–546.

Lee, C., Wei, X., Kysar, J.W., Hone, J., 2008. Measurement of the elastic properties and intrinsic strength of monolayer graphene. Science 321 (5887), 385–388.

Leenaerts, O., Peelaers, H., Hernández-Nieves, A.D., Partoens, B., Peeters, F.M., 2010. First-principles investigation of graphene fluoride and graphane. Phys. Rev. B 82 (19) 195436(6).

Liu, F., Ming, P., Li, J., 2007. Ab initio calculation of ideal strength and phonon instability of graphene under tension. Phys. Rev. B 76 (6) 064120(7).

Liu, X., Robinson, J.T., Wei, Z., Sheehan, P.E., Houston, B.H., Snow, E.S., 2010. Low temperature elastic properties of chemically reduced and CVD-grown graphene thin films. Diam. Relat. Mater. 19 (7–9), 875–878.

Liu, X., Metcalf, T.H., Robinson, J.T., Houston, B.H., Scarpa, F., 2012. Shear modulus of monolayer graphene prepared by chemical vapor deposition. Nano Lett. 12 (2), 1013–1017.

McWeeny, R., Coulson, C.A., 1980. Coulson's Valence, third ed. Oxford University Press, USA. ISBN-10: 0198551452, ISBN-13: 978-0198551454.

Muñoz, E., Singh, A.K., Ribas, M.A., Penev, E.S., Yakobson, B.I., 2010. The ultimate diamond slab: graphAne versus graphEne. Diam. Relat. Mater. 19 (5–6), 368–373.

Nair, R., Ren, W., Jalil, R., Riaz, I., Kravets, V., Britnell, L., Blake, P., Schedin, F., Mayorov, A., Yuan, S., Katsnelson, M., Cheng, H., Strupinski, W., Bulusheva, L., Okotrub, A., Grigorieva, I., Grigorenko, A., Novoselov, K., Geim, A.K., 2010. Fluorographene: a two-dimensional counterpart of teflon. Small 6 (24), 2877–2884.

Nemes-Incze, P., Yoo, K.J., Tapasztó, L., Dobrik, G., Lábár, J., Horváth, Z.E., Hwang, C., Biró, L.P., 2011. Revealing the grain structure of graphene grown by chemical vapor deposition. Appl. Phys. Lett. 99 (2) 023104(3).

Ogata, S., Shibutani, Y., 2003. Ideal tensile strength and band gap of single-walled carbon nanotubes. Phys. Rev. B 68 (16) 165409(4).

Rafiee, M.A., Rafiee, J., Yu, Z.-Z., Koratkar, N., 2009. Buckling resistant graphene nano-composites. Appl. Phys. Lett. 95 (22) 223103(3).

Rafiee, M.A., Rafiee, J., Srivastava, I., Wang, Z., Song, H., Yu, Z.-Z., Koratkar, N., 2010. Fracture and fatigue in graphene nanocomposites. Small 6 (2), 179–183.

Ranjbartoreh, A.R., Wang, B., Shen, X., Wang, G., 2011. Advanced mechanical properties of graphene paper. J. Appl. Phys. 109 (1) 014306(6).

Robinson, J.T., Zalalutdinov, M., Baldwin, J.W., Snow, E.S., Wei, Z., Sheehan, P., Houston, B.H., 2008. Wafer-scale reduced graphene oxide films for nanomechanical devices. Nano Lett. 8 (10), 3441–3445.

Ruiz-Vargas, C.S., Zhuang, H.L., Huang, P.Y., van der Zande, A.M., Garg, S., McEuen, P.L., Muller, D.A., Hennig, R.G., Park, J., 2011. Softened elastic response and unzipping in chemical vapor deposition graphene membranes. Nano Lett. 11 (6), 2259–2263.

Salvetat, J.-P., Briggs, G.A.D., Bonard, J.-M., Bacsa, R.R., Kulik, A.J., Stöckli, T., Burnham, N.A., Forró, L., 1999. Elastic and shear moduli of single-walled carbon nanotube ropes. Phys. Rev. Lett. 82 (5), 944–947.

Scarpa, F., Adhikari, S., Phani, A.S., 2009. Effective elastic mechanical properties of single layer graphene sheets. Nanotechnology 20 065709(11).

Schniepp, H.C., Li, J.-L., McAllister, M.J., Sai, H., Herrera-Alonso, M., Adamson, D.H., Prud'homme, R.K., Car, R., Saville, D.A., Aksay, I.A., 2006. Functionalized single graphene sheets derived from splitting graphite oxide. J. Phys. Chem. B 110 (17), 8535–8539.

Schniepp, H.C., Kudin, K.N., Li, J.-L., Prud'homme, R.K., Car, R., Saville, D.A., Aksay, I.A., 2008. Bending properties of single functionalized graphene sheets probed by atomic force microscopy. ACS Nano 2 (12), 2577–2584.

Sen, D., Novoselov, K.S., Reis, P.M., Buehler, M.J., 2010. Tearing graphene sheets from adhesive substrates produces tapered nanoribbons. Small 6 (10), 1108–1116.

Suk, J.W., Piner, R.D., An, J., Ruoff, R.S., 2010. Mechanical properties of monolayer graphene oxide. ACS Nano 4 (11), 6557–6564.

Tombler, T.W., Zhou, C., Alexseyev, L., Kong, J., Dai, H., Liu, L., Jayanthi, C.S., Tang, M., Wu, S.-Y., 2000. Reversible electromechanical characteristics of carbon nanotubes under local-probe manipulation. Nature 405, 769–772.

Walters, D.A., Ericson, L.M., Casavant, M.J., Liu, J., Colbert, D.T., Smith, K.A., Smalley, R.E., 1999. Elastic strain of freely suspended single-wall carbon nanotube ropes. Appl. Phys. Lett. 74 (25) 3803(3)

Wei, C., Cho, K., Srivastava, D., 2003. Tensile strength of carbon nanotubes under realistic temperature and strain rate. Phys. Rev. B 67 (11) 115407(6).

Xu, B., Guo, X., Tian, Y., 2011. Graphene Simulation, Chapter 11: Universal Quantification of Chemical Bond Strength and Its Application to Low Dimensional Materials. http://www.intechopen.com ISBN: 978-953-307-556-3.

Yu, M.-F., Lourie, O., Dyer, M.J., Moloni, K., Kelly, T.F., Ruoff, R.S., 2000a. Strength and breaking mechanism of multiwalled carbon nanotubes under tensile load. Science 287 (5453), 637–640.

Yu, M.-F., Files, B.S., Arepalli, S., Ruoff, R.S., 2000b. Tensile loading of ropes of single wall carbon nanotubes and their mechanical properties. Phys. Rev. Lett. 84 (24), 5552–5555.

Zhao, H., Aluru, N.R., 2010. Temperature and strain-rate dependent fracture strength of graphene. J. Appl. Phys. 108 (6) 064321(5).

Zhao, Q., Nardelli, M.B., Bernholc, J., 2002. Ultimate strength of carbon nanotubes: a theoretical study. Phys. Rev. B 65 (14) 144105(6).

Zhao, H., Min, K., Aluru, N.R., 2009. Size and chirality dependent elastic properties of graphene nanoribbons under uniaxial tension. Nano Lett. 9 (8), 3012–3015.

Zheng, Q., Li, Z., Geng, Y., Wang, S., Kim, J.-K., 2010a. Molecular Dynamics Study of the Effect of Chemical Functionalization on the Elastic Properties of Graphene Sheets. J. Nanosci and Nanotech., Carbon 48, 4315.

Zheng, Q., Li, Z., Geng, Y., Wang, S., Kim, J.-K., 2010b. Molecular Dynamics Study of the Effect of Chemical Functionalization on the Elastic Properties of Graphene Sheets. J. Nanosci. Nanotech 10, 1.

Zheng, Y., Wei, N., Fan, Z., Xu, L., Huang, Z., 2011. Mechanical properties of grafold: a demonstration of strengthened graphene. Nanotechnology 22 (40) 405701(9).

The Thermal Properties of Graphene

Mark H. Rümmeli

IFW Dresden, Germany

Carbon materials form a variety of allotropes with a remarkable range of properties, including their thermal properties. The thermal conductivity of the different carbon allotropes spans over five orders of magnitude, with the lowest being attributed to amorphous carbon ($\sim 0.01\ W\,mK^{-1}$) and the highest to diamond and graphene ($\sim 2000\ W\,mK^{-1}$). Graphene, though, unlike diamond is a two dimensional crystal and offers tantalising opportunities for heat management on various fronts, for example, graphene holds promise as a heat spreader in microelectronics systems. In this section, the remarkable properties of graphene, bilayer graphene and few layer graphene are examined. A full understanding and quantification of their various thermal properties has yet to be reached and the reader is urged to bear that in mind.

3.5.1. THERMAL CONDUCTIVITY

The thermal conductivity, K, of a material is the property of its ability to conduct heat and is usually introduced through Fourier's law for heat conduction. Fourier's law is written as:

$$\vec{q} = -k\nabla T$$

where q is the local heat flux and represents the amount of energy that flows through a unit area per unit time, K is the thermal conductivity and ∇T is the local temperature gradient. The thermal conductivity is often thought of as a constant, however this is not always true. Over a wide temperature range K is a function of T and in anisotropic materials it depends on crystal orientation and is represented by a tensor (Balandin, 2011).

Heat conduction in solid materials usually takes place through acoustic phonons (ion-core vibrations in the crystal lattice of the material) and by electrons. Thus K may be written as the sum of K_p and K_e where K_p and K_e are the phonon and electron contributions respectively.

In metals, heat conduction is dominated by the large concentrations of free carriers, electrons (viz. K_e) whereas in carbon materials heat conduction is dominated by phonons (K_p). This is also true for graphite even though it has metal-like properties. This is because the strong co-planar sp^2 bonds forming graphene's chicken wire lattice provide an efficient means for heat transfer by lattice vibrations. In the case of doped materials, K_e may become more relevant.

In solids, usually the phonons carrying heat can be scattered when encountering other phonons, lattice defects, impurities, conduction electrons and interfaces. Phonon transport can be either diffusive or ballistic. The distinction lies in whether the phonon mean free path (Λ) is smaller than the sample length (L) in which case many scattering events occur and the thermal transport is then diffusive. Fourier's law is based on this regime. When L is smaller than Λ, the phonon can effectively traverse the entire sample length without decaying and this regime is known as ballistic transport.

The *intrinsic* (or highest) thermal conductivity of a material occurs when phonons are only scattered by phonons (i.e. other scattering mechanisms such as defects or impurities do not exist). The fact that phonons scatter at other phonons (in other words they 'see' each other) occurs due to anharmonicity[1]. Umklapp scattering is the dominant process leading to finite K in 3 dimensions. In the Umklapp process the anharmonic phonon–phonon interaction, the total phonon momentum is changed. The degree of anharmonicty a material exhibits is characterised by the so called Gruneisen parameter (γ) and includes Umklapp scattering. In addition, the thermal conductivity can be affected by extrinsic affects such as phonon scattering at defects or rough boundaries (edges). This can obviously be more relevant at the nanoscale, e.g. GNRs.

The determination of the thermal conductivity of low dimensional materials, like graphene, is challenging due to handling difficulties and uncertainties in knowing the power dissipated to the sample. Moreover, the conventional 3ω method requires a significant temperature drop over the thickness of the examined film and since graphene is only 1 atom thick, the technique is not suitable. Measurement across a thermal bridge is possible, but difficult to accomplish technologically.

The first successful measurement of the thermal conductivity of graphene was accomplished using an optothermal Raman technique (see Fig. 3.5.1). The technique was developed by Alexander Balandin's team at the University of California – Riverside. They first investigated the temperature dependency of graphene's Raman sensitive G mode (elongation of the C–C bond). The G mode frequency shift decreases linearly from ~1584 cm^{-1} at $-200\,^\circ$C to ~1578 cm^{-1} at 100 $^\circ$C. This temperature shift is a result of anharmonic effects and can be, at a simplified level, argued to result from anharmonic coupling of the phonon modes and a contribution due to thermal expansion of the crystal.

---

1. Atomic dynamics can be described by a harmonic approximation in which the potential energy is expanded to a quadratic term in the atomic displacement which leads to a parabolic potential energy curve. Real materials are characterised by nonparabolic dependencies of the potential energy curve. This is known as anharmonic behaviour and helps explain how, for example, phonon–phonon interactions, thermal expansion, crystal vibration decay and why thermal conductivity is not infinite.

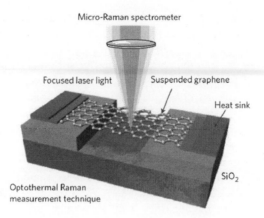

FIGURE 3.5.1. Experimental setup for thermal conductivity measurement of graphene and few layer graphene, using an opto-thermal Raman technique. *Image from Balandin (2011).*

Even though the thermal expansion itself results from anharmonicity, the physical mechanism is different and related to changes in the force constants of the graphene with volume (Calizo et al., 2007). The frequency shift of the G mode as a function of temperature in essence enables a Raman spectrometer to function as an optical thermometer. The Balandin team then capitalised on this dependency to measure the thermal conductivity of suspended graphene using a setup as shown in Fig. 3.5.1, in which the Raman excitation laser power is used to heat the sample. The amount of heat dissipated in the graphene sample can be measured through the integrated Raman intensity of the G peak (Balandin et al., 2008) or by placing a detector under the graphene layer (Cai et al., 2010). Correlating the change in temperature with the change in power for graphene samples with known geometry can lead to a solution for the thermal conductivity, K, through a heat-diffusion equation, so long as the graphene specimens are large enough to ensure a diffuse regime. The Raman technique for measuring K is a direct steady-state method. Using the technique, the Balandin team found a thermal conductivity, K, of ~3000 W mK^{-1} near room temperature, using mechanically cleaved graphite. Later, studies from a different group, again using the Raman technique, but this time using CVD grown graphene, obtained a K value of ca. 2500 W mK^{-1} near room temperature and ca. 1400 W mK^{-1} at 500 K (Cai et al., 2010). Another team modified the Raman route to also investigate the thermal conductivity. Here, the graphene membrane was formed with Corbino geometry – the graphene membrane was placed over a copper disk with a pinhole. The graphene edges (which extend outside the pinhole) were thermally short circuited to the copper heat sink with silver epoxy. The membrane had a well-defined geometry. The experimentalists were able to get a direct readout of the temperature from the intensity ratio of the Stokes to anti-Stokes Raman scattering signals from the low-energy phonon of graphene (the G band).

The steady state form of the heat diffusion equation is given by:

$$\kappa \cdot \nabla^2 T(\vec{r}) + q(\vec{r}) = 0$$

where $q(r)$ is the heat generated in a disk of radius R and thickness d, ∇T is the local temperature gradient and K the thermal conductivity. Based on this equation, they were able to extract the thermal conductivity of the graphene and found a K value of ~600 W mK^{-1} (Faugeras et al., 2010). This value is somewhat lower than the previously mentioned values. The authors of this work also commented on this fact and highlighted that they used an absorbed laser power of 2.3% based on more recent absorption studies, whereas previous studies used a laser power absorption of 13%. They pointed out that if they also used a power absorption of 13% they would get a K value of 3600 W mK^{-1}. In other words, understanding the laser light-absorption characteristics is important and can affect the extracted thermal conductivity. Other important technical aspects, such as strain in the membrane and quality of the graphene used can affect the final K value. These aspects are discussed later on.

Nonetheless, the actual values are very good. Bulk copper has a thermal conductivity of ~400 W mK^{-1} and highlights graphene as a competitive heat spreading material.

Copper thin films (for example as electrical interconnects) have reduced-thermal conductivity down to 250 W mK^{-1} and since the reliability of devices often strongly depend on temperature, any future use of graphene as a material for devices requires a good understanding of its heat generation and dissipation. Thus, it is important to understand graphene's thermal properties when residing on a substrate. The most logical substrate to investigate is of course Si/SiO$_2$ wafers, viz. an amorphous SiO$_2$ surface. To experimentally measure K of supported graphene a different approach to the Raman based measurement is required. Here a thermal measurement circuit is required (Seol et al., 2010) which consists of a set of resistance thermometer lines covering SLG deposited on a 300 nm SiO$_2$ beam. From the thermal circuit, the thermal resistance of the central beam for different electrical heating is measured. The process was repeated after etching away the surface graphene layer using an oxygen plasma. Since the thermal conductance is equivalent to the inverse of the thermal resistance, the difference between the measurements before and after O$_2$ etching allowed the thermal conductance of the graphene layer to be determined for various temperatures. Initially, the thermal conductivity increases to around 300 K after which it saturates between 300 K and 400 K. For the room temperature measurement, the thermal conductivity K was found to be approximatly 600 WmK^{-1}. This is lower than most of the values obtained from experimental determinations from freestanding graphene. The reduced thermal conductivity is argued to be related to phonons from the graphene monolayer leaking across the graphene-support interface and strong interface scattering of flexural modes. Indeed, since the thermal properties of graphene are dominated by phonons it is worth briefly examining the primary phonon modes in graphene before continuing.

3.5.1.1. Primary Phonon Modes Carrying Heat in Graphene

In bulk carbon materials the thermal conductivity is dominated by three acoustic[2] phonon modes: one longitudinal (LA) mode where the atomic vibration is in the direction of the wave propagation; and two transverse (TA) modes where the atomic vibration is perpendicular to the wave propagation direction. The frequencies of both acoustic modes are proportional to the wave vector (reciprocal wavelength) and so the phonon dispersion is linear (along the graphene basal plane). In graphene, there is a third mode the so called flexural (ZA) mode in which the vibrations are out of plane (see Fig. 3.5.2). The ZA phonon mode has a quadratic dispersion. Initially it was believed the ZA phonon mode did not contribute significantly to carrying heat in freestanding graphene because its group velocity is smaller than the group velocity of the LA and TA in-plane modes. (This arises because thermal conductivity is the product of specific heat, mean phonon scattering time and the square of the phonon group velocity integrated over the whole phonon frequency range (Prasher, 2010)). However, the studies described above in which the thermal conductivity of supported graphene was experimentally determined over a wide temperature range could not be explained, based solely on the LA and TA phonon modes. The researchers were able to explain their data by invoking the ZA phonon mode, indeed they showed that the flexural (ZA) mode can carry significant heat both in suspended graphene and supported graphene, because the specific heat for the ZA mode is higher than the LA and TA mode for temperatures up to ca. 360 K and that the scattering time of the ZA mode is much larger than either the LA or TA modes (Prasher, 2010; Seol et al., 2010). In the case of supported graphene the ZA mode contribution to thermal transport is reduced. It is argued that the ZA phonon mode leaks into the substrate far more efficiently than the LA and TA modes as the force constant of the ZA mode is much larger than that of the LA and TA modes (Prasher, 2010).

---

2. The collective vibration of atoms in crystals can be described by two types of phonons: acoustic and optical phonons. In the case of acoustic phonons the the movement of the atoms in the lattice with respect to their equilibrium positions are coherent such that their similarity to sound waves in air led to them being termed acustic. They can be longitudinal or transverse and are usually abbreviated to LA or TA respectively. In optical phonons atom movement in the lattice is out of phase. They are often excited by infrared light and so are termed 'optical'. The transverse and longitudinal modes are abbreviated as LO and TO respectively.

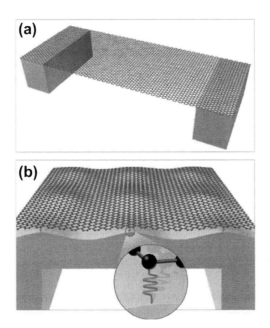

FIGURE 3.5.2 (a) Thermal conductivity measurements were performed on suspended graphene. (b) Seol et al., instead studied graphene supported on a substrate. The graphene layer does not conform to the nanoscale roughness of the substrate; rather, it makes contact on the summits of the rough surface, interacting with the substrate through van der Waals forces (red springs) *(After Prasher, 2010)*.

3.5.1.2. Thermal Conductivity in Few-layer Graphene

It is also interesting to investigate the evolution of the thermal conductivity with graphene-layer number. An opto-thermal Raman study investigating 1, 2, 3, and 4 layer graphene, as well as various graphite samples, showed that at room temperature, the thermal conductivity of free-standing graphene changes from ~2800 W mK^{-1} to ~1300 W mK^{-1} as the number of layers increases from 2 to 4. The change in K with layer number shows a dimensional crossover from 2D to bulk (3D) graphite and is ascribed to the coupling of cross plane coupling of low energy phonons and changes in the phonon Umklapp scattering (Ghosh et al., 2010). Theoretical studies comparing mono-layer and bi-layer graphene in which atomistic simulations of their structural and thermodynamic properties were implemented show significant changes as one goes from 1 to 2 layers. In the case of bi-layer, the bending rigidity is twice larger than that for monolayer graphene, which implies a reduction in out of plane fluctuations (ZA phonon mode). The heat capacity however is similar to that for SLG. Another study using non-equilibrium molecular dynamics simulations also found that the in-plane thermal conductivity of graphene at room temperature (300 K) decreases with increase of layer number. This is attributed, much as before, to interactions/constraints from neighbouring layers limiting the free vibration of the graphene. In other words phonon transport is hindered. These arguments are similar to that for supported SLG. In the case of encased graphene the opposite is observed. An investigation into different layer numbers encased or sandwiched graphene/FLG in-between SiO_2 layers using the 3ω technique at 310 K found a thermal conductivity of 160 WmK^{-1} for a single-layer graphene that increased to 1000 WmK^{-1} for a graphite film 8 nm high. The results show the strong effect of the encasing oxide, in essence, disrupting the thermal conductivity of adjacent graphene layers. This effect penetrates some 2.5 nm (ca. 7 layers) into the core and is attributed to phonon leakage to the oxide layers by the outermost layers and scattering by the inhomogeneous graphene oxide interface (Jang et al., 2010). Due to the layered nature of graphite, it is somewhat intuitive that its cross-layer (C axis) thermal conductivity is poor (i.e. vibrations will not couple easily across layers). Indeed, a Debye model employed to calculate the C-axis thermal conductivity shows it is four orders of magnitude smaller than in the graphite basal plane and it decreases as temperature increases (Sun et al., 2009).

3.5.1.3. Extrinsic Effects on the Thermal Conductivity of Graphene

Extrinsic affects such as phonon scattering at defects or rough boundaries (edges) can reduce thermal conductivity. Strain will also reduce thermal conductivity. In bulk materials, tensile strain decreases thermal conductivity by stiffening phonon modes while compressive strain enhances thermal conductivity since the phonon modes soften. In graphene and FLG, compressive strain leads to buckling. The deformation results in an increase in phonon scattering and thus reduces the thermal conductivity (Li et al., 2010; Sun et al., 2009). With tensile strain, the thermal conductivity decreases remarkably and is attributed to a softening of the phonon modes and an increase in lattice anharmonicity (Baimova eta l., 2012; Li et al., 2010). GNRs have also been shown (theoretically) to exhibit a strong reduction in thermal conductivity upon compressive or tensile strain (Guo et al., 2009; Li et al., 2010; Wei et al., 2011) due to the same underlying causes as for bulk graphene (phonon softening/hardening and scattering upon buckling). In addition, the edge termination, i.e. zig-zag or armchair, as well as the ribbon width can affect the thermal conductivity of GNRs according to non-equilibrium molecular dynamics simulations. Zig-zag terminated nanoribbons are found to initially show an increase in thermal conductivity and then decrease with increasing width. On the other hand, armchair terminated nanoribbons show a monotonous increase in thermal conductivity with width increase. The reason for this difference is attributed to a competitive mechanism in which on one side, upon increasing width, the number of phonon modes increases, while the number of edge localised phonons modes does not change.

Here, the edge effect decreases thermal conductivity while increasing the width increases thermal conductivity. On the other hand, the energy gap between different phonons increases with increasing width. This leads to an increase in phonon Umklapp processes and reduces a decrease in K. In the case of zig-zag terminated ribbons, edge dominated process are limited and so a reduction in thermal conductivity is not observed until the width is sufficiently large for Umklapp processes to dominate which then reduces K. Armchair configurations are edge sensitive and so show an immediate thermal conduction reduction with width increase (Guo et al., 2009). The length of GNRs can also affect thermal conductivity, such that thermal conductivity increases with increasing ribbon length (Guo et al., 2009; Wei et al., 2011).

The incorporation of a foreign atom, vacancy, defect or isotope in the graphene lattice, can affect thermal conductivity. Experimental and theoretical works show higher thermal conductivities for isotopicaly pure graphene as compared to $^{12}C/^{13}C$ mixtures (Wei et al., 2011; Zhang et al., 2010). Experimentally the thermal conductivity of isotopicaly pure ^{12}C (0.01% ^{13}C) graphene using the optothermal Raman technique yielded a K value of over 4000 WmK^{-1}, while for a 1:1 ratio of $^{12}C:^{13}C$ resulted in a K of less than half. Usually, the mass, size and various other factors are used to describe phonon scattering at point defects. In this case though, supporting molecular dynamics simulations suggest point defect phonon scattering in isotopicaly modified graphene is dominated by the mass difference (Chen et al., 2012). Molecular dynamics simulations on a different type of defect, namely, vacancies show a significant reduction of thermal conductivity with defect concentration such that with vacancy defect concentrations of 8.25% the conductivity can be reduced to a staggeringly low value of 3 W mK^{-1} (at a temperature of 300 K). Two primary causes are identified for the abrupt reduction in K. In the first, the broadening of phonon modes means a reduced mean free path, and in the second, an average increase of the density of states may also cause a reduction in phonon relaxation times and mean free paths. Stone–Wales defects also reduce the thermal conductivity but to a lesser degree than point defects (Hoa et al., 2011). The effect of grain boundaries on thermal conductivity is particularly pertinent given CVD grown graphene over polycrystalline catalysts usually consists of a stitched patchwork of grains with grain boundaries at their interfaces (see Section 6.5). Molecular dynamics simulations show a decrease in thermal conductivity with increasing grain orientation angle ($< 30°$) and decreasing grain size. In both cases, there is an increase in 5–7 defect pairs per unit area which leads to an increase in scattering of phonons (Bagri et al., 2011). In practice, polycrystalline graphene, in addition to grain boundaries has vacancies and voids which will further reduce the thermal conductance. Moreover, often the graphene needs to be transferred for use in whatever application has been chosen. The transfer process can leave behind residues. For example, polymer based transfer routes (e.g. using poly(methyl methacrylate) PMMA) often leave behind a polymeric residue. Experimental studies on suspended bilayer graphene samples suspended between two micro-resistance thermometers found K values between 560 and 620 W/mK at room temperature. The authors of this work argued residual polymeric material (confirmed through transmission electron microscopy investigations) scatters phonons and thus reduces the thermal conductivity (Pettes et al., 2011).

The effect of structural modifications in graphene forming graphene derivatives is also relevant. A detailed study using non-equilibrium molecular dynamics simulations investigated hydrogenated graphene (so called graphane). The study found that the thermal conductivity depends on the hydrogen distribution and coverage. In the case of random coverage the investigation showed a rapid decrease in thermal conductivity for coverage up to 30% after which little change is observed with continued coverage. An appealing aspect of the study was hydrogenated stripes were also investigated. Stripes parallel to the heat flux showed a gradual decrease from 0% to 100%. Stripes perpendicular to the heat flux cause a sharp decrease in K. For example, a 5% (perpendicular) coverage leads to a 60% drop on K. The leading cause behind the reduction in thermal conductivity is attributed to the sp^2-sp^3 bonding transition, which occurs in the hydrogenation process that softens the G-band phonon modes (Pei et al., 2011).

More sophisticated sp^2 carbon structures have also been investigated theoretically. Pillared graphene in which graphene layers interconnected by single wall carbon nanotube pillars were studied in-plane and out-of-plane with respect to the graphene plane. The thermal transport was shown to be governed by the minimum interpillar distance and the pillar length. This was shown to be due to scattering of phonons at the graphene–carbon nanotube junctions (Varshney et al., 2010).

3.5.1.4. Lattice Thermal Properties of Graphene

Prior to the isolation of SLG the stability of 2D layers and membranes was hotly debated, and the general consensus argued long wavelength fluctuations would destroy the long-range order of 2D crystals. This is similar to 2D membranes tending to crumple when in 3D space. The emergence of SLG clearly highlights 2D layers can be stable. It is generally argued that destructive fluctuations can be suppressed by anharmonic coupling between the intrinsic bending instability and the in-plane stretching modes as first predicted by Lifshitz (1952) over 50 years ago. This coupling prevents crumpling and stabilises the flat phase (Fasolino et al., 2007; Zakharchenko et al., 2009). Experimentally, ripples have been observed in graphene (Meyer et al., 2007). These ripples are obviously temperature dependant. In the case of graphene in the ground state (0 K) all the carbon bonds are equivalent (viz. conjugated bonds) at elevated temperatures, there is a significant probability of an asymmetric distribution of bond lengths, which implies curvature. Monte Carlo simulations show an anomalously broad distribution of first-neighbour bond lengths, going down to the length of double bonds even at 300 K (Fasolino et al., 2007). One can also engineer controlled ripple orientation, wavelength and amplitude in suspended graphene through thermally generated strains in which one takes advantage of graphene's negative thermal expansion coefficient (Bao et al., 2009). The in-plane thermal expansion of graphite is known to be negative in the low temperature region, with its lowest coefficient of thermal expansion around room temperature. A negative coefficient of thermal expansion means that the material will contract or shrink. The negative coefficient of thermal expansion (CTE) for graphene is predicted to be higher than graphene however the exact point at which the coefficient of thermal expansion changes from negative to positive with increasing temperature is not clear as is also the minimum value of the CTE. The same is also true for bi-layer graphene. Figure 3.5.3 highlights the large differences in the predicted CTE behaviours (shown through the relative expansion) from different groups or see refs (Jiang et al., 2009; Mounet and Marzari, 2005; Pozzo et al., 2011; Tsang et al., 2005; Zakharchenko et al., 2009, 2010). Experimentally, at 300 K the CTE, α, is ca. $-7 \times 10^{-6} K^{-1}$, and its magnitude decreases with increasing temperature (Bao et al., 2009). This is larger than that for graphite, $\alpha \sim -1 \times 10^{-6} K^{-1}$ (Kellett and Richards, 1964). Another experimental study estimated the CTE of SLG using temperature-dependant Raman spectroscopy between 200 K and 400 K. They observed the CTE to remain negative in this entire temperature range and their room temperature value was $\alpha \sim -8 \times 10^{-6} K^{-1}$ (Yoon et al., 2011). Another experiment in which suspended eletromechanical resonators were implemented to study the resonant frequency as a function of temperature between 30 K and 300 K allowed the thermal expansion of SLG to be extracted as a function of temperature. The thermal expansion was observed to remain negative between 30 K and 300 K (Singh et al., 2010).

The thermal contraction of graphene (and graphite) has been explained by the negative Grüneisen parameters which are dominant at low temperatures, since in this region, most optical modes with positive Grüneisen parameters are not excited. The negative Grüneisen parameters correspond to the lowest ZA modes in graphene and in the case of graphite, the "equivalent modes", the Z optical phonon (ZO)' modes. These phonon modes frequencies increase when the in-plane lattice parameter is increased (Mounet and Marzari, 2005). In essence this is a membrane effect predicted by Lifshitz (1952). At higher temperatures where the CTE becomes positive, the graphene lattice expands. Experimentally the expansion of a suspended graphene (bilayer) constriction has been directly observed in a transmission electron microscope. Here, whilst residing inside a TEM a current is applied across the constriction which in the process reduces the width of the constriction further, increasing the current density across the ribbon – very high current densities are achieved (ca. 28 mA/

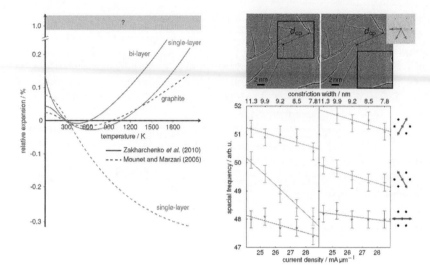

FIGURE 3.5.3 Left panel: sketch highlighting the uncertainty in predicted coefficients of thermal expansion for graphene (Börrnert et al., 2012). Right panel: (top) TEM images showing a bilayer graphene constriction. Right panel: (bottom) the spatial frequency showing the relative expansion of the constriction (left) and outside the constriction (right) with increasing current density (Börrnert et al., 2012).

μm). Information derived from the lattice reflexes indicate lattice expansion at the constriction which increases with increasing current density and is an indication of thermal expansion. However, the relative expansion is of the order of a few percent and cannot be explained by temperature extrapolations which suggest a temperature of around 600 to 700 K which would lead to a lattice expansion of not more than 0.3%. The additional lattice expansion was attributed to impact ionisation. In this mechanism, high-energy electrons scatter at valence electrons forming an exciton in the process. Effectively, this process transfers a valence electron (bonding orbital) into the conduction band (anti bonding orbital), which weakens the bonding and, thus, leads to further lattice expansion beyond that attributable to thermal expansion (Börrnert et al., 2012).

Classical atomistic molecular dynamics simulations investigating graphane (hydrogenated graphene) show graphane has a larger contraction with temperature as compared to graphene. The larger contraction of graphene is attributed to the larger amplitude of the ripples (as compared to graphene). This is somewhat contrary to what one might expect from a thicker material. The study also investigated the heat capacities for graphane as compared to graphene. It turns out that graphane has a larger heat capacity due to the extra storage of vibrational energy in the C–H bonds The study determined a heat capacity of 24.98 ± 0.15 J/molK for graphene and 29.32 ± 32 J/molK for graphane (Neek-Amal and Peeters, 2011).

3.5.1.5. Thermoelectric Aspects of Graphene

Graphene hints at having interesting thermoelectric properties in that it has a high Seebeck coefficient as compared to elemental semiconductors and its sign can be changed by gate bias rather than doping. Nonetheless, a clear picture of its thermoelectric characteristics is not available to date (Balandin, 2011). Experimentally thermoelectric power values between ~50 and 100 $\mu V \ K^{-1}$ have been obtained (Checkelsky and Ong, 2009; Wei et al., 2009; Zuev et al., 2009). Theory provides consistent results (Hwang et al., 2009). The thermoelectric figure of merit is given by ZT which is given by:

$$ZT = S^2 \sigma T / (K_e + K_p)$$

where S is the Seebeck coefficient, σ is the electrical conductivity and K_e and K_p are the thermal conductivity contributions from free carriers and phonons respectively. In state of the art, thermoelectrics ZT is around 1 at room temperature. Theoretical studies suggest that in GNRs ZT may be as high as 4 at room temperature. It is argued that this significant improvement in ZT arises from phonon-edge disorder scattering, while electron transport is not significantly compromised (Sevincli and Cuniberti, 2010). It may be that graphene with intentionally introduced disorder might be attractive for thermoelectric energy conversion (Balandin, 2011).

REFERENCES

Bagri, A., Kim, S.-P., Ruoff, R.S., Shenoy, V.B., 2011. Thermal transport across twin grain boundaries in polycrystalline graphene from nonequilibrium molecular dynamic simulations. Nano Lett. 11, 3917–3921.

Yu, A., Baimova, S.V., Dmitriev, A.V., Savin, Yu., Kivshar', S., 2012. Velocities of sound and the densities of phonon states in a uniformly strained flat graphene sheet. Phys. Solid State 54, 866.

Balandin, A.A., Ghosh, S., Bao, W., Calizo, I., Teweldebrhan, D., Miao, F., Lau, C.N., 2008. Superior thermal conductivity of single-layer graphene. Nano Lett. 8 (3), 902–907.

Balandin, A.A., 2011. Thermal properties of graphene and nanostructured carbon materials. Nat. Mater. 10, 569–581.

Bao, W., Miao, F., Chen, Z., Zhang, H., Jang, W., Dames, C., Lau, C.N., 2009. Controlled ripple texturing of suspemded graphene and ultrathin graphite membranes. Nat. Nanotechnol. 4, 562–566.

Börnert, F., Barreiro, A., Wolf, D., Katsnelson, M.I., Büchner, B., Vandersypen, L.M.K., Rümmeli, M.H., February 17, 2012. Lattice expansion in bilayer graphene constriction at high bias. Nano Lett.

Cai, W., Moore, A.L., Zhu, Y., Li, Y., Chen, S., Shi, L., Ruoff, R.S., 2010. Thermal transport in suspended and supported monolayer gaphene grown by chemical vapor deposition. Nano Lett. 10, 1645–1651.

Calizo, I., Balandin, A.A., Bao, W., Miao, F., Lau, C.N., 2007. Temperature dependance of the Raman spectra of graphene and graphene multilayers. Nano Lett. 7 (9), 2645–2649.

Checkelsky, J.G., Ong, N.P., 2009. Thermopower and Nernst effect in graphene in a magnetic field. Phys. Rev. B 80 081413(4).

Chen, S., Wu, Q., Mishra, C., Kang, J., Zhang, H., Cho, K., Cai, W., Balandin, A.A., Ruoff, R.S., 2012. Thermal conductivity of isotopically modified graphene. Nat. Mater.. http://dx.doi.org/10.1038/NMAT3207.

Fasolino, A., Los, J.H., Katsnelson, M.I., 2007. Insinsic ripples in graphene. Nat. Mater. 6, 858–861.

Faugeras, C., Faugeras, B., Orlita, M., Potemski, M., Nair, R.R., Geim, A.K., 2010. Thermal conductivity of graphene in corbino membrane geometry. Nano 4, 1889–1892.

Ghosh, S., Bao, W., Nika, D.L., Subrina, S., Pokatilov, E.P., Lau, C.N., Balandin, A.A., 2010. Dimensional crossover of thermal transport in few-layer graphene. Nat. Mater. http://dx.doi.org/10.1038/NMAT2753.

Guo, Z., Zhang, D., Gong, X.-G., 2009. Thermal conductivity of graphene nanoribbons. App. Phys. Lett. 95 163103(3).

Hao, F., Fang, D., Xu, Z., 2011. Mechanical and thermal properties of graphene with defects. App. Phys. Lett 99 041901(3).

Hwang, E.H., Rossi, E., Das Sarma, S., 2009. Theory of carrier transport in bilayer graphene. Phys. Rev. B 80 235415(5).

Jang, W., Chen, Z., Bao, W., Lau, C.N., Dames, C., 2010. Thickness-dependent thermal conductivity of encased graphene and ultrathin graphite. Nano Lett. 10, 3909–3913.

Jiang, J.-W., Wang, J.-S., Li, B., 2009. Thermal expansion in single-walled carbon nanotubes and graphene: nonequilibrium Green's function approach. Phys. Rev. B 80 205429(7).

Kellett, E.A., Richards, B.P., 1964. The thermal expansion of graphite within the layer planes. J. Nuc. Mater. 12 (2), 184–192.

Li, X., Maute, K., Dunn, M.L., Yang, R., 2010. Strain effects on the thermal conductivity of nanostructures. Phys. Rev. B 81 245318(11).

Lifshitz, I.M., 1952. Zh. Eksp. Teor. Fiz. 22, 475.

Lindsay, L., Broido, D.A., Mingo, N., 2010. Flexural phonons and thermal transport in graphene. Phys. Rev. B 82 115427(6).

Meyer, J.C., Geim, A.K., Katsnelson, M.I., Novoselov, K.S., Booth, T.J., Roth, S., 2007. The structure of suspended graphene membrane. Nature 446, 60–63.

Mounet, N., Marzari, N., 2005. First-principles determination of the structural, vibrational and thermodynamic properties of diamond, graphite, and derivates. Phys. Rev. B 71 205214(14).

Neek-Amal, M., Peeters, F.M., 2011. Lattice thermal properties of graphene: thermal contraction, roughness and heat capacity. Phys. Rev. B 83 235437(6).

Pei, Q.-X., Sha, Z.-D., Zhang, Y.-W., 2011. A theoretical analysis of the conductivity of hydrogenated graphene. Carbon 49, 4752–4759.

Pettes, M.T., Jo, I., Yao, Z., Shi, L., 2011. Influence of polymeric residue on the thermal conductivity of suspended bilayer graphene. Nano Lett. 11, 1195–1200.

Pozzo, M., Alfe, D., Lacovig, P., Hofmann, P., Lizzit, S., Baraldi, A., 2011. Thermal expansion of supported and freestanding graphene: lattice constant versus interatomic distance. Phys. Rev. Lett. 106 135501(4).

Prasher, R., 2010. Graphene spreads the heat. Science 328, 185–186.

Seol, J.H., Jo, I., Moore, A.L., Lindsay, L., Aitken, Z.H., Pettes, M.T., Li, X., Yao, Z., Huang, R., Broido, D., Mingo, N., Ruoff, R.S., Shi, L., 2010. Two-dimensional phonon transport in supported graphene. Science 328, 213–216.

Sevincli, H., Cuniberti, G., 2010. Enhanced thermoelectric figure of merit in edge-disordered zigzag graphene nanoribbons. Phys. Rev. B 81, 113401.

Singh, V., Sengupta, S., Solanki, H.S., Dhall, R., Allain, A., Dhara, S., Pant, P., Deshmukh, M.M., 2010. Probing thermal expansion of graphene and modal dispersion at low-temperature using graphene nanoelectromechanical systems resonators. Nanotechnol. 21 165204(8pp).

Sun, K., Stroscio, M.A., Dutta, M., 2009. Graphite C-axis thermal conductivity. Superlattices and Mircostructures 45, 60–64.

Tsang, D.K.L., Marsden, B.J., Fok, S.L., Hall, G., 2005. Graphite thermal expansion relationship for different temperature ranges. Carbon 43, 2902–2906.

Varshney, V., Patnaik, S.S., Roy, A.K., Froudakis, G., Farmer, B.L.., 2010. Modeling of thermal transport in pillard-graphene architectures. ACS Nano 4 (2), 1153–1161.

Wei, P., Bao, W., Pu, Y., Lau, C.N., Shi, J., 2009. Anomalous thermoelectric transport of Dirac particles in graphene. Phys. Rev. Lett. 102 166808(4).

Wei, N., Xu, L., Wang, H.-Q., Zheng, J.-C., 2011. Strain engineering of thermal conductivity in graphene sheets and nanoribbons: a demonstration of magic flexibility. Nanotechnol. 22 105705(11pp).

Yoon, D., Son, Y.-W., Cheong, H., 2011. Negative thermal expansion coefficient of graphene measures by Raman spectroscopy. Nano Lett. 11, 3227–3231.

Zakharchenko, K.V., Katsnelson, M.I., Fasolino, A., 2009. Finite temperature lattice properties of graphene beyond the quasiharmonic appoximation. Phys. Rev. Lett. 102 046808(4).

Zakharchenko, K.V., Los, J.H., Katsnelson, M.I., Fasolino, A., 2010. Atomistic simulations of structural and thermodynamic properties of bilayer graphene. Phys. Rev. B 81 235439(6).

Zhang, H., Lee, G., Fonseca, A.F., Borders, T.L., Cho, K., 2010. Isotope effect on the thermal conductivity of graphene. J. Nanomater. ID537657(5). http://dx.doi.org/10.1155/2010/537657.

Zuev, Y.M., Chang, W., Kim, P., 2009. Thermoelectric and magneto-thermoelectric transport measurements of graphene. Phys. Rev. Lett. 102 096807(4).

Methods for Obtaining Graphene

Mechanical Exfoliation

Imad Ibrahim and Mark H. Rümmeli

IFW Dresden, Germany

4.1.1. INTRODUCTION TO MECHANICAL EXFOLIATION

Mechanical exfoliation refers to the process where mechanical force is used to separate graphene layers from bulk graphite. This generally involves using either an adhesive tape to attach to the surface of graphite and using force to peel off the tape plus graphene layers attached or by rubbing the surface of graphite against another material to slide off graphene sheets from the bulk. The principle of writing with a pencil involves this process where the lead is actually graphite and as it is forced against paper thin graphite sheets are adhered to the paper leaving a black mark. At some stage in most of our lives we have all made graphene when putting pencil to paper.

The ease of production and low cost make exfoliation of graphite the most popular route to prepare graphene (Green and Hersam, 2010; Rümmeli et al., 2011). Many routes have been demonstrated for graphite exfoliation, including micromechanical exfoliation, (Novoselov et al., 2004) ultrasound treatment in solution, (Hernandez et al., 2008) milling (Zhao et al., 2010a) and intercalation steps (Zhu, 2008). Despite the enormous success achieved in this field, there are still some challenges to be overcome. Micromechanical exfoliation is achieved either by atomic force microscopy (AFM) tip-based techniques or adhesive tape exfoliation. The AFM tip-based approach is very limited due to the following: its complexity; it is time consuming; it provides a low output yield; and it has a lack of controllability on the produced graphene layer number. In contrast, the adhesive tape micromechanical

Graphene. http://dx.doi.org/10.1016/B978-0-12-394593-8.00004-7

exfoliation is easy and a quick one and provides high-quality large graphene sheets which are useful for the various experimental studies, including the preparation of device-grade graphene from highly oriented pyrolytic graphite (HOPG).

The main drawback of this technique is that it does not provide sufficient output yield for many applications. Moreover, the purity of the produced material often contains exfoliating agent remnants. In addition, micromechanical exfoliation may induce strain on the graphene layer during deposition on a substrate and introduce various types of defects, including atomic defects, wrinkles or ripples, and microscopic corrugation (Choi et al., 2011). These defects may lower the electrical performance of graphene devices since they break translational or rotational symmetry. Liquid-phase sonication for exfoliating graphene and ball milling are promising scale-up routes for producing high yield of single and few layer graphene. However, the small sheet size and the challenges in controlling the thickness as well as the possible introduction of defects due to sonication, in addition to the negative effect of residual material are all limitations. Nevertheless, exfoliation is still attractive as a facile method to obtain a single crystalline graphene due to its simplicity and the possibility to obtain graphene with little to no grain boundaries since high quality HOPG graphene can be used as a starting material.

4.1.2. MICROMECHANICAL EXFOLIATION

Since the invention of the scanning tunnelling microscope (STM), graphite has been highly valuable as an atomically smooth substrate for imaging. In order to achieve such nice atomic smoothness from the surface of graphite, mechanical cleaving of layers is required. This is the first step along the road to achieving graphene by scotch-tape mechanical exfoliation. Early attempts to micromechanically exfoliate graphene go back as far as when Ohashi et al. (1997) managed to mechanically exfoliate 20–100 layers of graphene. Later, another group managed to deposit graphitic flakes with a lateral size of 2 μm and thickness of 10–100 nm with micromechanical manipulations (Zhang et al., 2004). The flakes were originally extracted from bulk HOPG. Metallic electrodes were then fabricated on the flakes so that EFE-dependent magnetoresistance (MR) and Hall resistance measurements in mesoscopic graphite crystallites could be performed (Zhang et al., 2005a).

Micromechanical cleaving methods based on AFM tip manipulation of graphite flakes was first reported by Ruoff et al. (Lu et al., 1999a,b) They managed to prepare islands of HOPG with well-defined sizes and heights which were later transferred to silicon substrates. They used patterned SiO_2 layers deposited using plasma enhanced chemical vapour deposition (CVD) onto a freshly cleaved HOPG surface as a mask for oxygen plasma etching of HOPG. Afterwards, the SiO_2 residues were removed with dilute aqueous HF acid. Thereafter, the HOPG substrate with graphite islands on it was rubbed against other substrate. This results in graphite flakes being transferred on to the target substrate. The investigators showed that the transferred islands fanned out into flakes with reduced heights. The fabricated (fanned out) islands were manipulated carefully by an AFM tip. The subsequent manipulation involved the peeling of graphene sheets from the graphite islands. A variation of this method involves gluing an individual graphite pillar to an AFM tip and then scratching it on Si substrates (Zhang et al., 2005b). Again, the researchers first prepared graphite islands in a similar manner to that discussed above to yield arrays of graphite pillars. A precision micromanipulator fixed above an inverted optical microscope was used to detach an individual pillar and position it on a silicon cantilever and then glued with the help of a small amount of ultraviolet sensitive epoxy. The silicon cantilever loaded with a graphite pillar was used as an AFM tip working in contact mode by applying different normal forces and scanning speeds. As a result, very thin layers of HOPG are sheared off onto the substrate. The fabricated flakes were with size of ~2 μm and thicknesses ranging from 10 to 100 nm. However, with AFM based techniques it is difficult to control the separation and the number of generated graphene layers. In this context, there are other micromechanical exfoliation techniques that are better suited for the preparation of single and few graphene layer and offer better process controllability.

The graphene boom started in 2004 based on mechanical exfoliation of graphite using scotch-tape, when Novoselov et al. (2004) managed to separate and characterise few and single layer graphene. Previously, free standing atomic planes were often 'presumed not to exist' due to thermodynamical instability at the nanometre scale (Sakamoto et al., 2009; Shenderova et al., 2002). To prepare their graphene Geim and Novoselov used adhesive tape to repeatedly mechanically exfoliate (repeated peeling) 1 mm thick highly oriented pyrolytic graphite (HOPG) into increasingly thinner pieces. In more detail: a fresh piece of HOPG is pressed firmly on to the adhesive-side of adhesive tape (e.g. Scotch tape). The tape is then gently peeled away leaving thick and shiny layers of graphite stuck to it. Then, the part of the tape with HOPG fragments remaining on it is refolded onto a clean adhesive section of the same tape, and the two layers are firmly pressed together. The tape is then gently unfolded so that two mirrored graphite fragments remain on the tape. Repeating the folding and unfolding process enables one to thin down an HOPG flake until it is no longer shiny, but has dull dark grey lustre. The tape with the attached optically transparent flakes is then dissolved in acetone. After this a Si wafer is dipped in the solution and then washed in plenty of water and propanol. By following this process some of the exfoliated flakes can be found on the wafer's surface. Thick flakes were removed with ultrasound cleaning in propanol, while the thin films remained attached to the Si wafer, presumably due to van der Waals and/or capillary forces. Using this technique, Geim and Novoselov could reliably and reproducibly obtain few layer graphene films with lateral sizes up to 10 μm and thicknesses of less than 10 nm, shown in Fig. 4.1.1. The electrical properties of the prepared graphene films were then investigated (Novoselov et al., 2004).

The route was simplified further by avoiding the graphene floatation in liquid steps (Novoselov et al., 2005). Large crystallites greater than 1 mm (visible by the naked eye) were obtained using this modified approach. The few layer graphene flakes could be distinguished from bulk graphite through visible light microscopy. While multi-layer (1.5 nm < thickness < 50 nm) graphene flakes were visible, flakes with thicknesses of less than 1.5 nm (as later measured with AFM) were not visible. This natural marker was used to distinguish between the two groups of films; few- and multi-layer graphene. In this case, atomic force and high resolution scanning electronic microscopes were required to further characterise the thin flakes of graphene which were no longer identifiable in a visible light microscope.

A drawback of scanning probe and scanning electron microscopy routes to identify graphene flakes is that the techniques are time consuming. Moreover, because the graphene crystallites left on a substrate (when prepared by mechanical exfoliation routes) are extremely rare, finding them is tantamount to finding a needle in a haystack (Blake et al., 2007). In essence, AFM provides insufficient throughput for single layer graphene identification and in the case of scanning electron microscopy, clear signatures for the number of atomic graphene layers are hard to obtain without state-of-the-art apparatus. Fortunately it is possible to observe single and few layer graphene with a visible light microscope if an appropriate substrate is chosen. For example silicon wafers with an oxide layer with an approriate thickness provides contrast due to interference effects (see Section 5.1) or the discussion by Blake et al. (2007) on the origin of optical contrast differences. The possibility to observe few layer

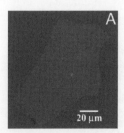

FIGURE 4.1.1 Photograph of a few tens of micrometre wide multilayer graphene flake with thickness ~3 nm on top of an oxidised Si wafer (Novoselov et al., 2004).

graphene flakes with optical microscopes is much easier than other techniques (Novoselov et al., 2005).

An improved route for the mechanical exfoliation of graphene was introduced by Dimiev et al. (2011), which implemented a 5 nm zinc layer instead of adhesive tape. The process starts by preparing multilayer flakes of graphite being attached to oxidised silicon substrates. Then a 5 nm layer of zinc is placed over the flakes using standard electron beam lithography and sputtering. After this step the zinc layer is removed by immersing the substrate in a solution of diluted HCl for 3 to 5 minutes which dissolves the zinc layer and results in removal of a single atomic layer of carbon material. Finally, the substrates are washed with water and dried in a stream of nitrogen. Repetition of these the steps allowed the controlled removal of additional carbon layers, one layer at a time. Another simple but effective procedure to prepare the few layer graphene involves rubbing a fresh surface of a layered crystal (graphite) against another solid surface which results in variety of flakes attaching to the solid surface. This procedure generally results in some single layer graphene and is referred to as drawing with graphene.

Mechanically exfoliated multilayered graphene that was prepared from commercial HOPG has been shown by Chang et al. (2010) to act as a saturable absorbers for passive mode-locking of a fibre lasers. The HOPG derived graphene was prepared by mechanical cutting in a ventilated hood to avoid possible contamination, followed by repeated peeling with scotch tape. The size of the resultant flakes ranged from subnanometre to nanometre with approximately two to eight layers. The graphene layers were fixed on the core region of an optical fibre by pressing them with fibre ferrule. Mechanically exfoliated single and few layer graphene are frequently used to study many basic physical properties of graphene due to the high-quality of HOPG-derived graphene (few defects and grain boundaries). Indeed, for these reasons electrical devices generally use HOPG. For example, exfoliated monolayer graphene prepared with peeling HOPG with Nitto tape and deposited into a Si/SiO$_2$ wafer was used to measure the spatial dependence of the local compressibility versus carrier density across a graphene sheet (Martin et al., 2008). In another report, a suspended graphene device was prepared with a single-layer graphene flake mechanically exfoliated with adhesive tape from natural flakes with an approximately rectangular shape, and then deposited on top of a silicon substrate covered with 300 nm of SiO$_2$ enabling single-layer graphene flakes to be identified based on their contrast in a visible light microscope. After fabricating metallic electrodes with standard electron beam lithography and thermal evaporation of metals, approximately 150 nm of SiO$_2$ across the substrate as well beneath the graphene flake was etched by dipping the entire device into 1:6 buffered oxide etch (BOE) for 90 s, while SiO$_2$ masked by the gold electrodes remains unetched. The mobility of suspended graphene was measured and found to exhibit a tenfold improvement in mobility as compared to the traditional graphene-based devices fabricated on a substrate (Bolotin et al., 2008).

4.1.3. MECHANICAL CLEAVAGE OF GRAPHITE

Mechanical cleavage using an ultra-sharp diamond wedge combined with ultrasonic oscillation can produce few layer graphene with thicknesses of tens of nanometres, from HOPG (Jayasena and Subbiah, 2011). In a study by Jayasena et al., an ultrasharp diamond wedge with a sharpness of less than 20 Å and an angle of 35° was used, as schematically shown in Fig. 4.1.2. It was shown that high-frequency oscillations applied along the wedge provides a smooth sliding motion for the cleaving of graphene layers at the diamond wedge surface. The smooth sliding motion leads to better quality layers. In a typical experiment, a small piece of HOPG was embedded into epofix-embedding medium, which was later trimmed into a pyramid shape, with the HOPG at the top. The diamond wedge mounted on an ultrasonic oscillation system capable of providing tunable frequencies and vibration amplitude of a few tens of nanometres was positioned and aligned with respect to the HOPG. The aligned wedge then sections off HOPG layers with defined layer thickness. The cleaved layers slide off the diamond wedge surface due to the applied high-frequency oscillations. The material is then floated in a water bath arrangement and then transferred to the target substrate. The main advantages of this strategy are the reproducibility of section thickness and chemical inertness. A simple hybrid route combining mechanical and chemical exfoliations of HOPG was developed by Warner et al. The facile route is suitable for the preparation of few layer graphene sheets. It involves repeatedly shaving off thin layers from the surface of HOPG, using a clean razor blade and then mixing with 1,2-dichloroethane, sonicated for 30–60 min and then left for one hour to allow large pieces to sediment. Finally, the supernatant is removed using a pipette and contains single and few layer graphene (Warner et al., 2010).

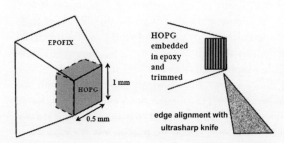

FIGURE 4.1.2 Schematic representation of graphite mechanical cleavage setup by positioning the graphite precursor in pyramid shape-trimmed epofix, and then cleaved with ultrasharp knife (Jayasena and Subbiah, 2011).

4.1.4. MECHANICAL MILLING OF GRAPHITE

One of the early studies on graphene milling was performed by Antisari et al. (2006) in which they managed to prepare carbon flakes with high aspect ratios and reduced thickness to a limited number of graphene layers. Furthermore, the lateral size of the graphene sheets covers a broad distribution up to micron sizes. To do this, they dispersed graphite powder in distilled water with continuous (wet) milling for 60 hours. They speculate that the water eases slip of graphene sheets and also prevents a back agglomeration of the graphene sheets. Ball milling, which is a common technique in the powder production industry, was demonstrated as a suitable route for the mechanical exfoliation of graphite flakes into graphene in various liquid media (Zhao et al., 2010a). Graphite nano-sheets with thickness of 30–80 nm were firstly dispersed in anhydrous N,N-dimethylformamide (DMF) solvent and then milled continuously for 30 hours on a planetary mill while keeping the rotating tray at low speeds. This ensures that the shear stress is dominant and avoids destroying the graphite along the in-plane crystal direction. Strong DMF–graphene interactions along with the weak van der Waals-like coupling between graphite layers results in efficient exfoliation of the graphite sheets into single and few layer graphene sheets (≤ 3 layers). The layer numbers were estimated using transmission electron microscopy and atomic force microscopy characterisation after centrifugation followed by DMF evaporation under vacuum and finally repeated washings with ethanol. As expected, the centrifugation helps to remove partially exfoliated and unexfoliated residual graphite sheets. The studies show that various milling parameters can be adjusted for the effective production of single and few layer graphene sheets. The primary parameters are the diameter and rotation speed of the milling ball, the milling time, the graphite sheets source and concentration in the DMF medium, and the speed of centrifugation. The ball milling approach was extended to include a wide variety of organic solvents, including ethanol, formamide, acetone, tetrahydrofuran (THF), tetramethyluren (TMU), N,N-dimethylformamide (DMF), and N-methylpyrrolidone (NMP) in order to create colloidal dispersions of unfuctionalised graphene sheets (Zhao et al., 2010b).

4.1.5. SUMMARY

The mechanical exfoliation of graphite into graphene using adhesive tape was instrumental in the rapid expansion of graphene research – widely acknowledged as being implemented first by the Manchester Group of Geim and Novoselov. The simplicity and effectiveness of this approach to obtaining graphene combined with the use of a thin oxide layer for ease of identification meant that many people could take this up and start investigating graphene. Unlike fullerenes and carbon nanotubes (CNTs), which required investment in substantial apparatus to obtain them, graphene required only purchasing graphite, adhesive tape, silicon wafers with oxide coating and an optical microscope, all of which are cheap, common and readily available. Although mechanical exfoliation is responsible for the graphene boom, it is not seen as the route towards industry uptake and commercial products that utilise graphene's outstanding properties. For this, new methods are required and these will now be discussed in the forthcoming sub-chapters.

REFERENCES

Antisari, M.V., Montone, A., Jovic, N., Piscopiello, E., Alvani, C., Pilloni, L., 2006. Low energy pure shear milling: a method for the preparation of graphite nano-sheets. Scr. Mater. 55, 1047–1050.

Blake, P., Hill, E.W., Castro Neto, A.H., Novoselov, K.S., Jiang, D., Yang, R., Booth, T.J., Geim, A.K., 2007. Making graphene visible. Appl. Phys. Lett. 91 063124:1–3.

Bolotin, K.I., Sikes, K.J., Jiang, Z., Klima, M., Fudenberg, G., Hone, J., Kim, P., Stormer, H.L., 2008. Ultrahigh electron mobility in suspended graphene. Solid State Commun. 146, 351–355.

Bourlinos, A.B., Georgakilas, V., Zboril, R., Steriotis, T.A., Stubos, A.K., 2009. Liquid-phase exfoliation of graphite towards solubilized graphenes. Small 5, 1841–1845.

Chang, Y.M., Kim, H., Lee, I.H., Song, Y.-W., 2010. Multilayered graphene efficiently formed by mechanical exfoliation for nonlinear saturable absorbers in fiber mode-locked lasers. Appl. Phys. Lett. 97 (211102), 1–3.

Choi, J.S., Kim, J.-S., Byun, I.-S., Lee, D.H., Lee, M.J., Park, B.H., Lee, C., Yoon, D., Cheong, H., Lee, K.H., Son, Y.-W., Park, J.Y., Salmeron, M., 2011. Friction anisotropy–driven domain imaging on exfoliated monolayer graphene. Science 333, 607–610.

Dimiev, A., Kosynkin, D.V., Sinitskii, A., Slesarev, A., Sun, Z., Tour, J.M., 2011. Layer-by-layer removal of graphene for device patterning. Science 331, 1168–1172.

Green, A.A., Hersam, M.C., 2010. Emerging methods for producing monodisperse graphene dispersions. Phys. Chem. Lett. 1, 544–549.

Hernandez, Y., Nicolosi, V., Lotya, M., Blighe, F.M., Sun, Z., De, S., McGovern, I.T., Holland, B., Byrne, M., Gun'Ko, Y.K., Boland, J.J., Niraj, P., Duesberg, G., Satheesh, K., Goodhue, R., Hutchison, J., Scardaci, V., Ferrari, A.C., Coleman, J.N., 2008. High-yield production of graphene by liquid-phase exfoliation of graphite. Nat. Nanotechnol. 3, 563–568.

Jayasena, B., Subbiah, S., 2011. A novel mechanical cleavage method for synthesizing few-layer graphenes. Nanoscale Res. Lett. 6 (95), 1–7.

Khan, U., O'Neill, A., Lotya, M., De, S., Coleman, J.N., 2010. High-concentration solvent exfoliation of graphene. Small 6, 864–871.

Li, X., Wang, X., Zhang, L., Lee, S., Dai, H., 2008. Chemically derived, ultrasmooth graphene nanoribbon semiconductors. Science 319, 1229–1232.

Lotya, M., Hernandez, Y., King, P.J., Smith, R.J., Nicolosi, V., Karlsson, L.S., Blighe, F.M., De, S., Wang, Z., McGovern, I.T., Duesberg, G.S., Coleman, J.N., 2009. Liquid phase production of graphene by exfoliation of graphite in surfactant/water solutions. J. Am. Chem. Soc. 131, 3611–3620.

Lu, X., Yu, M., Huang, H., Ruoff, R.S., 1999a. Tailoring graphite with the goal of achieving single sheets. Nanotechnology 10, 269–272.

Lu, X., Huang, H., Nemchuk, N., Ruoff, R.S., 1999b. Patterning of highly oriented pyrolytic graphite by oxygen plasma etching. Appl. Phys. Lett. 75, 193–195.

Martin, J., Akerman, N., Ulbricht, G., Lohmann, T., Smet, J.H., von Klitzing, K., Yacoby, A., 2008. Observation of electron–hole puddles in graphene using a scanning single-electron transistor. Nat. Phys. 4, 144–148.

Novoselov, K.S., Geim, A.K., Morozov, S.V., Jiang, D., Zhang, Y., Dubonos, S.V., Grigorieva, I.V., Firsov, A.A., 2004. Electric field effect in atomically thin carbon films. Science 306, 666–669.

Novoselov, K.S., Jiang, D., Schedin, F., Booth, T.J., Khotkevich, V.V., Morozov, S.V., Geim, A.K., 2005. Two-dimensional atomic crystals. PNAS 102, 10451–10453.

Ohashi, Y., TKoizumi, Yoshikawa, T., Hironaka, T., Shiiki, K., 1997. Size effect in the in-plane electrical resistivity of very thin graphite crystals. TANSO, 235–238.

Rümmeli, M.H., Rocha, C.G., Ortmann, F., Ibrahim, I., Sevincli, H., Börrnert, F., Kunstmann, J., Bachmatiuk, A., Pötschke, M., Shiraishi, M., Meyyappan, M., Büchner, B., Roche, S., Cuniberti, G., 2011. Graphene: piecing it together. Adv. Mater. 23, 4471–4490.

Sakamoto, J., van Heijst, J., Lukin, O., Schlüter, A.D., 2009. Two-dimensional polymers: just a dream of synthetic chemists? Angew. Chem. Int. Ed. 48, 1030–1069.

Shenderova, O.B., Zhirnov, V.V., Brenner, D.W., 2002. Carbon nostructures. Crit. Rev. Solid State Mater. Sci. 27, 227–356.

Warner, J.H., Rummeli, M.H., Bachmatiuk, A., Buchner, B., 2010. Examining the stability of folded graphene edges against electron beam induced sputtering with atomic resolution. Nanotechnology 21, 325702–325707.

Zhang, Y., Small, J.P., Pontius, W.V., Kim, P., 2004. Fabrication and electric-field-dependent transport measurements of mesoscopic graphite devices. Appl. Phys. Lett. 86 073104–1.

Zhang, Y., Tan, Y.-W., Stormer, H I.., Kim, P., 2005a. Experimental observation of the quantum Hall effect and Berry's phase in graphene. Nature 438, 201–204.

Zhang, Y., Small, J.P., Amori, M.E.S., Kim, P., 2005b. Electric Field Modulation of Galvano-magnetic Properties of mesoscopic Graphite. Phys. Rev. Lett. 94 (176803), 1–4.

Zhao, W., Fang, M., Wu, F., Wu, H., Wang, L., Chen, G., 2010a. Preparation of graphene by exfoliation of graphite using wet ball milling. J. Mater. Chem. 20, 5817–5819.

Zhao, W., Wu, F., Wu, H., Chen, G., 2010b. Preparation of colloidal dispersions of graphene sheets in organic solvents by using ball milling. J. Nanomater. 528235, 1–5.

Zhu, J., 2008. Graphene production: new solutions to a new problem. Nat. Nanotechnol. 3, 528–529.

Chapter 4.2

Chemical Exfoliation

Jamie H. Warner

Department of Materials, University of Oxford, Oxford, UK

4.2.1. INTRODUCTION TO CHEMICAL EXFOLIATION

The layers of graphene that are stacked together to form graphite can be separated from the bulk using solution based chemistry. In order to achieve layer separation, the strong van der Waals forces that stick graphene sheets together must be overcome. This generally requires some form of input energy and calculations suggest that 2 eV/nm^2 is needed to separate layers (Niyogi et al., 2006). Once graphene sheets have been separated, some means of preventing their restacking is also required. Polar solvents such as N-Methyl-2-pyrrolidone (NMP) are very effective at separating sheets of graphene from bulk graphite and maintaining their isolation to form a colloidal-like solution (Hernandez et al., 2008). A typical process involves placing graphite into a vial, adding a suitable solvent such as NMP, then providing mild sonication for around an hour or so using a bath sonicator. Micron sized single and few layer graphene sheets can then be isolated by centrifuging the sample to remove large graphite chunks floating in solution. This leaves a supernatant rich with micron sized sheets of graphene and few-layer graphene. This process is scalable and graphene in solution is ideal for spray coating onto substrates, ink-jet printing, mixing with polymers for composites and electrophoresis deposition in functional electronic devices. Reduced graphene oxide is another variant form of graphene obtained by chemical exfoliation, and will be discussed in detail in Chapter 4.3. Here we will focus on pristine graphene obtained from bulk graphite using polar solvents and aqueous surfactants, we will review some of the early key papers on the topic and provide some guidelines on how to prepare samples and effectively characterise them.

4.2.2. REVIEW OF CHEMICAL EXFOLIATION

4.2.2.1. Non-aqueous Solvents

The discovery of fullerenes and nanotubes prior to graphene meant that there were plenty of researchers with knowledge on how to disperse sp^2-carbon nanomaterials in solvents. Many of the early successes in graphene dispersions were a direct result of previous work on CNTs.

One of the earliest reports of direct chemical exfoliation of graphite into graphene without a graphene oxide step comes from the Nobel prize winners Prof. K. Novoselov and Prof. A. Geim (Blake et al., 2008). In 2008, realising the limitations of mechanically exfoliated graphene for large area transparent-conducting electrodes, they reported a simple method whereby graphite was placed in dimethylformamide and subjected to sonication to produce thin graphite pieces and some monolayer sheets. After centrifuging at 13000 rpm for 10 minutes the thick graphite pieces were removed, leaving monolayer and few layer graphene sheets in a DMF suspension that was then used to spray cast a thin transparent conducting film (Blake et al., 2008). Figure 4.2.1 shows (a) a TEM image of one of their graphene flakes along with the selected area electron diffraction pattern to its right, (b) an SEM image of the thin graphene film produced by spray casting from solution and (c) light transmission image of the glass with spray-coated graphene film on the right.

Li et al. (2008a) reported a simple process for obtaining graphene nanoribbons in solution by sonicating commercial expandable graphite. It

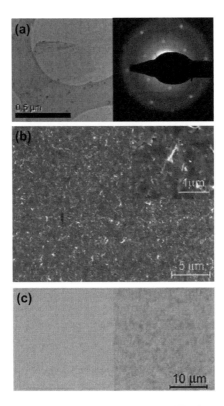

FIGURE 4.2.1   (a) TEM image (left panel) and electron diffraction pattern (right panel) of a graphene flake obtained by the chemical exfoliation method. Equal intensity of first- and second-order diffraction peaks confirm that the flake is exactly one monolayer thick. (b) Scanning electron micrograph of a thin graphitic film obtained by chemical exfoliation and spray-coating. Inset shows the same area under higher magnification. (c) Light transmission through an original glass slide (left) and the one covered with the graphitic film (right). *Reproduced from Blake et al. (2008), Fig. 4. Copyright (2008) American Chemical Society.*

involved heating expandable graphite in forming gas (3% Hydrogen in Ar) at 1000 °C for 60 seconds, followed by dispersing in a solution of 1,2-dichloroethane (DCE) and poly(*m*-phyenylenevinylene-co-2,5-dioctoxy-*p*-phenylenvinylene (PmPV). Dispersion was achieved using sonication and centrifuging removed large pieces of graphite. Graphene ribbons were produced with various widths ranging from 50 nm down to sub-10 nm and layer numbers from 1–3 (Li et al., 2008a). Figure 4.2.2(a) shows a schematic diagram of a graphene nanoribbon in a field effect transistor (FET), and Figures 4.2.2(b)-(c) show AFM images of graphene nanoribbons with various widths obtained by this process (Li, 2008; Wang, 2008). Field effect transistors (FETs) made from these graphene nanoribbons showed semi-conducting behaviour that was width dependent.

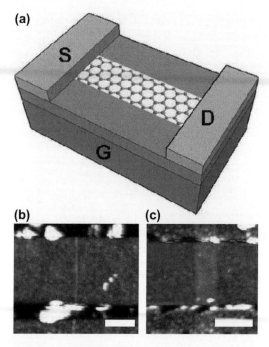

(a)

S

D

G

(b)

(c)

FIGURE 4.2.2 (a) Schematic diagram of a graphene nanoribbon in a field effect transistor (FET). (b) and (c) Atomic force microscopy images of graphene nanoribbons with two different widths in a FET. *Reproduced from Wang et al (2008). Copyright (2008) American Physical Society.*

However, this method in (Li et al., 2008a) was limited to producing ribbons, and a couple of months later, a detailed study on the liquid phase exfoliation of graphene for high yield production of graphene emerged (Hernandez et al., 2008). This was the first systematic study of a variety of organic solvents' capability for dispersing graphene and forming stable monolayer graphene suspensions. Solvents such as NMP, N, N-Dimethylacetamide (DMA), γ-butyrolactone (GBL), and 1,3-dimethyl-2-imidazolidinone (DMEU) were explored (Hernandez et al., 2008). Suspensions were prepared using sieved graphite powder in spectrophotometric grade solvent such as NMP by soni- cation in a bath sonicator to obtain a grey liquid. Aggregates were present and removed by gentle centrifugation leaving a brown homogeneous solution that remained stable for at least five months. Figure 4.2.3(a) shows an optical image of cuvettes with dispersed graphene in NMP at various concentrations, (b) the absorption spectra for graphene dispersed in a variety of solvents and (c) the optical absorption divided by the cell length as a function of concentration for different solvents, and finally (d) is the graphene concentration measured after centrifugation plotted against the surface energy of the solvents (Hernandez et al., 2008). The ability of the solvents to exfoliate graphene was explained in terms of the balance of graphene and solvent surface energies. It was found that solvents with surface tensions in the range of 40–50 mJ m^{-2} were good for dispersing graphene and that concentrations up to 0.01 mg·ml^{-1} could be obtained (Hernandez et al., 2008). This work has been further expanded upon with studies into improvements to the yield of material retained in solution (Khan et al., 2010).

Li et al. (2008b) followed on from their previous report in Li et al. (2008a) with a study on highly conducting graphene sheets prepared from expandable graphite. First, the expandable graphite was exposed to forming gas at 1000 °C for 60 seconds, then ground with NaCl crystallites for 3 minutes to form a grey mixture. Then pieces of graphite were collected and the NaCl removed by washing with water and filtration, followed by treatment with oleum at room temperature for a day. The treatment in oleum was stated as important for obtaining high quality graphene sheets without excessive chemical functionalisation. The sample was then ultrasonicated using a cup-horn sonicator in dimethylformamide (DMF) and tetrabutylammonium hydroxide (TBA) solution for 5 minutes, and then allowed to rest for 3 days so that the TBA could insert between the graphene layers, as shown in the schematic in Fig. 4.2.4(a) and (b) (Li et al., 2008b). Some fraction of the suspension was then bath-sonicated with DSPE-mPEG for 1 hour and then centrifuged to achieve a supernatant consisting black suspension with single layer graphene sheets. Figure 4.2.4(c) shows an atomic schematic of the graphene coated with DSPE-mPEG and an optical image of a vial containing the dispersed graphene in solution (Li et al., 2008b). Figure 4.2.4(d) shows an AFM image with topographic height of ~1 nm, (e) TEM image and (f) the selected area electron diffraction revealing single crystal structure (Li et al., 2008b). The suspension of graphene sheets was used to produce a large cm sized thin film on glass, Fig. 4.2.5(a), with transparencies ranging from 75–90% at 400 nm, Fig. 4.2.5(b), and a resistance of 150 kΩ at 92% transparency, Fig. 4.2.5(c) (Li et al., 2008b).

In late 2008, we showed how sonicating graphite in 1,2-dichloroethane resulted in exfoliation and few layer graphene sheets were dispersed (Warner et al., 2009). The advantage of this approach over solvents such as NMP, is the lower boiling point of DCE that results in easier removal of DCE at room temperature, which enabled TEM studies of clean graphene surfaces. A similar chlorine containing solvent ortho-dichlorobenzene has also shown to be effective at exfoliating graphene by covalent functionalisation (Hamilton et al., 2009). In early 2009, Bourlinos et al. (2009) showed the ability of a series of perfluorinated aromatic solvents (C_6F_6, $C_6F_5CF_3$, C_6F_5CN, and C_5F_5N) to exfoliate graphene and produce colloids that were stable for months, shown in Fig. 4.2.6. Concentrations ranging from 0.05–0.1 mg·mL^{-1} were achieved, depending on the solvent. The authors state that 'The mechanism of solubilisation most likely involves charge transfer through π–π stacking from the electron-rich carbon layers to the electron-deficient aromatic molecules, the

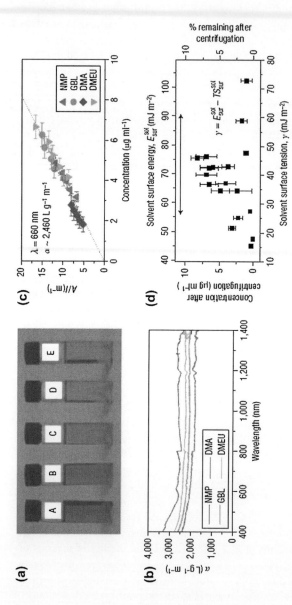

FIGURE 4.2.3 Optical characterisation of graphite dispersions. (a) Dispersions of graphite flakes in NMP at a range of concentrations ranging from 6 µg ml^{-1} (A) to 4 µg ml^{-1} (E) after centrifugation. (b) Absorption spectra for graphite flakes dispersed in NMP, GBL, DMA and DMEU at concentrations from 2 to 8 µg ml^{-1}. (c) Optical absorbance (λ_{ex} = 660 nm) divided by cell length (A/l) as a function of concentration for graphene in the four solvents NMP, GBL, DMA and DMEU, showing Lambert–Beer behaviour with an average absorption coefficient of $<\alpha_{660}>$ = 2,460 L g^{-1} m^{-1}. The x-axis error bars come from the uncertainty in measuring the mass of graphene/graphite in solution. (d) Graphite concentration measured after centrifugation for a range of solvents plotted versus solvent surface tension. The data were converted from absorbance (660 nm) using $A/l = <\alpha_{660}>C$ with $<\alpha_{660}>$ = 2,460 L g^{-1} m^{-1}. The original concentration, before centrifugation, was 0.1 mg ml^{-1}. The y-axis error bars represent the standard deviation calculated from five measurements. Shown on the right axis is the percentage of material remaining after centrifugation. On the top axis, the surface tension has been transformed into surface energy using a universal value for the surface entropy of $S_{sur}^{sol} \sim 0.1$ mJ K^{-1} m^{-2}. The horizontal arrow shows the approximate range of the reported literature values for the surface energy of graphite. (For colour version of this figure, the reader is referred to the online version of this book) *Reproduced from Hernandez et al. (2008), Fig. 1. Copyright (2008) Nature Publishing Group.*

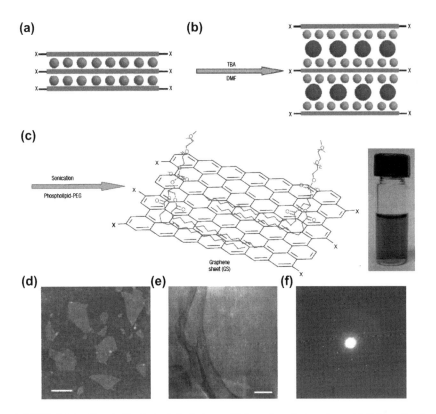

FIGURE 4.2.4   Chemically derived single-layer GS from the solution phase. (a) Schematic of the exfoliated graphite reintercalated with sulphuric acid molecules (teal spheres) between the layers. (b) Schematic of TBA (blue spheres) insertion into the intercalated graphite. (c) Schematic of GS coated with DSPE-mPEG molecules and a photograph of a DSPE-mPEG/DMF solution of GS. (d) An AFM image of a typical GS several hundred nanometres in size and with a topographic height of ~1 nm (see Supplementary Information for height details). The scale bar is 300 nm. (e) Low-magnification TEM images of a typical GS several hundred nanometres in size. The scale bar is 100 nm. (f) Electron diffraction (ED) pattern of an as-made GS as in e, showing excellent crystallisation of the GS. *Reproduced from Li et al. (2008b), Fig. 1. Copyright (2008) Nature Publishing Group*

latter containing strong electron-withdrawing fluorine atoms (Bourlinos et al., 2009).

4.2.2.2. Surfactant-assisted Dispersions

Research into dispersing carbon nanotubes showed that surfactants could wrap around them and effectively disperse them in water and other solvents. Aqueous solutions containing sodium dodecylbenzene sulfonate (SDBS) or

Graphene

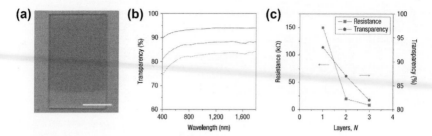

FIGURE 4.2.5 Large-scale Langmuir–Blodgett (LB) films of GS. (a) A photograph of a two-layer GS LB film on quartz with part of it left clear. The scale bar is 10 mm. (b) Transparency spectra of one- (black curve), two- (red curve) and three-layer (green curve) GS LB films. The transparency was defined as the transmittance at a wavelength of 1000 nm. (c) Resistances (red) and transparencies (blue) of one-, two- and three-layer LB films. The small percentage of bilayer and few-layer GS in our sample and GS overlapping in the LB film over the substrate contributed to the transparency loss. *Reproduced from Li et al. (2008b), Fig. 1. Copyright (2008) Nature Publishing Group.*

sodium cholate (SC) are able to disperse single-walled carbon nanotubes and enabled the separation of metallic and semiconducting chiralities (Arnold et al., 2006). Not surprisingly, this was quickly extended to graphene, where the surfactants helped stabilised exfoliated graphene in water (Lotya et al., 2009, 2010). Lotya et al. (2009) first reported using SDBS in water at concentrations between 5 and 10 mg/ml to exfoliate graphene from graphite with the assistance of sonication. Large quantities of multilayer graphene (<5 layers) were produced with some smaller amount of monolayer sheets at concentrations of at most 0.05 mg/ml. The presence of the surfactant attached to the graphene surface prevented reaggregation due to Coulomb repulsion. Figure 4.2.7 shows the concentration of graphite in solution after centrifugation (Lotya et al., 2009).

The group further improved their work by using SC instead of SDBS and achieved higher concentration loadings of up to ~0.3 mg/ml (Lotya et al., 2010). Long sonication times up to 430 h with mild power were important factors for the improved dispersion (Lotya et al., 2010). Figure 4.2.8 shows (a)

Solvent	C_6F_6	$C_6F_5CF_3$	C_6F_5CN	C_5F_5N

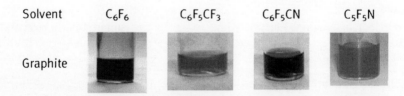

Graphite

FIGURE 4.2.6 Colloidal dispersions obtained after liquid-phase exfoliation of graphite using the perfluorinated aromatic solvents below. *Reproduced from Hamilton et al. (2009), Table 1. Copyright (2009) American Chemical Society.*

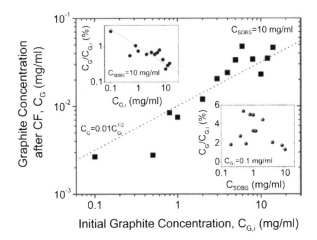

Initial Graphite Concentration, $C_{G,i}$ (mg/ml)

FIGURE 4.2.7   Graphite concentration after centrifugation (CF) as a function of starting graphite concentration (C_{SDBS}) 10 mg/ml). Upper inset: the same data represented as the fraction of graphite remaining after CF. Lower inset: fraction of graphite after centrifugation as a function of SDBS concentration ($C_{G,i}$) 0.1 mg/ml). *Reproduced from Lotya et al. (2009), Fig. 2. Copyright (2009) American Chemical Society.*

how the absorbance per unit length varies with the centrifuging speed, (b) an optical image of vials containing graphene with various centrifuging speeds and (c) the absorbance per unit length as a function of sonication time (Lotya et al., 2010). Centrifuging can also be used to separate out graphene sheets with different layer numbers, but the sheet size from this approach is often submicron (Green and Hersam, 2009). Figure 4.2.9 shows (a) a photograph of a centrifuge tube after the density gradient centrifugation (DGU), (b) and (c) AFM images taken from fractions f4 and f16, (c) height profiles showing different thickness of graphene flakes (Green and Hersam, 2009).

The addition of a stabilising polymer such as ethyl cellulose to an ethanol solution enabled graphene to be exfoliated in concentrations exceeding 1 mg/ml in ethanol (Liang and Hersam, 2010). In most cases the stabilising polymer relies on van der Waals forces to adhere to the graphene and is not covalently bound, but Das et al. (2011) performed an in situ polymerisation technique to coat graphene sheets with nylon in water, as shown in the schematic of Figure 4.2.10(a). Figure 4.2.10 (b) and (c) show photographs of the aqueous dispersions of surfactant-stabilised graphene and nylon coated surfactant-stabilised graphene, with a schematic of the polymerisation technique in Fig. 4.2.10(d) (Das et al., 2011). Functionalised pyrene molecules have also shown to be effective at stabilising graphene in water (Zhang et al., 2010). Removing the stabilising surfactant from the surface of graphene is significantly more of a challenge compared to chemical exfoliation based on polar solvents.

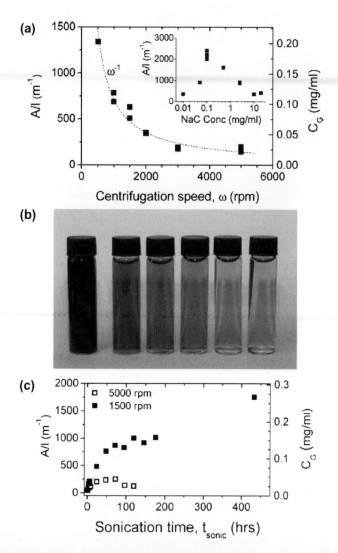

FIGURE 4.2.8 (a) *A/l* as a function of centrifugation speed ($C_{G,i}$ = 5 mg/ml, C_{NaC} = 0.1 mg/ml, t_{sonic} = 24 h, centrifugation: 90 min). Inset: Absorbance/cell length, *A/l*, as a function of surfactant concentration (*CG* = 5 mg/ml, t_{sonic} = 24 h, centrifugation: 1000 rpm for 30 min). Subsequently, a surfactant concentration of 0.1 mg/ml was used for all dispersions. (B) Photos of surfactant-stabilised graphene dispersions; $C_{G,i}$ = 5 mg/ml, C_{NaC} = 0.1 mg/ml and t_{sonic} = 24 h. Left to right: uncentrifuged, centrifuged for 90 min at 1000 rpm, 1500 rpm, 2000 rpm, 3000 rpm, 5000 rpm. Note that the centrifuged dispersions have been diluted by a factor of 10 to highlight the colour change. (C) *A/l* as a function of sonication time ($C_{G,i}$ = 5 mg/ml, C_{NaC} = 0.1 mg/ml, centrifugation: 90 min) for centrifugation speeds of 5000 and 1500 rpm. Note that in both A and C the right axis shows the graphitic concentration calculated using an absorption coefficient of 6600 L g^{-1} m^{-1}. *Reproduced from Lotya et al. (2010), Fig. 1. Copyright (2010) American Chemical Society.*

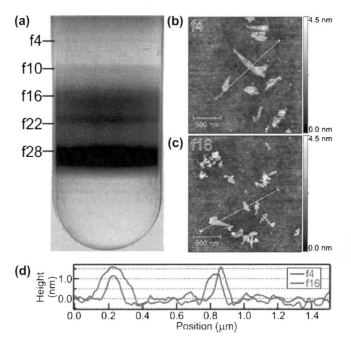

FIGURE 4.2.9 (a) Photograph of a centrifuge tube following the first iteration of density gradient ultracentrifugation (DGU). The concentrated graphene was diluted by a factor of 40 to ensure that all graphene bands could be clearly resolved in the photograph. Lines mark the positions of the sorted graphene fractions within the centrifuge tube. (b,c) Representative AFM images of graphene deposited using fractions f4 (B) and f16 (C) onto SiO_2. (d) Height profile of regions marked in panels B (blue curve) and C (red curve) demonstrating the different thicknesses of graphene flakes obtained from different DGU fractions. *Reproduced from Green and Hersam (2009), Fig. 2. Copyright (2009) American Chemical Society.*

4.2.3. DIFFERENT TYPES OF GRAPHITE

Graphite comes in a variety of forms and the choice of starting material for chemical exfoliation will be influenced by the application that the graphene will be used for. There are three main types of graphite, natural flake, kish and synthetic. Natural flake graphite is sourced by mining and generally has poor crystallinity in terms of rotational stacking faults. Impurities are often associated with natural flake graphite. Kish graphite is obtained as a byproduct of the steel making process, by skimming off waste from molten iron. Therefore Kish graphite contains iron impurities. Highly ordered pyrolytic graphite, known as HOPG, is a synthetic form that has excellent Bernal stacking of the graphene layers with less than 1° rotational mismatch and very low impurity content. It is easier to separate graphene layers that are not AB Bernal stacked, and therefore, graphite with high degree of rotational stacking faults, such as flake, can be exfoliated faster.

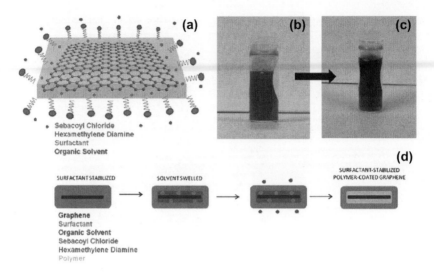

FIGURE 4.2.10 (a) Schematic representation of mechanism of wrapping of nylon around graphene. Photographs of (b) aqueous dispersion of surfactant-stabilised graphene and (c) aqueous dispersion of nylon-coated surfactant-stabilised graphene. (d) Schematic representation of the polymerisation technique to stabilise graphene. *Reproduced from Das et al. (2011), Fig. 1. Copyright (2011) American Chemical Society.*

Flake and Kish graphites contain iron impurities that result in paramagnetic properties. Therefore they should not be used as a starting source of graphite for magnetic or spintronic applications but are highly suitable for uses in mechanically strong composites and thin film conductors. Due to the high crystallinity of HOPG, it is difficult to obtain similar yields of graphene dispersed in solvents compared to using flake or Kish. Therefore HOPG should be used as the starting material for applications that require a very high quality and pure graphene source, such as electronics, magnetism and spintronics. Care should also be taken regarding which company to purchase HOPG from as quality is also variable.

4.2.4. DIFFERENT TYPES OF SOLVENTS

A wide range of solvents have been systematically studied to test their viability for exfoliation graphene. Early work indicated that a match between the solvent surface energy and that of graphene was an important indicator of exfoliation ability. Most solvents studied to date are polar, meaning that some electrostatic dipole forms within the molecule. This often arises when N or O is added to a hydrocarbon, such as in NMP. Care must be taken to use solvents free of water or moisture, which have a detrimental effect on exfoliation ability. We have recently studied a series of non-polar molecules based on a benzene ring with increasing length of single-side alkyl chains (Edson, 2010). In this work we

(a)

FIGURE 4.2.11 Photographs of vials containing solvents and graphite with 3 hours of sonication time: (1) cyclopentane, (2) cyclohexane, (3) toluene, (4) butylbenzene, (5) hexylbenzene, (6) phenylheptane, (7) phenyloctane, (8) phenyldodecane and (9) NMP for (a) 5-minute sedimentation and (b) 1 hour sedimentation (Edson, 2010).

(b)

found that hexylbenzene (6-atom side chain) and heptapentene (7-atom side chain), were able to produce excellent dispersions of graphene in solution, whilst shorter or longer side chains were ineffective, shown in Fig. 4.2.11. Measurements of the surface energy of the series of solvents showed little variation (Fig. 4.2.12), indicating that surface energy matching is not the only criterion that needs to be solved for exfoliation to occur.

Recipes have now been published that are simple, cheap and effective for chemical exfoliation of graphene. One can follow these step-by-step

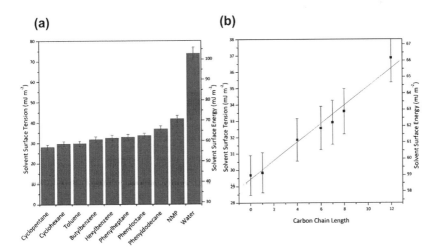

FIGURE 4.2.12 Surface tensions converted to surface energy (right axis) using a universal value for surface entropy (a) of all solvents used (b) against carbon chain length (Edson, 2010).

procedures to obtain gram scale quantities of graphene and few layer graphene. The dispersions remain stable for months and a quick sonication breaks apart any aggregates that may have formed. The utilisation of graphene suspensions is dependent upon the host solvent. Many of the effective solvents for exfoliation, such as DMF and NMP, have relatively high boiling points compared to alcohols, such as ethanol. High boiling point solvents are difficult to remove by evaporating in air. Therefore it is often the case that mild heating in an oven is used to help remove solvents when necessary. Alternatively, there are a couple of solvents that have a lower boiling point, such as 1,2 dichloroethane, which are able to exfoliate graphene from graphite. When deciding which solvent to use for exfoliation, think about the application and solvent compatibility. What are the key aspects of solvent removal required?

4.2.5. DIFFERENT TYPES OF SONICATION

Sonication is usually a crucial step that provides the energy to exfoliate graphene sheets from graphite. Without sonication, nothing much usually happens. It is possible to try to simply heat the solution containing graphite, but it is relatively ineffective compared to sonicating to obtain graphene. When recipes are published on how to produce graphene by chemical exfoliation, particular care must be taken to read the details of the sonication apparatus used. The final graphene product produced will be sensitive to the frequency, amplitude and power of the sonication system. In general, there are two main ways of applying ultrasonic power to a solution containing graphite, a sonication bath, or a tip sonicator, shown in Fig. 4.2.13. Sonication baths are usually rectangular metal dishes that can deliver ultrasonic waves through water that is filled in and come with different power capability and structural form. The sonication power transferred to the vial filled with solvent and graphite will be affected by how

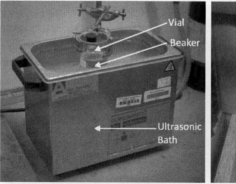

FIGURE 4.2.13 Photograph of the bath (left) and tip (right) sonicators.

much water you place in the dish, whether or not the vial is touching the base of the dish and the location of the vial within the dish. For reproducible results, it is recommended to keep these three parameters consistent. Tip sonicators contain a sharp end tip that is placed directly within the solution in a vial. The power in these systems is often larger than a bath, but the process is usually more reproducible. Both baths and tip sonicators offer either continuous or pulsed modes of sonication. During prolonged sonication, the water or solvent in the sample vial can heat, leading to chemical reactions with the sample. Therefore, it is common to have water circulation in a bath system and for a tip sonicator the sample vial is often placed in an ice bath. Pulsed mode operation helps to reduce heating.

4.2.6. HOW TO CHARACTERISE CHEMICALLY EXFOLIATED GRAPHENE

No doubt that if you followed a recipe to produce graphene, or you are developing new methods for graphene production, you need to know how to characterise the material you have produced and convince others you actually have graphene. Chapter 5 provides an overview of several key characterisation techniques. The first step in assessing whether you have exfoliated graphene from graphite is to visually inspect the vial after sonication. If the vial is clear and no suspension has formed, then exfoliation did not occur. If the vial has a grey, black or brown tinge to it, then it is likely some graphene has been exfoliated. The next step is to allow the vials to stand for 5 min and to repeat the visual inspection. If all the black material drops to the bottom leaving a clear solution then the exfoliation is not stable. If after 30 min, the solution still appears to have colour, then some graphene is suspended. Usually within the suspension there will be graphene, few layer graphene and thin graphite pieces. In order to enrich the sample with graphene it can be centrifuged. The speed of centrifugation and starting concentration will influence the final material. After centrifugation, if the sample has a brown colour, then it is likely you have formed a suspension of graphene. A quantitative analysis of the amount of graphene by the concentration by weight can be conducted. This involves measuring the optical absorbance of the sample using a UV–Vis absorption spectrometer. Care should be taken to subtract the background absorbance of the solvent. There are published graphs plotting the optical absorbance values at a specific wavelength as a function of concentration by weight.

The next characterisation step might involve transmission electron microscopy and electron diffraction. Because chemically exfoliated graphene is in a solution, TEM sample preparation is straight forward. Place a lacey carbon-coated copper-mesh TEM grid on a filter paper and with a pipette allow one drop of the graphene suspension to fall onto the TEM grid. If high boiling point solvents were used, such as NMP, then it is often a good idea to follow the first drop with a second drop of methanol. This will then wash off any excess NMP liquid and leave many of the graphene flakes adhered to the carbon grid. TEM imaging should be done at a low accelerating voltage such as 80 kV to obtain good contrast and minimal damage. On the lacey grid there should be many individual pieces of graphene or few layer graphene. Taking several low magnification images of a number of entire graphene pieces allows size distribution to be evaluated. A comparison should be made with the size distribution obtained from separate atomic force microscopy measurements on the same sample deposited on a substrate.

The number of layers of graphene within a single piece can then be evaluated using a tilt-series of electron diffraction. Using a selected area aperture, obtain a diffraction pattern from a graphene piece and repeat for about 10 different tilt angles, ranging between 0 and 20. Details of how to use electron diffraction for graphene layer evaluation are found in Chapter 5.5. There are several reports that only analyse the intensity of graphene SAED patterns for 0° tilt angle to infer the number of layers. This is not recommended, as it is not a robust method due to the relative intensity of spots in the SAED pattern being dependent upon tilt angle. Multilayered graphene samples can have similar SAED patterns to monolayer graphene, if they are tilted in projection. The edges of the graphene pieces should be directly imaged, using HRTEM to count the lines of contrast. Regions where back-folding has occurred are particularly useful for counting the number of layers, similar to carbon nanotube contrast, as shown in Fig. 4.2.14 for (a) bilayer graphene and (b) trilayer graphene. If you have access to an aberration-corrected HRTEM then it is also possible to image the lattice structure to obtain the layer number.

Raman spectroscopy is difficult to implement with the small micron sized pieces of graphene obtained by chemical exfoliation. Atomic force micron sized is great for determining the average size but is challenging for

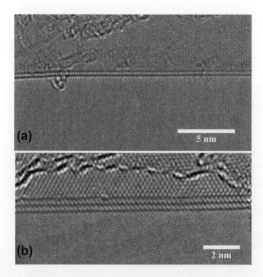

FIGURE 4.2.14 TEM images of back-folded (a) bilayer graphene and (c) trilayer graphene.

determining the distribution of layer numbers due to height profiles being influenced by solvent underneath the graphene flake as well as surface residues.

4.2.7. OTHER 2D CRYSTALS

The methods for exfoliating graphene from graphite were quickly extended to boron nitride, and thin BN sheets were produced by chemical exfoliation. One of the first reports of chemically exfoliated BN sheets was from Han et al. (2008) where 0.2 mg of a hexagonal BN single crystal sample was placed in a 5 ml of 1,2 dichloroethane solution containing 0.6 mg of poly(*m*-phenylenevinylene-co-2,5-dictoxy-*p*-phenylenevinylene) and sonicated for an hour to yield mono and few layer BN sheets in solution. Zhi et al. (2009) reported using DMF as the solvent to exfoliate BN. In their work, 1 g of BN powder was sonicated for 10 h in 40 ml of DMF using a tip sonicator. The sample was then centrifuged at 5000–8000 rpm to remove large BN pieces. The average BN sheet thickness was adjusted by the centrifuge speed and was as thin as ~1.2 nm, corresponding to ~3 BN layers. Micron sized sheet diameters were achieved. We have shown that few layer BN sheets can be exfoliated by using only 1,2 dichloroethane and sonication, enabling TEM studies of clean BN surfaces (Warner et al., 2010). Coleman et al. (2011) recently expanded the liquid exfoliation to a wide number of 2D crystals, such as MoS_2, WS_2, $MoSe_2$, $MoTe_2$, $TaSe_2$, $NbSe_2$, $NiTe_2$, BN and Bi_2Te_3. This provides a pathway to exploring many 2D materials.

4.2.8. SUMMARY

Chemical exfoliation is a great method for obtaining large amounts of micron sized graphene flakes in a wide range of solvent hosts. It is compatible with processing to produce polymer composites and for aqueous studies of biocompatibility. The advantage of direct exfoliation compared to using a graphene oxide step is that the quality of the graphene should be better, in terms of the pristine nature of the lattice. Efforts have been made to maximise the concentration loading of graphene in solution, whilst maintaining reduced sedimentation over time. Graphene in solution can be spray coated onto glass for an easy conductive thin film, but the challenge here lies in the fact that each sheet is poorly connected to one another and relies on weak van der Waals forces for adhesion. With the expansion of exfoliation now entering other 2D materials, new opportunities are available for exploring these materials in a wide range of applications where graphene is not ideal, for example in ultra-strong polymer composites. It is not ideal to have the film tarnished black due to the graphene inclusions, whereas BN should offer a transparent alternative for achieving the same goals in composites requiring only mechanical robustness.

REFERENCES

Arnold, M.S., Green, A.A., Hulvat, J.F., Stupp, S.I., Hersam, M.C., 2006. Sorting carbon nanotubes by electronic structure using density differentiation. Nat. Nanotechnol.

Blake, P., Brimicombe, P.D., Nair, R.R., Booth, T.J., Jiang, D., Schedin, F., Ponomarenko, L.A., Morozov, S.V., Gleeson, H.F., Hill, E.W., Geim, A.K., Novoselov, K.S., 2008. Graphene-based liquid crystal device. Nano Lett. 8, 1704–1708.

Bourlinos, A.B., Georgakilas, V., Zboril, R., Steriotis, T.A., Stubos, A.K., 2009. Liquid-phase exfoliation of graphite towards solubilized graphenes. Small 5, 1841–1845.

Coleman, J.N., et al., 2011. Two-dimensional nanosheets produced by liquid exfoliation of layered materials. Science 331, 568.

Das, S., Wajid, A.S., Shelburne, J.L., Liao, Y.-C., Green, M.J., 2011. Localized in situ polymerization on graphene surfaces for stabilized graphene dispersions. ACS Appl. Mater. Interfaces 3, 1844–1851.

Edson, C., 2010. Part II thesis titled 'Electron Spin Properties of Graphene'. Department of Materials, University of Oxford.

Green, A.A., Hersam, M.C., 2009. Solution phase production of graphene with controlled thickness via density differentiation. Nano Lett. 9, 4031–4036.

Hamilton, C.E., Lomeda, J.R., Sun, Z., Tour, J.M., Barron, A.R., 2009. High yield organic dispersions of unfunctionalized graphene. Nano Lett. 9, 3460–3462.

Han, W. Q., Wu, L., Zhu, Y., Watanabe, K., Taniguchi, T., 2008. Structure of chemically derived mono- and few-atomic-layer boron nitride sheets. Appl. Phys. Lett. 93 (223103).

Hernandez, Y., Nicolosi, V., Lotya, M., Blighe, F.M., Sun, Z., De, S., McGovern, I.T., Holland, B., Byrne, M., Gunko, Y.K., Boland, J.J., Niraj, P., Duesberg, G., Krishnamurthy, S., Goodhue, R., Hutchison, J., Scardaci, V., Ferrari, A.C., Coleman, J.N., 2008. High yield production of graphene by liquid-phase exfoliation of graphite. Nat. Nanotechnol. 3, 563.

Khan, U., O'Neill, A., Lotya, M., De, S., Coleman, J.N., 2010. High concentration solvent exfoliation of graphene. Small 6, 864–871.

Li, X., Wang, X., Zhang, L., Lee, S., Dai, H., 2008a. Chemically derived ultrasmooth graphene nanoribbon semiconductors. Science 319, 1229.

Li, X., Zhang, G., Bai, X., Sun, X., Wang, X., Wang, E., Dai, H., 2008b. Highly conducting graphene sheets and langmuir-blodgett films. Nat. Nanotechnol. 3, 538.

Liang, Y.T., Hersam, M.C., 2010. Highly concentrated graphene solutions via polymer enhanced solvent exfoliation and iterative solvent exchange. J. Am. Chem. Soc. 132, 17661–17663.

Lotya, M., Hernandez, Y., King, P.J., Smith, R.J., Nicolosi, V., Karlsson, L.S., Blighe, F.M., De, S., Wang, Z., McGovern, I.T., Duesberg, G.S., Coleman, J.N., 2009. Liquid phase production of graphene by exfoliation of graphite in surfactant/water solutions. J. Am. Chem. Soc. 131, 3611–3620.

Lotya, M., King, P.J., Khan, U., De, S., Coleman, J.N., 2010. High-concentration, surfactant-stabilized graphene dispersions. ACS Nano 4, 3155–3162.

Niyogi, S., Bekyarova, E., Itkis, M.E., McWilliams, J.L., Hamon, M.A., Haddon, R.C., 2006. Solution properties of graphite and graphene. J. Am. Chem. Soc. 128, 7720–7721.

Wang, X., Ouyang, Y., Li,, X, Wang, H., Guo, J., Dai, H., 2008. Room Temperature All Semiconducting Sub-10nm Graphene Nanoribbon Field-Effect Transistors, Physical Review Letters 100, 206803.

Warner, J.H., Ruemmeli, M.H., Gemming, T., Buechner, B., Briggs, G.A.D., 2009. Direct imaging of rotational stacking faults in few layer graphene. Nano Lett. 9, 102–106.

Warner, J.H., Ruemmeli, M.H., Bachmatiuk, A., Buechner, B., 2010. Atomic resolution imaging and topography of boron nitride sheets produced by chemical exfoliation. ACS Nano 4, 1299–1304.

Zhang, M., Parajuli, R.R., Mastrogiovanni, D., Dai, B., Lo, P., Cheung, W., Brukh, R., Chiu, P.L., Zhou, T., Liu, Z., Garfunkel, E., He, H., 2010. Production of graphene sheets by direct dispersion with aromatic healing agents. Small 6, 110–1107.

Zhi, C., Bando, Y., Tang, C., Kuwahara, H., Golberg, D., 2009. Large-scale fabrication of boron nitride nanosheets and their utilization in polymeric composites with improved thermal and mechanical properties. Adv. Mater. 21, 2889–2893.

⎯⎯⎯⎯⎯⎯⎯⎯⎯⎯⎯⎯⎯⎯⎯⎯⎯⎯(Chapter 4.3)

Reduced Graphene Oxide

Jamie H. Warner

Department of Materials, University of Oxford, Oxford, UK

4.3.1. GRAPHENE OXIDE

Graphene oxide is the name given to graphene that has been oxidised and as such the pristine nature of the graphene lattice disrupted. The oxidisation of graphite has been the subject of investigation since at least the mid 19th century (Brodie, 1860). One of the first reports was by Brodie (1860) where graphite was treated with potassium chlorate ($KClO_3$) and fuming nitric acid (HNO_3). Staudenmaier, (1898) later improved this approach for oxidising graphite by slowly added the potassium chlorate over the course of a week to a solution containing concentrated sulphuric acid, concentrated nitric acid (63%) and graphite. However, needing a 10:1 mass ratio of potassium chlorate to graphite, researchers found this method dangerous due to possibility of explosion and time-consuming. More than 56 years later Hummers and Offeman (1958) reported an alternative 'safer' method known as the Hummers method, which involved a water-free mixture of concentrated sulphuric acid, sodium nitrate, and potassium permanganate. Temperatures of only 45 °C were required and the entire reaction took only 2 hours to complete. The Hummers method has become the basis for many of today's recipes for graphene oxide production and is briefly summarised.

4.3.1.1. Hummers Method for Graphite Oxide (Hummers and Offeman, 1958)

50 g of sodium nitrate and 100 g of graphite are added to 2.3 L of concentrated sulphuric acid and cooled to 0 °C using an ice-bath. Vigorous agitation is applied as 300 g of potassium permanganate is added to the mixture at a rate such that the temperature does not reach above 20 °C. The ice-bath is removed and the temperature raised to 35 °C for 30 min at which point the mixture becomes pasty with a brown-grey colour. Next, 4.6 L of water is added, resulting in effervescence and a temperature rise to 98 °C. After 15 min at 98 °C, the solution is diluted with 14 L of warm water and 3% hydrogen peroxide to reduce permanganate and manganese dioxide to manganese sulfate. The solution is filtered whilst still warm to form a solid cake and then washed three times with a total of 14 L of water. Remaining salt impurities were removed by treating with resinous anion and cation exchangers. A dry graphite oxide powder was finally obtained by centrifugation, followed by drying at 40 °C over phosphorous pentoxide.

The Brodie, Staudenmaier, and Hummers methods are the three main approaches for oxidising graphite. Once graphite has been oxidised it can be exfoliated into individual graphene oxide sheets by either sonicating or prolonged stirring in water. Graphene oxide sheets disperse well in water due to the negative surface charges on the sheets that arise from the phenols and carboxylic acid groups that decorate graphene oxide and keep it from reaggregating (He et al., 1998). The hydrophilic nature of graphene oxide means that water molecules easily intercalate graphite oxide, leading to variable inter-sheet separations ranging from 0.6 nm to 1.2 nm (Buchsteiner et al., 2006).

For many years there was uncertainty regarding the specific atomic structure of graphite oxide and subsequently graphene oxide. (Szabo et al., 2006) looked at five existing proposed models of graphite oxide [Hofmann and Holst, 1939; Ruess, 1946; Scholz and Boehm, 1969; Nakajima et al., 1988) and Lerf-Klinowski (He et al., 1996, 1998; Lerf et al., 1997, 1998)] and in the end proposed a new model bringing together elements from both Scholz-Boehm's and Ruess' models. In many of the graphite oxide structure studies, solid-state NMR was used to provide insights, however the low nature abundance of ^{13}C leads to relatively poor signal to noise ratios. Cai et al. (2008) overcame this problem by using synthetic ^{13}C labelled graphite to prepare graphene oxide. Their solid-state NMR results revealed that only the Lerf-Klinowski (He et al., 1998) and Dekany models (Buchsteiner et al., 2006) were suitable structures for graphite oxide (Lerf et al., 1998). Figure 4.3.1 shows the proposed structural models of graphite oxide of (a) Lerf-Klinowski model (He et al., 1998) and (b) Dekany model (Szabo et al., 2006).

The majority of research on exfoliating graphite oxide into graphene oxide has focused on aqueous dispersions. Paredes et al. (2008) showed that graphene oxide can also be dispersed in a wide range of organic solvents, such as *N,N* dimethylformamide, *N*-methyl-2-pyrrolidone, THF and ethylene glycol. Figure 4.3.2 shows vials of solvents containing graphene oxide with 1 hour sonication time, just after sonication and also after three weeks of resting.

Graphene oxide can be converted to graphene by chemical methods involving reducing agents (i.e. sodium borohydride (NaBH$_4$) (Si and Samulski, 2008), hydrazine (Stankovich et al., 2007; Tung et al., 2008), dimethylhydrazine (Stankovich et al., 2006)), or by heat treatment (McAllister et al., 2007; Schniepp et al., 2006), or by UV assistance (Williams et al., 2008), and even electrochemically (Zhou et al., 2009). We shall discuss each of these separately.

4.3.2. CHEMICAL REDUCTION OF GRAPHENE OXIDE

One of the initial challenges of reducing graphene oxide to graphene in aqueous solutions is that graphene is not generally soluble in water and will aggregate.

(a)

graphite oxide

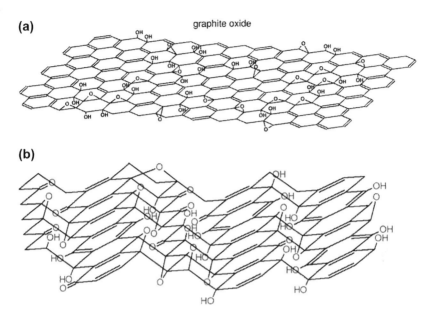

(b)

FIGURE 4.3.1 Proposed atomic models of graphene oxide (a) Lerf-Klinowski model (He et al., 1998) *Reprinted from He (1998). Copyright (1998) Elsevier.* (b) Dekany model (Szabo et al., 2006). *Reprinted from Szabo (2006). Copyright (2006) American Chemical Society.*

Stankovich et al. (2006) showed that reducing graphene oxide in the presence of a stabilising polymer poly(sodium-4-styrenesulfonate) enables stable aqueous dispersions to be achieved. Li et al. (2008) showed that graphene oxide could be converted to graphene in aqueous solutions using hydrazine if care was taken with the solute environment to control the charge on graphene's surface. They found that the metal salts and acids remaining from the oxidation

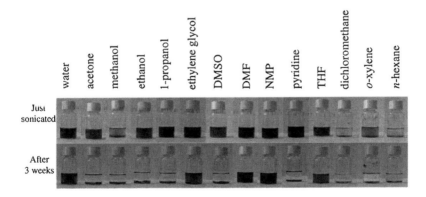

FIGURE 4.3.2 Photographs of vials containing graphene oxide dispersed in a variety of solvents, just after sonication and after 3 weeks of resting. Paredes et al. (2008). *Reprinted from Paredes (2008). Copyright (2008) American Chemical Society.*

of graphite neutralise the charge on graphene sheets that is critical for preventing aggregation. A key part of this work was the use of ammonia to control
the pH of the solution. The method of Li et al. for reducing graphene oxide to
graphene is summarised as follows (Stankovich et al., 2006):

*A modified Hummers method was used to obtained graphite oxide that was dispersed in
water at 0.05%wt by 30 min of sonication. Thirty minutes of 3000-rpm centrifugation
was then applied to remove aggregates, leaving a homogeneous graphene oxide solution. 5 ml of the graphene oxide solution was mixed with 5 ml of water, 5 μl of hydrazine
(35%wt in water), and 35 μl of ammonia solution (28%wt in water), then shaken for
a few minutes and placed in a water bath at 95 °C for 1 h.*

Tung et al. (2008) showed that immersing graphene oxide in pure hydrazine
resulted in both reduction of graphene oxide to hydrazinium graphene and also
dispersion. They suggest that the chemically converted graphene consists of
a negatively charged reduced graphene sheet surrounded by $N_2H_4^+$ counter ions.
Unlike the case of Li et al. (2008) where negatively charged sheets repel each
other in water, in pure anhydrous hydrazine, the majority of the carboxylic acid
groups of graphene oxide are reduced and the positive charge of the $N_2H_4^+$
counter-ions assists in the stabilisation. The use of pure anhydrous hydrazine
requires a dry-box, which may hinder future large-scale production, relative to
other methods. Figure 4.3.3(a) shows a photograph of graphene oxide solid in

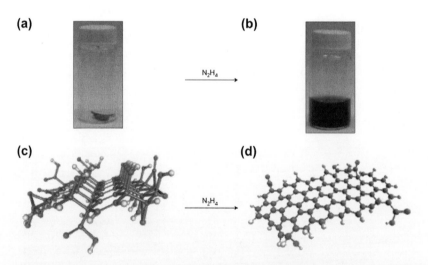

FIGURE 4.3.3 Photographs of (a) solid graphene oxide in a glass vial, (b) hydrazinium graphene
dispersed in hydrazine. *Reprinted from Tung (2008). Copyright (2008) Nature Publishing Group.*
(c) atomic model of graphene oxide (C – grey, O – red, H – white), (d) atomic model of hydrazinium graphene (Li et al., 2008). *Reprinted from Li (2008). Copyright (2008) Nature Publishing
Group.*

a glass vial and (b) after reducing and dispersion in pure hydrazine to form hydrazinium graphene. Figure 4.3.3(c) and (d) show atomic models of graphene oxide and hydrazinium graphene respectively.

The approach of Si and Samulski (2008) to achieve water-soluble graphene without stabilising polymers is slightly different in that chemical functionalisation was used. In their process, a small number of p-phenyl-SO$_3$H groups are introduced to the graphene oxide before it is fully reduced. After the final reduction step, the charged –SO$_3^-$ groups stop the graphene from aggregating. The resulting product is a lightly sulfonated graphene sheet that is stable in water. Their 3-stage method is comprised of (1) partial reduction of graphene oxide to graphene using sodium borohydride for 1 h at 80 °C, (2) sulfonation with the aryl diazonium salt of sulfanilic acid in an ice bath for 2 h, and (3) exposure to hydrazine for 24 h at a temperature of 100 °C, which removed the remaining oxygen groups. This approach has the advantage over Stankovich et al. (2006) in that (1) the graphene can be dispersed in a variety of solvents, (methanol, acetone, and acetonitrile), (2) it is not as sensitive to the pH of water, but has the disadvantage that the atomic structure of the graphene is perturbed from its pristine state due to the sulfonated groups attached.

4.3.3. HEAT TREATMENT OF GRAPHENE OXIDE

Thermal expansion of graphite oxide can be used to both exfoliate and produce functionalised graphene sheets. At a critical temperature of 550 °C the decomposition rate of epoxy and hydroxyl sites of graphene oxide becomes larger than the rate of diffusion of the evolved gases and this causes a build up of pressure (McAllister et al., 2007). The pressure becomes so great that it overcomes the van der Waals forces that bind the graphene sheets together and exfoliation occurs. In the work of McAllister et al. they first used the Staudenmaier method to produce graphene oxide, and placed 200 mg into a quartz tube flushed with argon and inserted into a tube furnace already at 1050 °C for 30 seconds. This resulted in functionalised graphene sheets with thickness between 1 and 3 nm and diameters between 0.3 and 2 microns. The functionalised graphene sheets were shown to be electrically conducting, indicating they were not graphene oxide, which is insulating (Schniepp et al., 2006).

4.3.4. ELECTROCHEMICAL REDUCTION OF GRAPHENE OXIDE

Chemical reducing agents, such as hydrazine, are classified as toxic and corrosive, and this was one motivating factor behind Zhou et al. (2009) to seek a greener, safer and more convenient method for reducing graphene oxide films. They report the use of an electrochemical method to reduce graphene oxide and couple it with a spray coating technique to achieve controllable synthesis and patterning of large area films on both insulating and conducting substrates. Figure 4.3.4 shows a schematic of their electrochemical setup. For

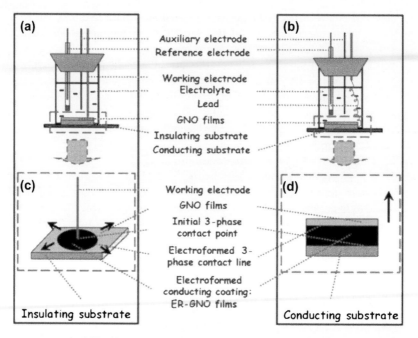

FIGURE 4.3.4 Schematic of the electrochemical reduction setup in Zhou et al. (2009) for (a) insulating and (b) conducting substrates. (c) and (d) electrolysing process for electrochemical reduction of graphene oxide based on three-phase-interlines (3PIs) model. Black arrows indicate the direction of film formation. Three phases: conductor/insulator/electrolyte; initial three phases: working electrode/graphene oxide/electrolyte; electro-formed three phases: electrochemically reduced graphene oxide/graphene oxide/electrolyte. *Reprinted from Zhou (2009). Copyright (2009) WILEY-VCH Verlag GmbH & Co.*

the insulating quartz substrate, the tip of the glassy carbon electrode was brought in contact with ~7-μm thick graphene oxide film on quartz in the presence of a sodium phosphate buffer solution (Na–PBS, 1 M, pH 4.12), and a voltage of −0.6 V initially applied. As the voltage increased, a sharp increase in current was observed as the applied voltage reached −0.87 V, indicating electrochemical reduction of the graphene oxide. Inspection of the sample using optical microscopy and SEM showed visible changes from brown to black (optical) and contrast changes in SEM, confirming reduction of graphene oxide, shown in Figure 4.3.5.

4.3.5. SUMMARY

Even though graphite oxide has been studied for more than 150 years there are still uncertainties surrounding the exact atomic structure and pathways in which reducing agents convert it to chemically functionalised graphene. Nonetheless, it is an excellent starting material for obtaining gram and possibly kilogram quantities of graphene sheets. It can be dispersed in a wide range of organic and

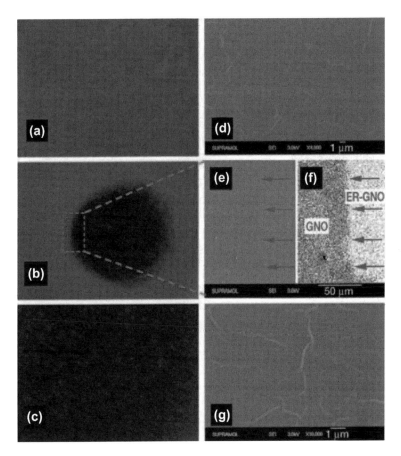

FIGURE 4.3.5   Photos (a–c) and SEM images (d–g) of ~7-μm-thick graphene oxide films on quartz (5 × 4 cm²) before (0 s; a, d), during (1000 s; b, e, f) and after electrolysis (5000 s; c and g). Image (f) was obtained from image (e) by using a contrast enhancement of ×100. The blue arrows indicate the boundary between the circular electro-reduced graphene oxide area and the unreduced graphene oxide in image (b). *Reprinted from Zhou (2009). Copyright (2009) WILEY-VCH Verlag GmbH & Co.*

aqueous solutions in high concentrations. The four main approaches for reducing graphene oxide were discussed, with particular emphasis on the popular chemical methods. Reduced graphene oxide is ideal for applications that are not so stringent on crystal quality, such as loading into polymers to produce light-weight ultrastrong materials. The inherent nature of starting with a highly defective graphene oxide sheet makes it less desirable compared with direct exfoliation in organic solvents for applications requiring pristine lattice structure. The advantage of having defect sites within the lattice of reduced graphene oxide is that they provide sites for chemical functionalisation, which is difficult in pristine-graphene structures. Functionalisation opens new avenues for incorporation graphene with other materials and surfaces to enhance its applicability.

REFERENCES

Brodie, B.C., 1860. Sur le poids atomique du graphite. Ann. Chim. Phys. 59, 466.

Buchsteiner, A., Lerf, A., Pieper, J., 2006. Water dynamics in graphite oxide investigated with neutron scattering. J. Phys. Chem. B 110, 22328–22338.

Cai, W., et al., 2008. Synthesis and solid-state NMR structural characterization of ^{13}C-labeled graphite oxide. Science 321, 1815–1817.

He, H., Riedl, T., Lerf, A., Klinowski, J., 1996. Solid-state NMR studies of the structure of graphite oxide. J. Phys. Chem. 100, 19954–19958.

He, H., Klinowski, J., Forster, M., Lerf, A., 1998. A new structural model for graphite oxide. Chem. Phys. Lett. 287, 53–56.

Hofmann, U., Holst, R., 1939. Ber. Dtsch. Chem. Ges. 72, 754.

Hummers, W.S., Offeman, R.E., 1958. Preparation of graphitic oxide. J. Am. Chem. Soc. 80, 1339.

Lerf, A., He, H., Riedl, T., Forster, M., Klinowski, J., 1997. ^{13}C and ^{1}H MAS NMR studies of graphite oxide and its chemically modified derivatives. Solid State Ionics 101–103, 857–862.

Lerf, A., He, H., Forster, M., Klinowski, J., 1998. Structure of graphite oxide revisited. J. Phys. Chem. B 102, 4477.

Li, D., Muller, M.B., Gilje, S., Kaner, R.B., Wallace, G.G., 2008. Processable aqueous dispersions of graphene nanosheets. Nat. Nanotechnol. 3, 101–105.

McAllister, M.J., Li, J.-L., Adamson, D.H., Schniepp, H.C., Abdala, A.A., Liu, J., Herrera-Alonso, M., Milius, D.L., Car, R., Prudhomme, R.K., Aksay, I.A., 2007. Single sheet functionalized graphene by oxidation and thermal expansion of graphite. Chem. Mater. 19, 4396–4404.

Nakajima, T., Mabuchi, A., Hagiwara, R., 1988. Carbon 26, 357.

Paredes, J.I., Villar-Rodil, S., Martinez-Alonso, A., Tascon, J.M.D., 2008. Graphene oxide dispersions in organic solvents. Langmuir 24, 10560–10564.

Ruess, G., 1946. Monatsch. Chem. 76, 381.

Schniepp, H.C., et al., 2006. Functionalized single graphene sheets derived from splitting graphite oxide. J. Phys. Chem. B 110, 8535–8539.

Scholz, W., Boehm, H.-P., 1969. Z. Anorg. Allg. Chem. 369, 327.

Si, Y., Samulski, E.T., 2008. Synthesis of water soluble graphene. Nano Lett. 8, 1679–1682.

Stankovich, S., Dikin, D.A., Dommett, G.H.B., Kohlhass, K.M., Zimney, E.J., Stach, E.A., Piner, R.D., S-B, T., Nguyen Ruoff, R.S., 2006. Graphene-based composite materials. Nature 442, 282–286.

Stankovich, S., Piner, R.D., Chen, X., Wu, N., Nguyen, S.T., Ruoff, R.S., 2006. Stable aqueous dispersions of graphitic nanoplatelets via the reduction of exfoliated graphite oxide in the presence of poly(sodium 4-styrenesulfonate). J. Mater. Chem. 16, 155–158.

Stankovich, S., et al., 2007. Synthesis of graphene-based nanosheets via chemical reduction of exfoliated graphite oxide. Carbon 45, 1558–1565.

Staudenmaier, L., 1898. Verfahren zur Darstellung der Graphitsaure. Ber. Deut. Chem. Ges. 31, 1481.

Szabo, T., Berkesi, O., Forgo, P., Josepovits, K., Sanakis, Y., Petridis, D., Dekany, I., 2006. Evolution of surface functional groups in a series of progressively oxidized graphite oxides. Chem. Mater. 18, 2740–2749.

Tung, V.C., Allen, M.J., Yang, Y., Kaner, R.B., 2008. High-throughput solution processing of large-scale graphene. Nat. Nanotechnol. 4, 25–29.

Williams, G., Serger, B., Kamat, P.V., 2008. TiO_2-graphene nanocomposites. UV-assisted photocatalytic reduction of graphene oxide. ACS Nano 2, 1487–1491.

Zhou, M., Wang, Y., Zhai, Y., Zhai, J., Ren, W., Wang, F., Dong, S., 2009. Controlled synthesis of large-area and patterned electochemically reduced graphene oxide films. Chem.–Eur. J. 15, 6116–6120.

Chapter 4.4

Bottom-up Synthesis of Graphene From Molecular Precursors

Alicja Bachmatiuk and Mark H. Rümmeli

IFW Dresden, Germany

4.4.1. INTRODUCTION

The bottom-up approaches for graphene synthesis include chemical vapor deposition and molecular based bottom-up routes. CVD methods to prepare graphene are discussed in the following sections (Sections 4.5 and 4.6). Here we focus on the implementation of molecules as building blocks for graphene. Of the various paths to accomplish this, the most developed is the solution-chemistry approach, however exciting chemothermal routes are now also emerging. For the most part, these routes are not suited for large-area graphene, however they do provide an attractive means to prepare graphene nano-ribbons and nano-dots (sometimes called nanoflakes) in large quantities which are well-defined and identical. The ability to produce these structures with tailored morphology is exciting because edge states in such structures significantly influence their properties. In the case of graphene nano-ribbons they can behave as metals, semiconductors, half-metals, feromagnets and antiferomagnets depending on their edge structures, chemical termination and width. Graphene nanoribbons with sufficiently reduced lengths such that confinement effects are now relevant in this direction (length) are essentially 0D structures (quantum dots). Their electronic structure can change from having discrete energy levels to a band-like structure as they increase in dimension. Moreover, apart from edge states they can also have corner states (see Figure 4.4.1 for some examples) that provide a further degree of engineering freedom.

4.4.2. SOLUTION-BASED APPROACHES

Since graphene is in effect part of the polycyclic aromatic hydrocarbon (PAH) class of molecules it makes sense that, hypothetically, it should be possible to synthesise small PAHs and ultimately small area graphene using solution based chemistry in which small molecule aromatic hydrocarbons are brought together

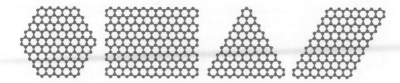

FIGURE 4.4.1 Schematic showing some corner states that could exist for graphene quantum dots. In addition, each of the edges could be different, for example, have zigzag or armchair termination.

to form larger structures through various coupling reactions. This differs from CVD reactions in which entirely new bonding arrangements are established. While solution based chemistry have much to offer, they suffer from a practical drawback, namely, the reduced solubility of large polycyclic systems makes solution-based methods for larger graphene domains challenging (Dreyer et al., 2010). In addition, the large number of chemical bonds that need to form necessitates efficient coupling reactions in order to achieve reasonable yields. Furthermore, solubilisation strategies need to be established to stabilise the graphene nanostructures so as to prevent aggregation. That said, the atomic precision afforded by solution-chemistry is superior to other routes (Yan, 2011).

The early synthesis routes (as far back as 1959) required rather harsh conditions that led to low production yields (Clar et al., 1959; Hendel et al., 1986; Kovacic and Jones, 1987). More recently less aggressive approaches have been developed. Typically well-designed polyphenylene dendric precursors are synthesised using stepwise chemical routes. Later, these structures are fused together in an oxidative environment to form PAHs. Examples of oxidative environments are $FeCl_3$ (Müller et al., 1998) and $CuCl_2$ (Stabel et al., 1995). Hyatt (1991) successfully synthesised hexa-peri-hex-abenzocoronene, which contains 42 conjugated carbon atoms. This molecule and many of its derivatives have become one of the best studied systems to form discotic[1] liquid crystalline phases (Fechtenkötter et al., 1999, 2001; Herwig et al., 1996; Ito et al., 2000; Kübel et al., 2000). Further work demonstrated that graphene flakes with up to 150 and 222 conjugated carbon atoms as well as ribbons reaching 12 nm in length could be synthesised (Simpson et al., 2002; Tomović et al., 2004; Wu, 2004).

4.4.3. SOLUBILISATION STRATEGIES

Often aliphatic side-chains are laterally attached to the edge of PAHs moieties to prevent aggregation, see for example the works of Wu et al. (2007) and

---

1. Flat-shaped liquid molecules can orient themselves in a layer-like fashion. This is known as the discotic nematic phase.

Sakamoto et al. (2009). The aim of this tactic is to overcome the inter-graphene attraction by the affinity established between the flexible aliphatic side-chains and the solvent. Nonetheless as one increases the number of conjugated carbon atoms the aliphatic side-chain strategy becomes less effective since the van der Waals energy, that leads to the aggregations, scales with the area. Since the area depends on the square of the diameter and the chain number scales with the diameter (circumference) it is easy to see that the use of flexible aliphatic side-chains can only be effective with small structures. Yan et al. (2010a) developed a novel stabilisation strategy for large graphene quantum dots to overcome this difficulty. In this technique, they created a three-dimensional cage around the graphene moieties by covalently attaching multiple $2',4',6'$-trialkyl-substituted phenyl moieties to the edges of the graphene moieties. This cage system is presented in Figure 4.4.2a and b. In essence the crowding at the graphene edges pushes the peripheral phenyl groups to twist from the graphene basal plane. This results in the alkyl chains at $2',6'$ positions extending out-of-plane while the chain at the $4'$ position extends laterally, providing a buffer between the graphene sheets and massively reducing agglomeration. Yan (2010) and Yan et al. (2010b) exploited the technique further to demonstrate the synthesis of highly soluble, large graphene nanoflakes with uniform and tunable sizes. The designed flakes contained 132, 168 and 170 conjugated carbon atoms as shown in Figure 4.4.2c. The developed graphene quantum dots are easily solubilised in common organic solvents, however, dynamic light scattering studies revealed the structures are dispersed in the form of reversible oligomers, which indicates the presence of

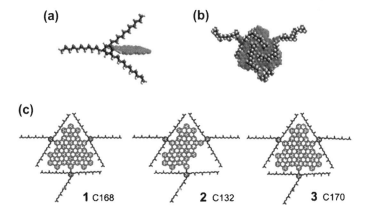

FIGURE 4.4.2   A new solubilisation strategy for graphene QDs and examples of colloidal graphene QDs. (a) A trialkyl-substituted phenyl moiety is attached to the edges of graphene. (b) An energy-minimised geometry of graphene QD 1 (C168) shows a 'cage' structure which effectively prevents the irreversible aggregation (Yan et al., 2010a). (c) Structures of three colloidal graphene QDs that have good solubility in common organic solvents (Yan et al., 2010b).

some residual inter-layer graphene attraction in solution, despite the presence of the solubilisation functionalities.

In addition to the controlled engineering of these structures with well-defined properties, they are also useful systems to understand coupling processes involving carbon materials, for example, the catalytic oxidative dehydrogenation of hydrocarbons. For graphene nano-flakes synthesiszed in this fashion to be fully developed there is still a need for new routes to create structures with larger areas as well as more efficient purification and structural characterisation routes.

4.4.4. SOLVOTHERMAL SYNTHESIS AND SONICATION

The bottom up synthesis of single and few layer graphene by solvothermal routes is attractive as the technique is generally simple and can lead to larger quantities of material as for example, solution based routes. However the technique does not lead to flakes of the same structure as found with solution based routes.

Kuang et al. (2004) developed a low temperature solvothermal route for production of crumpled carbon nanosheets (few layer graphene nanostructures). To achieve this tetrachloromethane was placed in an autoclave. Metallic potassium was then added to the autoclave in a nitrogen atmosphere. The autoclave was then sealed and rapidly heated to the desired temperature (60–100 °C) which was maintained for 10 h before being cooled to room temperature. The product was then filtered out and then thoroughly washed with toluene, then absolute ethanol and finally water. As a last step, the material was dried in vacuum for 4 h at a temperature of 60 °C. X-ray diffraction and Raman spectroscopic data indicated the material was graphitic in nature. Microscopy studies showed the morphology of the as-prepared material consisted of nanosheets interlaced like a cluster of crumpled paper. The thickness of the few-layer graphene nanosheets was <10 nm. Large specific area and BET measurements showed the material to have specific surface areas upto $97.2 \text{ m}^2 \text{ g}^{-1}$. This is tens of times more than graphite and significantly higher than exfoliated graphene.

The chemical route for the production of these structures occurs through the following reaction: first the tetrachloromethane is first reduced by the metallic potassium to diclorocarbon and forms free –c=c– units via a series of dechloronating and bonding processes. The free –c=c– units assemble into two-dimensional hexagonal clusters and then form a graphitic sheet and finally into potassium intercalated graphitic layered structures.

Shen and Feng (2008) reported a higher temperature (800 °C) solvothermal route in which ferrocene and carbon di-sulfide were reacted in a closed chamber for 10 h. The process leads to the formation of flower-like carbon nanosheet aggregations and is suitable for large scale production. The nanosheet thicknesses ranged between 5 and 10 nm and exhibited a specific surface area up to $93.6 \text{ m}^2 \text{ g}^{-1}$.

Choucair et al. (2009) developed a solvothermal route that leads to the formation of high-purity monolayer graphene sheets on the gram scale. Typically the synthesis consisted of heating a 1:1 molar ratio of sodium and ethanol in a sealed reaction vessel for 72 h at a temperature of 220 °C. This yields the solvethermal product in solid form, viz. the graphene precursor. The solvothermal product is then rapidly pyrolysed and the remaining material is then washed with deionised water. The suspended solid (graphene) is then vacuum filtered and then finally dried in a vacuum oven at 100 °C for 24 h. Postsynthesis characterisations revealed the nature of the final product. Transmission electron microscopy showed relatively large graphene sheets that have a tendency to coalesce into overlapped regions. Raman spectroscopic investigations showed the G and D modes to be of similar intensity suggesting defective graphene. AFM showed that the heights measured between the surface of the sheets and substrate to be consistently 4 ± 1 Å, confirming the sheets are one atomic layer thick. Selected area electron diffraction, no interplanar reflexes were observed in the diffraction pattern indicating no interplanar correlations. The observable reflexes were diffuse and of constant intensity indicating the sample consists of free sheets that may not lie wholly perpendicular to the incident electron beam (Meyer et al., 2007). The bulk conductivity of the material was found to be around 0.05 S m^{-1}, which is consistent with highly conductive sheets having multiple interfaces rather than amorphous carbon. X-ray photoemission spectroscopy and elemental analysis by combustion showed the presence of oxygen and hydrogen in the sample, which was attributed to adsorbed water in the samples.

4.4.5. CHEMO-THERMAL BASED APPROACHES

Zhang et al. (2009) developed a graphene production strategy based on a confined self-assembly approach within lamellar mesostructured silica. The technique is advantageous because the entire preparation process takes place under mild conditions and no hazardous chemicals are required. Moreover the synthesis approach is controllable and can produce graphene on a gram scale.

The overall production steps, in short; a lysine-based surfactant containing a terminal pyrrole moiety (PyC$_{12}$Lys) served as both the structure directing agent and the carbon source since polypyrrole is an ideal precursor to form graphite. The lamellar mesostructured silica was prepapred in a basic medium by hydrothermal synthesis using tetraethoxysilane (TEOS) as the silica source, PyC$_{12}$Lys. The actual synthesis consists of first dissolving PyC$_{12}$Lys in deionised water with sodium hydroxide. TEOS is then added to the solution dropwise at room temperature. This mixture is then stirred at room temperature for 2 h and then transferred into an autoclave where it is heated for three days at a temperature of 85 °C. After the heating step (hydrothermal treatment) the mixture is filtered leaving a white powder which is then washed with millipore water and acetone. The sample is then dried in air at room temperature. After this the polypyrrole thin films within a mesostructured silica framework is prepared by dispersing the as-synthesised lamellar mesoporous silica in chloroform through vigorous stirring at room temperature. After this step, a mix of chloroform, $FeCl_3$ and ethanol are added to the mixture to convert the densely packed pyrrole moieties within the silica into conjugated polypyrrole nanosheets. After stirring for 1 h, ammonium persulfate is added to the mixture to induce further poly-merisation of the pyrrole moieties. The mixture is filtrated which leaves a black solid that is then thoroughly washed with water and then dried in air at room temperature. The final steps then lead to graphene within confined silica pores. This is done by pyrolyzing the polypyrrole nanosheets in a nitrogen atmosphere at a temperature of 800 °C for 6 h. The silica framework is then easily dissolved away using aqueous hydrofluoric acid. The residue finally rinsed with Millipore water and acetone and dried in air. Transmission electron microscopy showed the graphene sheets have lateral dimensions of hundreds of nanometres. Selected area electron diffraction confirmed the crystalline structure of the sheets to be that of single layer graphene since the (1100) reflexes were more intense than the (2100) reflexes. AFM revealed an average height of 0.57 nm for the sheets which is concomitant with single layer sheets. The relative intensity of the D mode to the G mode obtained from Raman spectroscopic investigations was about 0.76. Studies by Jang and Yoon (2003) suggest the D/G mode ratio can be improved by increasing the pyrolysis temperature.

A team from Switzerland devised a facile route to form atomically precise graphene nanoribbons of different topologies and widths by taking advantage of surface–assisted coupling of molecular precursors to linear polyphenylenes, which can then be cyclodehydrogenated to yield graphene nanoribbons (Cai et al., 2010). The topology and width and edge periphery of the graphene nanoribbons can be tailored depending on structure of the precursor monomers. To highlight this, the researchers showed various ribbon designs on different surfaces. Their first example formed an armchair ribbon of width $N = 7$ and was derived from 10,10′-dibromo-9,9′-bianthryl precursor monomers, shown in Fig. 4.4.3. Initially the precursor monomers are deposted in an Au(111) surface by sublimation. This process removes their halogen substituents, leaving behind the molecular building blocks for the graphene nanoribbon. Next, intermolec-ular colligation through radical addition is achieved by thermal activation at 200 °C which allows the dehalogebated intermediates to diffuse along the surface and form single covalent C–C bonds between each monomer to yield polymer chains. The successive anthracene units forming the chain are tilted with respect to each other due to steric hinderance (i.e. blocked access to a reactive site by nearby groups) between hydrogen atoms of adjacent anthra-cene units. Indeed, STM investigations highlight the deviation from planarity. To then yield the graphene nanoribbon the sample is annealed at 400 °C to induce cyclodehydrogenation of the polymer chain. Figure 4.4.3 shows the different synthesis steps to achieve atomically precise graphene nanoribbons. Scanning

Precursor monomer   'Biradical' intermediate

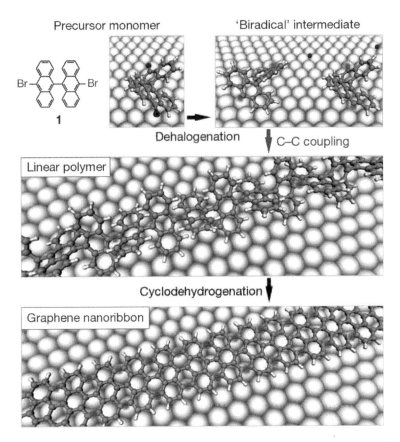

FIGURE 4.4.3 Bottom-up fabrication of atomically precise GNRs. Basic steps for surface-supported GNR synthesis, illustrated with a ball-and-stick model of the example of 10,10′-dibromo-9,9′-bianthryl monomers. (1) Grey, carbon; white, hydrogen; red, halogens; underlying surface atoms shown by large spheres. Top, dehalogenation during adsorption of the di-halogen functionalised precursor monomers. Middle, formation of linear polymers by covalent interlinking of the dehalogenated intermediates. Bottom, formation of fully aromatic GNRs by cyclo-dehydrogenation. *From Cai et al. (2010).*

tunnelling microscopy confirmed the reaction products were atomically precise $N = 7$ graphene nanoribbons with fully hydrogen-terminated armchair edges. The Raman spectroscopic data, as expected, showed the G and D modes. In addition various other peaks are present and are due to the finite width and low symmetry of the ribbons and their presence and spectral positions are in good agreement with calculations (Vandescuren et al., 2008).

The team also showed more complete structures could be fabricated by changing the functionality pattern of the precursor monomers. For example by using 6,11-dibromo-1,2,3,4-tetraphenyltriphenylene precursor monomers they could obtain chevron-type graphene nanoribbons with alternating widths of $N = 6$ and $N = 9$ and a periodicity of 1.7 nm. These chevron-type ribbons had pure armchair edge structure. X-ray photoemission spectroscopy showed the C1s core level to consist of a single sharp component at 284.5 eV binding energy characteristic of sp^2-bonded carbon. There were no signs of carbon in different bonding environments, for example, C–O, C=O and COOH components, which indicate that the samples are chemically pure and inert under ambient conditions. The technique can also yield tailored nanoribbon structures on Ag(111) surfaces, however there is still a major obstacle that this method needs to overcome for technological use, namely, the technique needs to be applicable on semiconducting surfaces or there needs to a be a facile route to transfer the nanoribbons from one surface to another (Sealy, 2010). Another team conducted theoretical investigations on the cooperative effects controlling the synthesis of graphene ribbons by this method on Au(111) surfaces, starting from an anthracene polymer using density functional calculations that include van der Waals interactions (Björk et al., 2011). They showed the Au(111) surface catalyses the cyclodehydrogenation in which H atoms are first pulled onto the surface. This is followed by a separate desorption into the vacuum. This reaction was show to have only a single-energy barrier, while other reactions pathways have an intermediate minimum followed by a high-energy barrier. The process leaves behind a C–C bond. The dehydrogenation was shown to occur successively from between anthracene units starting from one and through to the other end in a step by step propogation through the polymer, much like a domino-like manner.

4.4.6. SELF-ASSEMBLY OF GRAPHENE OXIDE NANOSHEETS

Graphene oxide is usually prepared using chemical exfoliation routes (see Section 4.3). A drawback for these routes is the need for strong oxidising agents and often the lateral size of the flakes are limited. Tang et al. (2012) presented a bottom-up approach for the self assembly of graphene oxide nanosheets with tunable thicknesses ranging from ca. 1 nm (monolayer) to 1500 nm. For the monolayer nanosheets, the lateral widths were around 20 μm while for the few layer material average widths were ca. 100 μm. To prepare the material glucose solutions in deionised water with concentrations ranging from 0.075 to 0.8 M were placed in a Teflon-lined autoclave and heated. For monolayer graphene a glucose solution of 0.5 M was heated for 70 min at a temperature of 160 °C, while for few layer samples upto 1500 nm in thickness, the temperature was increased to 180 °C and the growth time raised to 660 min. Their experiments suggest that most carbohydrates which contain C, H and O with ratios close to 1:2:1 may be used as carbon sources to prepare the graphene oxide nanosheets so long as the hydrogen and oxygen exist in hydroxyl, carboxyl or carbonyl form. They showed this by using sugar and fructose as carbon sources which successfully grew graphene oxide nanosheets, while the use of monosodium glutamate and paraffin did not yield any graphene oxide nanostructures. The chemistry involved in the synthesis of these graphene oxide nanosheets is as follows: under the applied hydrothermal conditions, at the interface of gas and liquid, the glucose molecules undergo cyclic polymerisation to form graphene oxide nanosheets. In the process, H_2O is released. Since the hydroxyl groups have been reduced the nanographene oxide is hydrophobic and so floats on the surface of the solution, in the meantime, glucose molecules diffuse from the solution to the interface due to the concentration gradient. At the interface, the glucose molecules undergo cyclic-polymerisation so that in essence the monolayer graphene oxide on the surface serves as a substrate for a new layer to form and provides a neat way to control the layer number. X-ray photoemission spectroscopy showed the presence of sp^2 carbon with $-OH$, $-CHO$, $-CH$, and various oxygenated groups like $C-O-C$, $C=O$ an $O=C-O-C$. The electrical resistivity of the material can be tuned by eight orders of magnitude by annealing (to chemically reduce the material), ranging from $10^6 \, \Omega$ cm to $10^{-2} \, \Omega$ cm. The team demonstrated the optoelectrical potential of the material by fabricating a photodetector using the graphene oxide material.

REFERENCES

Björk, J., Stafström, S., Hanke, F., 2011. Zipping up: cooperativity drives the synthesis of graphene nanoribbons. J. Am. Chem. Soc. 133 (38), 14884–14887.

Cai, J., Ruffieux, P., Jaafar, R., Bieri, M., Braun, T., Blankenburg, S., Muoth, M., Seitsonen, A.P., Saleh, M., Feng, X., Müllen, K., Fasel, R., 2010. Atomically precise bottom-up fabrication of graphene nanoribbons. Nature 466, 470–473.

Choucair, M., Thordarson, P., Stride, J.A., 2009. Gram-scale production of graphene based on solvothermal synthesis and sonication. Nat. Nanotechnol. 4 (1), 30–33.

Clar, E., Ironside, C.T., Zander, M., 1959. The electronic interaction between benzenoid rings in condensed aromatic hydrocarbons. 1: 12-2: 3-4: 5-6: 7-8: 9-10: 11-hexabenzocoronene, 1: 2-3: 4-5: 6-10: 11-tetrabenzoanthanthrene, and 4: 5-6: 7-11: 12-13: 14-tetrabenzoperopyrene. J. Chem. Soc., 142–147.

Dreyer, R., Ruoff, R.S., Bielawski, C.W., 2010. From conception to realization: an historial account of graphene and some perspectives for its future. Angew. Chem. Int. Ed. 49 (49), 9336–9344.

Fechtenkötter, A., Saalwächter, K., Harbison, M.A., Müllen, K., Spiess, H.W., 1999. Highly ordered columnar structures from hexa-peri-hexabenzocoronenes-synthesis, X-ray diffraction, and solid-state heteronuclear multiple-quantum nmr investigations. Angew. Chem. Int. Ed. 38 (20), 3039–3042.

Fechtenkötter, A., Tchebotareva, N., Watson, M., Müllen, K., 2001. Discotic liquid crystalline hexabenzocoronenes carrying chiral and racemic branched alkyl chains: supramolecular engineering and improved synthetic methods. Tetrahedron 57 (17), 3769–3783.

Hendel, W., Khan, Z.H., Schmidt, W., 1986. Hexa-peri-benzocoronene, a candidate for the origin of the diffuse interstellar visible absorption bands? Tetrahedron 42 (4), 1127–1134.

Herwig, P., Kayser, C.W., Müllen, K., Spiess, H.W., 1996. Columnar mesophases of alky-lated hexa-peri-hexabenzocoronenes with remarkably large phase widths. Adv. Mater. 8 (6), 510–513.

Hyatt, J.A., 1991. Synthesis of a hexaalkynylhexaphenylbenzene. Org. Prep. Proced. Int. 23 (4), 460–463.

Ito, S., Wehmeier, M., Brand, J.D., Kübel, C., Epsch, R., Rabe, J.P., Müllen, K., 2000. Synthesis and self-assembly of functionalized hexa-peri-hexabenzocoronenes. Chem.–Eur. J. 6 (23), 4327–4342.

Jang, J., Yoon, H., 2003. Fabrication of magnetic carbon nanotubes using a metal-impregnated polymer precursor. Adv. Mater. 15 (24), 2088–2091.

Kovacic, P., Jones, M.B., 1987. Dehydro coupling of aromatic nuclei by catalyst-oxidant systems: poly(p-phenylene). Chem. Rev. 87 (2), 357–379.

Kuang, Q., Xie, S.-Y., Jiang, Z.-Y., Zhang, X.-H., Xie, Z.-X., Huang, R.-B., Zheng, L.-S., 2004. Low temperature solvothermal synthesis of crumpled carbon nanosheets. Carbon 42 (8–9), 1737–1741.

Kübel, C., Eckhardt, K., Enkelmann, V., Wegner, G., Müllen, K., 2000. Synthesis and crystal packing of large polycyclic aromatic hydrocarbons: hexabenzo[bc, ef, hi, kl, no, qr]coronene and dibenzo[fg, ij]phenanthro[9,10,1,2,3-pqrst]pentaphene. J. Mater. Chem. 10 (4), 879–886.

Meyer, J.C., Geim, A.K., Katsnelson, M.I., Novoselov, K.S., Booth, T.J., Roth, S., 2007. The structure of suspended graphene sheets. Nature 446, 60–63.

Müller, M., Kübel, C., Müllen, K., 1998. Giant polycyclic aromatic hydrocarbons. Chem.–Eur. J. 4 (11).

Sakamoto, J., van Heijst, J., Lukin, O., Schlüter, A.D., 2009. Two-dimensional polymers: just a dream of synthetic chemists? Angew. Chem. Int. Ed. 48 (6), 1030–1069.

Sealy, C., 2010. Bottom-up approach to graphene nanoribbons. Nano Today 5 (5), 374–376.

Shen, J.-M., Feng, Y.-T., 2008. Formation of flower-like carbon nanosheet aggregations and their electrochemical application. J. Phys. Chem. C 112 (34), 13114–13120.

Simpson, C.D., Brand, J.D., Berresheim, A.J., Przybilla, L., Räder, H.J., Müllen, K., 2002. Synthesis of a giant 222 carbon graphite sheet. Chem.–Eur. J. 8 (6), 1424–1429.

Stabel, A., Herwig, P., Müllen, K., Rabe, J.P., 1995. Diodelike current–voltage curves for a single molecule–tunneling spectroscopy with submolecular resolution of an alkylated, peri-condensed hexabenzocoronene. Angew. Chem. Int. Ed. 34 (15), 1609–1611.

Tang, L., Li, X., Ji, R., Teng, K.S., Tai, G., Ye, J., Wei, C., Lau, S.P., 2012. Bottom-up synthesis of large-scale graphene oxide nanosheets. J. Mater. Chem. 22 (12), 5676–5683.

Tomović, Ž, Watson, M.D., Müllen, K., 2004. Superphenalene-based columnar liquid crystals. Angew. Chem. Int. Ed. 43 (6), 755–758.

Vandescuren, M., Hermet, P., Meunier, V., Henrard, L., Lambin, Ph., 2008. Theoretical study of the vibrational edge modes in graphene nanoribbons. Phys. Rev. B 78 (19) 195401(8).

Wu, J., Pisula, W., Müllen, K., 2007. Graphenes as potential material for electronics. Chem. Rev. 107 (3), 718–747.

Wu, J., Tomović, Ž, Enkelmann, V., Müllen, K., 2004. From branched hydrocarbon propellers to C3-symmetric graphite disks. J. Org. Chem. 69 (16), 5179–5186.

Yan, X., Cui, X., Li, B., Li, L.-S., 2010a. Large, solution-processable graphene quantum dots as light absorbers for photovoltaics. Nano Lett. 10 (5), 1869–1873.

Yan, X., Cui, X., L.-s., Li, 2010b. Synthesis of large, stable colloidal graphene quantum dots with tunable size. J. Am. Chem. Soc. 132 (17), 5944–5945.

Yan, X., L.-s., Li, 2011. Solution-chemistry approach to graphene nanostructures. J. Mater. Chem. 21, 3295–3300.

Zhang, W., Cui, J., Tao, C.-a., Wu, Y., Li, Z., Ma, L., Wen, Y., Li, G., 2009. A strategy for producing pure single-layer graphene sheets based on a confined self-assembly approach. Angew. Chem. Int. Ed. 48 (32), 5864–5868.

Chemical Vapour Deposition Using Catalytic Metals

Alicja Bachmatiuk and Mark H. Rümmeli

IFW Dresden, Germany

4.5.1. INTRODUCTION

The fabrication of graphene over metals by chemical vapour deposition (CVD) is one of the most popular synthesis routes, if not the most popular. The reasons for this are various and include the potential to scaleup fabrication, the technique is already well established in industrial settings and it is easy to setup in research laboratories amongst other attractive traits. CVD grown graphene over metals can also be established over large areas and this is important for applications, for example transparent conducting electrodes for solar cells, where a contiguous covering of graphene is required. This is fundamentally different to exfoliation routes that result in graphene flakes scattered randomly over a substrate.

The formation of graphitic material over transition metal surfaces is well known for over 50 years (Banerjee et al., 1961). Indeed the notion of combining carbon with other materials followed by dissociation of carbon can be traced back to 1896 (Acheson, 1896; Arsem, 1911). Historically graphitic layers were first formed over Ni by exposure to hydrocarbons or evaporated carbon. More or less at the same time graphite layers forming on the surface of single crystal platinum were found during catalyst experiments (Mattevi et al., 2011). In more recent years there has been strong interest in the use of non-transition metals for CNT growth (Rümmeli et al., 2011b). The current interest in graphene led to renewed interest in these deposition techniques for the controlled synthesis of graphene and has been accomplished on Ni, Co, Ru, Ir, Re, Pt, Pa and Cu among other metals (Mattevi et al., 2011). As we shall see, it is also possible to form graphene using CVD techniques over alloys.

4.5.2. CHEMICAL VAPOUR DEPOSITION (CVD) BASICS

The chemical vapour depositing of sp^2 carbon entails passing a carbon feed-stock over the surface of a catalyst substrate (e.g. transition metal) at elevated temperatures. The catalyst then catalytically decomposes the feedstock to provide a supply of carbon. The catalytic potential of transition metals is well established and is argued to arise from partially filled d orbitals or by the formation of intermediate compounds which can absorb and activate the reacting medium. In essence, the metals provide low energy pathways for reactions by changing oxidation states easily or through the formation of intermediates (Mattevi et al., 2011). Once the feedstock has decomposed and provided a source of carbon, the carbon can be absorbed by the metal and then later precipitate out to form graphene as for example Ni and Co, or if carbon solubility is limited, then sp^2-carbon formation can occur as a surface process as for example is the case for Cu. Although the basic experimental procedure is simple our understanding of the dynamics of carbon deposition and domain growth remain somewhat limited. The CVD parameters that most affect the graphene outcome are cooling rate, carbon exposure time and concentration, flow rate and carbon feedstock (source). In addition, the geometry of the oven reactor can affect the flow and deposition characteristics. Moreover, the role of impurities can also play a negative role in the final growth of the desired graphene. Hence finding the proper balance of all the CVD parameters is for the most part an experimental task and settings may vary considerably from reactor to another.

4.5.3. SUBSTRATE SELECTION

In the early days of graphene research since the first isolation of single layer graphene, Ni was explored as a catalyst for graphene formation in great detail. This was because it was already known to yield high quality graphite (Karu and Beer, 1966) and carbon nanotubes (Helveg et al., 2004). In addition, the lattice mismatch between Ni (111) surfaces and graphene is less than 1%, in other words the graphene is commensurate with the substrate lattice. The same is true for the Co (0001) surface. In these cases the graphene growth is often referred to as epitaxial. In cases where the lattice mismatch is greater than 1% the growth is incommensurate; examples are Pt (111), Pd (111), Ru (111) and Ir (111) (Mattevi et al., 2011).

Intensive early studies using nickel soon revealed a fundamental limitation with this catalyst, namely that single and few layer graphene is obtained over tens of microns and is not homogeneous across the substrate surface. In other words control over the number of layers is limited. This is argued to occur because Ni has a large carbon solubility (0.6 weight % at 1326 °C) (Mattevi et al., 2011). Above 800 °C carbon and nickel form a solid solution. The solubility of carbon decreases below 800 °C so that upon cooling, carbon diffuses out of the Ni. In short, carbon segregation is rapid within Ni grains and heterogeneous at grain boundaries. This means the number of graphene layers that form at grain boundaries exceeds that forming over Ni grains and leads to a variation in the number of graphene layers forming on the surface. To some degree this can be alleviated by using single crystalline Ni, however whilst this is attractive for producing graphene for fundamental studies, it is limited in practical terms for large area and cost. Co also forms a meta-stable carbide phase at higher temperatures so that upon cooling, the carbon segregates out. Fe, on the other hand has a stable Fe_3C phase and so graphite precipitation requires highly controlled cooling steps. The difficulties usually encountered with substrates of high carbon solubility seems to be less problematic with platinum (carbon solubility is 0.9 atomic % at 1000 °C (Siller et al., 1968)). Atmospheric pressure CVD studies show large hexagonal grain single crystal graphene formation formed over both polycrystalline and single crystalline Pt (Gao et al., 2012).

Copper has a low carbon solubility at high temperature (0.008 weight% at 1084 °C) (Oshima and Nagashima, 1997). The interest in copper as a substrate stems from its potential to catalyse various carbon allotropes such as graphite (Ong et al., 1992), diamond (Constant et al., 1997) and carbon nanotubes (Rümmeli et al., 2006). Unlike most substrates with high carbon solubility at elevated temperature, substrates with low carbon solubility allow more facile single graphene formation over large areas, for example on Cu foil. Graphene formation over ruthenium has also been demonstrated. Ru has a carbon solubility in between that of Ni and Cu, and can be termed a substrate with intermediate C solubility at high temperature. With Ru implementing a gradual decrease in temperature enables uniform graphene nucleation and growth (Sutter et al., 2009). Substrates based on CNT work in which noble metals, for example, Ag and Au (Takagi et al., 2006) and poor metals such as In and Pb (Rümmeli et al., 2006) may in the future also show promise for graphene formation, indeed graphene formation over Ag has already been demonstrated (Di et al., 2008).

4.5.4. SUBSTRATE PRE-TREATMENT

Prior to running a CVD reaction it is usual to expose the substrate to a flow of hydrogen at elevated temperature. The temperature, exposure time, flow and H concentration vary according to substrate and research group. The process fulfils several reactions to improve graphene formation. In the first, hydrogen helps reduced the native surface oxide layer that may have formed on the surface. Oxides are argued to reduce catalytic activity, although this point remains somewhat unclear, given that CNTs can be grown from a variety of oxides (Rümmeli et al., 2011b). The treatment also helps clean the surface and crystallise the surface. This is because the annealing process increases grain size and assists in surface rearrangement reducing surface defects (smoother surface) (Rümmeli et al., 2011a). This leads to fewer and more uniform graphene layers. It is also argued that hydrogen annealing helps remove unwanted impurities such as sulfur and phosphor since these species can induce carbon solubility variations in the substrate (Yu et al., 2008).

Other techniques to improve crystallisation are being developed, for example, pressing at high pressure. Wet chemical pre-treatments are also possible – for example, oxide reduction on Cu films can be achieved by submerging in acetic acid (Mattevi et al., 2011). In the case of Ni, optimising the thickness the film by depositing Ni on a support (e.g. Si/SiO$_2$ wafer) can help increase the size and quality of graphene sheets. This is because by controlling the Ni film thickness, one can control the amount of carbon absorbed in the Ni film, this in turn controls the carbon available to precipitate in the CVD reaction and hence control the number of layers formed (Kim et al., 2009).

4.5.5. GRAPHENE OVER NI AND CU

In this subsection Ni and Cu substrates for graphene formation using CVD are discussed separately to other support materials simply because these are the most popular support. Other substrates are discussed in the following section.

Continuous thin graphene films with graphene layer numbers ranging from 1 to 10 can be synthesised over polycrystalline nickel films (deposited on Si/SO₂ supports) at temperatures between 900 and 1000 °C using a highly diluted hydrocarbon feedstock at ambient pressure (Reina et al., 2009). The use of a diluted feedstock helps control the C supply. An alternative means to accomplish this is to use low pressures. Highly crystalline graphene formed over polycrystalline Ni can also be achieved at low pressures ($\sim 10^{-3}$ mbar) using temperatures below 1000 °C (Yu et al., 2008). In this case short growth times and a rapid quenching rate are implemented. This is because the quenching rate is related to the carbon solubility in Ni as well as the kinetics of carbon segregation. In short using very quick cooling rates, carbon atoms can rapidly precipitate to the surface and this leads to defective graphitic structures, on the other hand if cooling rates are too slow, carbon atoms tend to diffuse into the bulk rather than migrate to the surface. Hence optimising these competing processes during cooling enables one to control carbon segregation to yield crystalline graphene. Typically after growing single- or few-layer graphene, wrinkles can be observed on the as grown sheets. The wrinkle formation is not related to grain boundaries but is argued to arise from either defects formed at step edges and defect lines and more commonly due to thermal stress due to differences between the thermal expansion coefficients of graphene and the substrate (Rümmeli et al., 2011a). As mentioned above, to minimise defects and inhomogeneous layer formation due to the presence of grain boundaries in polycrystalline Ni substrates, single crystals can be employed. Zhang et al. (2010) conducted a comparative CVD study at ambient pressure between polycrystalline Ni and single crystalline Ni (111). They found that with polycrystalline Ni the best obtainable single layer graphene coverage using their setup was 72%, whilst with single crystalline Ni they were able to obtain greater than 90% coverage for both single-layer graphene and monolayer graphene. The researchers argued the improved coverage was due to single crystalline Ni having an atomically smooth surface and no grain boundaries. The growth of graphene over single crystalline Ni (111) at low pressure has also been demonstrated and in addition this was achieved at the low temperature of

460 °C (Lahiri et al., 2011). In this study ethylene (contrary to the more usual methane) was used as the feedstock at a pressure around 10^{-5} mbar. Upon exposure to the ethylene a nickel carbide layer forms after which the nucleation and growth of graphene is observed due to carbon segregation.

The great attraction with Cu as compared to Ni is that it has a low carbon solubility making monolayer graphene formation more facile and in addition grain sizes (after annealing) are larger than those found with Ni. In addition Copper has a low reactivity with carbon because of its electron configuration. Copper has a filled 3d electron shell $\{(Ar)3d^{10}4s^1\}$ which along with the half filling $3d^5$, is the most stable configuration as the electron distribution is symmetrical. This keeps electron–electron repulsions minimised and thus in the case of carbon, only weak bonds can be achieved by charge transfer of the empty Cu 4s electrons with antibonding electrons (π electrons) from sp^2 carbon. Thus despite copper's low affinity towards carbon, it is still able to stabilise carbon on its surface through weak bonds. To grow graphene over Cu in CVD, usually, methane is the feedstock of choice. CVD reaction temperatures with methane are usually between 800 and 1050 °C. As with nickel, the CVD reaction can be run at ambient pressure or low pressure. Usually some hydrogen is mixed into the reaction (see subsection below on the role of hydrogen).

In addition a variety of flow conditions can be used. Table 4.5.1, taken from reference (Mattevi et al., 2011) nicely highlights the variety of conditions. The table shows that in general 1, 2 and 3 layer graphene forms over CVD copper. This controllability to form so few layers is notably different to Ni where far more layers tend to form. This is because very little carbon diffuses in copper and so the formation of graphene is predominantly a surface process. Some argue graphene grown over Cu is a self-terminating process such that once a full coverage of graphene has formed no further growth can occur since the catalytic Cu surface is no longer available. However, the fact that various groups obtain 1, 2 and even 3 layer graphene (see Fig. 4.5.1) indicates the mechanism is more complex, and a clearer understanding of the different synthesis conditions to control few-layer graphene formation on Cu is still needed.

Another great advantage of Cu is that it can be grown over huge areas. Bae et al. (2010) demonstrated this beautifully in a roll to roll process for the production of predominantly 30-inch graphene films (see Fig. 4.5.2). The films were found to have a sheet resistance of around 150 Ω/square with 97.4% optical transmittance and exhibit half-integer quantum hall effect which highlights the high quality of the material making it suitable for transparent electrode fabrication on an industrial scale.

4.5.6. EARLY GROWTH

A variety of studies have focused on the early stages of graphene growth so as to better comprehend the nucleation and growth processes. An early and

TABLE 4.5.1 Summary of CVD Conditions Reported in the Literature

Pre-annealing	Cu thickness	C Precursor	Growth Pressure (mbar)	Temp. (°C)	Time (min)	Cooling rate	Graphene layer No.	Ref.
1. Ar (20 sccm, 0.55 mbar 12 min). 2. H₂ (20 sccm, 0.4 mbar 1.25 min) up to 766 °C	206 nm	C_2H_2	0.52	800	10	–	1,2,3	(Lee and Lee 2010)
900 °C 30 min H₂ 13.33 mbar	50 μm	CH_4	66.7	850–900	10	10 °C/s	Few layers	(Cai et al., 2009)
(pre-vacuum 0.01) Ar/H₂ 400 sccm 11–12 mbar up to 950 °C	25 μm	Hexane	0.667	950	4	–	1,2	(Srivastava et al., 2010)
(pre-vacuum) Heating up to 1000 °C H₂ (13 sccm, 0.1333 mbar) 30 min	25 μm 125 μm	CH_4	0.4	1000	0.5–30	9 °C/min	1,2	(Mattevi et al., 2011)
1000 °C H₂ (2 sccm, 0.053 mbar) 30 min	25 μm	CH_4	0.667	1000	1–60	40–300 °C/ min	>95% 1	(Li et al., 2009b)

(pre-vacuum) Heating in H₂ up to 1000 °C	100–450 nm	CH₄	0.133 –0.667	1000	15–420	—	1	(Ismach et al., 2010)
Heating up to 1000 °C H₂ (8 sccm, 0.24 mbar) 30 min	25 μm	CH₄	2.133	1000	30	10 °C/s	1	(Bae et al., 2010)
(base pressure C.133 mbar) Acetic acid + heating up to 1000 °C H₂ (50–200 sccm, 2.67 mbar 40 °C/min)	500 nm Cu foil 25 μm	CH₄	14.7	1000	10–20	20 °C/min	>93% 1	(Levendorf et al., 2009)
Heating up to 1000 °C in ambient pressure then, 30 min, 1000 °C He (1000 sccm) + I₂ (50 sccm)	700 nm	CH₄	1013	1000	5	10 °C/s	1,2	(Lee et al., 2010)
Heating up to 1000 °C H₂ and Ar, 30 min	Cu foil	CH₄	0.667	1000	2–10	Rapidly cool	Few layers	(Robertson and Warner, 2011)
(pre-vacuum) Heating in H₂ up to 1000 °C	25 μm foil	CH₄	2	850–1050	15	Rapidly cool	1,2,3	(Rümmeli, 2012)

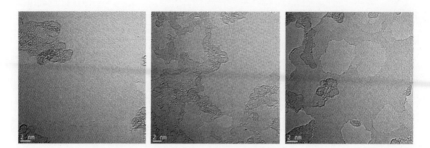

FIGURE 4.5.1 TEM micrographs showing single layer graphene, bi-layer and tri-layer graphene (left to right) grown over Cu using low-pressure (~2 mbar) CVD with methane as the feedstock. The 2 and 3 layer images show Moiré patterns due to rotational stacking faults between graphene sheets. Images provided courtesy of A. Bachmatiuk, S. Gorantla and M. H. Rümmeli – IFW Dresden, Germany.

important work was that by Li et al. (2009a). In this study they investigated graphene formation over both Ni and Cu. The study involved the use of carbon isotope labelling in the precursors as a means to track the mechanism and kinetics of the graphene CVD growth. Bond force constants change between ^{12}C and ^{13}C (cf. changing a mass on a spring) which leads to different frequencies between Raman modes for the two extremes and varying $^{12}C:^{13}C$ ratios in-between. By changing the isotope fraction during the reaction and combining this with Raman spectroscopy one can track the behaviour of the carbon species during growth. In the case of Ni, the carbon species would be expected to diffuse into the bulk and mix. This should lead to a random mix of ^{12}C and ^{13}C atoms that precipitate out of the Ni to form graphene on the surface. Spatial Raman mapping showed an even random spread of the two carbon isotopes confirming the C-diffusion process. In the case of Cu, spatial Raman spectroscopy mapping showed the ^{12}C and ^{13}C spatial distribution to follow the isotopic precursor time sequence. This shows growth on Cu occurs predominantly through surface adsorption. Linear growth rates were shown ranging from 1 to 6 μm/min depending on the Cu grain orientation.

In terms of the nucleation stages, Wofford et al. (2010) observed four-lobed, 4-fold symmetric islands to nucleate and grow. They found that each of the graphene lobes had a different crystallographic alignment with respect to the underlying Cu substrate. They argued the polycrystalline islands occur through complex heterogeneous nucleation processes at surface imperfections. The shape of the lobes was explained as being due to angularly dependant growth velocities. These studies were conducted at low pressure. Robertson and Warner (2011) investigated early growth of graphene on copper foils at atmospheric pressure with a high methane flow and high hydrogen:methane ratio. They found hexagonal-shaped single-crystal domains of few-layer

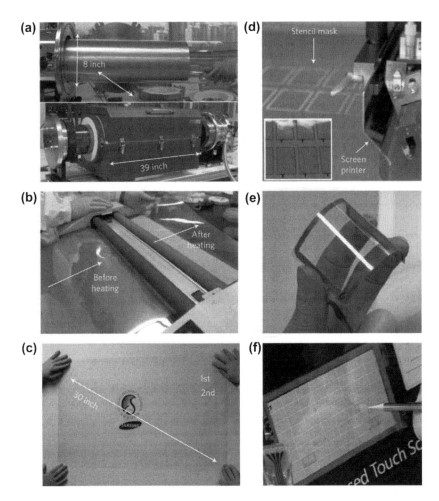

FIGURE 4.5.2 Photographs of the roll-based production of graphene films. (a) Copper foil wrapping around a 7.5-inch quartz tube to be inserted into an 8-inch quartz reactor. The lower image shows the stage in which the copper foil reacts with CH_4 and H_2 gases at high temperatures. (b) Roll-to-roll transfer of graphene films from a thermal release tape to a PET film at 120 °C. (c) A transparent ultralarge-area graphene film transferred on a 35-inch PET sheet. (d) Screen printing process of silver paste electrodes on graphene/PET film. The inset shows 3.1-inch graphene/PET panels patterned with silver electrodes before assembly. (e) An assembled graphene/PET touch panel showing outstanding flexibility. (f) A graphene-based touch-screen panel connected to a computer with control software. For a movie of its operation, see Supplementary Information. *After ref. Bae et al. (2010).*

graphene form. The hexagonal islands are randomly oriented on the copper foil, but sites of graphene nucleation were found to show some correlation in that they formed in linear rows. TEM showed the islands to consist of 5–10 layers in the central region which thin out toward the edges of the domain. Selected area

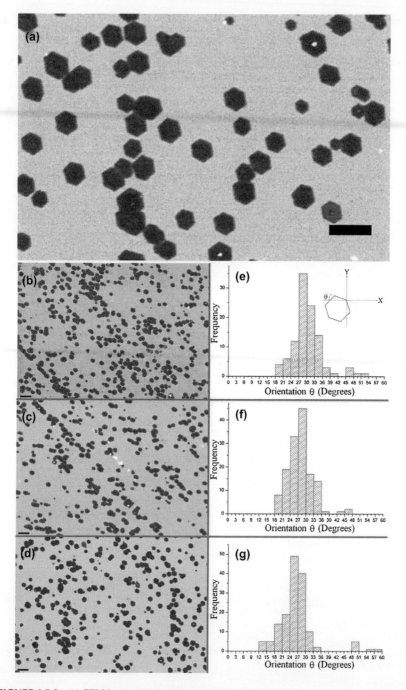

FIGURE 4.5.3 (a) SEM image showing alignment of the edges of few-layer graphene hexagons grown at 990 °C on Cu with smooth surface structure. (b) SEM image at position x, $y = 41.8$, 64.9 mm, (c) x, $y = 41.7$, 67.4 mm, and (d) x, $y = 45.4$, 69.8 mm. (e)–(g) show the histograms of the orientation of the edge of the hexagons (θ) from the respective SEM images in (b)–(d). Inset in (e) shows definition of (θ). Scale bar indicates 2 μm in all images.

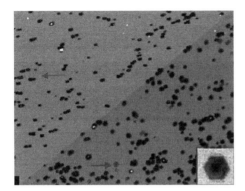

FIGURE 4.5.4 SEM image of a Cu grain boundary (diagonal) with graphene hexagons grown on the bottom Cu area and rectangular graphene domains grown on the top Cu area. Electron backscattered diffraction confirms the change in Cu lattice orientation with change in graphene domain shape.

electron diffraction revealed the layers were AB stacked and free from rotational stacking faults. The work also highlights an absence of self-limited monolayer graphene growth on Cu.

A later work by Li et al. (2011) again using low pressure CVD, but this time with copper-foil enclosures and using methane as the precursor and temperatures of 1035 °C was able to demonstrate large area single domain islands. The enclosure was formed by bending a piece of Cu foil and then crimping the three remaining sides. Graphene grew both on the outside and inside of the Cu enclosures. The graphene forming on the outside consisted of 1–3 layers. However the graphene formed on the inside of the enclosure consisted of single crystal graphene domains. In the early stages the domains grew as hexagons but with continued reaction exposure, the graphene domains grew into very large graphene domains with growing edges, resembling dendrites which upon closer examination were shown to be single-crystal graphene domains. They could grow as large as 0.5 mm. The reason for the different graphene formation inside the Cu foil enclosure was attributed to a lower partial pressure of methane and an improved growth environment in which the Cu vapour is in static equilibrium so that potentially there is a much lower pressure of unwanted species. The mobility of these domains was found to be better than $4000 \, \text{cm}^2\text{V}^{-1}\text{s}^{-1}$.

The work of Wu et al. (2011) showed that if care is taken to prepare a very smooth surface of Cu, an epitaxial relationship between the growing graphene domain and the Cu lattice emerges, shown in Fig. 4.5.3. As shown in Fig. 4.5.4, different Cu grains led to different shaped graphene domains and orientations, indicating that growth rates are influenced by the Cu surface structure. Monolayer graphene can also be produced using Cu in its molten state (Wu et al. 2012). Suprisingly the graphene domains were epitaxially aligned by the underlying Cu crystal and as the aligned domains merged together with longer growth time, large single crystals of monolayer graphene resulted. The method of epitaxial alignment is promising for achieving large single crystals of monolayer graphene in a reasonable time frame. If one calculates the time it would take to grow a $4''$ graphene single crystal from a single seed based on current area growth rates, it would come to nearly 2 years, which is unfeasible.

4.5.7. THE ROLE OF HYDROGEN IN THE CVD REACTION

As discussed previously, hydrogen is often employed for the pretreatment of the catalyst support (cleaning and improved crystallisation) for both Ni and Cu. A fraction of hydrogen along with the feedstock (and argon) is often implemented. It is sometimes argued the hydrogen is an "inactive" species that recombines and desorbs from the catalyst surface. Losurdo et al. (2011) investigated the matter in greater depth and found evidence to link hydrogen with the kinetics and structure of graphene. They found that H_2-dissociative chemisorption competes with CH_4-dissociative dehydrogenation on surface sites. In essence, these competitive processes play an inhibitor role on the kinetics of graphene formation on Cu. Moreover, hydrogen was shown to slow down the deposition kinetics of graphene on Cu by blocking sites on the Cu surface. Hydrogen was also demonstrated to have a negative role on the quality of the graphene because it can create point defects consisting of hybridised sp^3 C–H bonds. In the case of Ni, hydrogen resurfacing and surface combination helps keep sites on the Ni surface free for hydrocarbon dehydrogenation, yielding diffusion/segregation of carbon. Thus, with careful adjustment of hydrogen in the reaction one can grow graphene over Ni with very few defects (no D mode observed in the Raman spectrum).

4.5.8. GRAPHENE–OTHER METALS AND ALLOYS

Aside from Cu and Ni various other metals have been shown to be suitable as catalysts for CVD- based graphene fabrication. For example, graphene has been grown over iridium (111) surfaces at very low pressures (ultrahigh vacuum) (Coraux et al., 2009). Initial annealing of the Ir film led to terrace formations which extend several hundreds of nanometres. The Ir (111) surface was then exposed to ethane which served as the carbon precursor. Graphene formation was observed to initiate at substrate step edges. Moreover the graphene spanned both sides of the step edge with the largest fraction of graphene being found on the lower terrace. With careful temperature optimisation they could obtain coverage up to 100% and moreover the graphene was of high quality. Graphene formation over Ir (111) at low vacuum pressures has also been demonstrated (Coraux et al., 2008). In this study ethylene was used as the feedstock. The obtained graphene sheets were found to lie over the substrate like a blanket and wrinkles were seen over step edges. Sutter et al. (2008) demonstrated one can grow epitaxial graphene on Ru (0001) surfaces. Ru has a relatively high carbon solubility, so like Ni, one has to control the cooling rate to in turn manage the temperature-dependant solubility of interstitial carbon so as to successfully control the number of graphene layers formed. In addition they showed the early stages of growth (nucleation) involves the formation of lens shaped islands. Gao et al. (2012) investigated the use of platinum surfaces for graphene formation using ambient pressure CVD. Early growth consists of large hexagonal single-crystal graphene islands. With extended growth times these grains join up to form polycrystalline graphene. This same process remained true for both polycrystalline Pt and single-crystalline Pt substrates. Interestingly, despite Pt having a large carbon solubility (ca. 0.9 at % at 1000 °C), apparently, it is relatively easy to grow monolayer graphene. Moreover, as compared to graphene formed over Cu, growth rates are not only higher, but the growth window (with respect to synthesis parameters) is relatively large. Indeed, single-layer graphene can be obtained at temperatures as low as 750 °C with CH_4 serving as the carbon precursor. Both temperature and feedstock concentration (relative to H_2 concentration) affect the nucleation density and grain size of the resultant graphene. With careful adjustment of these parameters, they could obtain individual grain with a perfect hexagonal shape with smooth edges with lateral dimensions up to 1.3 mm on a polycrystalline Pt foil. With small CH_4/H_2 flow rates, small bilayer or few-layer regions can be observed. It is argued that this occurs because the amount of active carbon available exceeds that required for monolayer formation. The high quality of the graphene formed using this technique is reflected in mobility experiments (on SiO_2/Si) showed extracted mobilities of around 7100 $cm^2 V^{-1} s^{-1}$.

Dai et al. (2011) developed an interesting strategy to overcome difficulties associated with uncontrollable carbon precipitation as found, for example, with Ni. The technique involves the careful selection of a binary metal system that suppresses carbon precipitation and activates a self-limited growth mechanism for homogeneous monolayer graphene. The technique was first demonstrated using a Ni–Mo combination. In short, a thin Ni layer is deposited onto a Mo foil and after an initial pre-treatment in hydrogen the substrate is exposed to a $CH_4/H_2/Ar$ mixture. This leads to strictly single layer graphene forming on the surface. The basic argument of the technique is that molybdenum traps excess carbon species since molybdenum carbide is very stable. X-ray photoelectron spectroscopy (XPS) shows direct evidence for molybdenum carbide (predominately Mo_2C) in the Ni–Mo system. Thus, although graphene formation on Mo is difficult (at best), the synergistic combination of Mo and Ni works extremely well for uniform single-layer graphene fabrication. An exciting aspect of the technique is that growth window (in terms of the synthesis parameters) in which homogeneous single-graphene forms is huge as compared to other CVD routes. This combined with the fact that only single layer graphene is obtained makes it an attractive technique suitable for scaling up.

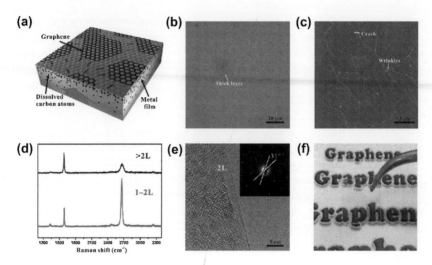

FIGURE 4.5.5 Segregation of graphene from bulk metal. (a) Schematic illustration of segregation technique, which demonstrates the surface accumulation of buried carbon atoms in bulk metal and formation of graphene at high temperature. (b–f) Characteristics of a graphene film segregated from polycrystalline Ni. (b) Optical microscope image of segregated graphene (shortly, s-graphene) transferred on 300 nm SiO$_2$/Si substrate. (c) Tapping mode AFM height image showing the existence of wrinkles and crack in monolayer or bilayer regions. (d) Raman spectra of 1–2 L (red) and >2 L (black) graphene. The excitation wavelength was 632.8 nm. (e) TEM image and SAED pattern revealing the nice crystallinity and disordered A–B stacking structure of bilayer s-graphene. (f) Photograph of graphene film transferred to quartz substrate (Liu et al., 2011).

4.5.9. SEGREGATION ROUTES

Carbon segregation routes are an alternative to CVD methods and are based on carbon precipitation from metals in which the carbon has already been supplied to the system before exposing the metal to elevated temperatures (see Fig. 4.5.5). Hence no C vapour needs to be supplied and thus helps avoid adsorption of extraneous organic molecules that arise during the catalytic decomposition of hydrocarbons (Amini et al., 2010; Liu et al., 2011). Amini et al. (2010) used solid graphite as a carbon source that dissolved into molten nickel at 1500 °C over 16 h at a pressure of ca. 10^{-6} mbar. Upon cooling, the dissolved carbon atoms segregate from the Ni and form graphene. Controlling the cooling rates allows one to form single-layer graphene. Another group explored the use of highly oriented pyrolitic graphite (HOPG) as the carbon source, again with Ni as the metal. Under optimised conditions in high vacuum and annealing at a lower temperature of 650 °C for 18 h, carbon atoms from the HOPG diffuse through the Ni and segregate on the free surface to form graphene (Xu et al., 2011). Liu et al. (2011) advanced the technique to yield wafer size graphene. They used commercially available metals that contained carbon impurities that had been evaporated directly onto Si/SiO_2 wafers. Few layer graphene (1 to 3 layers) were formed on the surface by annealing at 1100 °C in a vacuum around 10^{-3} Pa. During the annealing step, the embedded carbon diffuses to the surface, where it segregates to form graphene. Later Zhang et al. (2011) developed a similar route to fabricate nitrogen-doped graphene.

REFERENCES

Acheson E.G., 1896 United States Patent 568, 323.

Amini, S., Garay, J., Liu, G., Balandin, A.A., Abbaschian, R., 2010. Growth of large-area graphene films from metal-carbon melts. J. Appl. Phys. 108 (094321), 1–7.

Arsem, W.C., 1911. Transformation of other forms of carbon into graphite. Ind. Eng. Chem. 3, 799–804.

Bae, S., Kim, H., Lee, Y., Xu, X., Park, J.-S., Zheng, Y., Balakrishnan, J., Lei, T., Kim, H.R., Il Song, Y., Kim, Y.-J., Kim, K.S., Özyilmaz, B., Ahn, J.-H., Hong, B.H., Iijima, S., 2010. Roll-to-roll production of 30-inch graphene films for transparent electrodes. Nat. Nanotechnol. 5, 574–578.

Banerjee, B.C., Hirt, T.J., Walker, P.L., 1961. Pyrolytic carbon formation from carbon suboxide. Nature 192, 450–451.

Cai, W., Zhu, Y., Li, X., Piner, R.D., Ruoff, R.S., 2009. Large area few-layer graphene/graphite films as transparent thin conducting electrodes. Appl. Phys. Lett. 95 (123115), 1–3.

Constant, L., Speisser, C., Normand, F.L., 1997. HFCVD diamond growth on Cu(111). Evidence for carbon phase transformations by in situ AES and XPS. Surf. Sci. 387, 28–43.

Coraux, J., N'Diaye, A.T., Busse, C., Michely, T., 2008. Structural coherency of graphene on Ir(111). Nano Lett. 8, 565–570.

Coraux, J., N'Diaye, A.T., Engler, M., Busse, C., Wall, D., Buckanie, N., Heringdorf, F.-J.M.Z., Gastel, R.V., Poelsema, B., Michely, T., 2009. Growth of graphene on Ir(111). New J. Phys. 11 (039801).

Dai, B., Fu, L., Zou, Z., Wang, M., Xu, H., Wang, S., Liu, Z., 2011. Rational design of a binary metal alloy for chemical vapour deposition growth of uniform single-layer graphene. Nat. Commun. 2, 522–527.

Di, C.A., Wei, D., Yu, G., Liu, Y., Guo, Y., Zhu, D., 2008. Patterned graphene as source/drain electrodes for bottom-contact organic field-effect transistors. Adv. Mater. 20, 3289–3293.

Gao, L., Ren, W., Xu, H., Jin, L., Wang, Z., Ma, T., Ma, L.-P., Zhang, Z., Fu, Q., Peng, L.-M., Bao, X., Cheng, H.-M., 2012. Repeated growth and bubbling transfer of graphene with millimetre-size single-crystal grains using platinum. Nat. Commun. 3 (699), 1–7.

Helveg, S., Lopez-Cartes, C., Sehested, J., Hansen, P.L., Clausen, B.S., Rostrup-Nielsen, J.R., Abild-Pedersen, F., Norskov, J.K., 2004. Atomic-scale imaging of carbon nanofibre growth. Nature 427, 426–429.

Ismach, A., Druzgalski, C., Penwell, S., Schwartzberg, A., Zheng, M., Javey, A., Bokor, J., Zhang, Y., 2010. Direct chemical vapor deposition of graphene on dielectric surfaces. Nano Lett. 10, 1542–1548.

Karu, A.E., Beer, M.J., 1966. Pyrolytic formation of highly crystalline graphite films. J. Appl. Phys. 37, 2179–21781.

Kim, K.S., Zhao, Y., Jang, H., Lee, S.Y., Kim, J.M., Kim, K.S., Ahn, J.-H., Kim, P., Choi, J.-Y., Hong, B.H., 2009. Large-scale pattern growth of graphene films for stretchable transparent electrodes. Nature 457, 706–710.

Lahiri, J., Miller, T., Adamska, L., Oleynik, I.I., Batzill, M., 2011. Graphene growth on Ni(111) by transformation of a surface carbide. Nano Lett. 11, 518–522.

Lee, Y.-H., Lee, J.-H., 2010. Scalable growth of free-standing graphene wafers with copper(Cu) catalyst on SiO2/Si substrate: thermal conductivity of the wafers. Appl. Phys. Lett. 96 (083101), 1–3.

Lee, Y., Bae, S., Jang, H., Jang, S., Zhu, S.-E., Sim, S.H., Song, Y.I., Hong, B.H., Ahn, J.-H., 2010. Wafer-scale synthesis and transfer of graphene films. Nano Lett. 10, 490–493.

Levendorf, M.P., Ruiz-Vargas, C.S., Garg, S., Park, J., 2009. Transfer-free batch fabrication of single layer graphene transistors. Nano Lett. 9, 4479–4483.

Li, X., Cai, W., Colombo, L., Ruoff, R.S., 2009a. Evolution of graphene growth on Ni and Cu by carbon isotope labeling. Nano Lett. 9, 4268–4272.

Li, X., Cai, W., An, J., Kim, S., Nah, J., Yang, D., Piner, R., Velamakanni, A., Jung, I., Tutuc, E., Banerjee, S.K., Colombo, L., Ruoff, R.S., 2009b. Large-area synthesis of high-quality and uniform graphene films on copper foils. Science 324, 1312–1314.

Li, X., Magnuson, C.W., Venugopal, A., Tromp, R.M., Hannon, J.B., Vogel, E.M., Colombo, L., Ruoff, R.S., 2011. Large-area graphene single crystals grown by low-pressure chemical vapor deposition of methane on copper. J. Am. Chem. Soc. 133, 2816–2819. MC15.

Liu, N., Fu, L., Dai, B., Yan, K., Liu, X., Zhao, R., Zhang, Y., Liu, Z., 2011. Universal segregation growth approach to wafer-size graphene from non-noble metals. Nano Lett. 11, 297–303.

Losurdo, M., Giangregorio, M.M., Capezzuto, P., Bruno, G., 2011. Graphene CVD growth on copper and nickel: role of hydrogen in kinetics and structure. Phys. Chem. Chem. Phys. 13, 20836–20843.

Mattevi, C., Kim, H., Chhowalla, M., 2011. A review of chemical vapour deposition of graphene on copper. J. Mater. Chem. 21, 3324–3334.

Ong, T.P., Xiong, F., Chang, R.P.H., White, C.W.J., 1992. Nucleation and growth of diamond on carbon-implanted single crystal copper surfaces. J. Mater. Res. 7, 2429–2439.

Oshima, C., Nagashima, A., 1997. Ultra-thin epitaxial films of graphite and hexagonal boron nitride on solid surfaces. J. Phys.: Condens. Matter. 9, 1–20.

Reina, A., Jia, X., Ho, J., Nezich, D., Son, H., Bulovic, V., Dresselhaus, M.S., Kong, J., 2009. Large area, few-layer graphene films on arbitrary substrates by chemical vapor deposition. Nano Lett. 9, 30–35.

Robertson, A.W., Warner, J.H., 2011. Hexagonal single crystal domains of few-layer graphene on copper foils. Nano Lett. 11, 1182–1189.

Rümmeli, M.H., Grüneis, A., Löffler, M., Jost, O., Schönfelder, R., Kramberger, C., Grimm, D., Gemming, T., Barreiro, A., Borowiak-Palen, E., Kalbác, M., Ayala, P., Hübers, H.-W., Büchner, B., Pichler, T., 2006. Novel catalysts for low temperature synthesis of single wall carbon nanotubes. Phys. Status Solidi B 243, 3101–3105.

Rümmeli, M.H., Rocha, C.G., Ortmann, F., Ibrahim, I., Sevincli, H., Boerrnert, F., Kunstmann, J., Bachmatiuk, A., Poetschke, M., Shiraishi, M., Meyyappan, M., Buechner, B., Roche, S., Cuniberti, G., 2011a. Graphene: piecing it together. Adv. Mater. 23, 4471–4490.

Rümmeli, M.H., Bachmatiuk, A., Boerrnert, F., Schäffel, F., Ibrahim, I., Cendrowski, K., Simha-Martynkova, G., Placha, D., Borowiak-Palen, E., Cuniberti, G., Buechner, B., 2011b. Synthesis of carbon nanotubes with and without catalyst particles. Nanoscale Res. Lett. 6 (303), 1–9.

Siller, R.H., Oates, W.A., McLennan, R.B., 1968. The solubility of carbon in palladium and platinum. J. Less-Common Met. 16, 71–73.

Srivastava, A., Galande, C., Ci, L., Song, L., Rai, C., Jariwala, D., Kelly, K.F., Ajayan, P.M., 2010. Novel liquid precursor-based facile synthesis of large-area continuous, single, and few-layer graphene films. Chem. Mater. 22, 3457–3461.

Sutter, P.W., Flege, J.-I., Sutter, E.A., 2008. Epitaxial graphene on ruthenium. Nat. Mater. 7, 406–411.

Sutter, E., Albrecht, P., Sutter, P., 2009. Graphene growth on polycrystalline Ru thin films. Appl. Phys. Lett. 95 (133109), 1–3.

Takagi, D., Homma, Y., Hibino, H., Suzuki, S., Kobayashi, Y., 2006. Single-walled carbon nanotube growth from highly activated metal nanoparticles. Nano Lett. 6, 2642–2645.

Wofford, J.M., Nie, S., McCarty, K.F., Bartelt, N.C., Dubon, O.D., 2010. Graphene islands on Cu foils: the interplay between shape, orientation, and defects. Nano Lett. 10, 4890–4896.

Wu, Y.A., Robertson, A.W., Schäffel, F., Speller, S.C., Warner, J.H., 2011. Aligned rectangular few-layer graphene domains on copper surfaces. Chem. Mater. 23, 4543–4547.

Wu, Y.A., Fan, Y., Speller, S., Creeth, G.L., Sadowski, J.T., He, K., Robertson, A.W., Allen, C.S., Warner, J.H., 2012. Large single crystals of graphene on melted copper using chemical vapour deposition. ACS Nano 6, 5010–5017.

Xu, M., Fujita, D., Sagisaka, K., Watanabe, E., Hanagata, N., 2011. Production of extended single-layer graphene. ACS Nano 5, 1522–1528.

Yu, Q., Lian, J., Siriponglert, S., Li, H., Chen, Y.P., Pei, S.-S., 2008. Graphene segregated on Ni surfaces and transferred to insulators. Appl. Phys. Lett. 93 (113103), 1–3.

Zhang, Y., Gomez, L., Ishikawa, F.N., Madaria, A., Ryu, K., Wang, C., Badmaev, A., Zhou, C., 2010. Comparison of graphene growth on single-crystalline and polycrystalline Ni by chemical vapor deposition. J. Phys. Chem. Lett. 1, 3101–3107.

Zhang, C., Fu, L., Liu, N., Liu, M., Wang, Y., Liu, Z., 2011. Synthesis of nitrogen-doped graphene using embedded carbon and nitrogen sources. Adv. Mater. 23, 1020–1024.

Chapter 4.6

CVD Synthesis of Graphene Over Nonmetals

Alicja Bachmatiuk and Mark H. Rümmeli

IFW Dresden, Germany

4.6.1. INTRODUCTION

The use of non-metal catalysts to form graphene over using a CVD process has certain attractions. For example, when using graphene as channel material in field effect transistors, it needs to reside over a nonmetal to be isolated from the gate. Growth of the graphene directly on the insulator material, e.g. an oxide, removes the need for processing steps, such as transfer, which can damage the graphene. In addition, there is a drive to achieve the direct synthesis of graphene on non-metals at temperatures below 500 °C as temperatures above this would likely damage other devices being prepared on the substrate (microchip fabrication). However, the challenges to grow high quality graphene using CVD over nonmetal surfaces are by no means trivial. Moreover, because of the complexities, less research is conducted on this technique which further slows progress. Nonetheless, progress is being made and much of it comes on the back of work done on the growth of CNTs using non-metallic catalysts (Rümmeli et al., 2011a). Uchino et al. (2005a,b) first showed the potential of semiconducting catalyst particles for the growth of CNT using carbon doped SiGe islands on Si after exposure to various oxidation and annealing treatments. Later, Tagaki showed CNT could be grown directly from pure Ge or pure Si nanoparticles. Alumina nanoparticles have also been shown to be suitable to grow sp^2 CNTs through an alcohol CVD route (Liu et al., 2008). More recently various groups showed CNT can be grown from SiO_2 nanoparticles (Bachmatiuk et al., 2009,2010; Huang et al., 2009; Liu et al., 2008), however, the question as to whether the SiO_2 (or any oxide) is reduced in the reaction is controversial. In addition, it has been shown that oxide surfaces used as supports for nanoparticles in CNT growth can themselves be involved with the growth of the sp^2 carbon nanostructures (Rümmeli et al., 2007a, 2010a).

4.6.2. ASPECTS TO CONSIDER WITH NONMETAL CATALYSTS

The role and stability of the nonmetal catalysts/substrates for graphene formation in CVD are poorly understood. The question of the chemical stability of the catalyst during the reaction is often not even addressed. In the case of SiO_x, it is not clear whether silica remains in the oxide phase or reduces to a carbide phase. With the use of SiO_x for CNT production, Bachmatiuk et al. have shown it can undergo a carbo-thermal reduction process to SiC. The determination of whether an oxide remains in the same phase is often difficult to determine through usual routes. X-ray photoemission spectroscopy is often used as means to determine the phase, however the technique is highly surface sensitive and is not always reliable. Diffraction methods can also be tricky. For example, in the case of Al_2O_3 (sapphire), it reduces to a carbide phase through intermediate oxycarbides and it is difficult to distinguish the diffraction reflexes between these intermediate oxycarbide compounds and sapphire as they are very similar. It is also more challenging to remove graphene off non-metals so as to characterise the graphene material, say in TEM. In addition, in some catalytic reactions, e.g. metal oxides, the catalyst can undergo both a reduction and re-oxidation simultaneously by the loss and gain of surface lattice oxygen. The complexity continues in that many oxide surfaces are prone hydroxylation, moreover, many chemical reactions at the surface of catalysts are transient states which occur during the catalytic cycle and are difficult to probe experimentally. In short, the catalytic activity of nonmetals is complex and less understood as compared to metals, nonetheless they do have catalytic potential with regard to graphene formation.

4.6.3. NON-METALS AS CATALYSTS FOR CVD GROWN GRAPHENE

An early work in 2007 exploring the potential of oxides for graphitic formation in CVD showed a variety of oxides could be used including, SiO_2, Al_2O_3, MgO, Ga_2O_3 and ZrO_2 (Rümmeli et al., 2007b). Two feedstock systems were investigated, namely methane and ethanol at a temperature of 850 °C. The work also highlighted that some oxides were reduced in the process. However, the study focused mostly on the use of oxide nanoparticles to illustrate the graphitisation process due to the ease with which such samples can be investigated in TEM. In terms of large area metal-free graphene growth, a notable early work is that Hwang et al. (2010), where they grew graphitic films over Al_2O_3 and SiC by CVD at temperatures between 1350 and 1650 °C. Propane diluted in argon served as the carbon source. The number of layers formed could be varied from a few to several tens of layers – be either adjusting the propane flow rate or growth time. Raman spectroscopy showed the I_d/I_g ratio improved with increasing temperature, viz. higher degrees of crystallisation are obtained with higher reaction temperatures. Synchrotron based X-ray studies confirmed that the graphitic layers are oriented parallel to the substrate surface and that their in-plane orientations were only weakly constrained to the substrate. They also observed that the layers were predominantly rhombohedral stacked (ABC/ACB). This is in contrast to natural bulk graphite which is about 80% Bernal stacked (AB) and 14% Rhombohedral. The basis for Rhombohedral stacking was not clear however the authors suggest this may depend on growth conditions (e.g. temperature and pressure). A more detailed analysis on graphene film formation over sapphire using metal-free CVD (Fanton et al., 2011). They were able to optimise the reaction conditions to yield monolayer graphene. They did this by first implementing thermochemical modelling to determine appropriate boundary conditions for C deposition and to determine the thermodynamically stable reaction products. The modelling which took into account Al, O, C, H and Ar suggested no covalent bonding between the graphene film and substrate. Moreover it predicts that two undesirable effects may take place during the CVD of graphene over sapphire. The first effect is no solid carbon will be available if the reaction to form gaseous CO proceeds at a rate significantly faster than the rate at which CH_4 can decompose to solid carbon. The second effect is that C (gas or solid) may etch the surface of the sapphire. This latter point is not desirable for semiconductor device processing. Their experimental studies confirmed that the surface roughness did deteriorate with increasing C in the growth environment suggesting C does etch the surface. Raman studies corroborated the findings by Hwang, that the I_d/I_g ratio (degree of crystallisation) improves with increasing temperature. Moreover, the

Raman data confirmed most of the film consisted of monolayer graphene and that little strain remains on the graphene film after cool down suggesting minimal interaction between the sapphire and graphene film, however some wrinkling was observed on the film indicating some degree of coupling. Core level XPS investigations showed no altered sapphire phases like Al–C or Al–O–C bonds suggesting a stable surface and no interfacial (buffer) layer. This also suggests chemical bonding between the substrate and film does not exist in agreement with the initial thermodynamic calculations. Carrier mobility studies revealed the quality of graphene grown in this manner is comparable with graphene found on SIC (0001) grown by SiC decomposition (see Section 4.7). One of the disadvantages of the discussed graphene growth over sapphire is that it seems to require very high CVD reaction temperatures. However, one can grow nanographene flakes at significantly lower temperatures, as will be discussed later. At the intermediate temperatures of 900 to 1100 °C it does seem possible to grow large-area graphene films of reasonable quality. In this temperature, as with higher temperatures, the primary decomposition mechanism of the carbon feedstock can be expected to predominantly arise from thermal cracking, viz. a noncatalytic process. Sun et al. (2011, 2012a) showed that one can form graphene-like films by CVD at 1000 °C, using methane as the feedstock. The technique was demonstrated on silicon nitride (deposited on Si substrates), quartz and sapphire. Raman spectroscopic measurements showed I_d/I_g ratios around 1, suggesting rather poor crystallisation. Indeed, TEM studies indicate the samples are comprised primarily of small monolayer crystallites around 10 nm in size. The thickness of the graphene-like films can be controlled by modifying the growth conditions. In essence, the self-assembly of the graphene crystallites form the resultant film. Similar graphene films were demonstrated by another group on SiO_2, BN, AlN and Si substrates at slightly higher temperatures (1100–1200 °C) (Sun et al., 2012b). Although the electrical characteristics from such films are not suitable for transistor fabrication they are potentially attractive for transparent electrodes which is attractive for photovoltaic applications.

Chen et al. (2011) improved the process and were able obtain polycrystalline graphene on silicon dioxide (Si/SiO$_2$ wafers or quartz). What the investigators did was to introduce a pre-CVD reaction step in which the substrate is annealed in flowing air prior to the CVD reaction. They argue this step removes any organic residue and activates growth sites. After the pretreatment step a conventional CVD reaction is applied using a feed gas mix of CH$_4$:H$_2$:Ar (14:50:65) and a temperature of 1100 °C with a reaction time of 3 h. Systematic studies investigating the different growth stages show initially small graphene islands form which grow and eventually merge to form a polycrystalline graphene layer. The technique also affords a layer number control from single-layer graphene to few-layer graphene, depending on the selected carbon flow, reaction temperature or growth time (see Fig. 4.6.1 panels a through e). Low-angle (30°) XPS studies showed no SiC for samples

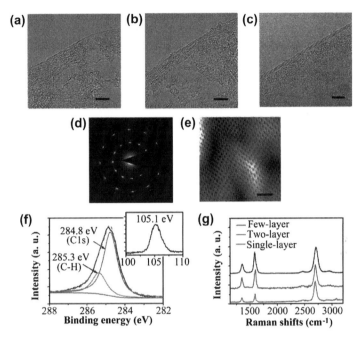

FIGURE 4.6.1   Analyses of graphene films produced by oxygen-aided CVD. (a–c) HRTEM images of (a) monolayer, (b) two-layer, and (c) three-layer graphene. Scale bars = 5 nm. (d) SAED pattern of graphene, showing two sets of hexagonal patterns. (e) STM image (It = 0.5 nA, Vs = 25 mV) of the graphene honeycomb lattice. Scale bar = 1 nm. (f) C 1s and (inset) Si 2p XPS spectra of a graphene film. (g) Raman spectra (514-nm laser wavelength) of a graphene film. *(Image from Chen et al. (2011))*

produced on Si/SiO$_2$ substrates, while on quartz substrates, SiC signatures were observed. This indicates that a carbothermal reduction process in which SiO$_2$ is reduced to SiC, as first highlighted by Bachmatiuk et al. (2009), may take place in the reaction. More in-depth studies are required to better elucidate this point. To evaluate the quality of the graphene films, field effect transistors were fabricated on Si/SiO$_2$ substrates. On–off ratios of around three were demonstrated. Moreover, carrier mobilities of ca. 500 cm^{2}V^{-1}s^{-1} were obtained indicating there is still plenty of room for improvement for graphene formed over SiO$_2$ using the above described CVD route. One can also form graphene at lower temperatures, for example, graphene growth of hafnia nanoparticles has been demonstrated at 900 °C (Kidambi et al., 2011) and magnesia nanoparticles at 700 °C (Rümmeli et al., 2009); however, when considering CVD growth within a microelectronic setting temperatures of at least below 500 °C so as to maintain the mechanical integrity of other devices residing on the wafer. Current research suggests one can form graphene at temperatures below 500 °C.

Sub-500 °C CVD of nanographene was first demonstrated by Rümmeli et al. (2010b). In this work, both cyclohexane and acetylene/argon mixtures were used to form graphene over MgO nanoparticles. With acetylene/argon mixtures, they were able to grow graphene at temperatures as low as 325 °C. They noted that the number of graphene layers that encapsulated the MgO nanoparticles would reach an upper limit of 10 layers regardless of reaction time – in other words, there seemed to be some limiting process. In addition, TEM investigations revealed the layers aligned themselves with the magnesia crystal lattice fringes as if to anchor or bond to the crystal. They suggested this might indicate carbon incorporation to the lattice occurs predominantly at the graphene/nanoparticle interface. Subsequent studies by the same group involved density functional theory investigations to help elucidate this point, namely, was there catalytic activity at the step sites of MgO? (since at the low temperatures used, (<400 °C) thermal decomposition of the carbon feedstock is highly unlikely (Scott et al., 2011)). The study suggested step sites provided regions for catalytic activity in which four distinct subprocesses could be identified. First, carbon feedstock molecules are adsorbed on the substrate surface followed by hydrogen dissociation with carbon remaining on the surface (surface diffusion) and finally the addition of carbon atoms adding into the graphene network (growth). The process is illustrated in Fig. 4.6.2 (Rümmeli et al., 2011b). The technique was nicely developed further by another group which demonstrated one can directly grow large-area flat

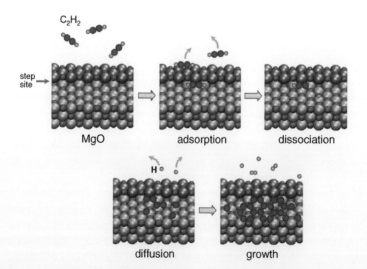

FIGURE 4.6.2 Proposed growth processes for graphene at step sites on oxides. C_2H_2 is preferentially adsorbed at step sites, where it dissociates, thereafter H diffuses away and finally C addition leads to graphene growth. (Blue = carbon, gray = hydrogen, red = oxygen, green/gold = Mg). *(From Rümmeli et al. (2011b)).*

nano-graphene films over silica at 400 °C (Medina et al., 2012). The domain sizes of the graphene were estimated between 3 and 5 nm in a continuous film. Both Raman spectroscopy and XPS data confirmed SiC formed between the graphene film and the SiO_2 substrate, indicating the carbothermal reduction of silica is relevant.

4.6.4. METAL-ASSISTED ROUTES

Aside from the metal free routes there exist a few modified CVD techniques in which metals are employed to assist in the growth of graphene over dielectrics. Probably, the most well known route is that in which a Cu sacrificial layer dewets and evaporates during synthesis (Ismach et al., 2010). In other words the graphene growth occurs over the copper, which then evaporates away. In this sense the technique is only a pseudo direct route to grow graphene over a dielectric since the substrate does not play a role in growth. Similarly, graphene films can be grown in the vicinity of an edge of a lithographic pattern on a SiO_2 substrate (Suzuki et al., 2010). Teng et al. (2012) developed an intriguing route in which a Cu catalyst assists remotely from the substrate such that the graphene growth truly takes place on an oxide substrate. In essence, what they did was to place a Cu foil ring in the upstream of the CVD reactor. At sufficiently high temperatures, Cu atoms sublime off the Cu foil surface and mix with the reaction gases. The Cu atoms assist in the catalytic decomposition of the hydrocarbon providing a rich source of C species to deposit and graphitise on oxide substrates a few centimetres downstream. The technique allows for the fabrication of relatively low defect graphene. Moreover within the detectable resolution limit, no Cu was found on the samples. Interestingly, XPS studies found the presence of a SiC peak when using a SiO_2 substrate. They found the SiC peak appears at the beginning of growth which later was hard to discern as the graphene grew. This highlights the difficulty of using XPS to unequivocally determine that SiO_2 has not reduced to a carbide. Transport studies to investigate the quality of the graphene showed mobilities in the range 100–600 $cm^2V^{-1}s^{-1}$ which is comparable to that found with graphene grown on Cu foils with small domain sizes (Huang et al., 2011).

4.6.5. NON-METALS AS CATALYSTS FOR CARBON NANOWALL FABRICATION (VERTICAL GRAPHENE)

The fabrication of 2D carbon nanostructures forming so-called carbon nanowalls or nanosheets has a history dating back to 1997, which is long before conventional CVD routes for graphene began (Hiramatsu et al., 2010). As a consequence, carbon nanowall synthesis routes are rather well developed. A common aspect of their growth is that they all require a plasma based CVD route. That said a disparate plasma techniques are available and include; Microwave Plasma Enhanced CVD

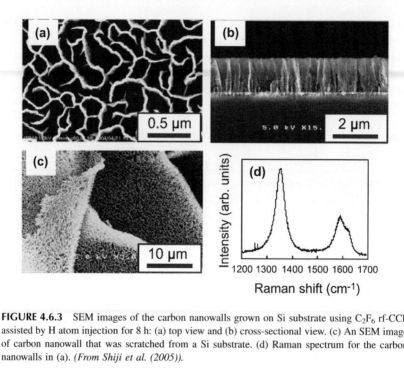

FIGURE 4.6.3 SEM images of the carbon nanowalls grown on Si substrate using C_2F_6 rf-CCP assisted by H atom injection for 8 h: (a) top view and (b) cross-sectional view. (c) An SEM image of carbon nanowall that was scratched from a Si substrate. (d) Raman spectrum for the carbon nanowalls in (a). *(From Shiji et al. (2005)).*

(MPCVD), Inductively Coupled Plasma Enhanced CVD (IC-PECVD), Capacitively Coupled Plasma Enhanced CVD (CC-PECVD), RF Plasma-Enhanced CVD (RF-PECVD), VHF Plasma-Enhaced CVD (VHF-PECVD), Electron Beam Excited Plasma Enhanced CVD (EBE-PECVD), Hot Filament CVD, Atmospheric Pressure Plasma and sputtering.

Carbon nanowalls are two dimensional few layer graphene domains that are typically vertically oriented on a substrate. Although they can be grown on metallic substrates the plasma enhanced routes are particularly apt for carbon nanowall fabrication on nonmetallic substrates and so are relevant to the current discussion (Fig. 4.6.3).

4.6.6. THE BASICS OF PLASMA-ENHANCED CHEMICAL VAPOUR DEPOSITION

Plasma Enhanced Chemical Vapour Deposition (PECVD) is a widely accepted technique within industry, particularly for thin-film production. The primary reason for its acceptance is its capability to operate at lower temperatures than thermally driven CVD. In PECVD a plasma which consists of ionised gas species (ions), electrons and some neutral species in both ground and excited states. The plasmas are usually ignited and sustained by applying a high frequency voltage (microwave frequencies, ultrahigh frequencies or radio frequencies) to a low pressure gas. Atmospheric pressure systems are also available however they are less common as high pressure plasmas are more difficult to sustain. In the plasma inelastic collitions take place between electrons and gas molecules forming reactive species, such as excited neutrals and free radicals, as well as ions and electrons. In essence the electrons acquire suffient energy from the applied electric field to create highly reactive species without significantly raising the gas temperature. PECVD takes advantage of these reactive species to deposit thin films enabling lower temperatures to be used. In the case of carbon nanowalls no catalysts are required however large amounts of H atoms (viz. decomposed H_2 molecules) are required in the reaction, which is similar to the case for diamond growth (Hiramatsu et al., 2010). Carbon nanowalls can be deposited on virtually any substrate including Si, SiO_2, Al_2O_3, Ni and stainless steel, at substrate temperatures of 400–700 °C.

4.6.7. NANOWALL OR NANOSHEET SYNTHESIS

One of the earlieast works demonstrating the potential of plasma enhanced CVD for carbon nanowall synthesis was by Obraztsov et al. (2003). Their system employed a dc discharge PECVD, which successfully grew nanowalls on Si as well as various metal substrates. They used gas pressures between 10 and 150 Torr and a various gas-mixing ratios of CH_4 and H_2. The produced material was, in general, quite thick. The first report of single- to few-layer graphene nanowalls by PECVD emerged in 2004 (Wang et al., 2004a,b). Here a radio frequency PECVD system was implemented and it was shown to successfully yield the vertical single and few layer graphene nanowalls on a variety of metallic substrates as well as Si, SiO_2 and Al_2O_3. Again, gas mixtures of CH_4 and H_2 were used. The pressure was 12 Pa (~0.1 Torr). The simplicity and versatility of the process triggered interest in the community. Shang et al. modified the process using microwaves to generate the PECVD (Malesevic et al., 2008). The technique worked very well on Si substrates yielding highly graphitised knife edge structures. Their findings showed high growth rates of 1.6 µm/min, which is about ten times faster than other processes. They went on to show that the produced material had excellent biosensing properties (dopamine). Detailed quantitative studies on the role of hydrogen atoms or radicals on the growth of the graphene nanowalls showed the morphologies of the nanowalls to be dependant on the type of carbon source and applied power. The main role for H radicals was argued to be an agent for removing undesirable amorphous phases during growth, this in turn allows improved surface morphology and crystalline quality of the nanowalls (Kondo et al., 2008; Shiji et al., 2005). Compared to microwave and radio-frequency PECVD, ultra high frquency PECVD is reported to lead to smaller and denser nanoflakes (Shang et al., 2002) and nanowalls (Dikonimos et al., 2007).

In terms of growth, the mechanisms for graphene or carbon nanowall formation poorly understood. Nonetheless, some ideas exist. One of the more commonly accepted routes is that described by Zhu et al. (2007). In their model, the graphene (nanosheet) initially grows parallel to the substrate with thicknesses of 1–15 nm before the onset of vertical growth. X-ray scattering studies confirm the formation of this initial parallel layer (French et al., 2005). Takeoff into the vertical position may occur due to a build up of the upward curling force at the grain boundaries so as to curl the leading edge of the top layers upward, viz. the electric field in the plasma induces an orientation of the growing planes that is perpendicular to the substrate. Once vertical growth is established the high surface mobility of the carbon bearing species and the induced polarisation of the graphitic layers (due to the electric field sheath layer) combine to encourage the graphene nanosheets to grow upward rather than outward (thicker).

4.6.8. SUBSTRATE-FREE PECVD SYNTHESIS OF GRAPHENE SHEETS

A nice variation to the more usual PECVD synthesis routes was developed by Dato et al. (2008). In this study an atmospheric microwave (2.45 GHz) plasma reactor was used. A running argon plasma was ignited and maintained, and then, ethanol droplets at a rate of 4×10^{-4} L/min were injected directly into the plasma. The residence time of the droplets within the plasma was 10^{-1} s. During the short time the ethanol droplets evaporate and dissociate in the plasma and form solid carbon matter, namely graphene sheets that are collected downstream on filter. Removal of the sheets is achieved by sonicating the filters in methanol. Examination of the graphene sheets showed they were predominantly single and bilayer. Electron energy loss spectroscopy (EELS) showed no detectable traces of hydrogen, oxygen or OH, indicating the sheets consist of pure sp^2 carbon.

4.6.9. GRAPHENE FORMATION FROM SOLID-CARBON SOURCES ON SURFACES

It is also possible to fabricate graphene using solid-carbon sources using a variety of techniques that although not in widespread use, are steadily on the increase. One of the earlier examples was presented in 2010 by the tour group at Rice University (Sun et al., 2010) in which they demonstrated a technique that leads to controlled layer number formation and can also yields doped graphene. In the case of synthesising undoped graphene poly(methyl methacrylate) (PMMA) was spin coated over a Cu film. By annealing the samples at low pressure at temperatures between 800 and 1000 °C in a reductive gas environment (H_2/Ar) for 10 min, they could obtain single-layer graphene. Raman spectroscopy confirmed monolayer graphene and also revealed a very weak D (defect) mode indicating the graphene has relatively few defects. The material's electrical properties were evaluated by fabricating field effect transistor (FET) devices. The estimated carrier mobility was around $410 \ cm^2 \cdot V^{-1} \cdot s^{-1}$ at room temperature and the on/off ratios about 2. Control of the number of graphene layers obtained could be achieved by varying the Ar and H_2 flow rates to yield single-layer, bilayer and few-layer graphene. Their interpretation for this was that hydrogen serves both as a reducing reagent and as a carrier gas to remove C atoms extruded from the PMMA during growth, Thus, lower H_2 flows leaves more carbon for growth and so favours few-layer growth. For doped graphene melamine ($C_3N_6H_6$) was mixed with PMMA and again spin-coated onto a Cu surface. This time the reaction was accomplished at atmospheric pressure to maintain a good nitrogen-atom concentration in the reaction. The N content in the graphene was estimated at 2–3.5%. The same team went on to develop the technique to fabricate bi layer graphene on insulating substrates (Yan et al., 2011). PMMA was used as the feedstock, or for N doped graphene poly (acrylonitrile-co-butadiene-co-styrene) (ABS) was used. To prepare the samples, either the PMMA or ABS was spin-coated on the insulating substrate (SiO_2, h-BN, Si_3N_4 or Al_2O_3) and then a capping layer of Ni was deposited on the surface. The samples were then annealed at low pressure in a reducing environment (Ar/H_2) at a temperature of 1000 °C.

After the reaction the Ni layer is dissolved away leaving bi-layer graphene on the surface. No remnant polymer material is detected. In the reaction, the polymer film decomposed and dissolves into the Ni film. As the sample cools (by removing it from the hot zone) carbon precipitates and forms graphene. Because the Ni is subsaturated with carbon, bilayer graphene is facilitated (Garaj et al., 2010). Increasing the film thickness to conversely increase the C supply leads to multilayer graphene with increased defects. Reducing the C supply leads discontinuous graphene film formation. The concept to decompose the solid carbon based material in the presence of a reducing atmosphere at elevated temperature was pushed further when the Tour team demonstrated they could obtain graphene from food, insects and waste. More specifically the carbon sources used without any prepurification were cookies, chocolate, grass, plastics, roaches and dog feces (Ruan et al., 2011). The detailed chemistry occurring during the graphene growth over the Cu foils from solid carbon feedstocks is limited. A preliminary study drew some insight (Ji et al., 2011). In the study, amorphous carbon was used as the solid carbon feedstock. The investigators confirmed that the inclusion of H_2 in the reaction was necessary to obtain graphene. A mass spectrometer was used to probe the reaction atmosphere during the reaction – CH_4 was shown to be present and forms from H_2 reacting with the amorphous carbon. The data points to gaseous hydrocarbons and/or their intermediates forming the graphene, somewhat like conventional CVD reactions.

The decomposition of of polymers to yield graphene is also possible using hydrogen flame synthesis. The technique has been shown to successfully yield

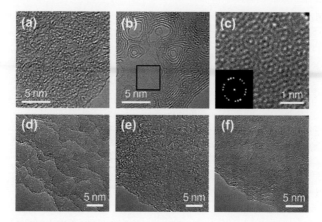

FIGURE 4.6.4 (a) Through (c) The graphitisation of amorphous carbon on graphene (a) to graphitised carbon on graphene is clearly observed by the Moiré patterns (b) and (c). Panel (d) shows the h-BN layer serving as the template substrate. Panel (b) shows the h-BN after amorphous carbon deposition and (c) shows the same region after crystallisation of the carbon on h-BN (Börrnert et al., 2012).

few-layer graphene on h-BN (Lin et al., 2012). h-BN crystallites were dispersed in PMMA and then dried in a vacuum oven at 100 °C for 6 h. The resultant material was then placed in an evacuated quartz tube and then the material was rapidly heated at 1100–1200 °C for 20 s by a hydrogen flame and then rapidly cooled down to room temperature. After the reaction graphene is found on the h-BN, as confirmed by Raman spectroscopic studies. By adjusting the PMMA concentration, one can adjust the ratio of graphene to h-BN. Apparently the graphene layers initially anchor to h-BN ($\bar{1}100$) surface plane, and then, growth continues to form graphene layers. This route, unlike those described previously, required no metal catalyst. The implementation of SiC as a carbon source has also been explored. In this route an amorphous SiC layer sandwiched between a SiO_2 substrate and a Ni capping layer. Upon rapid thermal annealing the SiC dissolves in the Ni. Upon cooling carbon segregates to the surface forming a graphene layer. Wet etching of the Ni allows the graphene to settle on the substrate below (Hofrichter et al., 2010). Similar segregation routes using amorphous carbon or carbon already embedded in the metal are also possible e.g. (Liu et al., 2011).

A rather interesting completely metal free route to fabricate graphene on graphene or h-BN has been demonstrated in a TEM. Here, amorphous carbon on the surface of a graphene or h-BN support crystallises to graphene or few layer graphene upon exposure to the electron beam. Exposing the electron beam to the amorphous carbon leads to radiation induced diffusion (similar to thermal induced diffusion) which allows the carbon to crystallise. Without a graphene of h-BN support, the crystallisation process leads to a carbon onion formation. However with a graphene or h-BN support, the crystallisation process is templated and planar crystallisation occurs, viz. graphene and few-layer graphene domains form as shown in Fig. 4.6.4 (Börrnert et al. 2012).

REFERENCES

Bachmatiuk, A., Börrnert, F., Grobosch, M., Schäffel, F., Wolff, U., Scott, A., Zaka, M., Warner, J.H., Klingeler, R., Knupfer, M., Büchner, B., Rümmeli, M.H., 2009. Investigating the graphitization mechanism of SiO_2 nanoparticles in chemical vapor deposition. ACS Nano 3 (12), 4098–4104.

Bachmatiuk, A., Börrnert, F., Schäffel, F., Zaka, M., Martynkowa, G.S., Placha, D., Schönfelder, R., Costa, P.M.F.J., Ioannides, N., Warner, J.H., Klingeler, R., Büchner, B., Rümmeli, M.H., 2010. The formation of stacked-cup carbon nanotubes using chemical vapor deposition from ethanol over silica. Carbon 48 (11), 3175–3181.

Börrnert, F., Avdoshenko, S.M., Bachmatiuk, A., Ibrahim, I., Büchner, B., Cunniberti, G., Rümmeli, M.H., 2012. Amorphous carbon under 80 kV irradiation: a means to make or break graphene. Adv. Mater.

Chen, J., Wen, Y., Guo, Y., Wu, B., Huang, L., Xue, Y., Geng, D., Wang, D., Yu, G., Liu, Y., 2011. Oxygen-aided synthesis of polycrystalline graphene on silicon dioxide substrates. J. Am. Chem. Soc. 133 (44), 17548–17551.

Dato, A., Radmilovic, V., Lee, Z., Phillips, J., Frenklach, M., 2008. Substrate-free gas-phase synthesis of graphene sheets. Nano Lett. 8 (7), 2012–2016.

Dikonimos, T., Giorgi, L., Giorgi, R., Lisi, N., Salernitano, E., Rossi, R., 2007. DC plasma enhanced growth of oriented carbon nanowall films by HFCVD. Diamond Relat. Mater. 16 (4–7), 1240–1243.

Fanton, M.A., Robinson, J.A., Puls, C., Liu, Y., Hollander, M.J., Weiland, B.E., LaBella, M., Trumbull, K., Kasarda, R., Howsare, C., Stitt, J., Snyder, D.W., 2011. Characterization of graphene films and transistors grown on sapphire by metal-free chemical vapor deposition. ACS Nano 5 (10), 8062–8069.

French, B.L., Wang, J.J., Zhu, M.Y., Holloway, B.C., 2005. Structural characterization of carbon nanosheets via X-ray scattering. J. Appl. Phys. 97 114317(8).

Garaj, S., Hubbard, W., Golovchenko, J.A., 2010. Graphene synthesis by ion implantation. Appl. Phys. Lett. 97 183103(3).

Hiramatsu, M., Hori, M., 2010. Carbon Nanowalls Synthesis and Emerging Applications. Springer Verlag.

Hofrichter, J., Szafranek, B.N., Otto, M., Echtermeyer, T.J., Baus, M., Majerus, A., Geringer, V., Ramsteiner, M., Kurz, H., 2010. Synthesis of graphene on silicon dioxide by a solid carbon source. Nano Lett. 10 (1), 36–42.

Huang, S., Cai, Q., Chen, J., Qian, Y., Zhang, L., 2009. Metal-catalyst-free growth of single-walled carbon nanotubes on substrates. J. Am. Chem. Soc. 131 (6), 2094–2095.

Huang, P.Y., Ruiz-Vargas, C.S., van der Zande, A.M., Whitney, W.S., Levendorf, M.P., Kevek, J.W., Garg, S., Alden, J.S., Hustedt, C.J., Zhu, Y., Park, J., McEuen, P.L., Muller, D.A., 2011. Grains and grain boundaries in single-layer graphene atomic patchwork quilts. Nature 469 (7330), 389–392.

Hwang, J., Shields, V.B., Thomas, C.I., Shivaraman, S., Hao, D., Kim, M., Woll, A.R., Tompa, G.S., Spencer, M.G., 2010. Epitaxial growth of graphitic carbon on C-face

SiC and sapphire by chemical vapor deposition (CVD). J. Cryst. Growth. 312 (21), 3219–3224.

Ismach, A., Druzgalski, C., Penwell, S., Schwartzberg, A., Zheng, M., Javey, A., Bokor, J., Zhang, Y., 2010. Direct chemical vapor deposition of graphene on dielectric surfaces. Nano Lett. 10 (5), 1542–1548.

Ji, H., Hao, Y., Ren, Y., Charlton, M., Lee, W.H., Wu, Q., Li, H., Zhu, Y., Wu, Y., Piner, R., Ruoff, R.S., 2011. Graphene growth using a solid carbon feedstock and hydrogen. ACS Nano 5 (9), 7656–7661.

Kidambi, P.R., Bayer, B.C., Weatherup, R.S., Ochs, R., Ducati, C., Vinga Szabó, D., Hofmann, S., 2011. Hafnia nanoparticles – a model system for graphene growth on a dielectric. Phys. Status Solidi (RRL) 5 (9), 341–343.

Kondo, S., Hori, M., Yamakawa, K., Den, S., Kano, H., Hiramatsu, M., 2008. Highly reliable growth process of carbon nanowalls using radical injection plasma-enhanced chemical vapor deposition. J. Vac. Sci. Technol. B 26, 1294–1300.

Lin, T., Wang, Y., Bi, H., Wan, D., Huang, F., Xie, X., Jiang, M., 2012. Hydrogen flame synthesis of few-layer graphene from a solid carbon source on hexagonal boron nitride. J. Mater. Chem. 22, 2859–2862.

Liu, H., Takagi, D., Ohno, H., Chiashi, S., Chokan, T., Homma, Y., 2008. Growth of single-walled carbon nanotubes from ceramic particles by alcohol chemical vapor deposition. Appl. Phys. Express 1 014001(3).

Liu, X., Fu, L., Liu, N., Gao, T., Zhang, Y., Liao, L., Liu, Z., 2011. Segregation growth of graphene on Cu–Ni alloy for precise layer control. J. Phys. Chem. C 115 (24), 11976–11982.

Malesevic, A., Kemps, R., Zhang, L., Erni, R., Van Tendeloo, G., Vanhulsel, A., Van Haesendonck, C., 2008. A versatile plasma tool for the synthesis of carbon nanotubes and few-layer graphene sheets. J. Optoelect. Adv. Mater. 10 (8), 2052–2055.

Medina, H., Lin., Y.-C., Jin, C., Lu, C.-C., Yeh, C.-H., Huang, K.-P., Suenaga, K., Robertson, J., Chiu, P.-W., 2012. Metal-free growth of nanographene on silicon oxides for transparent conducting applications. Adv. Funct. Mater. 22 (10), 2123–2128.

Obraztsov, A.N., Zolotukhin, A.A., Ustinov, A.O., Volkov, A.P., Svirko, Yu., Jefimovs, K., 2003. DC discharge plasma studies for nanostructured carbon CVD. Diamond Relat. Mater. 12 (3–7), 917–920.

Ruan, G., Sun, Z., Peng, Z.i, Tour, J.M., 2011. Growth of graphene from food, insects, and waste. ACS Nano 5 (9), 7601–7607.

Rümmeli, M.H., Schäffel, F., Kramberger, C., Gemming, T., Bachmatiuk, A., Kalenczuk, R.J., Rellinghaus, B., Büchner, B., Pichler, T., 2007a. Oxide-driven carbon nanotube growth in supported catalyst CVD. J. Am. Chem. Soc. 129 (51), 15772–15773.

Rümmeli, M.H., Kramberger, C., Grüneis, A., Ayala, P., Gemming, T., Büchner, B., Pichler, T., 2007b. On the graphitization nature of oxides for the formation of carbon nanostructures. Chem. Mater. 19 (17), 4105–4107.

Rümmeli, M.H., Schäffel, F., Bachmatiuk, A., Trotter, G., Adebimpe, D., Simha-Martynková, G., Plachá, D., Rellinghaus, B., McCormick, P.G., Borowiak-Palen, E., Ayala, P., Pichler, T., Klingeler, R., Knupfer, M., Büchner, Bernd, 2009. Oxide catalysts for carbon nanotube and few layer graphene formation. Phys. Status Solidi B 246 (11–12), 2530–2533.

Rümmeli, M.H., Schäffel, F., Bachmatiuk, A., Adebimpe, D., Trotter, G., Börrnert, F., Scott, A., Coric, E., Sparing, M., Rellinghaus, B., McCormick, P.G., Cuniberti, G., Knupfer, M., Schultz, L., Büchner, B., 2010a. Investigating the outskirts of Fe and Co catalyst particles in alumina-supported catalytic CVD carbon nanotube growth. ACS Nano 4 (2), 1146–1152.

Rümmeli, M.H., Bachmatiuk, A., Scott, A., Börrnert, F., Warner, J.H., Hoffman, V., Lin, J.-H., Cuniberti, G., Büchner, B., 2010b. Direct low-temperature nanographene CVD synthesis over a dielectric insulator. ACS Nano 4 (7), 4206–4210.

Rümmeli, M.H., Bachmatiuk, A., Börrnert, F., Schäffel, F., Ibrahim, I., Cendrowski, K., Simha-Martynkova, G., Plachá, D., Borowiak-Palen, E., Cuniberti, G., Büchner, B., 2011a. Synthesis of carbon nanotubes with and without catalyst particles. Nanoscale Res. Lett. 6, 303.

Rümmeli, M.H., Rocha, C.G., Ortmann, F., Ibrahim, I., Sevincli, H., Börrnert, F., Kunstmann, J., Bachmatiuk, A., Pötschke, M., Shiraishi, M., Meyyappan, M., Büchner, N., Roche, S., Cuniberti, G., 2011b. Graphene: piecing it together. Adv. Mater. 23, 4471–4490.

Scott, A., Dianat, A., Börrnert, F., Bachmatiuk, A., Zhang, S., Warner, J.H., Borowiak-Paleń, E., Knupfer, M., Büchner, B., Cuniberti, G., Rümmeli, M.H., 2011. The catalytic potential of high-κ dielectrics for graphene formation. Appl. Phys. Lett. 98 073110(3).

Shang, N.G., Au, F.C.K., Meng, X.M., Lee, C.S., Bello, I., Lee, S.T., 2002. Uniform carbon nanoflake films and their field emissions. Chem. Phys. Lett. 358 (3–4), 187–191.

Shiji, K., Hiramatsu, M., Enomoto, A., Nakamura, M., Amano, H., Hori, M., 2005. Vertical growth of carbon nanowalls using rf plasma-enhanced chemical vapor deposition. Diamond Relat. Mater. 14 (3–7), 831–834.

Shiji, K., Hiramatsu, M., Enomoto, A., Nakamura, M., Amano, H., Hori, M., 2005. Vertical growth of carbon nanowalls using rf plasma-enhanced chemical vapor deposition. Diamond Relat. Mater. 14 (3–7), 831–834.

Sun, Z., Yan, Z., Yao, J., Beitler, E., Zhu, Y., Tour, J.M., 2010. Growth of graphene from solid carbon sources. Nature 468 (7323), 549–552.

Sun, J., Lindvall, N., Cole, M.T., Teo, K.B.K., Yurgens, A., 2011. Large-area uniform graphene-like thin films grown by chemical vapor deposition directly on silicon nitride. Appl. Phys. Lett. 98 (25) 252107(3).

Sun, J., Cole, M.T., Lindvall, N., Teo, K.B.K., Yurgens, A., 2012a. Noncatalytic chemical vapor deposition of graphene on high-temperature substrates for transparent electrodes. Appl. Phys. Lett. 100 (2) 022102(3).

Bi, H., Sun, S., Huang, F., Xie, X., Jiang, M., 2012b. Direct growth of few-layer graphene films on SiO₂ substrates and their photovoltaic application. J. Mater. Chem. 22, 411–416.

Suzuki, S., Kobayashi, Y., Mizuno, T., Maki, H., 2010. Non-catalytic growth of graphene-like thin film near pattern edges fabricated on SiO₂ substrates. Thin Solid Films 518 (18), 5040–5043.

Teng, P.-Y., Lu, C.-C., Akiyama-Hasegawa, K., Lin, Yung-C., Yeh, C.-H., Suenaga, K., Chiu, P.-W., 2012. Remote catalyzation for direct formation of graphene layers on oxides. Nano Lett. 12 (3), 1379–1384.

Uchino T., Bourdakos K.N., de Groot C.H., Ashburn P., Kiziroglou M.E., Dilliway G.D., Smith D.C., 2005a. Catalyst free low temperature, direct growth of carbon nanotubes. In: Proceedings of 2005 Fifth IEEE Conference on Nanotechnology 5:1.

Uchino, T., Bourdakos, K.N., de Groot, C.H., Ashburn, P., Kiziroglou, M.E., Dilliway, G.D., Smith, D.C., 2005b. Metal catalyst-free low-temperature carbon nanotube growth on SiGe islands. Appl. Phys. Lett. 86 233110(3).

Wang, J., Zhu, M., Outlaw, R.A., Zhao, X., Manos, D.M., Holloway, B.C., Mammana, V.P., 2004a. Free-standing subnanometer graphite sheets. Appl. Phys. Lett. 85 1265(3).

Wang, J., Zhu, M., Outlaw, R.A., Zhao, X., Manos, D.M., Holloway, B.C., 2004b. Synthesis of carbon nanosheets by inductively coupled radio-frequency plasma enhanced chemical vapor deposition. Carbon 42 (14), 2867–2872.

Yan, Z., Peng, Z., Sun, Z., Yao, J., Zhu, Y., Liu, Z., Ajayan, P.M., Tour, J.M., 2011. Growth of bilayer graphene on insulating substrates. ACS Nano 5 (10), 8187–8192.

Zhu, M., Wang, J., Holloway, B.C., Outlaw, R.A., Zhao, X., Hou, K., Shutthanandan, V., Manos, D.M., 2007. A mechanism for carbon nanosheet formation. Carbon 45 (11), 2229–2234.

Chapter 4.7

Epitaxial Growth of Graphene on SiC

Alicja Bachmatiuk and Mark H. Rümmeli

IFW Dresden, Germany

4.7.1. INTRODUCTION

The attraction in the use of silicon carbide for the formation of graphene stems from the fact that graphene films can be grown epitaxially on commercial SiC subtrates. Moreover, the grown graphene can be patterned using standard nanolithography methods without the need for transfer. This makes the technique compatible with current semiconductor technology (Berger et al., 2004). In addition, the technique is very clean because no metal or hydrocarbons are involved since the eptaxially matching support itself provides the carbon. Epitaxial graphene grown on SiC can exhibit long phase coherence lengths and mobilities exceeding $25,000 \text{ cm}^2/\text{V s}$, further adding attraction to graphene grown in this manner (Berger et al., 2006). These aspects have made it a widely used technique. The technique is based on the controlled sublimation of Si from single-crystalline SiC surfaces. The sublimation process, in which a direct transformation of material in the solid phase to the vapour phase occurs, does not preserve stochiometry for binary compounds. This is related to the binding energy between atoms such that less tightly bound atoms in the solid sublimate first. In the specific case of SiC, Si sublimates first leaving behind a few layers of nearly free carbon species. These layers reconstruct (rearrange) on the surface so as to minimise energy forming graphene in the process (Camara et al., 2010). Calculations on molar densities show that approximately three bi layers of SiC are required to free sufficient carbon atoms for the formation of a single-graphene layer (Hass et al., 2008). To form epitaxial graphene in this manner the (0001) (silicon terminated) and (000$\overline{1}$) (carbon terminated) faces of 4H and 6H α-SiC wafers are usually used.

The technique itself dates back to the early 1960s when Badami during X-ray scattering studies found graphite on SiC after heating it to $2150 \,^{\circ}\text{C}$ in UHV (Badami, 1962). He suggested that the graphite formed through preferential surface Si-out diffusion (sublimation). Over a decade later, Van Bommel and co-workers found single layers of graphite (what we now term graphene) on SiC after annealing at $800 \,^{\circ}\text{C}$ and enhanced graphitisation around $1500 \,^{\circ}\text{C}$.

They also identified that the observed differences in the crystallinity of the graphene layers has a dependance on the surface termination, namely, the Si-terminated (0001) face and the C-terminated (000$\bar{1}$) face. They found that the epitaxial alignment of graphene on the C-terminated (000$\bar{1}$) face is rotated 30° with respect to the Si-face unit cell (Van Bommel et al., 1975). Later in the 1980s and 1990s, Muehlhoff et al., Forbeaux et al., and Charrier et al. conducted more detailed studies on the graphitisation process, which, in short, confirmed the results of Van Bommel (Charrier et al., 2000; Forbeaux et al., 1998, 1999; Muehlhoff et al., 1986). Recently, the parallel publication of the electrical response of graphene in 2004 by Novoselov et al. and Berger et al. (who used graphene grown from SiC) provided new impetus to optimise the growth conditions of graphene on SiC (Al-Temimy et al., 2009; Berger et al., 2004).

4.7.2. REACTION PROTOCOL

Normally a hydrogen treatment is applied to the SiC substrate prior to the reaction as this removes any surface oxides and if the surface has been polished (usually it is) and resultant remnant scratches are removed. Typically the sample is heated in a mixture of H_2 and Ar (5%/95%) at temperatures around 1600 °C for approximately 60 min. The cooling rates should be slow (ca. 50 °C/min) so as to avoid silicon crystallisation on the surface. After the treatment the surface is highly uniform with atomically flat terraces. The distance between terraces or the terrace width depends on the angular deviation of the cut with respect to the crystallographic orientation. After this treatment, the samples are cleaned, usually with ethanol or acetone. Sometimes other pre-treatments to the surface may be applied, for example, a sacrificial oxide layer can be thermally grown on the surface, which is then chemically etched in HF to remove any remaining sub-surface damage (Camara et al., 2010). Another pre-treatment is to anneal the sample in UHV at temperatures ranging from 950 °C to 1100 °C for several minutes to remove surface oxides. It is worth noting that the importance of the pre-synthesis surface structure with respect to the resultant graphene has not been demonstrated, however a general argument is that smoother surfaces yield larger graphene sheets, albeit unsubstantiated. This issue is further complicated by the fact that the reaction occurs at rather high temperatures, which may also lead to significant surface degradation/restructuring.

The pre-cleaning/surface smoothing treatments can lead to a depletion of surface Si. To compensate for this an external Si flux is applied before the SiC substrate is heated to a higher temperature to grow graphene. The reaction to grow graphene can be achieved in UHV or at atmospheric pressure. Historically, the sublimation process was achieved in UHV, however, more recently the use of an unreactive gas (Ar) at ambient pressure. The reasoning behind the use of an unreactive ambient gas is as follows: roughness, excess surface steps and small-grained graphene are obtained if the reaction conditions don't allow the most favourable morphology because inherent microscopic processes, namely the detachment, diffusion and reattachment of surface atom are surpressed. One can increase the temperature to minimise these constraints. Under UHV conditions increasing the temperature leads to an increase in sublimation rates and higher graphene growth rates, but this promotes surface roughening and the nucleation of small graphene flakes which is not the goal. To reduce these limitations, one needs to decouple Si evaporation away from the surface from mass transport on the surface (carbon) (Sutter, 2009). Emtsev et al. (2009) implemented a neat trick to accomplish this. Instead of running the reaction in UHV, they ran the reaction in Ar, an unreactive gas, close to ambient pressure. This ensures a dense cloud of gas atoms resides over the SiC substrate surface which slows down the transport of Si atoms away from the surface, viz. the sublimation rate is reduced. This allows one to increase the reaction temperature by several hundred degrees, which favours the formation of large flat terrace formation some tens of microns long and several microns wide. This is significantly larger than obtained with UHV-grown epitaxial graphene which tend not to grow beyond 100 nm (Yu et al., 2011). Thus, an Ar atmosphere is attractive for the formation of large single-crystalline monolayer graphene regions. Moreover, mobility experiments comparing the Ar environment grown graphene with UHV-grown graphene showed the Ar process leads to a nearly two-fold improvement in mobility.

4.7.3. NUCLEATION AND GROWTH

The mechanism of the nucleation and growth of graphene from SiC has, for the most part, been well explored (Rümmeli et al., 2011). Hannon and Tromp (2008) observed the formation of deep pits in the substrate during annealing in vacuum in an in situ study. The pits form because domains from the buffer layer in essence pin decomposing surface steps. Graphene has been shown to nucleate in these pits where the step density is high. Hupalo et al. (2009) found that Si desorption from steps is the main controlling factor for graphene nucleation. They also showed that different desorption rates for different step directions. In addition, faster heating rates can suppress these different speeds leading to significantly larger graphene sheets. From this last statement, it is obvious how the use of an Ar atmosphere helps. TEM studies by Norimatsu and Kusunoki (2010) and Robinson et al. (2010) revealed the preferential nucleation site at steplike facets on the SiC surface as illustrated in Fig. 4.7.1. In-situ Raman studies show small domains of graphene covering the substrate during the early stages of growth. These domains grow, and the number of layers can increase with increasing growth time. The G and 2D modes were observed to shift to higher frequencies with the growth time and was attributed to a strong compressive strain from the substrate (Kamoi et al., 2012).

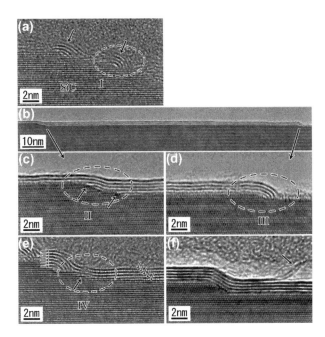

FIGURE 4.7.1 HRTEM images of typical graphene layers formed on stepped SiC. (a) Initial growth occurs on steps. (b) Wide-area scan at a later stage of growth. The images in (c) and (d) are enlargements of part of (b). Images (e) and (f) show graphene around the step. *Reproduced with permission from Norimatsu and Kusunoki (2010) Copyright (2010), ACS.*

4.7.4. EPITAXIAL GRAPHENE ON THE SiC (0001) FACE

Before the graphitisation process begins, a number of surface reconstructions occur. They are usually observed and studied using low-energy electron diffraction (LEED) – (see Section 5.5). Figure 4.7.2 shows LEED patterns collect from different stages of the growth of graphene through a typical UHV

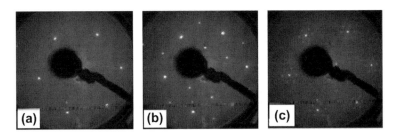

FIGURE. 4.7.2 LEED patterns with a primary energy of 180 eV, obtained at four different stages during the growth of sample A. (a) 1×1 spots of SiC, after a 5 min anneal around 1000 °C, followed by the initial cleaning procedure under Si flux. (b) $(\sqrt{3} \times \sqrt{3})$ R30 reconstruction, after 5 min around 1100 °C. (c) $(6\sqrt{3} \times 6\sqrt{3})$ R30 reconstruction, after 10 min around 1250 °C (Yu et al., 2011).

grown process (Yu et al., 2011). First, the initial Si-rich surface (3 × 3) phase is obtained by exposing a well outgassed SiC substrate to a silicon flux at a temperature around 800 °C. The substrate is subsequently exposed to an annealing step at around (for 5 min). This gives rise to sharp pattern, as shown in Fig. 4.7.2 panel a, which corresponds the 1 × 1 spots of SiC. Further annealing at 1100 °C, again for 5 min, produces the ($\sqrt{3}$ × $\sqrt{3}$) R30 pattern (see panel b in Fig. 4.7.2). The next and final annealing step at 1250–1350 °C for ca. 10 min yields the complex ($6\sqrt{3}$ × $6\sqrt{3}$) R30 reconstruction (panel c) which indicates the formation of graphene.

In the case of graphene growth on the SiC (0001) face, the first carbon layer to form on the Si-face of the SiC is a buffer layer, viz. it is not graphene. This arises because although the atomic arrangement is identical to that of graphene the atom bonding is not because about one third of the carbon atoms are covalently bound to the underlying Si atoms residing on the topmost SiC layer. The interactions between the first graphene layer (buffer layer) and the substrate influence the electrical characteristics, because the covalent bonds disturb the linear π bands. Various techniques exist to decouple the graphene from the substrate. Riedl et al. developed a route in which they intercalated hydrogen between the graphene and the substrate. This saturates the topmost Si layer and, as a consequence, frees the graphene π bands (Riedl et al., 2009). However, the intercalation effect is reversible, and the long-term stability of devices fabricated from decoupled graphene in this manner remains to be demonstrated. Another technique developed by Oida et al. (2010), involves the introduction of an oxide buffer layer. This can be achieved by heating the sample to 250 °C in 1 atm O_2 for 5 s. With this technique, one must be careful as, at higher temperatures the graphene is easily etched. An alternative approach is to reduce the process temperature by introducing a partial pressure of HCl mixed in an argon atmosphere during the graphene growth to assist in the removal of the first Si layers (Fanton et al., 2010).

4.7.5. FACE TO FACE GROWTH

In order to improve the size of the graphene flakes obtained when processing in UHV the so-called face-to-face method was developed (Yu et al., 2011). The technique is simple and economical and yields high quality graphene in large length scales. In this route, two pieces of SiC are stacked together (face to face) and separated by a 25 µm gap, which is obtained by using a Ta-foil spacers between the two inner surfaces. The samples are then heated (the initial experiment used resistive heating). At temperatures below 1500 °C before the graphene begins to grow, both pieces act as sources and sinks of SiC on opposing surfaces viz. material exchange. This step leads to the formation of large atomically flat terraces without the need for careful hydrogen etching steps that are usually required. These large flat terraces encourage the forma-tion of graphene sizes similar to that of the terraces, hence large graphene regions. Moreover the close proximity of the two surfaces partially traps sublimed Si atoms within the gap. The elevated Si pressure at the surface in turn restricts the Si sublimation rate, again encouraging large graphene pieces to form. At the edges where Si is not trapped the resultant graphene is thicker and of poorer quality. In general the graphene consists of 1 to 3 layers.

4.7.6. LASER-INDUCED GROWTH OF EPITAXIAL GRAPHENE

The use of a laser beam to grow epitaxial graphene is extremely attractive, as it provides the opportunity to simultaneously synthesise and pattern graphene. Moreover, studies show this can be achieved rapidly, with easily available commercial equipment and that the successful fabrication of graphene does not need SiC pretreatments or high vacuum. In addition, the heating process is localised, which means surrounding regions remain cool and are therefore not exposed to the risk of damage. This is because the laser can be considered as a surface-heating source and as a result the substrate is held essentially at room temperature (Lee et al., 2010). All these factors make the technique industrially attractive and make it a green solution for epitaxial graphene formation (Yannopoulos et al., 2012).

An early work demonstrating the basic principle that SiC exposed to a laser beam can yield sp^2 carbon is that from Ohkawara et al. (2003). In short they fired a pulsed Nd:YAG laser beam ($\lambda = 1064$ nm, pulse width 5 ms) onto polycrystalline silicon carbide in various gaseous atmospheres, namely, argon, carbon dioxide and air. The central region of the resultant spot (due laser irritation) showed an enhanced C presence from EDX studies. Subsequent Raman spectroscopy and X-ray diffractometry investigations confirmed the presence of graphite. The authors did not investigate the number of layers formed (presumably, as the study was conducted in 2003, before graphene became such a hot topic). It is also remarkable that they were able to obtain graphite in the presence of air and carbon dioxide atmospheres. The concept of laser-based epitaxial graphene growth by laser beam was taken later in 2011 by Lee et al. (2010). They employed a pulsed excimer laser ($\lambda = 248$ nm) and were able to confirm graphene growth from single-crystal SiC (0001), using reflection high-energy electron diffraction (RHEED), Raman spectroscopy, synchrotron X-ray diffraction, TEM and STM. By adjusting the laser fluence, they were able to control the thickness of the graphene film down to single layer graphene. They found the structural quality of their graphene to match that from thermally grown graphene on SiC (0001), however, they found that for few-layer graphene, it did not conform to AB Bernal stacked graphene. They also demonstrated the potential to directly pattern graphene on SiC. More recently it was shown CO_2 lasers ($\lambda = 10.6$ μm) can be used to grow epitaxial graphene (Yannopoulos et al., 2012). They nicely demonstrated that no SiC pretreatment or high vacuum was required and showed growth on the second scale. SEM, XPS, SMS and

Raman spectroscopy confirmed the presence of graphene after irradiating the surface of SiC (0001) substrates. Their studies confirmed the graphene was of low strain. Moreover, it seems the high heating rates achieved with the CO_2 laser seems to avoid different Si desorption rates from adjacent SiC steps. They also found the fast cooling rates affect the stacking order of the different graphene layers.

4.7.7. EPITAXIAL GRAPHENE ON THE SIC (000$\bar{1}$) FACE

As compared to the Si-terminating face (SiC (0001)), studies on graphene formation from the SiC (000$\bar{1}$) face is generally less studied. This is because graphene films obtained from the C terminating face are of poor quality and rotationally disordered. The rotational disorder occurs because when growing graphene on the C face of a SiC substrate there is almost no coupling between the first graphene layer and the substrate and as a consequence the layers are somewhat randomly oriented viz. turbostratic material is obtained (Camara et al., 2010). Nonetheless a detailed LEED study indicates that the azimuthal disorder is not random. For multilayer graphene grown on the SIC (000$\bar{1}$) face, epitaxial layers can orient themselves either in the 30° phase or in the 2° phase with respect to the substrate. The different orientations between adjacent layers result in them decoupling from each other. This in effect forms a system which keeps the transport and electronic properties of free-standing monolayer graphene (Berger et al., 2006; de Heer et al., 2007). Because of this, there is an increasing interest in graphene obtained on the C face of SiC. However, the goal of large area, high quality graphene with precise graphene layer number control is no trivial matter. Indeed, a systematic study by Camara et al. (2010) suggested that using the standard growth parameters (low-pressure conditions and a temperature range between 1500 and 1600 °C), large areas of monolayer graphene will be difficult to reach because graphitisation is not an intrinsic process. The reason for this is that any existing structural defect (e.g. threading dislocations) creates a nucleation centre that is much more efficient than the spontaneous growth process which requires more energy to take place. Detailed STM studies also suggested large area graphene from SiC (000$\bar{1}$) faces are difficult to obtain (Biedermann et al., 2009). In other words, to grow large and homogeneous graphene layers this process (defect-induced nucleation centres) needs to be controlled. To achieve this a radically different growth technique was developed (Camara et al., 2009). In this new technique a graphite cap is used to cover the SiC sample, as this increases the carbon and silicon partial pressure at the surface of the SiC substrate. This lowers the Si sublimation rate but still allows a rather large diffusion length for both the Si and C species. The procedure to obtain graphene in this manner is temperature-sensitive. At 1500 °C graphite growth is fully quenched and so no graphene layer forms. One simply obtains a large reconstruction of the substrate surface. By raising the temperature to ca. 1700 °C for about 15 min epitaxial graphene layers start to form. Full wafer coverage is not obtained, but instead a large number of long, self-organised graphene ribbons are obtained. On average the ribbons are around 100 μm long and 4 μm wide. By increasing the reaction time the lengths of the ribbons increase without changing the widths much. For example, for a reaction time of 30 min, ribbon lengths of 300 μm with widths still close to 4 μm are found. In many, but not all, cases a dislocation defect can be seen at the starting point (centre) of a ribbon. Many ribbons without any identified defect at the origin are also found. In addition, dislocation defects in regions devoid of any graphene are also found. AFM studies reveal that the reaction leads to a surface reorganisation of the SiC face into long and uniform terraces (see Fig. 4.7.3). The terracing effect is a standard effect known as step bunching. It arises because of a small initial miscut of the wafer surface with respect to the nominal 6H–SiC surface.

The step bunching is similar to the facet nucleation mechanisms reported for <111> silicon. In essence a minimum width of terrace (W_c) is required for surface reconstruction to take place as discussed in (Camara et al., 2009; Jeong and Weeks 1998). Surface reconstruction expands rapidly once a seed has been formed by a dislocation or other nucleation centre. The graphene grows on the reconstructed parts of terraces and this explains the observed anisotropic growth rates as well as the unusual lengths of the ribbons. So long as the width of a terrace is below the critical value W_c, no surface reconstruction can occur. This means a graphene layer can expand preferentially on one isolated terrace and explains why at the end of the process the graphene ribbons occupy a single terrace surrounded by high and sharp edges. However, occasionally a graphene

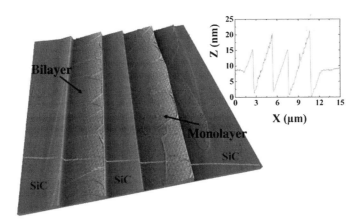

FIGURE 4.7.3   AFM picture of sample graphitised at 1700 °C for 15 min with a graphitic cap covering the sample. Two long graphene layers are seen on single large terraces. Above one of the terraces, a monolayer lies, while a Bernal bilayer AB lies on the other one. The graphene layers are surrounded by step-bunched bare SiC. The corresponding profile indicates that very high steps in the range 10–20 nm high are detected while in this example the ribbons are 3–4 μm large. *From Camara et al. (2010).*

layer can expand over two or three terraces. This probably depends on the size of the initial defect that initiates growth (Camara et al., 2010).

4.7.8. GRAPHENE GROWTH BY MOLECULAR BEAM EPITAXY OF SiC

Molecular beam epitaxy (MBE) as a technique is based on the interaction of one or more molecular or atomic beams that occurs on the surface of a heated crystalline substrate. MBE can deliver high quality and homogeneous wafer-scale epitaxial layers. Moreover, the layer thickness control is high (atomic) and is highly reproducible hence it is an attractive technique for graphene growth. One of the earliest works in this vein is by Al-Temimy et al. (2009) – the used 6H–SiC (0001) and 4H–SiC(000-1) samples that were first hydrogen etched to remove polishing damage and generate regular arrays of atomically flat terraces. Thereafter, they were placed in the UHV chamber and then cleaned with a Si flux. After further surface reconstruction steps, the graphitisation step included a carbon deposition source. The carbon-assisted approach enabled them to obtain graphene on SiC(0001) at a temperature of 950 °C, which is lower than for conventional vacuum annealing routes (ca. 1200 °C). The addition of a carbon source modifies the growth dynamics of the graphene such that the initial surface morphology of the SiC samples is not altered (since carbon is not required from the SiC). When using SiC ($000\bar{1}$) crystals, the carbon source-induced growth of epitaxial graphene leads to a predominant surface termination of 0° with respect to the substrate, instead of the more usual 30° as found with conventional UHV thermal preparation. Moreau et al. (2010), explored the technique using a solid carbon source on SiC ($000\bar{1}$). In agreement with the work by Al-Temimy et al., they found that the graphene layers grow without altering the initial atomically flat step and terrace structures on the substrate surface. This combined with the fact that the technique yields precise control of the thickness, particularly at the beginning to obtain single- or bilayer graphene makes it a promising technique. Park et al. (2010) used graphite and C_{60} as solid-carbon sources and again confirmed the stability of the terraces with the MBE technique. Their AFM studies showed the graphene on the surface has wrinkles along and across step edges. Raman spectroscopy mapping showed homogeneous growth over the substrate. In addition, they also noted the high degree of control for layer thickness.

4.7.9. GRAPHENE SYNTHESIS ON CUBIC SiC/Si Wafers

Cubic 3C–SiC (β-SiC) is generally not considered useful for the epitaxial growth of graphene due to its large lattice mismatch. However if graphene growth on cubic SiC could be achieved it would be attractive as it is readily grown with large sizes (>300 mm) and is commercially available on cheap Si wafers. In 2010, Aristov et al. achieved this. In short, they exposed their samples with a Si-rich surface of SiC (001) to a series of annealing cycles with increasing temperatures from 1200 K to 1500 K (Aristov et al., 2010). LEED investigations showed the surface to go through all known SiC (001) reconstructions finishing in the β-SiC (001) 1 × 1 carbon rich surface. Photo-emission studies showed a very weak interaction with the substrate. Despite the mismatch they found the graphene growth to be guided along the (110) crystallographic direction of the SiC (001) substrate. The technique is attractive not only because it is economically viable but it is also appropriate for industrial mass production.

4.7.10. GRAPHENE FROM THE CARBOTHERMAL REDUCTION OF SiO$_x$

One of the main synthesis methods for SiC is a carbothermal reduction process, known as the Acheson process. The general reaction is:

$$3C + SiO_2 \rightarrow SiC + 2CO$$

Silica has previously been shown to grow sp^2 carbon nanostructures, and more detailed studies show that at the root of the nanostructures is SiC, viz. the silica reduced to SiC in the CVD reaction (Bachmatiuk et al., 2009). Moreover, the study revealed carbon formation on the surface of the resultant SiC nanoparticles. Continued studies in progress in which sandwiched solid carbon layers in SiO$_x$ are treated to annealing steps show graphene formation (Rümmeli et al., 2011) and the presence of SiC. Further studies are required to clarify the growth route.

4.7.11. SiC/METAL HYBRID SYSTEMS FOR GRAPHENE FORMATION

In this route a metallic film, e.g. Ni, is deposited on the surface of SiC. Under heating, the metal extracts and absorbs carbon. During the cooling stage the absorbed carbon precipitates on the free surface of the metal and forms single layer or few layer graphene. An advantage of the route is that the graphene is continuous over the entire surface which is not the case for convetional epitaxial graphene synthesis. The technique was first developed by Juang et al. (2009) in which they deposited Ni (200 nm) on single crystalline 6H–SiC (1000) and 3C–SiC coated on Si substrates. Their data pointed to the heating rate controlling the carbon absorption and the cooling rate controlling the precipitation. Playing with both these parameters allows one to control the graphene thickness (number of layers). In addition, they were able to accomplish graphene formation at the reduced temperature of 750 °C. Hofrichter et al. (2010) modified the technique by depositing amorphous SiC on Si and then depositing a Ni film (500 nm) on the SiC. They used processing temperatures of 1100 °C to obtain graphene. After the reaction, the Ni film was removed by wet etching. This leads to the graphene film settling directly on the Si/SiO$_2$ substrate, since the SiC dissolves in the Ni and is removed along with the Ni in the wet etching step.

REFERENCES

Al-Temimy, A., Riedl, C., Starke, U., 2009. Low temperature growth of epitaxial graphene on SiC induced by carbon evaporation. App. Phys. Lett. 95 231907(3).

Aristov, V. Yu., Urbanik, G., Kummel, K., Vyalikh, D.V., Molodtsova, O.V., Preobrajenski, A.B., Zakharov, A.A., Hess, C., Hänke, T., Büchner, B., Vobornik, I., Fujii, J., Panaccione, G., Ossipyan, Y.A., Knupfer, M., 2010. Graphene synthesis on cubic SiC/Si wafers. Perspectives for mass production of graphene-based electronic devices. Nano Lett. 10, 992–995.

Bachmatiuk, A., Börrnert, F., Grobosch, M., Schäffel, F., Wolff, U., Scott, A., Zaka, M., Warner, J.H., Klingler, R., Knupfer, M., Büchner, B., Rümmeli, M.H., 2009. Investigating the graphitization mechanism of SiO$_2$ nanoparticles in chemical vapor deposition. ACS Nano 3 (12), 4098–4104.

Badami, D.V., 1962. Graphitization of alpha-silicon carbide. Nature 193 (4815), 569–570.

Berger, C., Song, Z., Li, T., Li, X., Ogbazghi, A.Y., Feng, R., Dai, Z., Marchenkov, A.N., Conrad, E.H., First, P.N., de Heer, W.A., 2004. Ultrathin epitaxial graphite: 2D electron gas properties and a route toward graphene-based nanoelectronics. J. Phys. Chem. B 108 (52), 19912–19916.

Berger, C., Song, Z., Li, X., Wu, X., Brown, N., Naud, C., Mayou, D., Li, T., Hass, J., Marchenkov, A.N., Conrad, E.H., First, P.N., de Heer, W.A., 2006. Electronic confinement and coherence in patterned epitaxial graphene. Science 312 (5777), 1191–1196.

Biedermann, L.B., Bolen, M.L., Capano, M.A., Zemlyanov, D., Reifenberger, R.G., 2009. Insights into few-layer epitaxial graphene growth on 4H–SiC(0001) substrates from STM studies. Phys. Rev. B 79 125411(19).

Camara, N., Huntzinger, J.-R., Rius, G., Tiberj, A., Mestres, N., Pérez-Murano, F., Godignon, P., Camassel, J., 2009. Anisotropic growth of long isolated graphene ribbons on the C face of graphite-capped 6H–SiC. Phys. Rev. B 80 (12) 125410(8).

Camara, N., Tiberj, A., Jouault, B., Caboni, A., Jabakhanji, B., Mestres, N., Godignon, P., Camassel, J., 2010. Current status of self-organized epitaxial graphene ribbons on the C face of 6H–SiC substrates. J. Phys. D. Appl. Phys. 43 (37) 374011(13).

Charrier, A., Coati, A., Argunova, T., Thibaudau, F., Garreau, Y., Pinchaux, R., Forbeaux, I., Debever, J.-M., Sauvage-Simkin, M., Themlin, J.-M., 2000. Solid-state decomposition of silicon carbide for growing ultra-thin heteroepitaxial graphite films. J. Appl. Phys. 92 (5), 2479–2484.

de Heer, W.A., Berger, C., Wu, X., First, P.N., Conrad, E.H., Li, X., Li, T., Sprinkle, M., Hass, J., Sadowski, M.L., Potemski, M., Martinez, G., 2007. Epitaxial graphene. Solid State Commun. 143 (1–2), 92–100.

Emtsev, K.V., Bostwick, A., Horn, K., Jobst, J., Kellogg, G.L., Ley, L., McChesney, J.L., Ohta, T., Reshanov, S.A., Röhrl, J., Rotenberg, E., Schmid, A.K., Wandmann, D., Weber, H.B., Seyller, T., 2009. Towards wafer-size graphene layers by atmospheric pressure graphitization of silicon carbide. Nat. Mater. 8, 203–207.

Fanton, M.A., Robinson, J.A., Hollander, M., Weiland, B.E., Trumbull, K., LaBella, M., 2010. Synthesis of thin carbon films on 4H–SiC by low temperature extraction of Si with HCl. Carbon 48 (9), 2671–2673.

Forbeaux, I., Themlin, J.-M., Debever, J.-M., 1998. Heteroepitaxial graphite on 6H–SiC(0001): interface formation through conduction-band electronic structure. Phys. Rev. B 58 (24), 16396–16406.

Forbeaux, I., Themlin, J.-M., Debever, J.-M., 1999. High-temperature graphitization of the 6H–SiC (000(1)over-bar) face. Surf. Sci. 442 (1), 9–18.

Hannon, J.B., Tromp, R.M., 2008. Pit formation during graphene synthesis on SiC(0001): in situ electron microscopy. Phys. Rev. B 77 (24) 241404(R)(4).

Hass, J., de Heer, W.A., Conrad, E.H., 2008. The growth and morphology of epitaxial multilayer graphene. J. Phys.: Condens. Matter 20 323202(27).

Hofrichter, J., Szafranek, B.N., Otto, M., Echtermeyer, T.J., Baus, M., Majerus, A., Geringer, V., Ramsteiner, M., Kurz, H., 2010. Synthesis of graphene on silicon dioxide by a solid carbon source. Nano Lett. 10, 36–42.

Hupalo, M., Conrad, E.H., Tringides, M.C., 2009. Growth mechanism for epitaxial graphene on vicinal 6H–SiC(0001) surfaces: a scanning tunneling microscopy study. Phys. Rev. B 80 041401(R)(4).

Jeong, H.-C., Weeks, J.D., 1998. Two-dimensional dynamical model for step bunching and pattern formation induced by surface reconstruction. Phys. Rev. B 57 (7), 3939–3948.

Juang, Z.-Y., Wu, C.-Y., Lo, C.-W., Chen, W.-Y., Huang, C.-F., Hwang, J.-C., Chen, F.-R., Leou, K.-C., Tsai, C.-H., 2009. Synthesis of graphene on silicon carbide substrates at low temperatures. Carbon 47, 2026–2031.

Kamoi, S., Kisoda, K., Hasuike, N., Harima, H., Morita, K., Tanaka, S., Hashimoto, A., Hibino, H., 2012. A Raman imaging study of growth process of few-layer epitaxial graphene on vicinal 6H–SiC. Diamond Relat. Mater. 25, 80–83.

Lee, S., Toney, M.F., Ko, W., Randel, J.C., Jung, H.J., Munakata, K., Lu, J., Geballe, T.H., Beasley, M.R., Sinclair, R., Manoharan, H.C., Salleo, A., 2010. Laser-synthesized epitaxial graphene. Nano 4 (12), 7524–7530.

Moreau, E., Godey, S., Ferrer, F.J., Vignaud, D., Wallart, X., Avila, J., Asensio, M.C., Bournel, F., Gallet, J.-J., 2010. Graphene growth by molecular beam epitaxy on the carbon-face of SiC. App. Phys. Lett. 97 241907(3).

Muehlhoff, L., Choyke, W.J., Bozack, M.J., Yates, J.T., 1986. Comparative electron spectroscopic studies of surface segregation on SiC(0001) and SiC(0001). J. Appl. Phys. 60 (8) 2842–1853.

Norimatsu, W., Kusunoki, M., 2010. Formation process of graphene on SiC (0001). Physica E 42 (4), 691–694.

Ohkawara, Y., Shinada, T., Fukada, Y., Ohshio, S., Saitoh, H., Hiraga, H., 2003. Synthesis of graphite using laser decomposition of SiC. J. Mater. Sci. 30, 2447–2453.

Oida, S., McFeely, F.R., Hannon, J.B., Tromp, R.M., Copel, M., Chen, Z., Sun, Y., Farmer, D.B., Yurkas, J., 2010. Decoupling graphene from SiC(0001) via oxidation. Phys. Rev. B 82 (4) 041411(R)(4).

Park, J., Mitchel, W.C., Grazulis, L., Smith, H.E., Eyink, K.G., Boeckl, J.J., Tomich, D.H., Pacley, S.D., Hoelscher, J.E., 2010. Epitaxial graphene growth by molecular beam epitaxy (CMMMBE). Adv. Mater. 22, 4140–4145.

Riedl, C., Coletti, C., Iwasaki, T., Zakharov, A.A., Starke, U., 2009. Quasi-free-standing epitaxial graphene on SiC obtained by hydrogen intercalation. Phys. Rev. Lett. 103 (24) 246804(4).

Robinson, J., Weng, X., Trumbull, K., Cavalero, R., Wetherington, M., Frantz, E., LaBella, M., Hughes, Z., Fanton, M., Snyder, D., 2010. Nucleation of epitaxial graphene on SiC(0001). Nano 4 (1), 153–158.

Rümmeli, M.H., Rocha, C.G., Ortmann, F., Ibrahim, I., Sevincli, H., Börrnert, F., Kunstmann, J., Bachmatiuk, A., Pötschke, M., Shiraishi, M., Meyyappan, M., Büchner, N., Roche, S., Cuniberti, G., 2011. Graphene: piecing it together. Adv. Mater. 23, 4471–4490.

Sutter, P., 2009. How silicon leaves the scene. Nat. Mater. 8, 171–172.

Van Bommel, A.J., Crombeen, J.E., Van Tooren, A., 1975. LEED and Auger electron observations of the SiC(0001) surface. Surf. Sci. 48 (2), 463–472.

Yannopoulos, S.N., Siokou, A., Nasikas, N.K., Dracopoulos, V., Ravani, F., Papatheodorou, G.N., 2012. CO_2-laser-induced growth of epitaxial graphene on 6H–SiC(0001). Adv. Funct. Mater. 22, 113–120.

Yu, X.Z., Hwang, C.G., Jozwiak, C.M., Köhl, A., Schmid, A.K., Lanzara, A., 2011. New synthesis method for the growth of epitaxial graphene. J. Electron Spectrosc. Relat. Phenom. 184 (3–6), 100–106.

Chapter 4.8

Transfer to Arbitrary Substrates

Sandeep Gorantla and Mark H. Rümmeli

IFW Dresden, Germany

4.8.1. INTRODUCTION

The fundamental research investigations and the potential applications of graphene often demand the need to transfer graphene from its source material/ substrate to other arbitrary substrates such as insulating SiO_2/Si and sapphire substrates, polymer substrates used for organic and transparent electronics, and metals. A variety of transfer routes have been proposed in the literature which are mostly specific to a particular graphene synthesis route such as commercially available highly-oriented pyrolytic graphite (HOPG), CVD growth of graphene and epitaxial growth of graphene on SiC. Notably, the realisation of the growth of large-area graphene through CVD process on different metal substrates triggered the development of several different routes for transferring graphene to arbitrary substrates. However, it is important to note that no transfer approach is completely free of transfer process-induced artifacts (defects, contamination). At present the graphene transfer routes are being further investigated and optimised and is one of actively researched areas of graphene science with newer methods being added to the list. Considering this, here, we provide a general overview of the most widely practiced graphene transfer protocols. They are classified in accordance to the different graphene synthesis routes.

4.8.2. TRANSFER OF MECHANICALLY EXFOLIATED GRAPHENE TO ARBITRARY SUBSTRATES

Reina et al. achieved the transfer of graphene from SiO_2/Si substrate to arbitrary substrates in 2008, by extending an earlier reported transfer approach applied for transferring CNTs (Jiao et al., 2008; Reina et al., 2008). This led to the possibility of fabricating graphene devices on other substrates. Figure 4.8.1 schematically depicts their approach for transferring graphene from prepared graphene deposited SiO_2/Si substrates from microcleaved HOPG.

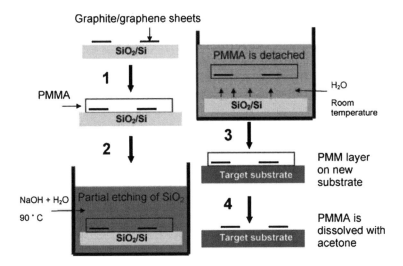

FIGURE 4.8.1   Schematic diagram of the transferring process. The graphene sheets are deposited on SiO₂/Si substrates (300 nm thermal oxide) via HOPG micro-cleaving and are finally transferred to a nonspecific substrate (Reina et al., 2008).

The underlying principle of this method involves the initial spin coating (e.g. 600 rpm at 6 s, 2000 rpm at 60 s) of PMMA polymer over the graphene already deposited on an SiO₂/Si substrate, followed by subsequent curing (e.g. 170 °C at 10 min) of the polymer. The polymer here serves as a dispensable support and carrier layer for transferring graphene. Please note that in the literature different conditions for spin coating and curing of the polymer are reported by different research groups. After curing the graphene preferentially adheres to the overlying polymer compared to the underlying SiO₂ layer, owing to the stronger carbon–carbon bonding affinity. After the curing process, the PMMA-coated substrate stack is submerged in 1 M NaOH solution which acts as medium to etch away (in this case only partially etched) the SiO₂ layer, thereby releasing the graphene with PMMA as a PMMA/graphene (PMMA/Gr) film. The PMMA/Gr film can now be either released manually with tweezers or can be transferred into water (preferably de-ionised water), where the surface tension of the water will keep the PMMA/Gr film afloat as the denser partially etched SiO₂/Si substrate sinks into water. The PMMA/Gr film can now be fished out/transferred onto any desired target substrate with the graphene side overlying the target substrate. Subsequently, by the use of acetone the PMMA can be dissolved and removed leaving the graphene transferred onto the arbitrary substrate. Notably, since this approach preserves the initial location of the graphene flakes on the starting SiO₂/Si substrate, by comparison of the optical images of graphene flakes before transfer on SiO₂/Si substrate and after transfer on the new substrate, the desired flakes can be easily identified and isolated for further investigations.

An alternative, simpler approach for transferring graphene/FLG layers from the SiO₂/Si wafers to perforated substrates for obtaining freestanding graphene membranes was reported by Zettl's group (Meyer et al., 2008). They used perforated holey carbon coated TEM grids (note: carbon is usually coated only on one-side of a TEM grid) as the target substrates for this purpose. At first, the regions on the SiO_2/Si substrate with single-layer graphene flakes are identified with an optical microscope. After this a holey carbon TEM grid is placed on the flake such that carbon side of the TEM grid is on the graphene flake. This is followed by putting just a drop of isopropanol (IPA) directly on the TEM grid and leaving the IPA drop to evaporate. As the IPA evaporates, the surface tension of the drop pulls the carbon on the TEM grid into close contact with the graphene and the surrounding substrate. After the IPA drop is completely evaporated, an additional drop of IPA is placed just next to the TEM grid. As this drop expands and wets the now adhered TEM grid, the IPA liquid passes between the graphene-attached carbon of the TEM grid and the substrate and thereby lifting up the TEM grid from the substrate. The thus-separated and suspended TEM gird in the IPA drop can now be picked up manually with tweezers. Although this transfer approach is simple, it is important to note the yield of FLG/Gr flakes sticking to the perforated carbon on the TEM grid is reportedly only about 25% (Meyer et al., 2008). With the aid of some additional steps the transfer yield in this approach is found to be improved by enhancing the initial adherence between the TEM grid carbon film and graphene-deposited substrate. For this, after the first IPA drop is allowed to evaporate completely in air, the substrate with the attached TEM grid is heated at 200 °C for 5 min on a hot plate to evaporate the IPA more effectively than in just air drying. Then, after cooling down to room temperature, the TEM grid is separated from substrate by placing the ensemble in a 30% potassium hydroxide (semiconductor grade) solution at room temperature. The potassium hydroxide partially etches SiO_2 slowly at room temperature reportedly ranging in time from a few minutes to a few hours leading to the separation of the graphene-transferred TEM grid (Meyer et al., 2008). The TEM grid with FLG/graphene flakes may now be cleaned by rinsing in a water bath followed by rinsing in IPA. Finally, the TEM grid is dried in air resulting in freestanding single- and few-layer graphene across the holes of the holey carbon TEM grid. If the use of bases/acid in the transfer process needs to be avoided, the starting SiO_2/Si substrate can be spin-coated with about 10–30 nm PMMA prior to depositing FLG/Gr from microcleaved HOPG. This helps in easily separating the TEM grid attached with the FLG/Gr flakes through IPA drop evaporation, by dissolving the PMMA in acetone. It is important to note that this transfer approach of the freestanding graphene is seemingly limited to arbitrary substrates coated with perforated amorphous carbon film on them. To date, we found no reports for the transfer of freestanding graphene using this approach to noncarbon coated or polymer free arbitrary substrates. This is probably due to the stronger van der Waals forces of attraction between graphene and carbon compared to the adhesion between graphene and metal oxide substrates.

The mechanical exfoliation of graphene from HOPG is a tedious approach, yielding relatively small flakes (few tens of micrometres or less) of monolayer graphene, and is not feasible to accomplish transfer of large area monolayer graphene. Therefore, the mechanical exfoliation of graphene from HOPG is limited in scalability. These challenges led to development of methods to synthesise large areas of monolayer graphene through CVD process on metal or metal-oxide substrates.

4.8.3. TRANSFER OF CVD-GROWN GRAPHENE ON METALS TO ARBITRARY SUBSTRATES

The commonly used approach for the transfer of CVD grown on Cu foil reported by the Ruoff's group, is similar to the approach developed by Reina et al. (2008) as described in the previous subsection with a few modifications (Li et al., 2009a). In a typical CVD process for fabricating graphene over Cu foil, the graphene grows on both sides of the foil. For the transfer of graphene from a chosen side of the foil, initially a polymer that serves as a graphene support/carrier material (either PMMA or polydimethylsiloxane (PDMS)) is spin coated and cured. After this the polymer/graphene/Cu foil stack is suspended on the surface of an aqueous iron nitrate solution to etch away Cu and thereby leaving behind PMMA/graphene film afloat on the solution. The Cu etching time depends on the etchant concentration and the area and thickness of Cu foil; reportedly a 25-μm thick Cu foil with 1 cm^2 area can be dissolved in 0.05 g/ml iron-nitrate solution overnight (Li et al., 2009a). Another possible etchant for dissolving Cu is aqueous iron chloride (0.03 g/ml) solution (Lin et al., 2011). After the Cu is completely etched away, the PMMA/Cu film can be removed from the iron nitrate solution and subsequently rinsed in de-ionised (DI) water to clean off remnant etchant solution and transferred to the desired arbitrary substrate. Finally, the PMMA can be dissolved in acetone leaving the graphene on the substrate and, thus, completing the transfer process. A similar wet-etching of the metal approach can be followed even in the case of CVD-grown graphene on Ni metal films (Reina et al., 2009). However, this approach is not a perfect, and there are still some associated issues that are currently being researched so as to achieve defect free (such as tears and cracks) and contamination free (such as carrier polymer residue and nanoparticular metal contamination) transfer of clean monolayer graphene to arbitrary substrates.

Graphene transferred using this simple approach has been observed to cause cracks in the graphene layer after the removal of the polymer carrier. This is, because the conformal graphene grown on the surface of the metal foil, which is usually rough, when transferred to an arbitrary flat surface there are always gaps between the graphene and target substrate surface (Li et al., 2009a).

Moreover, the PMMA transforms into a hard coating after it is cured and does not aid in relaxing the graphene after being placed on a target substrate. Owing to this, when the PMMA is dissolved in acetone the unattached regions of the graphene are highly susceptible to breakage leading to cracks in the transferred monolayer graphene. Two possible paths to overcome this problem were proposed by Ruoff's group: after the transfer of the PMMA/Gr film onto the target substrate, an additional drop of PMMA is placed on to this film. This was reported to help in relaxing the PMMA/Gr film by partially or fully dissolving the previously cured hard PMMA layer by the freshly deposited wet PMMA drop. This can help improve the contact of graphene with the target substrate surface (Li et al., 2009a). An alternative approach is the softening of the PMMA layer of the PMMA/Gr film after transferring it to the target substrate by heating the PMMA above its glass transition temperature (T_g) at 150 °C for longer than 12 h (Suk et al., 2011). In a comparative study of these two methods, as reported by Pham (2010), in improving the contact between the PMMA/Gr film and the substrate (300 nm SiO_2/Si substrate) it is found that heating of the PMMA above its T_g and holding at this temperature resulted in better contact compared to the additional wet PMMA drop approach.

The final dissolution step of removing the PMMA using acetone is often observed to leave some residual polymer contamination on the graphene as revealed by HRTEM investigations. Commonly, after the PMMA removal, the graphene on substrate is rinsed in IPA to reduce carrier polymer residue. In our labs, an alternative approach to substantially reduce residual polymer material from graphene transferred onto TEM lacey or holey carbon grids is to heat them at 170 °C under dynamic vacuum ($\sim 10^{-6}$ Torr) for about 8 h. Recently, Lin et al. (2011) have shown that replacing PMMA carrier polymer with poly(bisphenol A carbonate) (PC) and the subsequent dissolution of the PC from the PC/Gr film using acetone-buffered chloroform resulted in much cleaner transferred graphene without the need for any postannealing step. However, a wider consensus on this approach by other research groups is yet to be achieved.

As an alternative approach devoid of the use of a carrier polymer coating, Zettl's group have shown that their direct transfer method applied in the case of transferring HOPG microcleaved graphene on SiO_2/Si wafers to perforated carbon coated TEM grids, as described in the previous subsection, can be extended for the transfer of CVD grown graphene over Cu (Regan et al., 2010). In this case, after the IPA drop, surface tension and evaporation are used to pull the carbon on the TEM grid into intimate contact with the graphene on the Cu foil, the cu foil is etched away in 0.1 g/ml iron chloride solution leaving behind graphene adhered on the perforated carbon of the TEM grid. This approach is schematically shown in comparison with the more commonly adopted PMMA-etching-acetone approach illustrated in Fig. 4.8.2. In this case, the graphene transferred TEM grid is further rinsed in DI water, followed by rinsing in IPA

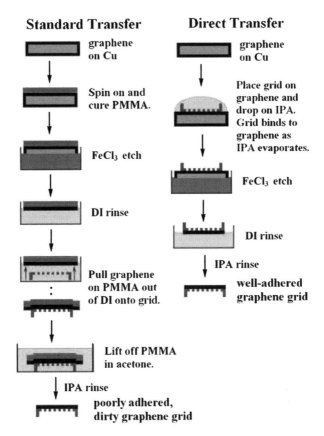

Standard Transfer   **Direct Transfer**

graphene on Cu — graphene on Cu

Spin on and cure PMMA. — Place grid on graphene and drop on IPA. Grid binds to graphene as IPA evaporates.

FeCl₃ etch — FeCl₃ etch

DI rinse — DI rinse

Pull graphene on PMMA out of DI onto grid. — IPA rinse / well-adhered graphene grid

Lift off PMMA in acetone.

IPA rinse / poorly adhered, dirty graphene grid

FIGURE 4.8.2 A comparison of the standard (e.g. PMMA) and direct transfer of layer-area graphene to holey a-C TEM grids (Regan et al., 2010).

for improved removal of any remaining Cu etchant and any possible organic contaminants. However, it was reported that while the graphene transferred TEM grid is being removed with tweezers from DI water, after rinsing, the strong surface tension under-pull by the DI water potentially induce tears in the transferred graphene, particularly, if the transferred graphene area is larger than the diameter of the TEM grid. Rinsing in ethanol instead of DI water by manually holding the graphene-transferred TEM grid with tweezers on the ethanol surface helps minimise this problem due to the lower surface tension of ethanol as compared to DI water (O'Hern, 2011).

For certain specific graphene based sensor device applications, such as freestanding graphene-sealed microchambers, there is a need for the dry transfer of large area CVD grown graphene over Cu (Bunch et al., 2008; Cha et al., 2008). To accomplish this, Ruoff's group developed a semi-dry transfer method for CVD grown graphene on Cu (Suk et al., 2011).

This approach however involves two initial wet chemical steps followed by a dry thermal approach for removing the carrier PMMA support. After the deposition of PMMA on the graphene through spin-coating and curing an additional PDMS block with a through hole in the centre is pressed onto this manually. Reportedly, the PDMS block that serves as an additional handling support stays fixed through natural adhesion, as shown in Fig. 4.8.2 (Suk et al., 2011). After this, the copper is etched away in 0.1 M ammonium persulfate and the PDMS/PMMA/Gr stack is rinsed in DI water followed by air drying. After this step the stack is transferred onto the desired substrate with holes in it. Now the whole composite structure is heated above the glass transition temperature of the PMMA (at 180 °C for about 3 h) to improve the contact between the graphene and the substrate by minimising the gaps between them. After this heating step, the PMMA/Gr adheres well with the substrate, and the PDMS handle support can be easily peeled off while the sample was still hot. Unlike the use of acetone solution to dissolve PMMA away, in this route, the PMMA is thermally removed by heating in a furnace at 350 °C with Ar and H_2 gas flow for approx. 2 h.

In strong contrast to above mentioned approaches, a novel electrochemical exfoliation pathway was reported to separate CVD grown graphene over Cu foils by Wang et al. This approach avoids etching of the Cu substrate with acids (Wang et al., 2011). After the spin-coating and curing of a PMMA film on the graphene, the PMMA/Gr/Cu foil stack is immersed in an electrolyte and used as a cathode in a simple electrochemical cell. An aqueous solution of 0.05 M $K_2S_2O_8$ was used the electrolyte, with a glassy carbon rod acting as anode. It was found that when the voltage is applied across the electrodes (cathode polarised at -5 V), the hydrogen bubbles release at the cathode due to decomposition of water in the electrolyte. Here, they claim that the bubbles of H_2 provide a gentle yet persistent force to gradually lift and separate PMMA/Gr film from the Cu substrate and this further aided by the permeation of the electrolyte at this interface as the PMMA/Gr film delaminates (Wang et al., 2011). The separated PMMA/Gr film can be transferred to any arbitrary substrate followed by the removal of the PMMA. However, they observed that during the delamination of the PMMA/Gr film, in one transfer cycle, a thin Cu substrate layer of about 40 nm is etched away. In further support of the efficacy of this approach recently Gao et al. (2012) have shown that graphene grown on Pt metal can be electrochemically delaminated through bubbling due to release of H_2 in the electrolyte (aq. NaOH) as a consequence of the electrolytic dissociation of H_2O. The Fig. 4.8.3 shows a schematic of the electrochemical exfoliation of PMMA spin-coated and cured graphene from the Pt-growth substrate. Owing to the chemical inertness of the Pt substrate, no etching of the Pt was observed and it supports the premise that the H_2 bubbles intercalate between the graphene and the Pt substrate and that this serves as the sole driving force acting as a gentle peeling force for separating the PMMA/Gr film. This approach holds promise for separating graphene without the need for

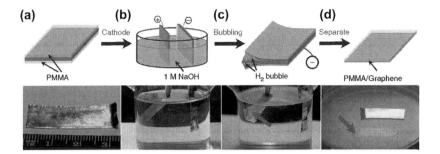

FIGURE 4.8.3 Illustration of the bubbling transfer process of graphene from a Pt substrate. (a) A Pt foil with grown graphene covered by a PMMA layer. (b) The PMMA/graphene/Pt in (a) was used as a cathode, and a Pt foil was used as an anode. (c) The PMMA/graphene was gradually separated from the Pt substrate driven by the H_2 bubbles produced at the cathode after applying a constant current. (d) The completely separated PMMA/graphene layer and Pt foil after bubbling for tens of seconds. The PMMA/graphene layer is indicated by a red arrow in (c) and (d) (Gao et al., 2012).

etching, which is relevant when inert or noble metals are used as the growth substrate. However, for now, this method seems to be restricted to electrically conductive metal substrates alone.

4.8.4. TRANSFER OF GRAPHENE GROWN ON SiC

The growth of epitaxial graphene layers on carbon-terminated face of SiC has received attention, since such grown graphene films are found to be of high quality. Hence, this is anticipated as a means to grow large area high quality graphene (Tedesco et al., 2008). The SiC is highly resistant to chemical etchants; due to this the wet-chemical etching approach for transfer as implemented in the case of metal-substrate grown graphene cannot be applied. Owing to this, thus far, the transfer of epitaxial graphene on SiC is limited to dry transfer approaches involving the use of adhesive tapes. Lee et al. (2008) have shown that the use of conventional scotch tape, as used for the mechanical cleavage of HOPG, can also be applied to transfer epitaxial graphene grown on SiC. However, even after repeated attaching of an adhesive tape on the graphene films the yield of monolayer graphene transferred to arbitrary substrates, though possible, is very low. In most cases, the sizes of the transferred graphene flakes are $1 \mu m^2$ or less (Lee et al., 2008). This is attributed to the relatively weak bonding strength between the adhesive tape and the graphene layers compared to their bonding with the SiC growth substrate. To overcome this problem and achieve better adhesion before peeling off, Unarunotai et al. (2009) proposed a modified dry transfer approach. Initially, a 100-nm layer of Au is deposited through electron beam evaporation on the epitaxial graphene/SiC substrate. Then, a layer of polyimide (PI) polymer (thickness ~ 1.4 μm) is spin-coated and cured to serve as an additional support layer for ease of handling during subsequent mechanical peeling. The PI/Au/graphene film is mechanically peeled off the SiC substrate manually and can then be transferred to any arbitrary substrate. After this, the PI and the Au support layers are etched away using oxygen plasma reactive etching and a wet chemical etching step respectively (Unarunotai et al., 2009). Though transfer of large area epitaxial graphene films are reported to be possible using this approach it is important to note that the deposition of support layers and their later removal were found to induce defects in the otherwise high-quality graphene films (Caldwell et al., 2010). A modified dry transfer route is reported by Caldwell et al. involving the use of a special kind of adhesive tape, namely, thermal release tape (TRP). They investigated the effect of applying an external force to improve the adhesion between the tape and epitaxial graphene layers for transferring graphene grown on SiC (Caldwell et al., 2010). In addition, they implemented a specific pretreatment of their target transfer substrate SiO2-coated Si substrate to further improve the adhesion of the transferred graphene. Figure 4.8.4 schematically illustrates their approach using TRP.

In the first step, a piece of TRP is pressed on to the epitaxial graphene. After this, the TRP/EG/SiC stack is transferred into a wafer bonding apparatus and

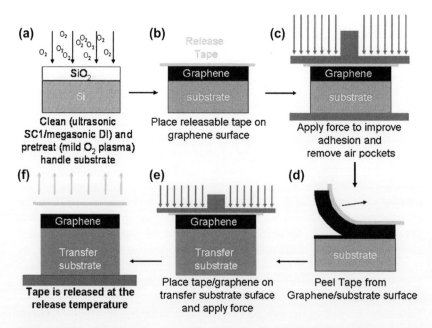

FIGURE 4.8.4 Schematic illustration of the transfer of EG grown on SiC to SiO2/Si substrate using modified thermal release tape as the support and transfer carrier for graphene (Caldwell et al., 2010).

a 2″ stainless steel pressure plate mounted on the stack. After evacuating to around 5×10^{-4} Torr pressure, a compression force ranging from 3–6 N/mm^2 is applied to the steel plate to uniformly press the TRP onto the graphene/SiC for 5 min. The authors observed an improvement in the quality of the transferred graphene (reduced number of voids) by increasing the force from 3 to 5 N/mm^2, but when further higher force of 6 N/mm^2 was applied only small islands of EG were found to be transferred; thus, 5 N/mm^2 was found to be the optimum bonding force to apply (Caldwell et al., 2010). After the pressing step, the TRP was mechanically peeled off the SiC substrate along with the epitaxial graphene, thereby achieving dry separation of graphene from the SiC substrate. The peeled off film can then be transferred to an arbitrary substrate (in this case, SiO$_2$/Si substrate). The transferred film is then pressed using the steel plate onto the transferred substrate as before. To release the TRP, the TRP/graphene/SiO$_2$-Si stack is heated at 1–2 °C above the release temperature of the TRP (120 °C) at which the adhesive capability of the tape is inhibited facilitating in easy removal of the TRP tape. However, the TRP tape was found to leave some contamination on the transferred graphene and requires further cleaning steps, using organic solvents (e.g. 1:1:1 toluene/methane/acetone) to remove the tape residue, followed by annealing at 250 °C in air for about 10 min.

Though the dry transfer methods seems, so far, the only viable means to transfer epitaxial graphene grown on SiC it is important to note that none of the above discussed methods claim the complete removal of all the EG layers grown on the SiC (mainly owing to mechanical peeling). Usually, the top graphene layers are easy to extract through these approaches leaving behind at least 1 to 2 EG layers closest to the SiC top growth layer. This emphasises the need for further investigation, optimisation and development of alternative transfer routes epitaxial graphene grown on SiC.

4.8.5. TOWARDS A UNIVERSAL TRANSFER ROUTE FOR GRAPHENE GROWN ON ARBITRARY SUBSTRATES

The current research trends in the CVD growth of graphene for its application into electronic devices is headed towards the direct growth of graphene on insulating metal oxide substrates (e.g. SiO$_2$, sapphire) for seamless integration into current semiconducting industry process technologies without the need for transferring graphene grown on a different substrate. However, to realise this, a fundamental understanding of the graphene grown directly on metal oxides is pivotal. For carrying out such investigations it is necessary to transfer the graphene directly grown on metal oxides to arbitrary substrates, for example TEM grids. Such a transfer in the case of graphene grown on SiO$_2$ is feasible using a polymer support and carrier film coating and subsequently etching away SiO$_2$ in HF solution. However, it is not so straight forward in the case of growth over sapphire substrates because of the high chemical inertness of sapphire similar to SiC. In our current attempts to overcome this challenge (in particular for the case of sapphire) we have developed a simple wet chemical separation route with which one can successfully isolate monolayer graphene directly grown on sapphire by CVD. Moreover, through this approach one can successfully separate graphene from Cu and Mo/Ni metal foils without the need for conventional metal etching. In addition, one can easily separate graphene deposited on SiO_2/Si substrates without the need for etching away SiO_2. Hence, the route holds promise as a universal transfer route for graphene grown on arbitrary substrates. In brief, the procedure requires one to deposit a PMMA layer over the graphene/ support substrate through spin-coating and subsequent curing. Then the PMMA/graphene/sapphire stack is placed in a solution of 1:1:3 NH_4OH:- H_2O_2:H_2O (vol %). The solution is then heated to 80 °C on a hot plate. The dissociation of H_2O_2 in this solution results in the release of O_2 gas, which causes bubbling in the solution. This bubbling in the solution serves as the peeling force to naturally separate PMMA/graphene films from the underlying sapphire substrate. The mechanism of bubbling separation we observe is similar to that observed in the case of electrochemical separation reported by Gao et al. (2012); Wang et al. (2011) but is much simpler and dispenses the need for an electrochemical reaction. After the complete decomposition of the peroxide the bubbling action in the solution ceases and now the PMMA/ graphene film can be manually peeled off with tweezers from the sapphire substrate or can be gently immersed in DI water allowing the surface tension of water to separate PMMA/graphene from the substrate.

4.8.6. SUMMARY

Often, it is the case that the substrate used to produce high quality graphene is not the final substrate we are interested in having graphene residing on. Transfer to a different substrate enables new applications and measurement capabilities. Graphene typically needs some form of support (usually polymer based) when being transferred and also the use of an etch solution to separate it from its current substrate. Removing the support and all contamination from the etchant solution is often a major challenge in obtaining clean undoped graphene. There is nothing to be gained by producing outstanding synthetic graphene on a metal catalyst substrate and then inducing defects and dopants during the transfer phase. From our experience in examining the atomic structure of graphene using TEM, even if graphene is cleaned, if it is exposed to ambient atmosphere it can re-accumulate hydrocarbon contamination within days, possibly hours. The choice of polymer used to support graphene during transfer is important as well, since some polymers are easier than others to remove by dissolving in solvents. For synthetic graphene to make further inroads in applications, not only do large area sheets need to be produced but the capabilities of transferring these large sheets reproducibly as well.

The element of human uncertainty, such as shaky hands, need to be eliminated from the process line and replaced with semi-automated systems that can reliably transfer graphene each time.

REFERENCES

Blake, P., Hill, E.W., Neto, A.H.C., Novoselov, K.S., Jiang, D., Yang, R., Booth, T.J., Geim, A.K., 2007. Making graphene visible. Appl. Phys. Lett. 91 063124-(1-3).

Bunch, J.S., Verbridge, S.S., Alden, J.S., van der Zande, A.M., Parpia, J.M., Craighead, H.G., McEuen, P.L., 2008. Impermeable atomic membranes from graphene sheets. Nano Lett. 8, 2458–2462.

Caldwell, J.D., Anderson, T.J., Culbertson, J.C., Jernigan, G.G., Hobart, K.D., Kub, F.J., Tadjer, M.J., Tedesco, J.L., Hite, J.K., Mastro, M.A., Myers-Ward, R.L., Eddy, C.R., Campbell, P.M., Gaskill, D.K., 2010. Technique for the dry transfer of epitaxial graphene onto arbitrary substrates. ACS Nano 4 (2), 1108–1114.

Cha, M., Shin, J., Kim, J.H., Kim, I., choi, J., Lee, N., Kim, B.G., Lee, J., 2008. Biomolecular detection with a thin membrane transducer. Lab Chip 8, 932–937.

Gao, L., Ren, W., Xu, H., Jin, L., Wang, Z., Ma, T., Ma, L., Zhang, Z., Fu, Q., Peng, L., Bao, X., Cheng, H., 2012. Repeated growth and bubbling transfer of graphene with millimetre-size single-crystal grains using platinum. Nat. Commun. 3 699-(1-7).

Jiao, L., Fan, B., Xian, X., Wu, Z., Zhang, J., Liu, Z., 2008. Creation of nanostructures with poly(methyl methacrylate)-mediated nanotransfer printing. J. Am. Chem. Soc. 130, 12612–12613.

Lee, D.S., Riedl, C., Krauss, B., Klitzing, K., Starke, U., Smet, J.H., 2008. Raman spectra of epitaxial graphene on SiC and of epitaxial graphene transferred to SiO_2. Nano Lett. 8 (12), 4320–4325.

Li, X., Cai, W., An, J., Kim, S., Nah, J., Yang, D., Piner, R., Velamakanni, A., Jung, I., Tutuc, E., Banerjee, S.K., Colombo, L., Ruoff, R.S., 2009a. Large-area synthesis of high-quality and uniform graphene films on copper foils. Science 324, 1312–1314.

Li, X., Zhu, Y., Cai, W., Borysiak, M., Han, B., Chen, D., Piner, R.D., Colombo, L., Ruoff, R.S., 2009b. Transfer of large-area graphene films for high-performance transparent conductive electrodes. Nano Lett. 9 (12), 4359–4363.

Lin, Y., Jin, C., Lee, J., Jen, S., Suenaga, K., Chiu, P., 2011. Clean transfer of graphene for isolation and suspension. ACS Nano 5 (3), 2362–3268.

Lu, X., Huang, H., Nemchuk, N., Ruoff, R.S., 1999. Patterning of highly oriented pyrolytic graphite by oxygen plasma etching. Appl. Phys. Lett. 75, 193–195.

Meyer, J.C., Girit, C.O., Crommie, M.F., Zettl, A., 2008. Hydrocarbon lithography on graphene membranes. Appl. Phys. Lett. 92 123110-(1-3).

Novoselov, K.S., Geim, A.K., Morozov, S.V., Jiang, D., Zhang, Y., Dubonos, S.V., Grigorieva, I.V., Firsov, A.A., 2004. Electric field effect in atomically thin carbon films. Science 306, 666–669

O'Hern S.C., 2011. Development of Process to Transfer Large Areas of LPCVD Graphene from Copper foil to Porous Support Substrate, Master Thesis, MIT-USA.

Pham P., 2010. Transferring chemical vapor deposition grown graphene. Proc. 2010 NNIN REU Convocation, 192–193.

Regan, W., Alem, N., Alemán, B., Geng, B., Girit, C., Maserati, L., Wang, F., Crommie, M., Zettl, A., 2010. A direct transfer of layer-area graphene. Appl. Phys. Lett. 96 113102-(1-3).

Reina, A., Son, H., Jiao, L., Fan, B., Dresselhaus, M.S., Liu, Z.F., Kong, J., 2008. Transferring and identification of single- and few-layer graphene on arbitrary substrates. J. Phys. Chem. C 112 (46), 17741–17744.

Reina, A., Jia, X., Ho, J., Nezich, D., Son, H., Bulovic, V., Dresselhaus, M.S., Kong, J., 2009. Large area, few-layer graphene films on arbitrary substrates by chemical vapor deposition. Nano Lett. 9 (1), 30–35.

Suk, J.W., Kitt, A., Magnuson, C.W., Hao, Y., Ahmed, S., An, J., Swan, A.K., Goldberg, B.B., Ruoff, R.S., 2011. Transfer of CVD-grown monolayer graphene onto arbitrary substrates. ACS Nano 5 (9), 6916–6924.

Tedesco, J.L., vanMil, B.L., Myers-ward, R.L., McCrate, J.M., Kitt, S.A., Campbell, P.M., Jernigan, G.G., Culbertson, J.C., Eddy, C.R., Gaskill, D.K., 2008. Hall effect mobility of epitaxial graphene grown on silicon carbide. Appl. Phys. Lett. 95 122102-(1-3).

Unarunotai, S., Murata, Y., Chialvo, C.E., Kim, H., MacLaren, S., Mason, N., Petrov, I., Rogers, J.A., 2009. Transfer of graphene layers grown on SiC wafers to other substrates and their integration into field effect transistors. Appl. Phys. Lett. 95 202101-(1-3).

Wang, Y., Zheng, Y., Xu, X., Dubuisson, E., Bao, Q., Lu, J., Loh, K.P., 2011. Electrochemical delamination of CVD grown graphene film: toward the recyclable use of copper catalyst. ACS Nano 5 (12), 9927–9933.

Characterisation Techniques

Optical Microscopy

Alicja Bachmatiuk

IFW Dresden, Germany

The optical light microscope is a very important tool for the study of micro-structure, and it is still a very useful technique even against other imaging microscopes, such as, e.g. scanning or transmission electron microscopes. Reflected light microscopy, which is commonly used for imaging metallic microstructures and transmitted light microscopy which can be efficiently used in the imaging of non-metals, for example, polymers. The choice of method depends on the measurement requirements such as magnification, resolution and depth of view. In an optical microscope the object is illuminated with focused light from the condenser lens which is placed perpendicularly to the axis of the objective lens. Part of the light is transmitted through the sample and some is reflected back to the objective lens which forms a magnified view of the sample. Light coming from the objective is diverted by a beam splitter/prism combination either into the eyepieces to form a virtual image, or through to the projection lens, where it can then form an image on a CCD photodiode array inside a digital camera. Usually the magnification of an optical microscope in the visible light range is up to $\times 1500$ with a theoretical resolution limit of around 0.2 micrometers due to diffraction limits. The image seen in the microscope depends not only on how the specimen is illuminated and positioned, but also on the characteristics of the specimen.

With the expansion of graphene technology, it became important to determine or verify the number of the deposited graphene layers. Optical microscopy is a well-established technique which in the first place can be used for surveying large area graphene samples due to the technique's simplicity. In fact optical microscopy allows for a quick thickness inspection before using more

Graphene. http://dx.doi.org/10.1016/B978-0-12-394593-8.00005-9

precise methods such as Raman spectroscopy, atomic force microscopy (AFM) or scanning and transmission electron microscopy (Roddaro et al., 2007). Information on optical imaging of thin graphite layers have sometimes been controversial as well: in some cases single-layer graphene is said to be invisible (Novoselov et al., 2004,2005), while other studies imply that it can be seen by straightforward optical means (Zhang et al., 2005).

Graphene visibility using optical microscopy is usually explained by the change of the interference colour of reflected light from graphene with respect to the empty substrate as well as by graphite's opacity which is sometimes considered as the key element in explaining this effect (Roddaro et al., 2007). Additionally, the substrate supporting graphene and the microscope equipment (e.g. objective lens, employed light source or filters) explored for the actual graphite visibility, play a significant role for graphene observations. A typical value for the numerical aperture (NA) of the objective used for graphene observations is around 0.8. A smaller NA value (e.g. 0.2) will increase the contrast of thin graphitic samples even though the image details can be negatively affected (Roddaro et al., 2007).

It takes a fair bit of effort to find monolayer graphene using an optical microscope on most substrates. The most popular substrate surface on which single-layer graphene becomes visible is a silicon wafer with a silicon dioxide layer (usually 300-nm thick). Preparation of graphene devices used for experimental studies rely on the fact that graphene layers can be observed using optical microscopy if prepared on top of Si wafers with a certain thickness of SiO_2. It is well-known that the thickness of a SiO_2 film grown on top of a Si substrate can be determined with some precision merely by evaluating its apparent colour (Henrie et al., 2004). This fact is due to the interference between the reflection paths that originate from the two air-to-SiO_2 and SiO_2-to-Si interfaces. Depending on their spacing, the interfering paths will experience relative phase shifts: thickness variations of a fraction of wavelength lead to colour shifts that can be easily appreciated by eye (Roddaro et al., 2007).

However two main facts, namely thickness of the silicon dioxide layer and incident light wavelength, depend strongly on graphene's visibility. By using monochromatic light, graphene can be isolated for any SiO_2 thickness, although 300 nm SiO_2 and, particularly, ~100 nm are most appropriate for its visual detection when using a white light source and 'naked eye' during the observations (Blake et al., 2007). Different thicknesses of SiO_2 for example 200 nm are not so suitable for graphene observation with a simple optical microscope (graphene is completely invisible). Only flakes thicker than ten layers can be found in white light on top of a 200-nm SiO_2 layer. In this case it is necessary to use a blue filter for single layer graphene observations as it enables it to be more easily observed even on top of 200-nm SiO_2 (Blake et al., 2007). In comparison to a monochromatic light source, the use of white light allows fast classification of graphene thickness regions due to the fact that graphene with different ranges of thickness can exhibit different colour bands that can easily be appreciated by the naked eye (Roddaro et al., 2007). Nevertheless, only minor colour differences between monolayer graphene and the substrate and between samples with different numbers of graphene layers are observed even for the well-accepted optimum substrate (Blake et al., 2007). The visibility of thin graphite on SiO_2/Si is generally agreed to allow for a few monolayers, with a total thickness of the order of 1 nm to still be identified (Novoselov et al., 2005). Most of the visibility effect is due to a modulation of the relative amplitude of the interfering paths. This is a result of the fact that graphite transparency depends on thickness, since graphite is a good conductor. Relative amplitude modulations are strongly visible by a favourable combination of permittivity values in the SiO_2/Si multilayer. This leads to a resonant cancellation of reflection by destructive interference at specific wavelengths for finely tuned and relatively thin graphite layers (Roddaro et al., 2007). Figure 5.1.1 shows the optical images using white light source on different thickness of graphene sheets on a Si wafer with a 280 nm SiO_2 capping layer (left panel), together with the contrast spectra (right panel) (Ni et al., 2007). From optical images using white-light illumination and the 'naked eye' (left panel) one can observe different colours and contrasts for different thicknesses

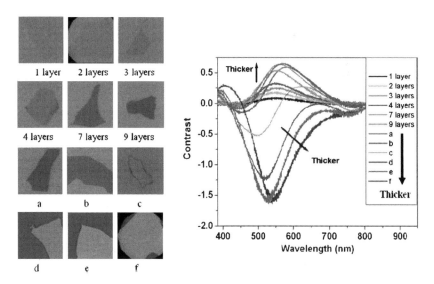

FIGURE 5.1.1 *Left panel:* the optical images of the samples with different thicknesses of graphene sheets, *Right panel:* the contrast spectra. Besides the samples with one, two, three, four, seven, and nine layers, samples a, b, c, d, e, and f are more than 10 layers and the thickness increases from a to f. The arrows in the graph show the trend of curves in terms of the thicknesses of graphene sheets (Ni et al., 2007).

of graphene due to the number of sheets on the Si wafer with a 280-nm SiO_2 layer. In this case, the characteristic visibility of the graphene is mainly due to an interference effect between the graphene and the thin layer of SiO_2 (Blake et al., 2007; Ni et al., 2007). Even though one can observe different colours and contrasts for graphene sheets of different thickness using the optical microscopy and the 'naked eye', still graphene's visibility strongly changes from one laboratory to another and it depends on experience of the observer and utilised equipment.

Therefore, one can use the contrast spectra to obtain a quantitative and accurate thickness determination, albeit different for each research laboratory (Fig. 5.1.1-right panel). The contrast spectra for different thicknesses of graphene sheets are obtained using the following calculation:

$$C(\lambda) = (R_0(\lambda) - R(\lambda))/R_0(\lambda) \tag{5.1.1}$$

where $R_0(\lambda)$ is the reflection spectrum from the SiO_2/Si substrate and $R(\lambda)$ is the reflection spectrum from graphene sheet on top of the substrate. With this method the contrast across the whole visible range (with a spectrum resolution higher than 1 nm) can be measured without any bandpass filter (Ni et al., 2007). For mono layer graphene the contrast spectrum has a peak centred at 550 nm, which is in the green–orange range and makes single-layer graphene visible. In addition, the position of the contrast peak is almost stable (~550 nm) with increasing numbers of graphene layers up to 10. The contrast value for single-layer graphene is about 0.09 and it increases with the number of layers, for example, 0.175, 0.255 and 0.330 for two, three and four layers, respectively. For graphene with around 10 layers the contrast of the sample saturates and the contrast peak shifts toward higher wavelengths. Additionally for samples with a larger number of layers negative contrast occurs due to their thickness. The reflections from their surface become stronger than that from the SiO_2/Si substrate, causing negative value contrast (Ni et al., 2007).

The existence of the contrast can be explained by Fresnel's equations, where the incident light from air ($n_0 = 1$) onto a trilayer system (graphene, SiO_2 and Si) should be considered. The reflected light intensity from such a trilayer system can then be described by the following equations (Anders, 1967; Blake et al., 2007; Ni et al., 2007):

$$R(\lambda) = r(\lambda)r * (\lambda) \tag{5.1.2}$$

$$r(\lambda) = r_a/r_b \tag{5.1.3}$$

$$r_a = \left(r_1 e^{i(\beta 1 + \beta 2)} + r_2 e^{-i(\beta 1 - \beta 2)} + r_3 e^{-i(\beta 1 + \beta 2)} + r_1 r_2 r_3 e^{i(\beta 1 - \beta 2)} \right) \tag{5.1.4}$$

$$r_b = \left(e^{i(\beta 1 + \beta 2)} + r_1 r_2 e^{-i(\beta 1 - \beta 2)} + r_1 r_3 e^{-i(\beta 1 + \beta 2)} + r_2 r_3 e^{i(\beta 1 - \beta 2)} \right) \tag{5.1.5}$$

where:

$$r_1 = (n_0 - n_1)/(n_0 + n_1) \qquad (5.1.6)$$

$$r_2 = (n_1 - n_2)/(n_1 + n_2) \qquad (5.1.7)$$

$$r_3 = (n_2 - n_3)/(n_2 + n_3) \qquad (5.1.8)$$

are the reflection coefficients for different interfaces, and

$$\beta_1 = 2\Pi n_1 (d_1/\lambda) \qquad (5.1.9)$$

$$\beta_2 = 2\Pi n_2 - (d_2/\lambda) \qquad (5.1.10)$$

are the phase differences when light passes through the media, which is determined by the path difference of two neighbouring interfering light beams. The thickness of the graphene sample can be estimated as:

$$d_1 = N\Delta d \qquad (5.1.11)$$

where N represents the number of layers and Δd is the thickness of mono-layer graphene ($\Delta d = 0.335$ nm) (Dresselhaus et al., 1996; Kelly, 1981). As the refractive index of graphene (n_1) the value for graphite can be used (2.6–1.3i) (Blake et al., 2007). d_2 is the thickness of silicon dioxide layer, n_2 – refractive index of silicon dioxide which is wavelength dependent, n_3 – refractive index of silicon is also wavelength dependent. However, $R_0(\lambda)$ – the reflection from silicon dioxide background is calculated by using $n_1 = n_0 = 1$, and $d_1 = 0$ (Ni et al., 2007). By using a Fresnel-law-based model, it is possible to investigate the dependence of the contrast on SiO_2 thickness and light wavelength, and it is possible to quantitatively describe the experimental data (Blake et al., 2007). Understanding the source of the observed contrast is essential for optimizing the graphene detection technique and transferring it to different substrates, aiding experimental progress in the research area. Another fact that, under technically the same observation conditions, graphene's visibility strongly changes between the laboratories, is a reason caused by different cameras providing better imaging for graphene observations and also relies on experience of the observer (Blake et al., 2007).

Figure 5.1.2 provides a guide for the investigation of graphene on top of SiO_2/Si wafers as a colour plot for the expected contrast as a function of SiO_2 thickness and wavelength (Blake et al., 2007). The presented plot shows the most appropriate filters for graphene observation on the top of a known silicon dioxide thickness.

By using filters almost any thickness of SiO_2 (except for around 150 nm and below 30 nm) one can visualise graphene samples. For 300-nm SiO_2 layers, the main contrast appears using a green filter, and graphene is undetectable in blue light. Additionally, the use of a blue filter makes graphene visible on top of 200 nm SiO_2. From different filters the green filter is the most comfortable to an observer's eyes (Blake et al., 2007) and this makes SiO_2 thicknesses of

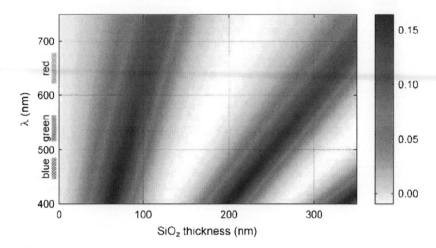

FIGURE. 5.1.2 Colour plot of the contrast as a function of wavelength and SiO$_2$ thickness according to Eqn. (5.1.1). The colour scale on the right shows the expected contrast (Blake et al., 2007).

approximately 90 and 280 nm most appropriate with the use of green filters as well as without any filters, in white light. Additionally, the lower thickness of ca. 90 nm provides a better choice for graphene's detection, and it is suggested as a substitute for the present benchmark thickness of ca. 300 nm (Blake et al., 2007). Moreover, the changes in the light intensity due to graphene are relatively minor, which allows the observed contrast to be used for measuring the number of graphene layers.

The current focus on SiO$_2$/Si is not so much given by any unique physical property but simply by the fact that graphene sits on a thin dielectric, which makes it easy to identify even with standard optical microscopy using white light illumination. As already stated, the optimum SiO$_2$ thickness to make graphene visible calculated using the Fresnel formula is 90 nm. This results in a maximum contrast of C of around 12%, which is easily visible in a standard optical microscope. However even a small change of SiO$_2$ thickness, say from 300 nm to 315 nm can significantly lower the contrast of visible graphene flakes (Blake et al., 2007; Geim and Novoselov, 2007).

Other substrates such as crystalline insulators (e.g. CaF$_2$, Al$_2$O$_3$, SrTiO$_3$ and TiO$_2$,) can also be used to prepare (by mechanical exfoliation and transfer) and identify (by optical microscopy) ultra-thin sheets of graphene on their surfaces – Fig. 5.1.3 (Akcoltekin et al., 2009).

The optical contrast on all the above mentioned substrates is rather poor in comparison to graphene on a 90 nm SiO$_2$ layer (a few percent instead of 12%), but in all cases it is still sufficient to identify regions with single layer flakes

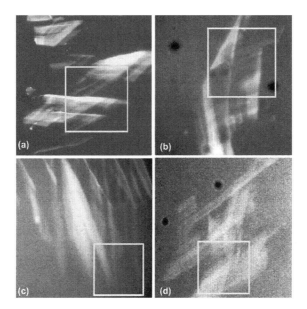

FIGURE. 5.1.3 Images taken by optical microscopy of graphene crystallites consisting of different numbers of graphene sheets found on the surface regions marked by the squares (a) CaF_2, (b)Al_2O_3, (c) $SrTiO_3$ and (d) TiO_2 (Akcoltekin et al., 2009).

(Akcoltekin et al., 2009). This fact shows that considerable contrast can be achieved also on bulk dielectrics without the need for thin films. Furthermore, other insulators, for example, 50-nm Si_3N_4 using blue light or 90-nm poly(methyl methacrylate) (PMMA) using white light can also be used to visualise graphene (Blake et al., 2007).

An interesting way to quantitatively study the effect of the light source and substrate on the optical imaging of graphene is the total colour difference (TCD) technique, which is based on a combination of the reflection spectrum calculation and international commission on illumination (CIE) colour space (Gao et al., 2008; Janos, 2007). In the visible light range the image contrast of graphene is a mixture of all the contrasts for each wavelength part. Thus, a colourful image is acquired in this case instead of the black and white image for narrow band illumination, therefore a new standard as a replacement for of the commonly used monochrome contrast is required due to the colour factor. The CIE colour space contains both brightness and colour and is based on the colour sensitivity and perception of human eyes (Janos, 2007). Therefore, the TCD is proposed as a colour standard based on a combination of the reflection spectrum calculation and CIE colour space, in order to quantitatively describe the contrast of colour image. By employing this method one can verify the most suitable dielectric layer (and its thickness) for maximizing the visibility of a graphene multilayer. It has been proven, using this technique that the best substrate for graphene characterisation using optical microscopy is a 72 nm thick layer of Al_2O_3 deposited on top of a silicon wafer, which is even better than deposited on silicon SiO_2 or Si_3N_4 films (Gao et al., 2008). The fact that a Si wafer with 72 nm Al_2O_3 film is the best for quick and accurate optical microscopy characterisation under visible light is related to the high sensitivity of the colour bands on the number of layers and larger total colour difference between different layers of graphene. What is more is that the image contrast of graphene layers can be improved significantly if the wavelength range of the light source is narrowed properly, instead of the commonly recommended monochromatic light (Blake et al., 2007; Jung et al., 2007).

It was shown that TCD between graphene samples with one layer difference is markedly increased when the wavelength range is decreased properly. As an example, the TCD values among single layer graphene and substrate and between graphene with one layer difference are raised from 2.3 to 4.6 and from 1.7 to 3.1, respectively, when the wavelength of light source is changed from 380–720 nm to 430–570 nm (Gao et al., 2008). Consequently, the visibility of single layer graphene and the contrast between graphene with diverse layers were significantly improved. This is important for the precise determination of graphene layers. Nevertheless, if the wavelength range of the light is too narrow (close to monochromatic light), the TCD turns out to be smaller. This suggests that monochromatic light is not a perfect choice for the measurements of graphene using the optical method. It has also been shown that the objective lens magnification has no significant influence on the TCD values of the graphene samples deposited on top of the Si wafer with 72 nm Al_2O_3 capping layer. In the case of the numerical aperture (NA) value of the objective lens it was found that the smaller the NA the larger the TCD values for less than 15 graphene layers. For layer numbers ranging between 16 and 35, the TCD values for the graphene drop.

Regarding characterisation using an optical microscope of grapheneoxide, a different behaviour when deposited on top of the substrates can be exhibited (Gao et al., 2008; Jung et al., 2007). The reflectance of grapheneoxide differs depending on the wavelength of the incident light and the material thickness. For the thickness region close to 0 nm, the smallest reflectance (R_{min}) appears near a wavelength of 530 nm and the highest reflectance (R_{max}) appears near the shortest given wavelength (ca. 380 nm). With increasing thickness, the wavelength for the smallest and highest reflectance also increases. The ratio between the two extreme reflectances (R_{max}/R_{min}) is 5.1 for the thickness region near 0 nm and becomes even higher, upto 11.5, when the thickness reaches around 300 nm, demonstrating that the interference effect is active over a range of material thicknesses. In comparison to graphene-oxide, the variation in the reflectance of graphene reduces significantly as the material thickness increases. For a thickness of 50 nm, the R_{max}/R_{min} is around 2.2. This value decreases close to 1.38 for a thickness of 100 nm and further reduces to 1.01 when the thickness reaches 300 nm. These values show that the interference effect in the graphene multilayer diminishes as the material thickness increases (Jung et al., 2012). The reflectance of graphene-oxide notably changes over a wide range of material thicknesses, due to the fact that graphene-oxide has a lower extinction coefficient than graphene. Thus, the resulting colours of graphene-oxide continually change as a function of the material thickness, but the colours of the graphene multilayer samples became saturated. Thus colour can potentially be employed as a tool for easy recognition and simple evaluation of the thicknesses of both graphene and graphene-oxide multilayer samples (Jung et al., 2012).

To conclude, optical microscopy is a very useful tool as a direct and complimentary microscopic technique for the imaging of graphene samples with single and higher numbers of layers.

REFERENCES

Akcoltekin, S., El Kharrazi, M., Kohler, B., Lorke, A., Schleberger, M., 2009. Nanotechnology 20, 155601.

Anders, H., 1967. Thin Films in Optics. The Focal Press, London.

Blake, P., Hill, E.W., Castro Neto, A.H., Novoselov, K.S., Jiang, D., Yang, R., Booth, T.J., Geim, A.K., 2007. Appl. Phys. Lett. 91, 063124.

Dresselhaus, M.S., Dresselhaus, G., Eklund, P.C., 1996. Science of Fullerenes and Carbon Nanotubes. Academic Press, San Diego, CA. 965.

Gao, L., Ren, W., Li, F., Cheng, H.M., 2008. ACS Nano 2, 1625.

Geim, A.K., Novoselov, K.S., 2007. Nat. Mater. 6, 183.

Henrie, J., Kellis, S., Schultz, S.M., Hawkins, A., 2004. Opt. Express 12, 1464.

Janos, S., 2007. Colorimetry: Understanding the CIE System. John Wiley & Sons, New Jersey, pp 25–88.

Jung, I., Pelton, M., Piner, R., Dikin, D.A., Stankovich, S., Watcharotone, S., Hausner, M., Ruoff, R.S., 2007. Nano Lett. 7, 3569.

Jung, I., Rhyee, J.S., Son, J.Y., Ruoff, R.S., Rhee, K.Y., 2012. Nanotechnology 23 025708.

Kelly, B.T., 1981. Applied Science London.

Ni, Z.H., Wang, H.M., Kasim, J., Fan, H.M., Yu, T., Wu, Y.H., Feng, Y.P., Shen, Z.X., 2007. Nano Lett. 7, 2758.

Novoselov, K.S., Geim, A.K., Morozov, S.V., Jiang, D., Zhang, Y., Dubonos, S.V., Grigorieva, I.V., Firsov, A.A., 2004. Science 306, 666.

Novoselov, K.S., Jiang, D., Schedin, F., Booth, T.J., Khotkevich, V.V., Morozov, S.V., Geim, A.K., 2005. Proc. Natl. Acad. Sci. U.S.A. 102, 10451.

Roddaro, S., Pingue, P., Piazza, V., Pellegrini, V., Beltram, F., 2007. Nano Lett. 7, 2707.

Zhang, Y., Tan, Y.W., Stormer, H.L., Kim, P., 2005. Nature 438, 201.

———————————————————————————————— (Chapter 5.2)

Raman Spectroscopy

Alicja Bachmatiuk

IFW Dresden, Germany

5.2.1. INTRODUCTION

Raman spectroscopy is a spectroscopic technique where the inelastic scattering of monochromatic light, typically from a laser source interacts with a sample. During Raman measurements the sample (solid, liquid or gaseous) is irradiated with a laser source in the visible, near infrared or ultraviolet light range. The scattered light is then gathered with a system of lenses and sent through spectrophotometer to obtain a Raman spectrum. The frequency of the photons forming the monochromatic light source can change when interacting with matter. In this process the sample absorbs the incoming photons and then re-emits them with the same or different frequency. Most of the scattered light has the same frequency as the incident light and hence, is very strong and is termed Rayleigh (elastic) scattering. The second scattering process is inelastic and has an altered or shifted frequency and is known as Raman scattering. It is very weak ($\sim 10^{-5}\%$ of the incident beam). Two different types of Raman scattering can be distinguished, namely Stokes scattering where the frequency of scattered light is reduced and Anti-Stokes scattering where the resulting frequency of the scattered light increases. These characteristic (Raman) shifts provide valuable information about vibrational, rotational and other low frequency changes in the samples (Ferraro et al., 2003).

Historically, Raman spectroscopy has always played an important role in the structural characterisation of graphitic materials (Cancado et al., 2004; Dresselhaus et al., 2005). Moreover it is a powerful tool to better understand the interaction of electrons and phonons in graphene samples. Due to the intrinsic dispersion of the π electrons in graphene, Raman spectroscopy of this carbon material is always resonant which makes it an effective tool with which to characterise sample vibrations as well as its electronic properties (Cancado et al., 2008a). Common to all carbon based systems (e.g. fullerenes, graphite or conjugated polymers), the Raman spectra contain only a few significant modes (including second-order peaks) in the spectral region 1000–2000 cm^{-1} (Ferrari and Robertson, 2004).

In the specific case of graphene, the Raman signal variations observed for different numbers of graphene layers not only demonstrate changes in the electron bands but also provide an easy and non-destructive means to determine single, bi-layer and few-layer graphene when stacked in the

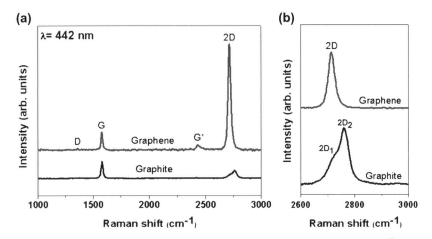

FIGURE 5.2.1 (a) Raman spectra of graphene and graphite, showing the main Raman features, the D, G, G* and 2D bands taken with a 442 nm laser. (b) Zoomed 2D region of graphene and graphite.

Bernal (AB) configuration (Ferrari et al., 2007; Malard et al., 2009). The most prominent Raman features from graphene are the so-called G mode and 2D mode as easily seen in Fig. 5.2.1, which shows a typical Raman spectrum for graphene and graphite respectively obtained using a 442 nm excitation laser.

The G mode resides around 1580 cm^{-1} and the 2D peak sits around 2700 cm^{-1}. Two further peaks can also be observed; the D peak may appear ca. 1350 cm^{-1} and G* peak at ca. 2450 cm^{-1}. In terms of their vibrational character, the G mode corresponds to bond stretching of all pairs of sp^2 atoms in both rings and chains, whereas the D mode arises due to the breathing modes of sp^2 atoms in rings. Usually the D mode is a forbidden transition, however in the presence of disorder (e.g. a defect) symmetry is broken and the transition is allowed. (Castiglioni et al., 2001; Ferrari and Robertson, 2000; Tuinstra and Koenig, 1970). Of all the graphene based peaks observed in its Raman spectrum, it is only the G mode, which originates from a first order Raman scattering process and is linked to a doubly degenerate in-plane, zone centre phonon mode (TO transverse optical and LO longitudinal optical) with E$_{2g}$ symmetry (Tuinstra and Koenig, 1970). The D and 2D modes originate from a second-order double resonant process between nonequivalent K points in graphene's first Brillouin zone (BZ), connecting two zone-boundary phonons (TO-derived) for the 2D mode and in the D mode, a single phonon and a defect (Maultzsch et al., 2004). Hence, in well-ordered (viz. defect free) graphene and graphite, the D band is absent (Dresselhaus et al., 2005). The G* mode originates from a combination of the zone-boundary in-plane longitudinal acoustic phonon and the in-plane transverse optical phonon modes

(Mafra et al., 2007; Shimada et al., 2005). The position of this mode red-shifts with increasing number of graphene layers from 2455 to 2445 cm^{-1}. Within different number of graphene layers the shape of the band remains the same, except for single-layer graphene, where it can be quite sharp. The 2D mode is a second order two-phonon process and corresponds to the overtone of the D band. Sometimes the 2D mode is also referred to as the G' mode. This labelling is rooted in the fact that it is the second most prominent peak in Raman spectrum of graphite (Dresselhaus et al., 2010; Malard et al., 2009). Since the 2D peak comes from the second-order of zone-boundary phonons which do not meet Raman fundamental selection rules and has nothing in common with the G peak, henceforth we refer to it as the 2D mode. If one compares the two spectra in Fig. 5.2.2, a noticeable difference in the shape and intensity of the 2D mode of graphene compared to graphite can be observed. This difference lies in the fact that in graphite the 2D mode is composed of two elements, namely, 2D$_1$ and 2D$_2$ (Saito et al., 1993; Yoon et al., 2008), which are in general about 1/4 and 1/2 the G mode amplitude, respectively. However, unique to graphene, the 2D peak is narrower and its intensity is around four times that of the G peak (Ferrari and Robertson, 2000; Ferrari et al., 2007; Malard et al., 2009).

An interesting as well as useful aspect of the G and 2D Raman modes is that they change their position, shape and relative intensity depending on the number of graphene layers in the sample. Figure 5.2.2 shows regions of G and

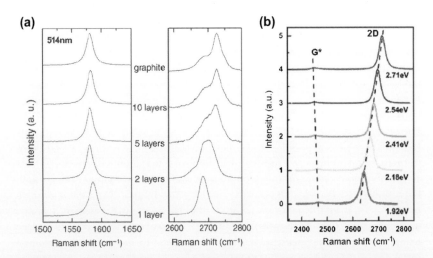

FIGURE 5.2.2 (a) Raman spectra of graphene with different number of layers and graphite taken with 514 nm laser length, showing the G and 2D bands (Ferrari et al. (2007) Copyrights). (b) Laser energy-dependent spectra of monolayer graphene sample, showing shifts of the G* and 2D bands (Malard (2009) Copyrights).

2D mode regions as a function of the number of layers in the graphene samples measured using a 514 nm excitation laser. When we compare the Raman spectrum from bi-layer graphene with that from a single-layer, it is easy to see that the 2D mode is much broader and its position up-shifted. The 2D band can be seen to evolve as the number layers increases to about 10 layers, where upon its profile matches that from graphite. The 2D mode is composed of sub-components; in the case of bi-layer graphene it can be broken down into four components, $2D_{1B}$, $2D_{1A}$, $2D_{2A}$, $2D_{2B}$, where $2D_{1A}$ and $2D_{2A}$ have the highest relative intensities. Occurance of these sub-peaks is related to a splitting of the valence and conduction bands. As the number of layers increases a large reduction of the relative intensity for the $2D_1$ mode is observed. Consequently, for the samples with five and more layers the Raman spectra are difficult to distinguish from graphite spectra. This feature in effect enables layer number metrology up to about 5 layers of graphene (Ferrari et al., 2007).

It is worth noting that choice of excitation laser energy can influence the position of both the G* and 2D modes (Fig. 5.2.2). This effect is related to a double resonance process which is connected with the relationship of the phonon wave vectors to the electronic band structure (Pimenta et al., 2007). When comparing turbostratic graphite with graphene the frequencies of the 2D and G* modes also change but the changes are not as obvious as for single layer graphene samples. Quantitavely, the 2D mode exhibits a highly dispersive behaviour with a slope of 88 cm^{-1} per 1eV for graphene and 95 cm^{-1} per 1eV for turbostratic graphite. The frequency changes in the G* mode are less pronounced both in turbostratic graphite as well as for monolayer graphene. Its frequency decreases with laser energy by about -18 cm^{-1} per 1eV for monolayer graphene and by about -31 cm^{-1} per 1eV for turbostratic graphite (Mafra et al., 2007).

An important characteristic to bear in mind with regards identifying the number of layers in graphene samples by Raman spectroscopy is that this technique is well established only for graphene with AB Bernal stacking (Ferrari et al., 2007; Malard et al., 2009). Such samples are usually produced by the mechanical exfoliation of natural or highly oriented pyrolytic graphite (HOPG). However graphene samples obtained by other techniques (e.g. CVD), or thermal decomposition of SiC) are not always characterised homogeneously with AB stacking of layers. For example, in turbostratic graphite the stacking of the layers with respect to c-axis are rotationally random, however it has single Lorentzian 2D band as found with monolayer graphene, however, it has a broader line width. This makes it quite difficult to distinguish between spectra from monolayer graphene and turbostratic graphite (Malard et al., 2009). The reason for this is that the electronic structure for turbostratic graphite as well as for monolayer graphene can be illustrated in the same way, however in the case of turbostratic graphite the broadening of 2D mode occurs due to a relaxation of the double resonance Raman selection rules due to the arbitrary orientation of the graphene layers with respect to each other. The width of 2D mode in monolayer graphene is around 24 cm^{-1}, while for turbostratic graphite it varies from 45 to 60 cm^{-1} (Cancado et al., 2002, 2008b). Additionally, the relative intensity of 2D and G modes – I_{2D}/I_G in turbostratic graphite is much smaller than in monolayer graphene, and the frequency of 2D is blue-shifted (Malard et al., 2009).

Since the graphene samples or flakes are usually quite small typically most Raman measurements are performed using an optical microscope, so called micro-Raman devices. Micro-Raman allows for better localisation of the graphene sample on substrate surfaces. Additionally, graphene samples are usually deposited on silicon wafers with a surface of silicon dioxide layer (~300 nm thick) for better localisation, since light reflects differently from the surface with graphene than from the pristine surface. Hence, it is easier to localise graphene on Si/SiO$_2$ surfaces under an optical microscope.

With regard to the choice of laser excitation energy for graphene measurements, typically visible light lasers (usually from 633 to 488 nm) are employed, since nice spectra using near infrared (NIR) sources (e.g. 780 or 785 nm) or ultraviolet (UV) are difficult to collect. Graphene samples deposited on silicon wafers with an oxide surface layer when measured using a NIR source have strong fluorescence signals. On the other hand, graphene, measured using a UV excitation laser leads to a Raman spectrum has differences with regards relative intensities of the characteristic graphene modes. However, they are useful as a complimentary method for layer metrology of graphene samples (Calizo, 2009). For example, relative to a spectrum obtained from a visible excitation laser, the G peak remains pronounced in a UV excited Raman spectrum and the 2D band is blue-shifted to the value around 2825 cm^{-1}. In addition, the intensity ratio of the 2D and G modes (I_{2D}/I_G) between different numbers of layers in graphene samples differs between UV and visible light excitation; with a visible source I_{2D}/I_G is around 4 while with a UV source the ratio is ca. 0.1. In the case of bi-layer graphene measured with UV source the ratio is even smaller (ca. 0.04) whereas with a visible source it is around 1.4. Table 5.2.1 presents typical values of the I_{2D}/I_G ratios for different number of

TABLE 5.2.1 Measured I_{2D}/I_G Ratios for Different Number of Graphene Layers n, Using Visible Source $\lambda = 488$ nm and UV Source $\lambda = 325$ nm (Calizo, 2009)

No. of layers, n	1	2	3	4	5
VIS ($\lambda = 488$ nm)	4.17	1.35	0.88	0.67	0.48
UV ($\lambda = 325$ nm)	0.11	0.44	0.40	0.44	0.28

graphene layers n from 1 to 5 for both a UV and visible excitation laser source. From the table it is clear the additional use of a UV source in conjunction with a visible laser strengthens layer metrology.

The selection of laser power is also important. Usually, powers between 0.04 mW and 4 mW are used. Within this range no significant changes in the Raman spectra or sample damage are observed. At higher power local heating from the laser can damage or burn the sample or lead to spectral variations (discussed below) and at powers below this range the signal to noise ratio can be very poor (Ferrari and Robertson, 2004).

With temperature increase the G and 2D modes from mono-layer graphene shift relatively linearly to lower wavenumbers, while the position of D mode remains stable (Calizo et al., 2008a; Late et al., 2011). The temperature coefficient from Raman spectra is also related to the number of graphene layers in the sample, such that the positions of the G and 2D modes red-shift with the increasing number of layers (for a given power) (Late et al., 2011). The number of layers also affects the temperature coefficients which are reflected in the full width at half maximums (FWHMs) of these bands. The width of the observed spectral responses from single-layer graphene remains stable or increase with temperature, and shrinks with increasing numbers of layers. The observed changes with temperature can be explained by the electron–phonon coupling, anharmonic phonon–phonon interactions and the negative thermal expansion of graphene (Late et al., 2011).

While graphene as a structure exists as a single atomic layer of carbon, it is also important to investigate the interaction between graphene and its substrate particularly with regards any potential applications and device fabrication. Possible interactions arise due to defects, coupling or surface charges between the two surfaces and, importantly, can be quite well monitored with Raman spectroscopy. The most popular method to obtain graphene samples for further tests is by the micromechanical cleavage graphene layers from natural graphite or from HOPG then as a next step, their deposition on a substrates surface (usually Si with a 300 nm SiO_2 layer) (Casiraghi et al., 2007a; Ferrari et al., 2006; Graf et al., 2007; Ni et al., 2008a). However depending on the application different substrates may be required, e.g. transparent substrates for optical applications – sapphire, single crystal quartz, glass or polymers (e.g. PDMS, PTFE) to name a few. Graphene can be also deposited on doped silicon, metals alloys (e.g. NiFe) or semiconductors (e.g. GaAs) (Ni et al., 2008; Wang et al., 2008). During the Raman measurements on graphene over substrates like quartz, polymers, silicon or metals one can observe that the G band positions fluctuate between 1580 cm^{-1} and 1588 cm^{-1}, which is similar to that obtained on standard Si/SiO_2 wafers (1580 cm^{-1}). These small changes are attributed to accidental electron or hole doping effects (Casiraghi et al., 2007b). However the graphene spectrum obtained over sapphire, glass or GaAs shows much bigger G position shifts and in the case of glass may even split into a double peak. Nonetheless,

Raman spectroscopy can still be used here to determine number of graphene layers on these surfaces (Calizo et al., 2007, 2008b). Thus, one can conclude from Raman measurements that the interaction between graphene and a substrate (van der Waals force) from samples prepared by micro-cleaving is not strong enough to influence the physical structure of graphene. However, things are very different for monolayer graphene prepared by epitaxial growth from SiC substrates, where one could expect a stronger interaction between the graphene layer and the SiC surface. Typically the G and 2D modes are strongly blue-shifted (~11 cm^{-1} and ~34 cm^{-1}, respectively) (Ni et al., 2008a). The observed shifts might be related to strain, caused by the strong interaction between the graphene and SiC substrate due to covalent bonding between the SiC substrate and epitaxial graphene through the interfacial carbon layer which has a graphene-like honeycomb lattice structure (Calizo et al., 2007, 2008b; Kim et al., 2008; Varchon et al., 2007). The strong interaction is caused by the lattice mismatch between the graphene lattice and interfacial carbon layer, where compressive stress on graphene causes changes in the lattice constant, atomic and electronic structures of graphene and consequently the Raman features (Ni et al., 2008b).

The Raman spectrum from $1 + 1$ folded single layer graphene is also worth mentioning since its spectrum differs from that of double layer graphene. In the case of $1 + 1$ folded graphene, the 2D signal is a single sharp peak, similar to that from monolayer graphene whereas the 2D band in double layer graphene is much wider and can be fitted with four sub peaks due to a splitting between the valence and conduction bands (Ferrari et al., 2006). In essence, the electronic structure of folded graphene and monolayer graphene should be similar with no splitting in energy bands. Even though the shape of 2D modes are similar in single layer and folded graphene there are differences in the 2D peak position where for folded graphene, the frequency blue-shifts (around 12 cm^{-1}) from monolayer graphene and also the intensity ratio I_{2D}/I_G in folded graphene is higher due to the different resonance conditions (Ni et al., 2008a).

Graphene oxide (GO) is another graphene based structure where Raman spectroscopy can be utilised as a characterisation tool. The G mode in GO samples is wider and blue-shifted to ca. 1590 cm^{-1}, the D mode is significantly more intense (sometimes, even stronger than G mode) due to the disorder in the sp^2 structure induced by oxidative synthesis of GO and also due to the attachment of hydroxyl and epoxide groups on the carbon basal plane (Kudin et al., 2008; Wilson et al., 2009; Yang et al., 2009). The 2D mode with respect to the D and G modes is very small, but it can be enhanced after temperature annealing of GO which also affects the frequency position. It can shift by around 6 cm^{-1} and is related to the reduction of GO (Yang et al., 2009).

Raman spectroscopy of graphene nanoribbons is useful to determine the properties of graphene nanoribbons with lengths in the micrometre range and few nanometre width, in a fast and non-destructive way (Malard et al., 2009). The observed modes in the Raman spectrum of nanoribbons are the defect induced D-mode ($\sim$1340 cm^{-1}), the G-mode ($\sim$1580–1590 cm^{-1}), the 2D-mode ($\sim$2680 cm^{-1}) and for the narrower ribbons an additional defect induced D'-mode ($\sim$1620 cm^{-1}) (Bischoff et al., 2011). Additionally, for single-layer graphene nanoribbons the width of 2D mode at half maximum is below 40 cm^{-1} (Graf et al., 2007; Han et al., 2007). The D and D' modes intensities do not depend on the ribbon width but depend on the edge-region and its roughness, which is related to the fabrication method of the graphene nanoribbons (Bischoff et al., 2011). In comparison, the intensities of G and 2D modes do depend on the ribbon area and width. The observation of the G mode behaviour is related to the size of the illuminated area of sp^2 bound carbon atoms by laser spot (Malard et al., 2009). The same irradiation dependance is observed for the 2D mode (Castiglioni et al., 2001; Lazzeri et al., 2008; Tuinstra and Koenig, 1970). These Raman observations can be very useful for designing further experiments and future graphene nanoelectronic devices based on graphene nanoribbons. Another important aspect for electronic device fabrication is the clear determination of the orientation of graphene edges (zigzag, armchair or disordered edges) since their orientation determines their electronic properties. The G and D modes in the Raman spectrum can be usefully utilised to identify edge orientations (Cong et al., 2010). The G peak at the edges of monolayer graphene shows polar behaviour and exhibits stiffening at zigzag-dominated edges, while it is softened for armchair-dominated edges. Moreover, the intensity of the D peak is weak at the zigzag edge due to momentum conservation (Cancado et al., 2004a,b) and strong at the armchair edge of graphene (Casiraghi et al., 2009; You et al., 2008). In addition, the intensity of the D to G responses depends on polarisation, relative position of the laser spot with respect to the edge and the amount of edge disorder.

Raman spectroscopy can be also used to determine doping levels (e.g. nitrogen) in graphene. Typically the Raman modes in graphene are altered after doping, viz. the G mode is blue-shifted to ca. 1585 cm^{-1} and is an indication for atomic insertion in graphene structure (Ferrari et al., 2007; Yan et al., 2007). Also, the D and 2D modes shift to 1345 cm^{-1} and 2684 cm^{-1}, respectively. In addition, the D mode at 1624 cm^{-1} appears (Jin et al., 2011; Panchokarla et al., 2009). The I_D/I_G ratio for doped graphene has a value between 0.3 and 0.4, since doping of sp^2 structure introduces defects (Ferrari et al., 2007). However the intensity ratio I_{2D}/I_G can vary with doping level (Zhang et al., 2011), where, for example, the ratio $I_{2D}/I_G = 0.6$ corresponds to a doping level around 4×10^{13} cm^{-2} (Das et al., 2008).

To summarise, Raman spectroscopy is a remarkably powerful tool for the characterisation of graphene and plays very important role as a direct or complimentary technique in any laboratory working with graphene.

REFERENCES

Ferrari, Andrea C., 2007. Solid State Commun. 143, 47–57.

Bischoff, D., Güttinger, J., Dröscher, S., Ihn, T., Ensslin, K., Stampfer, C., 2011. J. Appl. Phys. 109, 073710.

Calizo, I., Bao, W.Z., Miao, F., Lau, C.N., Balandin, A.A., 2007. Appl. Phys. Lett. 91, 201904.

Calizo, Irene, Ghosh, Suchismita, Teweldebrhan, Desalegne, Bao, Wenzhong, Miao, Feng, Ning Lau, Chun, Balandin, Alexander A., 2008a. SPIE 7037, 70371B.

Calizo, I., Teweldebrhan, D., Bao, W., Miao, F., Lau, C.N., Balandin, A.A., 2008b. Spectroscopic Raman nanometrology of graphene and graphene multilayers on arbitrary substrates. J. Phys.: Conf. Ser. 109, 012008.

Calizo, Irene, Bejenari, Igor, Rahman, Muhammad, Liu, Guanxiong, Balandinc, Alexander A., 2009. J. Appl. Phys. 106, 043509.

Cançado, L.G., Pimenta, M.A., Saito, R., Jorio, A., Ladeira, L.O., Grüneis, A., Souza-Filho, A.G., Dresselhaus, G., Dresselhaus, M.S., 2002. Phys. Rev. B 66.

Cancado, L.G., Pimenta, M.A., Neves, B.R.A., Dantas, M.S.S., Jorio, A., 2004a. Phys. Rev. Lett. 93, 247401.

Cancado, L.G., Pimenta, M.A., Neves, B.R.A., Medeiros-Ribeiro, G., Enoki, T., Kobayashi, Y., Takai, K., Fukui, K., Dresselhaus, M.S., Saito, R., Jorio, A., 2004b. Phys. Rev. Lett. 93, 047403.

Cancado, L.G., Reina, A., Kong, J., Dresselhaus, M.S., 2008a. Geometrical approach for the study of G' band in the Raman spectrum of monolayer graphene, bilayer graphene, and bulk graphite. Phys. Rev. B 77, 245408. 035415.

Cançado, L.G., Takai, K., Enoki, T., Endo, M., Kim, Y.A., Mizusaki, H., Speziali, N.L., Jorio, A., Pimenta, M.A., 2008b. Carbon 46, 272.

Casiraghi, C., Hartschuh, A., Lidorikis, E., Qian, H., Harutyunyan, H., Gokus, T., Novoselov, K.S., Ferrari, A.C., 2007a. Rayleigh imaging of graphene and graphene layers. Nano Lett. 7, 2711–2717.

Casiraghi, C., Pisana, S., Novoselov, K.S., Geim, A.K., Ferrari, A.C., 2007b. Raman fi ngerprint of charged impurities in graphene. Appl. Phys. Lett. 91, 233108.

Casiraghi, C., Hartschuh, A., Qian, H., Piscanec, S., Georgi, C., Fasoli, A., Novoselov, K.S., Basko, D.M., Ferrari, A.C., 2009. Nano Lett. 9, 1433.

Castiglioni, C., Negri, F., Rigolio, M., Zerbi, G., 2001. J. Chem. Phys. 115, 3769.

Cong, Chunxiao, Yu, Ting, Wang, Haomin, 2010. ACS Nano 4 (6), 3175–3180.

Das, A., Pisana, S., Chakraborty, B., Piscanec, S., Saha, S.K., Waghmare, U.V., Novoselov, K.S., Krishnamurthy, H.R., Geim, A.K., Ferrari, A.C., Sood, A.K., 2008. Nat. Nanotechnol. 3, 210.

Dresselhaus, M.S., Dresselhaus, G., Saito, R., Jorio, A., 2005. Phys. Rep. 409, 47.

Dresselhaus, M.S., Jorio, A., Hofmann, M., Dresselhaus, G, Saito, R., 2010. Nano. Lett. 10, 751–758.

Ferrari, A.C., Robertson, J., 2000. Phys. Rev. B 61, 14095.

Ferrari, A.C., Robertson, J. (Eds.), 2004, Raman spectroscopy in carbons: from nanotubes to diamond. Philos. Trans. R. Soc. Ser. A, vol. 362, pp. 2267–2565.

Ferrari, A.C., Meyer, J.C., Scardaci, V., Casiraghi, C., Lazzeri, M., Mauri, F., Piscanec, S., Jiang, D., Novoselov, K.S., Roth, S., Geim, A.K., 2006. Raman spectrum of graphene and graphene layers. Phys. Rev. Lett. 97, 187401.

Graf, D., Molitor, F., Ensslin, K., Stampfer, C., Jungen, A., Hierold, C., Wirtz, L., 2007. Nano Lett. 7, 238.

Han, M.Y., Özyilmaz, B., Zhang, Y., P. Kim, 2007. Phys. Rev. Lett. 98, 206805.

Jin, Zhong, Yao, Jun, Kittrell, Carter, Tour, James M., 2011. ACS Nano 5 (5), 4112–4117.

Kim, S., Ihm, J., Choi, H.J., Son, Y.W., 2008. Origin of anomalous electronic structures of epitaxial graphene on silicon carbide. Phys. Rev. Lett. 100, 176802.

Kudin, K.N., Ozbas, B., Schniepp, H.C., Prud'homme, R.K., Aksay, I.A., Car, R., 2008. Raman spectra of graphite oxide and functionalized graphene sheets. Nano Lett. 8, 36–41.

Late, Dattatray J, Maitra, Urmimala, Panchakarla, L.S., Waghmare, Umesh V, Rao, C.N.R., 2011. J. Phys.: Condens. Matter. 23, 5. 055303.

Lazzeri, M., Attaccalite, C., Wirtz, L., Mauri, F., 2008. Phys. Rev. B 78, 081406.

Mafra, D.L., Samsonidze, G., Malard, L.M., Elias, D.C., Brant, J.C., Plentz, F., Alves, E.S., Pimenta, M.A., 2007. Phys. Rev. B 76, 233407.

Malard, L.M., Pimentaa, M.A., Dresselhaus b, G., Dresselhaus, M.S., 2009. Phys. Rep. 473, 51–87.

Maultzsch, J., Reich, S., Thomsen, C., 2004. Double-resonant Raman scattering in graphite: interference effects, selection rules, and phonon dispersion. Phys. Rev. B 70, 155403.

Dresselhaus, Mildred S., Jorio, Ado, Hofmann, Mario, Dresselhaus, Gene, Saito, Riichiro, 2010. Nano Lett. 10, 751–758.

Nakamoto, Kazuo, Brown, Chris W., 2003. Introductory Raman Spectroscopy, second ed., ISBN 978-0-12-254105-6.

Ni, Z.H., Wang, H.M., Kasim, J., Fan, H.M., Yu, T., Wu, Y.H., Feng, Y.P., Shen, Z.X., 2007. Graphene thickness determination using refl ection and contrast spectroscopy. Nano Lett. 7, 2758–2763.

Ni, Zhenhua, Wang, Yingying, Yu, Ting, Shen, Zexiang, 2008a. Nano Res. 1, 273–291.

Ni, Z.H., Chen, W., Fan, X.F., Kuo, J.L., Yu, T., Wee, A.T.S., Shen, Z.X., 2008b. Raman spectroscopy of epitaxial graphene on a SiC substrate. Phys. Rev. B 77, 115416.

Panchokarla, L.S., Subrahmanyam, K.S., Saha, S.K., Govindaraj, A., Krishnamurthy, H.R., Waghmare, U.V., Rao, C.N.R., 2009. Synthesis, structure, and properties of boronand nitrogen-doped graphene. Adv. Mater. 21, 4726–4730.

Pimenta, M.A., Dresselhaus, G., Dresselhaus, M.S., Cançado, L.G., Jorio, A., Saito, R., 2007. Phys. Chem. Chem. Phys. 9, 1276–1291.

Saito, R., Dresselhaus, G., Dresselhaus, M.S., 1993. J. Appl. Phys. 73, 494.

Shimada, T., Sugai, T., Fantini, C., Souza, M., Cancado, L.G., Jorio, A., MaPimenta, A., Saito, R., Gruneis, A., Dresselhaus, G., Dresselhaus, M.S., Ohno, Y., Mizutani, T., Shinohara, H., 2005. Carbon 43, 1049.

Tuinstra, F., Koenig, J.L., 1970. Raman spectrum of graphite. J. Chem. Phys. 53, 1126–1130.

Varchon, F., Feng, R., Hass, J., Li, X., Nguyen, B.N., Naud, C., Mallet, P., Veuillen, J.Y., Berger, C., Conrad, E.H., Magaud, L., 2007. Electronic structure of epitaxial graphene layers on SiC: effect of the substrate. Phys. Rev. Lett. 99 126805.

Wang, Y.Y., Ni, Z.H., Yu, T., Wang, H.M., Wu, Y.H., Chen, W., Wee, A.T.S., Shen, Z.X., 2008. Raman studies of monolayer graphene: the substrate effect. J. Phys. Chem. C 112, 10637–10640.

Wilson, Neil R., Pandey, Priyanka A., Beanland, Richard, Young, Robert J., Kinloch, Ian A., Gong, Lei, Liu, Zheng, Suenaga, Kazu, Rourke, Jonathan P., York, Stephen J., Sloan, Jeremy, 2009. ACS Nano 3 (9), 2547–2556.

Yan, J., Zhang, Y., Kim, P., Pinczuk, A., 2007. Electric field effect tuning of electron-phonon coupling in graphene. Phys. Rev. Lett. 98, 166802.

Yang, Dongxing, Velamakanni, Aruna, Bozoklu, Gülay, Park, Sungjin, Stoller, Meryl, Piner, Richard D., Stankovich, Sasha, Jung, Inhwa, Field, Daniel A, Ventrice Jr., Carl A., Ruoff, Rodney S., 2009. Carbon 47, 145–152.

Yoon, D., Moon, H., Son, Y.-W., Samsonidze, G., Park, B.H., Kim, J.B., Lee, Y., Cheong, H., 2008. Nano Lett. 8, 4270.

You, Y.M., Ni, Z.H., Yu, T., Shen, Z.X., 2008. Appl. Phys. Lett. 93, 163112.

Zhang, C., Fu, L., Liu, N., Liu, M., Wang, Y., Liu, ., Z., 2011. Synthesis of nitrogen-doped graphene using embedded carbon and nitrogen sources. Adv. Mater. 23, 1020–1024.

Scanning Electron Microscopy

Franziska Schäffel

University of Oxford, Oxford, UK

Due to the combination of high magnification, large depth of focus, high resolution and ease of applicability scanning electron microscopy (SEM) is one of the most frequently used methods in materials characterisation. An electron beam is focussed to a small probe and scanned across the surface of the sample. The interaction of the primary electrons with the sample results in the generation of secondary electrons, which are collected for image formation giving a three-dimensional impression of the sample's surface topography. The resolution in a SEM is limited by the diameter of the electron probe and the probe's interaction volume within the sample.

In graphene research, SEM is predominantly employed to study CVD graphene grown on a conductive substrate. It allows for straightforward determination of graphene domain size (sometimes also referred to as grain size), domain morphology, sample coverage, nucleation density and growth rates. SEMs are often equipped with additional detectors, e.g. for energy-dispersive X-ray analysis (EDX), allowing for chemical analysis of the sample. Some SEMs are equipped with an electron backscatter diffraction (EBSD) detector that allows for the examination of the crystallographic orientation of the substrate and is used to study the correlation of graphene domain shape and orientation of the catalytic substrate.

Due to the ease with which SEM images can be acquired, SEM is often employed to study the progress of graphene growth. To do this, the samples are exposed to a hydrocarbon source at a specific growth temperature for different growth durations, and subsequent SEM analysis is then carried out to compare the domain sizes and determine a growth rate. For example, Weatherup et al. investigated graphene growth using catalytic Ni films with an additional Au layer and compared their results to graphene grown on pure Ni films (Weatherup et al., 2011). Figure 5.3.1a–c show graphene domains grown on pure Ni at 450 °C in C_2H_2 for 120 s, 270 s and 420 s, respectively. SEM does not allow for the determination of the exact number of layers; however, it does provide some information on the graphene thickness grown on the catalytic substrate. A strong contrast variation between areas with different numbers of graphene layers is observed in Fig. 5.3.1a–c. The SEM image contrast is a result of the amount of secondary electrons that are generated in the upper few nanometres of the sample surface. Further, a higher secondary electron yield is expected for Ni as compared to C; thus, Ni appears brighter in the SEM

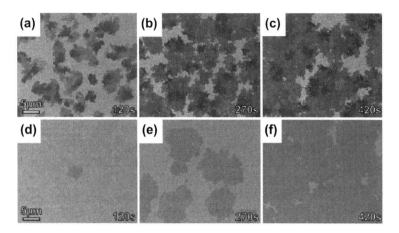

FIGURE 5.3.1   Comparative study of graphene grown at 450 °C on a 550 nm Ni film (a–c) and a Ni film covered with a 5 nm Au layer (d–f) after different growth durations of 120 s (a,d), 270 s (b,e) and 420 s (c,f). *Reprinted with permission from Weatherup et al. (2011). Copyright (2011) American Chemical Society.*

image (Seiler, 1983). Therefore, areas that are dark in the SEM image are covered by a larger number of graphene layers, and areas that appear bright are covered by fewer graphene layers. The bright contrast arises from secondary electrons that can still escape from the underlying Ni substrate. For comparison, Figures 5.3.1d–f show graphene domains grown on Ni with an additional 5 nm Au layer synthesised under the same conditions (Weatherup et al., 2011). In contrast to Ni catalysed growth, graphene domains grown using the Au–Ni system appeared to be larger and more homogeneous in thickness and layer number. Additional Raman spectroscopy confirms that ~74% of the domains were graphene monolayers. Further, the SEM images revealed that the graphene nucleation density on Au–Ni is significantly reduced as compared to pure Ni, which was attributed to Au decoration of highly reactive Ni surface sites, such as step edges, where graphene nucleation predominantly occurs. In both cases, the domain shape appeared to be irregular (Weatherup et al., 2011).

The question of domain shape is of particular interest for graphene domains grown from Cu catalysts which are found to exhibit very peculiar shapes. To study the shape of the graphene domains SEM is the technique that has predominantly been employed. The domain shape is dependent on various parameters as, for example, the catalyst material, the crystallographic orientation of the catalytic grains, the growth temperature, the growth atmosphere and the partial pressures of the gases involved (Fan et al., 2011; Gao et al., 2012; Li et al., 2010; Sutter et al., 2008; Vlassiouk et al., 2011; Wu et al., 2011).

Li et al. (2010) carried out low-pressure CVD growth of graphene on Cu foil and studied the influence of temperature, methane flow rate and methane partial pressure on the graphene domains. From SEM analysis they deduced that the nucleation density decreases with increasing temperature or decreasing methane flow rate and methane partial pressure, respectively. Thus larger graphene domains could be obtained controlling these parameters. They further observed the graphene islands to exhibit a four-lobed shape (Li et al., 2010). Such four-lobed, fourfold symmetric islands are found to nucleate and grow on the (100)-textured surface of Cu that forms during annealing of the foil (Wofford et al., 2010). The graphene islands are polycrystalline with each lobe being an individual graphene crystal differently oriented on the underlying Cu grain. The shape evolution of the lobes is attributed to a growth mode dominated by edge kinetics with an angularly dependent growth velocity (Wofford et al., 2010).

In contrast to the four-lobed domain shapes obtained in low pressure CVD experiments, graphene domains grown on Cu foil using atmospheric pressure CVD often exhibit six-lobed or even hexagonal shapes. Vlassiouk et al. (2011) carried out atmospheric pressure graphene synthesis from Cu foil using Ar as buffer gas and a very low partial pressure of methane (30 ppm) as carbon source. They reported that shape and size of the graphene domains change drastically with different partial pressures of hydrogen. Figure 5.3.2a shows the evolution of graphene grain size as a function of the partial pressure of H_2 ($P(H_2)$) as determined from SEM analysis. The maximum graphene growth rate was observed at intermediate hydrogen partial pressures, where the ratio of the partial pressures of H_2 and CH_4 was approximately 200–400. Vlassiouk et al. (2011) showed that this value also holds if other buffer gases (e.g. He) are used or if graphene growth is carried out under low-pressure CVD conditions without buffer gas. In Fig. 5.3.2b, SEM micrographs of graphene domains obtained at different hydrogen pressures are depicted. Clearly, strong variations in domain shape can be seen. At low and intermediate hydrogen partial pressures ($P(H_2)/P(CH_4) < 400$) the grains appeared irregular with poorly defined

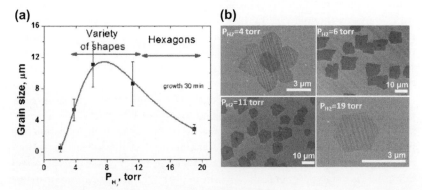

FIGURE 5.3.2 (a) Grain size of graphene grains grown for 30 min at 1000 °C on Cu foil at atmospheric pressure using 30 ppm methane in Ar buffer gas as a function of the partial pressure of H_2 ($P(H_2)$); (b) SEM images illustrating the different domain shapes observed for different partial pressures of H_2. *Reprinted with permission from Vlassiouk et al. (2011). Copyright (2011) American Chemical Society.*

edges; sometimes they exhibited lobes with sixfold symmetry giving a flower-like appearance. At higher hydrogen pressures ($P(H_2)/P(CH_4) > 400$, $P(H_2) =$ 19 torr) distinct hexagonal domains with well-defined zigzag edges were observed (Vlassiouk et al., 2011). Robertson and Warner (2011) also observed domains with hexagonal shape in their atmospheric pressure CVD studies of graphene grown on Cu. Here, the high concentration of CH_4 in the gas mixture led to the formation of FLG domains; the high partial pressure of H_2 affected the hexagonal shape of the domains.

Wu et al. (2011) carried out atmospheric pressure graphene synthesis on high purity Cu foils (99.999%) in order to reduce the artificial nucleation of graphene domains on impurities. At growth temperatures of 990 °C and 1000 °C and relatively high partial pressures of H_2 they predominantly observed the growth of hexagonal domains, as reported by others (Robertson and Warner, 2011; Vlassiouk et al., 2011; Wu et al., 2011). This indicates that the growth is dominated by the sixfold symmetry of the graphene lattice and the Cu lattice does not affect the domain shape. However, at a reduced growth temperature of 980 °C Wu et al. (2011) frequently detected rectangular graphene domains. Further, within individual Cu grains the rectangular graphene domains appeared to be aligned with respect to each other which suggests that this is a result of the relationship between the atomic structure of the Cu and the growing graphene, along with the role hydrogen plays in forming faceted structures. Wu et al. (2011) carried out electron backscatter diffraction (EBSD) analysis to examine the crystallographic orientation of the underlying Cu grains. They found that rectangular graphene domains only grow on Cu grains with (111) orientation, while hexagonal domains can form on nearly all non-(111) Cu grains. This was surprising since both Cu(111) and graphene are sixfold symmetric, but the rectangular shaped graphene domains are twofold symmetric. The observation of rectangular domains was attributed to the lattice mismatch of ~4% between the Cu(111) interatomic spacing and the C–C distance in graphene, which leads to strain in the system. This may result in faster growth kinetics along the elongated direction of the rectangle. Wu et al. (2011) suggested that, alternatively, graphene growth may also be slowed along one direction due to hydrogen being more effective at etching away carbon atoms from the respective edges. Rectangular graphene domains were also reported from low-pressure CVD at 500 °C and 600 °C using electropolished Cu foils as substrate and toluene as the carbon source (Zhang et al., 2012). The SEM image depicted in Fig. 5.3.3a shows typical rectangular graphene domains grown at 500 °C. However, in this case EBSD mapping predominantly revealed Cu grains with (001) orientation. Here, the rectangular shape of the graphene domains was solely attributed to relatively high partial pressure of hydrogen resulting in strong faceting of the graphene domains through etching (Zhang et al., 2012).

Low magnification SEM can also be employed to study the grain size of the Cu foil. Different Cu grains often appear with different contrast levels in an SEM image. Depending on the crystallographic orientation of the grains, the depth of

FIGURE 5.3.3 (a) SEM image of typical rectangular graphene domains grown at 500 °C via low-pressure CVD using toluene. *Reprinted with permission from Zhang et al. (2012). Copyright (2012) American Chemical Society.* (b) SEM image of graphene domains grown on Cu foil via atmospheric pressure CVD at 1000 °C. A Cu grain boundary runs through the image. The neighbouring Cu grains exhibit different domain shapes and coverage.

penetration of the incoming electrons varies. If the incoming electrons encounter a low density of atoms, e.g. on a low index surface, the scattering probability is reduced, and thus the grain appears darker. Using this channelling effect Cu grain sizes have been deduced from SEM images (Huang et al., 2011).

Further, the positions of Cu grain boundaries can be derived with SEM taking advantage of the channelling effect. These defect sites often give rise to different growth behaviour. For example, the nucleation density is often found to be enhanced at grain boundaries and scratches due to the presence of reactive step edges (Han et al., 2011; Reina et al., 2009). In the SEM image in Fig. 5.3.3b, the grain boundary can easily be made out as the line of tightly arranged graphene domains. Further, different graphene coverage as well as different domain shapes are observed in the two neighbouring Cu grains in Fig. 5.3.3b. SEM analysis also revealed that graphene can grow across metal steps (Sutter et al., 2008) and grain boundaries (Li et al., 2009). Often wrinkles of the graphene layers are observed in SEM; these wrinkles are associated with the difference of the coefficients of thermal expansion between the substrate and graphene (Li et al., 2009).

Ismach et al. (2010) studied graphene growth on thin sacrificial Cu films that have been deposited on dielectric substrates. Due to the high growth temperature of 1000 °C and the low pressure (100–500 m Torr) employed, the Cu films dewetted and evaporated during or after graphene growth. Thus direct deposition of graphene on the dielectric substrate has been achieved. Using EDX analysis in a SEM, Ismach et al. (2010) could illustrate the dewetting process. Figure 5.3.4a shows an SEM image of a typical sample with a 450 nm Cu film on a quartz substrate after 2 h growth time. The finger-like (white) structures correspond to the dewetted Cu film, as confirmed by the EDX elemental mapping in Fig. 5.3.4b, where bright areas correspond to Cu. Figure 5.3.4c and d correspond to elemental maps of O and Si, respectively.

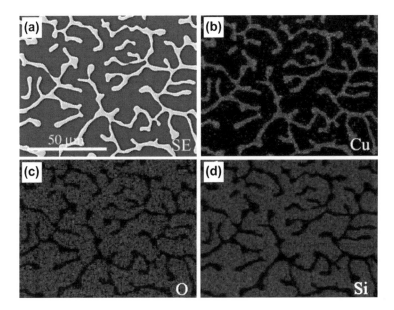

FIGURE 5.3.4   (a) SEM image of Cu (white) on quartz after a growth time of 2 hours (b–d) EDX elemental maps of Cu, O and, Si, respectively. The brighter areas are the areas from which an elemental signal is obtained. *Reprinted with permission from Ismach et al. (2010). Copyright (2010) American Chemical Society.*

These maps confirm the absence of Cu between the fingers where the oxygen and silicon signals of the underlying quartz substrate can escape through the emerging graphene layer (Ismach et al., 2010).

To summarise, SEM is a versatile and easily applicable technique which, in the graphene community, is predominantly used to gain information on the graphene domain size, domain shape, nucleation density and sample coverage. Using special detectors also elemental (EDX) or orientational (EBSD) information can be obtained to study correlations of graphene evolution with the underlying substrate.

REFERENCES

Fan, L., Li, Z., Li, X., Wang, K., Zhong, M., Wei, J., Wua, D., Zhu, H., 2011. Controllable growth of shaped graphene domains by atmospheric pressure chemical vapour deposition. Nanoscale 3, 4946–4950.

Gao, L., Ren, W., Xu, H., Jin, L., Wang, Z., Ma, T., Ma, L.-P., Zhang, Z., Fu, Q., Peng, L.-M., Bao, X., Cheng, H.-M., 2012. Repeated growth and bubbling transfer of graphene with millimetre-size single-crystal grains using platinum. Nat. Commun. 3, 699.

Han, G.H., Günes, F., Bae, J.J., Kim, E.S., Chae, S.J., Shin, H.-J., Choi, J.-Y., Pribat, D., Lee, Y.H., 2011. Influence of copper morphology in forming nucleation seeds for graphene growth. Nano Lett. 11, 4144–4148.

Huang, P.Y., Ruiz-Vargas, C.S., van der Zande, A.M., Whitney, W.S., Levendorf, M.P., Kevek, J.W., Garg, S., Alden, J.S., Hustedt, C.J., Zhu, Y., Park, J., McEuen, P.L., Muller, D.A., 2011. Grains and grain boundaries in single-layer graphene atomic patchwork quilts. Nature 469, 389–392.

Ismach, A., Druzgalski, C., Penwell, S., Schwartzberg, A., Zheng, M., Javey, A., Bokor, J., Zhang, Y., 2010. Direct chemical vapor deposition of graphene on dielectric surfaces. Nano Lett. 10, 1542–1548.

Li, X., Cai, W., An, J., Kim, S., Nah, J., Yang, D., Piner, R., Velamakanni, A., Jung, I., Tutuc, E., Banerjee, S.K., Colombo, L., Ruoff, R.S., 2009. Large area synthesis of high-quality and uniform graphene films on copper foils. Science 324, 1312–1314.

Li, X., Magnuson, C.W., Venugopal, A., An, J., Suk, J.W., Han, B., Borysiak, M., Cai, W., Velamakanni, A., Zhu, Y., Fu, L., Vogel, E.M., Voelkl, E., Colombo, L., Ruoff, R.S., 2010. Graphene films with large domain size by a two-step chemical vapor deposition process. Nano Lett. 10, 4328–4334.

Reina, A., Jia, X., Ho, J., Nezich, D., Son, H., Bulovic, V., Dresselhaus, M.S., Kong, J., 2009. Large area, few-layer graphene films on arbitrary substrates by chemical vapor deposition. Nano Lett. 9, 30–35.

Robertson, A.W., Warner, J.H., 2011. Hexagonal single crystal domains of few-layer graphene on copper foils. Nano Lett. 11, 1182–1189.

Seiler, H., 1983. Secondary electron emission in the scanning electron microscope. J. Appl. Phys. 54, R1–R18.

Sutter, P.W., Flege, J. I., Sutter, E.A., 2008. Epitaxial graphene on ruthenium. Nat. Mater. 7, 406–411.

Vlassiouk, I., Regmi, M., Fulvio, P., Dai, S., Datskos, P., Eres, G., Smirnov, S., 2011. Role of hydrogen in chemical vapor deposition growth of large single-crystal graphene. ACS Nano 5, 6069–6076.

Weatherup, R.S., Bayer, B.C., Blume, R., Ducati, C., Baehtz, C., Schlögl, R., Hofmann, S., 2011. In situ characterization of alloy catalysts for low-temperature graphene growth. Nano Lett. 11, 4154–4160.

Wofford, J.M., Nie, S., McCarty, K.F., Bartelt, N.C., Dubon, O.D., 2010. Graphene islands on Cu foils: the interplay between shape, orientation, and defects. Nano Lett. 10, 4890–4896.

Wu, Y.A., Robertson, A.W., Schäffel, F., Speller, S.C., Warner, J.H., 2011. Aligned rectangular few-layer graphene domains on copper surfaces. Chem. Mater. 23, 4543–4547.

Zhang, B., Lee, W.H., Piner, R., Kholmanov, I., Wu, Y., Li, H., Ji, H., Ruoff, R.S., 2012. Low-temperature chemical vapor deposition growth of graphene from toluene on electropolished copper foils. ACS Nano 6, 2471–2476.

Chapter 5.4

Transmission Electron Microscopy

Franziska Schäffel

University of Oxford, Oxford, UK

5.4.1. INTRODUCTION

With respect to the morphological and structural characterisation of graphene, the technological platform that is provided by a state of art transmission electron microscope (TEM) enables researches to analyse their samples from micron scale down to atomic resolution and further allows to observe in-situ growth or transformation processes.

To give a brief introduction, in a TEM electrons are emitted from an electron gun by thermionic or field emission under ultra-high vacuum (UHV) conditions, injected into the microscope column by virtue of accelerating voltages typically between 200 and 300 kV, and directed through a double or triple electromagnetic condenser lens system to illuminate the nanometre thin specimen (Williams and Carter, 2009). As the electrons pass the sample, they are scattered by the electrostatic potentials of its atoms. Image contrast arises from incoherent electron scattering as well as from (coherent) electron diffraction (ED). Depending on the thickness and the atomic number of the atoms in the sample, the electrons are scattered under specific angles giving rise to the so-called mass–thickness contrast. Additional diffraction contrast arises from coherent superposition of scattered waves in a crystalline sample; here, sample areas which fulfil Bragg's law scatter the electrons stronger and, thus, appear darker (in bright-field TEM micrographs). In high resolution transmission electron microscopy (HRTEM) interference of the scattered and unscattered waves leads to the formation of an image of the crystal lattice. Using conventional HRTEM, structures with lateral distances of 1–2 Å can be resolved.

TEMs can often be operated in different imaging modes, i.e. conventional TEM, where a large area of the sample is illuminated under almost 'parallel' beam conditions, or scanning transmission electron microscopy (STEM), where a very small probe of electrons is formed and scanned across the sample by deflection coils (Christenson and Eades, 1986; Williams and Carter, 2009). Further, TEMs are often equipped with analytical tools, such as detectors for energy-dispersive X-ray analysis (EDX) and electron energy loss spectroscopy (EELS) allowing for additional chemical analysis being carried out inside a TEM.

The electron collisions, that provide the useful signals for (S)TEM imaging and chemical analysis can also cause electron-beam damage which affects the structure and/or the chemistry of the sample. This so called knock-on damage depends on the energy of the incident electron beam (Williams and Carter, 2009). The emergence of aberration corrected electron microscopes makes it possible to examine samples that are sensitive to knock-on damage at a lower acceleration voltage where knock-on damage is significantly reduced while still retaining a very high resolution. These new generation electron microscopes allow for delocalisation-free imaging and reach sub-Å resolution (Sasaki et al., 2010). This is particularly interesting for the characterisation of carbon based nanomaterials, such as graphene. To remove a carbon atom from the lattice a minimum incident energy of 86 keV is required (Smith and Luzzi, 2001). Only with the availability of aberration correctors, atomic resolution imaging of carbon nanostructures, including graphene and its defects, became accessible. For carbon nanostructure investigations the first aberration corrected microscopes were typically operated at 80 kV accelerating voltage. Yet, structural modifications have still been observed since lattice defects and curvature play an important role (Girit et al., 2009; Warner et al., 2009c). At present, further development of aberration correctors enabling sub-Å imaging at even lower accelerating voltages down to 30 kV is pursued and promises novel insights into graphene at the sub-Å scale (Sasaki et al., 2010; Suenaga and Koshino, 2010).

This section will summarise the observations that can be made using TEM imaging and spectroscopic techniques, such as determination of the number of graphene layers, detection of atomic scale defects (vacancies, dislocations, grain boundaries, etc.), and chemical characterisation. It will especially focus on the novel insights that can be gained via new-generation aberration corrected low voltage high resolution transmission electron microscopes (LV-HRTEMs).

5.4.2. ATOMIC RESOLUTION IMAGING (TEM/STEM) AND ATOMIC SCALE SPECTROSCOPY (EELS)

The versatility of the analytical platform provided by a TEM becomes particularly clear when concerning atomic-scale spatial resolution as well as chemical analysis with atomic resolution. TEM, STEM and EELS can all be carried out using an analytical TEM.

Using the aberration-corrected, monochromated TEAM 0.5 microscope operated at 80 kV, Meyer et al. (2008a) were able to resolve lattice structure from carbon atoms in suspended graphene via direct imaging (Kisielowski et al., 2008). Figure 5.4.1a shows an atomic resolution TEM micrograph acquired under imaging condition where the carbon atoms of the monolayer appear white (cf. reference (Meyer et al., 2008a) for details). The image is an

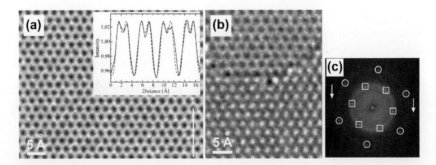

FIGURE 5.4.1 (a) Atomic resolution TEM image of monolayer graphene. The carbon atoms appear with white contrast. Inset: measured (solid line) and simulated (dashed line) contrast profile of monolayer graphene. (b) Atomic resolution TEM micrograph of a monolayer (top) next to a bilayer graphene region (bottom). (c) FFT of a graphene bilayer region with the reflections corresponding to an information transfer of 2.13 Å, 1.23 Å and 1.06 Å marked with squares, circles and arrows, respectively. *Adapted with permission from Meyer et al. (2008a). Copyright (2008) American Chemical Society.*

average of seven exposures to improve the signal-to-noise ratio. The inset shows the bright-field phase-contrast profile measured along the white line in Fig. 5.4.1a (solid line) along with the simulated contrast profile (dashed line). Monolayer graphene has a unique signature in the image, which becomes more apparent when directly comparing mono- and bi-layer graphenes, as shown in Fig. 5.4.1b. A mono-layer region is imaged in the upper part of the figure, while an AB stacked bi-layer region with qualitatively different contrast is observed in the lower part. White-atom contrast in the mono-layer switches to black atom contrast in the bi-layer. However, when directly interpreting the number of layers from a bright-field phase contrast image, one has to take great care, since an atomic column can show up as either a white or a black dot, depending on how the phase transfer function is tuned. For example, imaging with a negative defocus (as opposed to the positive defocus used for the acquisition of Fig. 5.4.1a,b) results in black-atom contrast for the monolayer and white atom contrast for the bi-layer (Warner, 2010). Therefore, other approaches, such as e.g. electron diffraction, should additionally be carried out for the unambiguous determination of the number of layers, as will be detailed in the next section. Figure 5.4.1c shows a fast Fourier transform (FFT) of a bi-layer graphene region (Meyer et al., 2008a). The inner hexagon of diffraction spots (marked with squares) corresponds to the $\{10\bar{1}0\}$ planes of graphene with a lattice spacing of $d = 2.13$ Å, whereas the larger hexagon (marked with circles) reflects the $\{11\bar{2}0\}$ planes with a shorter lattice spacing of $d = 1.23$ Å (cf. Section 2.1). In order to be able to resolve individual carbon atoms in graphene an information transfer of the 2.13 Å reflections is not sufficient; the reflections at 1.23 Å have to be transferred to resolve every single carbon atom. Further, a set of peaks corresponding to an information transfer of 1.06 Å is observed (marked with arrows), which demonstrates the microscope's capability to resolve even finer structural details (Meyer et al., 2008a).

As mentioned above, TEM imaging using bright-field phase contrast is highly sensitive on the focussing conditions. In contrast, images acquired with a high-angle annular dark-field (HA-ADF) detector in a STEM are directly interpretable and are particularly useful for determining the number of graphene layers as the contrast intensity varies monotonically with sample thickness in thin samples (Gass et al., 2008; Li et al., 2008). Krivanek et al. (2010) gave an impressive example of atomic resolution imaging using dark-field imaging in an aberration-corrected STEM operated at 60 kV. They studied single layer hexagonal boron nitride sheets which are, as discussed in Section 2.4 the structural analogue of graphene with alternating boron and nitrogen atoms substituting the carbon atoms within the 2D honeycomb lattice. Thus, in principle their results are also applicable to graphene imaging. The intensity I of the dark-field signal is proportional to the sample thickness and increases with the atomic number Z by $I \sim Z^{1.7}$ (Hartel et al., 1996). Therefore, heavier atoms appear brighter than lighter atoms. Although boron and nitrogen are both very light atoms, they can clearly be distinguished in the raw

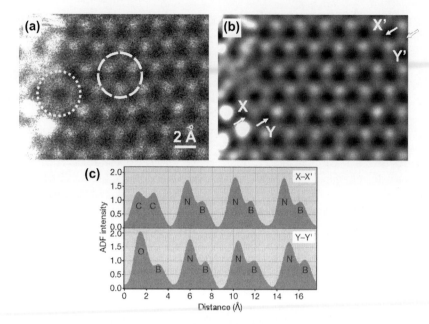

FIGURE 5.4.2 Atomic resolution annular dark field STEM imaging of a monolayer of hexagonal boron nitride: (a) raw image revealing darker boron atoms and brighter nitrogen atoms (dashed circle) as well as atoms with intermediate intensity (dotted circle). (b) The same image after removing the tail contribution of nearest-neighbour sites. (c) Contrast intensity profiles measured along the two lines in (b) with tentative elemental assignment. *Reprinted by permission from Macmillan Publishers Ltd: Nature (Krivanek et al., 2010). Copyright (2010).*

high-magnification ADF-STEM image in Fig. 5.4.2a (Krivanek et al., 2010). As an example, the dashed ring highlights a hexagonal ring with three darker boron and three brighter nitrogen atoms. However, there are some deviations from this contrast pattern, e.g. the hexagon highlighted by the dotted ring, showing atoms with intermediate intensities as compared to boron and nitrogen. After image deconvolution, that removed the tail contribution from the nearest-neighbour sites as well as the pixel-to-pixel statistical noise, Fig. 5.4.2b is obtained (Krivanek et al., 2010). Figure 5.4.2c shows contrast profiles acquired along the two lines X–X' and Y–Y' in Fig. 5.4.2b, where the intensities have been tentatively attributed to boron ($Z = 5$), carbon ($Z = 6$), nitrogen ($Z = 7$) and oxygen atoms ($Z = 8$). Thorough quantitative statistical analysis confirmed the elemental assignment with >99% confidence (Krivanek et al., 2010). It becomes apparent that carbon atoms only substitute B–N pairs whereas single-oxygen atoms substitute single-nitrogen atoms in the hexagonal boron-nitride (h-BN) lattice. Further, in-plane distortions around the substitutional defects have been detected, which can be tracked with a precision of about 0.1 Å. These studies demonstrate the feasibility of atom-by-atom structural and chemical analysis using ADF-STEM (Krivanek et al., 2010).

While Krivanek et al. (2010) derive their elemental information from contrast intensities in images, Suenaga and Koshino (2010) take chemical characterisation at the atomic scale further in that they demonstrate atom-by-atom spectroscopy via ADF-STEM in combination with electron energy loss spectroscopy (EELS). They acquired ADF images of graphene monolayers at 60 kV with a resolution better than 1.06 Å and use the same electron probe to obtain EELS spectra to carry out energy-loss near edge fine structure analysis (ELNES) at specific atomic sites within the graphene lattice and at the graphene edge. ELNES spectra of the carbon K (1s)-edge from atoms in the 'bulk' graphene lattice showed the typical features of sp^2-coordinated carbon, i.e. the π^* peak at ~286 eV and the excitation peak σ^* at ~292 eV (Garvie et al., 1994). Edge atoms within a carbon hexagon at the edge showed an extra peak around ~282.6 eV and reduced π^* and σ^* intensities (Suenaga and Koshino, 2010). Single-coordinated edge atoms, the so-called Klein edges, are difficult to analyse due to perpetual electron-beam damage that primarily affects the graphene edge (Klein, 1994). Still, spectra have been obtained from such a Klein edge and showed that the extra peak was shifted to ~283.6 eV and the π^* and σ^* peaks showed reduced intensities as in the case of the double-coordinated carbon atoms in an edge hexagon. From ELNES simulation and further experiments performing EELS in the spectrum-line mode across a graphene edge, Suenaga and Koshino (2010) concluded that open graphene edges exhibit single- and double-coordinated carbon atoms with edge states that are completely localised at the atomic level. Thus, site-specific spectroscopy is feasible with atomic resolution to explore electronic states of unknown materials.

5.4.3. SURFACE CONTAMINATION

Unambiguous sample characterisation via (S)TEM requires clean specimen devoid of surface adsorbates. However, TEM inspection of graphene is often impaired by residual species partially or fully covering the graphene surface after graphene synthesis and/or TEM sample preparation. As an example, after CVD growth of graphene residual catalytic nanoparticles, e.g. Cu or Ni, may remain attached to the graphene layer (cf. Section 4.5). More often hydrocarbon contaminations are found to reside on the highly reactive graphene surface. Graphene transfer from the growth substrate onto a TEM grid often results in additional contamination with residual polymers, predominantly Poly(methyl methacrylate) (PMMA), from the temporary transfer scaffold as well as traces from the solvents or the ferric chloride etchant employed (Lin et al., 2011, 2012). Further, also synthesis methods which do not involve graphene transfer, e.g. liquid phase exfoliation (cf. Section 4.2), do not necessarily yield graphene with clean surfaces. Here, residual solvent species from the exfoliation treatment often remain attached to the graphene surface and may obscure the TEM image (Schäffel et al., 2011). In TEM inspection graphene surfaces often show only small patches on the order of several 10s of nanometres where the surface

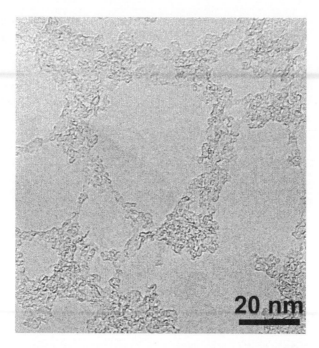

FIGURE 5.4.3 TEM micrograph of hydrocarbon contamination typically obtained on the graphene membrane after sample preparation leaving only small clean patches of the surface for TEM
characterisation. *Reprinted by permission from Macmillan Publishers Ltd: Nature (Meyer et al.,
2008b). Copyright (2008).*

is devoid of contaminants. A typical TEM overview is shown in Fig. 5.4.3
(Meyer et al., 2008b).

Prior to inspection of graphene in TEM intensive efforts are often put into
cleaning the graphene sheet. Generally, it is assumed that annealing in air at
moderate temperatures (up to 300 °C) or argon/hydrogen atmosphere at 400 °C
burns off residual hydrocarbon to leave a clean graphene surface for imaging or
electronic characterisation (Ishigami et al., 2007; Lin et al., 2012). However,
this assumption is being questioned nowadays. A thorough graphene cleaning
study has been carried out by Lin et al. (2011, 2012) combining side by side
Raman and TEM characterisation. Using PMMA as a transfer scaffold or as
electron-beam resist for lithography in device fabrication is known to leave
a thin layer of residual polymer on the sample even after an intensive rinse with
various organic solvents. This residual PMMA obscures the image in TEM, and
the atomic structure of graphene cannot be revealed. Lin et al. (2011, 2012)
found annealing at 200–250 °C in air followed by additional annealing in an
argon/hydrogen atmosphere provides the cleanest graphene surface. However,
adsorbates are not removed completely. Even at higher temperatures to which
the graphene sheet is still susceptible before breaking (~300 °C), the PMMA

networks fail to decompose completely, and prolonged annealing treatments at moderate temperatures lead to partial removal of graphene layers, since the presence of oxygen causes the formation of defects. Annealing-induced tears and holes are found in these samples even after prolonged annealing in air at only 200 °C. From their inspection with Raman, Lin et al. (2012) observe a significant blue-shift of the 2D mode up to $23\,\text{cm}^{-1}$ after annealing which is indicative of an annealing-induced band-structure modulation and thus holds significant implications for electronic characterisation of graphene membranes. This demonstrates that contamination always has to be considered when analysing graphene membranes with TEM and other characterisation methods.

5.4.4. DETERMINING THE NUMBER OF LAYERS THROUGH (SCANNING) TRANSMISSION ELECTRON MICROSCOPY

There are several ways to determine the number of graphene layers using a TEM. Frequently, the number of graphene layers is simply and reliably analysed by counting lines of contrast along a backfolded edge of a graphene sheet, allowing for a cross-sectional view of the membrane, similar to determining the number of walls of a carbon nanotube. This is schematically illustrated in Fig. 5.4.4a for a mono- and a bi-layer membrane. HRTEM micrographs of a mono- and a trilayer graphene membrane are shown in Fig. 5.4.4b (Reina et al., 2009).

Counting the number of layers on the basis of graphene edges around a sputtered area or hole in the membrane is an alternative option. In the graphene membranes shown in Fig. 5.4.4c and d, holes have formed after continuous electron beam irradiation. The membrane in Fig. 5.4.4c can clearly be identified to be a monolayer (Wu et al., 2012). In few-layer membranes, one layer at a time is selectively removed through irradiation and discrete step edges can be made out around a hole (Warner et al., 2009a). The HRTEM micrograph shown in Fig. 5.4.4d presents a few-layer graphene (FLG) flake consisting of a total of six graphene layers (Schäffel et al., 2011). This approach is more reliable than counting contrast patterns produced at edges of a graphene sheet; however, it does involve some sample damage.

The number of layers of a graphene membrane can also be extracted from contrast patterns formed in an atomic resolution image in HRTEM. Warner (2010) performed detailed HRTEM simulation of mono-layer graphene and FLG and compared the contrast of simulated images to experimentally obtained HRTEM micrographs. Figure 5.4.5a shows a set of simulated images using simulation parameters where black atom contrast is obtained for mono-layer graphene (1L, top left panel), i.e. a negative spherical aberration coefficient $C_S = -0.05$ mm and a negative defocus of -5 nm. As demonstrated above, when imaging a bi-layer membrane under the same conditions, atom contrast switches. This can be also derived from the image simulations; the graphene bilayer now shows white atom contrast (2L, Fig. 5.4.5a). Interestingly, a distinctly different contrast pattern is produced of graphene membranes with

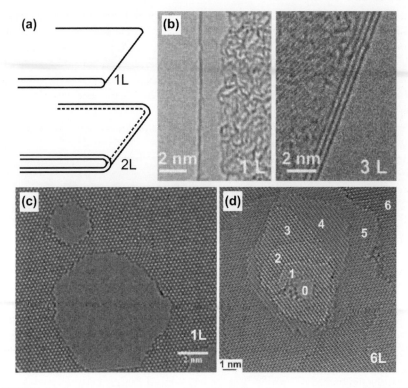

FIGURE 5.4.4 Determination of the number of graphene layers via conventional HRTEM imaging: (a) schematic of backfolded edges of monolayer (1L) and bilayer (2L) graphene. (b) HRTEM micrograph of the backfolded edge of a mono- and a trilayer graphene membrane. *Reproduced with permission from (Reina et al., 2009), Copyright (2009) American Chemical Society.* (c, d) HRTEM micrographs of holes in a c) monolayer and d) in a six-layer membrane allowing for counting the discrete step edges for layer number determination. *Reproduced with permission from Wu et al. (2012) and Schäffel et al., (2011). Copyright (2011) and (2012) American Chemical Society.*

an odd number of layers. Alternating black and white triangles are observed instead of the hexagonal pattern which makes is possible to distinguish tri-layer and five-layer graphene from a mono- and a bi-layer (Warner, 2010). The contrast pattern of the even numbered membranes is similar to the pattern derived from the bilayer membrane. This is supported by the line profiles shown in Fig. 5.4.5b, taken at the position marked with a line in Fig. 5.4.5a (6L). While the contrast maxima marking the atom positions in membranes with an even number of layers are at the same level, they are asymmetric for membranes with an odd number of layers (Warner, 2010). In Fig. 5.4.5c, an atomic resolution HRTEM micrograph of an FLG membrane is shown, where electron beam sputtering has resulted in selective removal of the graphene layers leaving distinct mono-, bi- and tri-layer area, as marked in the image.

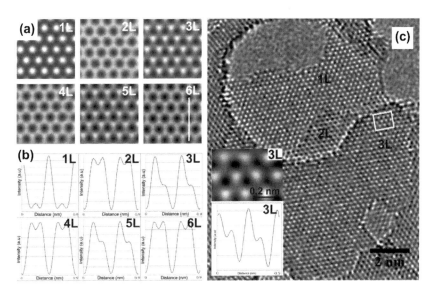

FIGURE 5.4.5 Determination of the number of graphene layers via contrast pattern recognition: (a) simulated TEM images of graphene membranes with one to six layers. (b) Contrast line profiles for the respective graphene membranes shown in a). (c) Atomic resolution HRTEM micrograph of a graphene membrane with a distinct mono-, bi- and trilayer area. Inset: magnified view of the trilayer area highlighted with a rectangle, together with the respective line profile, clearly showing the typical contrast features (i.e. alternating black and white triangles) of trilayer graphene. *Reproduced with permission from Warner (2010). Copyright (2010) IOP Publishing Ltd.*

The inset in Fig. 5.4.5c shows a magnified view of the tri-layer area highlighted with a rectangle in the image, together with the respective line profile (Warner, 2010). Clearly, the pattern of the alternating triangles can be made out. Also the line profile resembles that of odd layered graphene membranes. This study demonstrates that contrast pattern recognition can be employed to determine the number of graphene layers.

A further route to determine the number of layers through transmission electron microscopy has been presented by Meyer et al. (2007). They recorded a dark field TEM image where the primary electron beam has been tilted to just outside the objective aperture in a way that no Bragg reflection is selected and only electrons that are scattered by a small angle contribute to the intensity of the dark field image (Meyer et al., 2007). In this way an image of the thickness of a single element sample, i.e. graphene, can be obtained and the intensity is proportional to the number of graphene layers. Figure 5.4.6a shows the small angle dark field TEM image acquired under these imaging conditions (Meyer et al., 2007). The areas marked with arrows correspond to single-layer graphene as verified by electron diffraction, which will be discussed in the following Section 5.5. Towards the right side the graphene flake appears to be folded

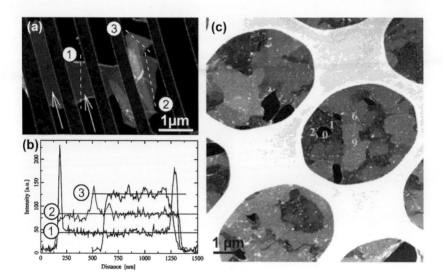

FIGURE 5.4.6 (a) Small-angle dark-field TEM micrograph of a graphene flake. The areas marked with arrows correspond to single-layer graphene as confirmed by electron diffraction. The right part of the flake shows higher intensities which can be attributed to backfolded regions. The recorded intensities in the folded areas are precisely integer multiples of the intensity in the monolayer area. (b) Intensity profiles taken along the dashed lines in a) indicated with the respective number of graphene layers. *Adapted from Meyer et al. (2007). Copyright (2007), with permission from Elsevier.* (c) ADF-STEM image of an FLG membrane spanned across 3.5 μm holes in the TEM grid. The numbers in the top-right hole indicate the number of graphene layers at the respective positions as directly determined from the ADF signal intensity. *Reprinted from Park et al. (2010). Copyright (2010), with permission from Elsevier.*

back. In Fig. 5.4.6b intensity profiles taken along the dashed lines in Fig. 5.4.6a are plotted. It can be recognised that the intensity levels in the folded area are integer multiples of that in the monolayer area. The profile acquired along the top right of the graphene flake (cf. Fig. 5.4.6a) can be attributed to a triple folded region, while the intensity obtained from the profile below corresponds to a double-fold. Since this imaging mode is highly thickness sensitive, it is also very sensitive to any surface adsorbates which should be kept in mind during such analysis (Meyer et al., 2007).

As indicated above, STEM in conjunction with the use of a high-angle annular dark-field (HA-ADF) detector represents a further tool to determine the number of graphene layers since the contrast intensity varies monotonically with sample thickness in thin samples (Gass et al., 2008; Li et al., 2008). As an example an ADF-STEM image of a FLG flake is shown in Fig. 5.4.6c (Park et al., 2010). Here, electron diffraction was used to first unambiguously identify a mono- and a bi-layer (cf. Section 5.5). The respective ADF intensities were then used as reference for the analysis of the graphene membrane thickness (Park et al., 2010). This example also demonstrates that ADF-STEM provides a straightforward platform for characterisation of the layer number on a larger scale, up to a few micron (cf. Fig. 5.4.6c), as compared to conventional HRTEM imaging where the layer number can be determined only locally and for only one graphene flake at a time.

5.4.5. CHARACTERISATION OF DEFECTS IN GRAPHENE

As any other material graphene exhibits structural defects that can strongly influence its properties. Especially with regard to its potential application in nanoelectronics, effects such as electron scattering at point defects, grain boundaries and membrane edges have to be considered. Defects often originate from synthesis, device fabrication or transfer procedures. However, they can also be induced intentionally by chemical functionalisation or irradiation. LV-HRTEM has been employed to study a vast amount of defect structures in graphene, including point defects such as vacancies and adatoms, dislocations and grain boundaries, with atomic resolution which will be covered in this section. The evolution and dynamics of larger defect structures, such as holes and edge structures, will be discussed in the next section, where a stronger focus is put on in-situ transformations.

A recent review by Banhart et al. (2011) focuses on structural point and line defects in graphene. With regard to defect formation graphene is quite unique due to the ability to reconstruct the lattice around intrinsic defects which is partly due to the different possible hybridisations of carbon allowing a different number of nearest neighbours thus leading to the formation of various polygons other than the typical hexagons. The review by Banhart et al. (2011) does not only summarise the defect structures that have been observed via imaging techniques, but also provides a thorough discussion of experimentally and theoretically derived defect formation and migration energies. In this section of the book we will, however, primarily focus on the observations that have been made by (S)TEM techniques.

Since low-voltage aberration corrected TEM now enables researchers to study graphene with atomic resolution it also provides a platform to investigate structural defects. The formation and dynamics of topological defects in graphene have been observed by Meyer et al. (2008a). As an example, the formation and healing of a so called Stone–Wales (SW) defect is shown in Fig. 5.4.7a, which may both have been initiated through electron impact in the TEM (Meyer et al., 2008a; Stone and Wales, 1986). The formation of this defect does not involve atom addition or removal; it is obtained by a 90° in-plane rotation of one of the C–C bonds and, thus, is a perfect example of graphene's ability to reconstruct its lattice by forming non-hexagonal polygons (Banhart et al., 2011). In the left panel in Fig. 5.4.7a, the unperturbed graphene lattice is shown. In the second panel a Stone–Wales defect is apparent and its atomic configuration is highlighted in the third panel of Fig. 5.4.7a. It consists

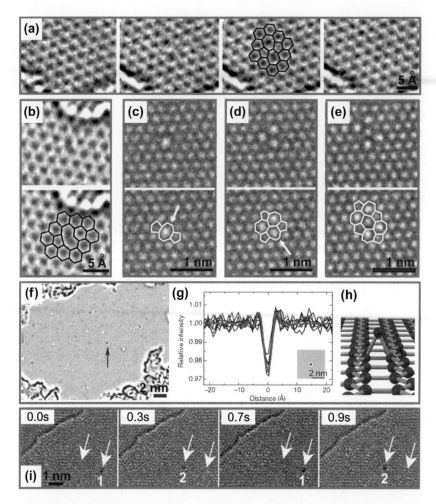

FIGURE 5.4.7   Point defects in graphene: (a) formation and healing of a Stone–Wales defect. Left panel: unperturbed graphene lattice; middle panels: Stone–Wales defect with the atomic configuration highlighted; right panel: unperturbed graphene lattice after relaxation (~4 s later). (b) Single vacancy. *Reproduced with permission from Meyer et al. (2008a). Copyright (2008) American Chemical Society.* (c) DV in 5-8-5 reconstruction. (d) DV in 555–777 reconstruction. (e) DV in 5555-6-7777 reconstruction. *Reprinted with permission from Kotakoski et al. (2011). Copyright (2011) by the American Physical Society.* (f) Carbon atom adsorbed on graphene. (g) Measured (black) and simulated (grey) contrast profiles across a carbon adatom. (h) Atomic model of a carbon adatom in bridge configuration. *Reprinted by permission from Macmillan Publishers Ltd: Nature copyright (2008).* (i) Time sequence of oscillations of a tungsten atom on an FLG sheet observed at 480 °C. The tungsten atom hops back and forth between two distant trapping sites 1 and 2 as highlighted with arrows. *Reprinted with permission from Cretu et al. (2010). Copyright (2010) by the American Physical Society.*

of two pentagons and two hexagons (SW 55-77). After 4 s, the defect has disappeared and the unperturbed lattice is visible again (right panel in Fig. 5.4.7a) (Meyer et al., 2008a). Further, also vacancies, i.e. missing lattice atoms, have been observed via LV-HRTEM. A TEM micrograph of a single vacancy (SV) together with an image highlighting its atomic configuration is presented in Fig. 5.4.7b (Meyer et al., 2008a). It consists of a nine- and a five-membered ring (SV 5-9). Further, double vacancies (DVs) can form through coalescence of two single vacancies or the removal of two neighbouring carbon atoms. A TEM image of a fully reconstructed double vacancy together with its structural representation is shown in Fig. 5.4.7c (Kotakoski et al., 2011). This double vacancy consists of two pentagons and an octagon (DV 5-8-5). However, two other lattice reconstructions that are energetically more favourable have also been observed. Figure 5.4.7d shows the TEM micrograph of a reconstruction that can be obtained by rotating the C–C bond marked with an arrow in Fig. 5.4.7c. The defect now consists of three pentagons and three heptagons (DV 555-777) (Kotakoski et al., 2011). A further reconstruction (Fig. 5.4.7e) is formed by rotating the C–C bond highlighted with an arrow in Fig. 5.4.7d. In this defect structure, a hexagon is symmetrically surrounded by four pentagons and four heptagons (DV 5555-6-7777) (Kotakoski et al., 2011). In general, a full reconstruction of the lattice is possible if an even number of lattice atoms is missing. These types of defects are energetically more favourable than vacancies with an odd number of missing atoms, since in the latter case a dangling bond remains due to geometrical reasons (Banhart et al., 2011). Single and double vacancies have also been detected using ADF-STEM (Gass et al., 2008).

Not only the above described intrinsic defects have been investigated with TEM but also extrinsic defects, such as adatoms or adsorbed molecules, which can be pinned by structural defects. For example, Meyer et al. (2008b) observed light atoms such as carbon and even hydrogen that were adsorbed on the graphene membrane. They collected TEM images with improved signal-to-noise ratio by summing multiple frames over a few minutes. Figure 5.4.7f shows a TEM micrograph with the carbon atom (black arrow) adsorbed on the graphene membrane. In Fig. 5.4.7g, intensity profiles collected from several carbon adatoms (thin black lines) exhibit a good match to a simulated contrast profile (grey line) derived from a carbon atom adsorbed in bridge configuration as schematically shown in Fig. 5.4.7h. Meyer et al. (2008b) could also detect the formation and healing of defects. Further, they obtained larger elongated adsorbates, possibly hydrocarbon chains (Meyer et al., 2008b). Schättel et al. (2011) showed that such hydrocarbon chains get trapped at defect sites and can swing around these anchor points. Further, Schäffel et al. (2011) also observed in situ growth of molecules into longer chains to be possible even under electron irradiation. In a combined theoretical and experimental study Erni et al. (2010) demonstrated that small molecules, such as $-CH_3$, can attach to a carbon atom in the graphene lattice and remain stable at room temperature if the adjacent carbon atom is decorated with a hydrogen atom. If hydrogen attachment to graphene could be achieved in a controlled manner, this mechanism could be used for specific graphene functionalisation.

TEM is also suited to study the diffusion behaviour of metal atoms. This is, in part, easier since metal atoms are stronger scatterers as compared to the light carbon and hydrogen atoms studied by Meyer et al. (2008b). The diffusion of individual gold and platinum atoms on graphene at elevated temperatures has been studied with in situ HRTEM (Gan et al., 2008). The metal atoms were found to reside on voids in the graphene layer, such as single or multiple vacancies. Further, the activation energy for in-plane migration of both gold and platinum atoms was determined to be ~2.5 eV, which indicates that the metal and carbon atoms are covalently bonded (Gan et al., 2008). Cretu et al. (2010) evaporated tungsten atoms onto FLG. At elevated temperature they could observe the tungsten atoms to repeatedly hop between two positions as depicted in the time sequence in Fig. 5.4.7i. This led to the assumption that there is an attraction over a distance larger than 1 nm between the tungsten atom and the trapping site. They concluded that defects such as 555-777 DVs can effectively act as trapping sites due to the increased reactivity of the distorted π-electron system in the strained graphene lattice (Cretu et al., 2010).

A further example employing aberration-corrected LV-HRTEM was presented by Sloan et al. (2010) who studied the dynamics of polyoxometalate anions on GO. The TEM contrast of these anions is dominated by the tungsten atoms within the structure of the molecule allowing for precise determination of the molecules' orientation with respect to the graphene-oxide support and capturing its dynamical motion (Sloan et al., 2010).

Also ADF-STEM was employed to study metal atom adsorption to graphene (Zan et al., 2011). While gold and iron atoms are found to bond to FLG where they reside on top of the carbon atoms and on top of C–C bonds, respectively, chromium atoms bond more strongly to monolayer graphene and are found to dissociate C–C bonds and thus induce vacancy formation (Zan et al., 2011).

Dislocations and grain boundaries in the graphene lattice have also been studied using TEM. While in a 3D crystal an edge dislocation constitutes an extra half-plane of atoms that is inserted into the crystal leading to distortion of neighbouring planes of atoms, in 2D graphene a dislocation is a one-dimensional defect. Only edge dislocation can exist in 2D graphene, since the Burgers vector **b**, which represents the magnitude and direction of the distortion in a crystal lattice generated by a dislocation, is forced to lie within the graphene plane (Yazyev and Louie, 2010). The Burger's vector $\mathbf{b} = r\mathbf{a}_1 + s\mathbf{a}_2$ is a translational lattice vector; thus a pair of intergers (r,s) can be used to describe dislocations in graphene (Yazyev and Louie, 2010).

Hashimoto et al. (2004) imaged an individual edge dislocation where an extra chain of zigzag atoms is inserted in the graphene lattice. Dislocation dipoles, comprising of two pentagon-heptagon pairs, have been observed in reduced graphene oxide (Gómez-Navarro et al., 2010). In a recent study by Warner et al. (2012) dislocation dynamics in graphene were investigated. Single-atom sensitivity was achieved using aberration corrected LV-HRTEM in combination with monochromation of the electron beam with a double Wien filter. Thus individual atomic positions could be accurately mapped and an unambiguous image of the atomic structure could be obtained in real time. In Fig. 5.4.8a, a HRTEM micrograph of a graphene monolayer with two (1,0) edge dislocations is shown. The core of a (1,0) dislocation consists of an edge-sharing pentagon–heptagon pair as high-lighted in Fig. 5.4.8b and schematically shown in Fig. 5.4.8c (Yazyev and Louie, 2010). This dislocation effects the insertion of an extra carbon chain along the armchair direction; the Burgers vector **b** is oriented along the zigzag direction as indicated in Fig. 5.4.8c (Yazyev and Louie, 2010). The presence of the two dislocations in the graphene layer causes strain on the lattice, as indicated by the dotted line in Fig. 5.4.8a, as well as significant bond contraction and elongation around the dislocation cores (Warner et al., 2012). From geometric phase analysis, Warner et al. (2012) determined the strain field around the dislocation cores, which corresponds to shear strain with values from 8–77% at a distance from 0.1–0.02 nm from the dislocation core (Hÿtch et al., 1998; Warner et al., 2012). Further, they observed the two dislocations to be stable over a significant time (>100 s) before a rapid step-wise dislocation movement occurred, induced through energy from electron beam irradiation. Structural transformations similar to dislocation creep and dislocation climb have been detected by

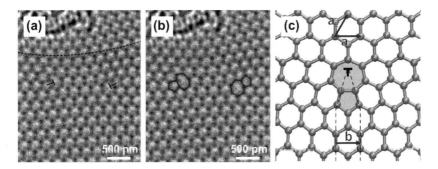

FIGURE 5.4.8   Dislocations in graphene: (a) graphene monolayer with two (1,0) edge disloca-tions that induce significant strain in the graphene lattice together with (b) an image highlighting the atomic configuration of the dislocation dipole. Courtesy of J. H. Warner, Oxford University. (c) Schematic representation of a (1,0) edge dislocation. *Reprinted with permission from Yazyev and Louie, (2010). Copyright (2010) by the American Physical Society.*

Warner et al. (2012). While dislocation creep corresponds to single bond rotation along the zigzag direction, a dislocation climb requires the loss of two carbon atoms (Warner et al., 2012).

It is well-known that mechanical and electronic properties in 3D materials are strongly influenced by the grain size and the atomic structure of the grain boundaries. Also in graphene, where grain boundaries are another type of linear defect, they have strong effects on the electronic (Kim et al., 2009; Reina et al., 2009; Tsen et al., 2012), thermal (Cai et al., 2010) and mechanical performance (Grantab et al., 2010). In a 3D material, grain boundaries are the two-dimensional interfaces between 3D grains or crystallites of different crystallographic orientation and are commonly regarded as an array of dislocations. In 2D graphene grain boundaries are composed of edges dislocations that are aligned into a one-dimensional chain that connects 2D domains of different lattice orientation (Yazyev and Louie, 2010). They generally are tilt boundaries with the tilt axis normal to the graphene plane (Banhart et al., 2011). Grain boundaries in graphene frequently arise during the CVD growth process due to simultaneous nucleation of graphene domains at different sites on the catalytic metal surface. The misfit between the metal support and graphene can lead to different orientations of the emerging graphene domains on the support thus leading to the formation of grain boundaries once the domains grow into one another (Gao et al., 2010; Huang et al., 2011; Park et al., 2010). As an example, a STM image of an extended one-dimensional defect in graphene is shown in Fig. 5.4.9a (Lahiri et al., 2010). This defect results from the lattice mismatch of graphene grown on Ni(111); a domain boundary

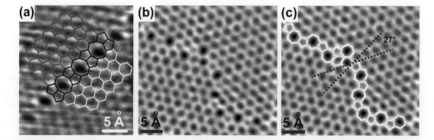

FIGURE 5.4.9 Domain boundaries in graphene. (a) STM image of a domain boundary consisting of alternating pairs of pentagons and octagons arranged along a straight line. *Reprinted by permission from Macmillan Publishers Ltd: Nat. Nanotechnol. (Lahiri et al., 2010), copyright (2010).* (b,c) ADF-STEM image of a curved tilt boundary with a 27° relative misorientation between the domains, predominantly composed of alternating pentagon–heptagon pairs. This type of tilt boundary has been repeatedly observed from graphene grown via CVD, using copper foil. *Reprinted by permission from Macmillan Publishers Ltd: Nature (Huang et al., 2011), copyright (2011).*

consisting of alternating pairs of pentagons and octagons arranged along a straight line is observed (Lahiri et al., 2010).

Grain boundaries have also been inspected using S(TEM) techniques (An et al., 2011; Huang et al., 2011; Kim et al., 2011). Huang et al. (2011) examined the grain boundaries of graphene CVD-grown on copper foil using ADF-STEM with a low accelerating of 60 kV to avoid electron beam-induced structural changes of the grain boundaries. Figure 5.4.9b shows a low-pass filtered ADF-STEM image of a tilt boundary where two graphene domains meet with a relative misorientation of 27°. In the same image in Fig. 5.4.9c, the polygons connecting the graphene domains are highlighted. In contrast to the domain boundary shown in Fig. 5.4.9a, this tilt boundary is not straight and does not consist of a periodic arrangement of polygons. It is composed of a series of pentagons, heptagons and a few distorted hexagons with the predominant feature being alternating pentagon–heptagon pairs (Huang et al., 2011). These results are supported by a study from Kim et al. (2011). Using aberration-corrected TEM at 80 kV they observed a curved 26.6° tilt boundary to consist of an alternating array of pentagons and heptagons exhibiting a mixture of (1,0) and (1,1) dislocations. Huang et al. (2011) and Kim et al. (2011) report that the grain boundaries are stable under the electron beam. Only under high electron beam doses bond rotations along the grain boundary have been observed (Huang et al., 2011). Further, lines of surface contamination were often found to mark grain boundaries (Huang et al., 2011), which is indicative of a higher chemical reactivity of these defects as theoretically predicted (Malola et al., 2010).

In order to obtain reliable information on grain sizes, dark field TEM (DF-TEM) is a useful technique. Huang et al. (2011) collected complete maps of the graphene grain structure by acquiring a dark-field image from a diffraction spot corresponding to grains with (almost) identical lattice orientation and over-laying this image with dark-field images derived from other diffraction spots, using a specific colour code for different lattice orientations. The resulting images reveal grains with a mean size of ~250 nm and complex shapes as well as a preference for low-angle grain boundaries with a tilt angle of ~7° and high angle grain boundaries with ~30° tilt angles. The latter has been attributed to a preferential alignment of the graphene domains with respect to the copper support (Huang et al., 2011). From a study combining DF-TEM and electrical measurements on individual grain boundaries, Tsen et al. (2012) concluded that well-connected grain boundaries, such as the one shown in Fig. 5.4.9b, have minimal electrical impact on the transport properties.

Examination of the domain boundaries and the interconnection of few-layer graphene domains has also been carried out by aberration corrected HRTEM (Robertson et al., 2011). In this study two distinct types of interfaces became apparent, i.e. an atomically bonded graphene–graphene interface as well as overlapping sheets with the termination of a graphene layer on top of another. Differentiation between the two types of domain boundaries was found to be possible by examining the HRTEM contrast profiles across the interface (Robertson et al., 2011).

5.4.6. CHARACTERISATION OF GRAPHENE EDGES

The edge structure of graphene has been extensively studied using S(TEM) techniques due to its large impact especially on the electronic and chemical properties of graphene (nano-)structures, as already highlighted in Section 2.5. Armchair and zigzag edges are the simplest and preferred edge structures most likely because they minimise the number of dangling bonds within the edge. Theoretical calculations, however, show that zigzag edges are metastable and that they can form other fully reconstructed arrangements, e.g. two hexagons in a zigzag edge can spontaneously transform into a pentagon–heptagon arrangement at room temperature (Koskinen et al., 2008; Lee et al., 2010). The edge atoms can either be pristine or passivated with hydrogenated and oxygenated groups. In practice, passivation may be employed intentionally to stabilise the edges or is an inherent result of the preparational procedures (Enoki et al., 2007; Guisinger et al., 2009; Kobayashi et al., 2006; Kosynkin et al., 2009). However, any chemical passivation also has an influence on the transport properties of, e.g., graphene nanoribbons (cf. Section 2.5) and therefore has to be critically considered.

To characterise graphene edges, Raman spectroscopy (Casiraghi et al., 2009; Jiao et al., 2010), STM (Neubeck et al., 2010; Ritter and Lyding, 2009) and HRTEM (Girit et al., 2009; Liu et al., 2009; Warner et al., 2009a) have been widely used. In this context TEM provides advantages in that it allows for fast and direct imaging of the edge. In comparison to the scanning probe techniques where image acquisition times are on the order of minutes or even hours, HRTEM can probe graphene edges in real time, thus, enabling the examination of edge dynamics, edge reconstructions and roughness.

Liu et al. (2009) studied edge structures of graphite powder thermally treated at 2000 °C in vacuum and provide evidence for the coexistence of closed and open edges in graphene. Before heat treatment the edge structure was wavy (Fig. 5.4.10a), after heat treatment, straight edges with well-developed facets along armchair and zigzag directions with intersecting angles of 30° are observed (Fig. 5.4.10b). Similar results have been reported from resistive Joule heating of suspended FLG (Huang et al., 2009; Jia et al., 2009). Liu et al. (2009) determined that the edges of thermally treated graphite are predominantly closed in order to minimise the number of dangling bonds. Closed edges give rise to a line of contrast in the TEM image similar to (back-)folded edges that are often used to determine the number of graphene layers as discussed above. A single line corresponds to one closed edge (Fig. 5.4.10c, left panel). Paired edge lines are also observed; they correspond to two parallel closed edges. Small regions of open, i.e. non-folded, edges have also been found. In Fig. 5.4.10d, an armchair edge is shown where a small region (on the

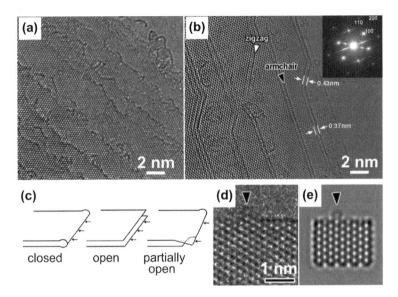

FIGURE 5.4.10 (a) HRTEM micrograph of graphite powder before heat treatment; the edges appear wavy. (b) After heat treatment at 2000°C in vacuum straight edge lines are observed. (c) Schematic illustration of closed, open and partially open edges in a graphene bilayer. (d) HRTEM image and (e) simulated TEM image of a partially open armchair edge with an isolated hexagon extruding from the edge. *Reprinted with permission from Liu et al. (2009). Copyright (2009) by the American Physical Society.*

left side of the image) is open. While the line contrast of the closed edge is apparent in the right part of the image, this strong contrast is absent along the left part of the edge. An isolated hexagon extrudes from this edge (indicated with a black arrow) which has been confirmed by TEM simulation (Fig. 5.4.10e) and further highlights the presence of an open edge. Open zigzag edges with additional carbon atoms protruding from the edge have also been reported. A trace of oxygen in supporting electron energy-loss spectra is indicative of –OH or –COOH termination of the open edges (Liu et al., 2009).

A study of the stability and dynamics of graphene edges was presented by Girit et al. (2009). Using aberration corrected HRTEM, they examined the dynamics of carbon atoms on the edge of a hole formed through prolonged electron beam irradiation in a graphene monolayer. The morphological changes of the edge are a result of the interaction with the electron beam through a combination of electron-beam induced displacement of atoms at the edge and their subsequent migration. Girit et al. (2009) observed armchair and zigzag edges, expanding over multiple hexagons to be stable under electron-beam irradiation for periods of at least 1 s (i.e. the acquisition time), which is indicative that these are stable-edge configurations. From analysing all their microscopy data, Girit et al. (2009) observe that the zigzag edge is the most prominent edge structure. Chuvilin et al. (2009) and Song et al. (2011) also report the observation of zigzag and armchair edges. Figure 5.4.11a and b show an example of a zigzag and an armchair edge, respectively (Chuvilin et al., 2009; Song et al., 2011).

Koskinen et al., (2009) further point out that unconventional edge structures, which have been theoretically predicted, have also been found to be stable. As indicated above, theory predicts that a zigzag edge can spontaneously transform into an edge structure consisting of alternating pentagons and heptagons (Koskinen et al., 2008). As shown in Fig. 5.4.11c, such reconstructions can indeed be observed experimentally; they have been detected with fair abundance with stabilities of sometimes more than 10 s (Girit et al., 2009; Koskinen et al., 2009; Chuvilin et al. 2009). Also other theoretically considered edge structures were experimentally detected, e.g. an armchair edge with pentagons or an armchair edge with two neighbouring heptagons; however, these types of edge segments were only rarely observed (Koskinen et al., 2008, 2009).

Further, Warner et al. (2009a) studied the interaction of high-intensity electron beam irradiation with few-layer graphene and observed a selective removal of monolayers from the back surface. The displacement energy is lower for atoms situated within the back surface of a FLG stack as compared to

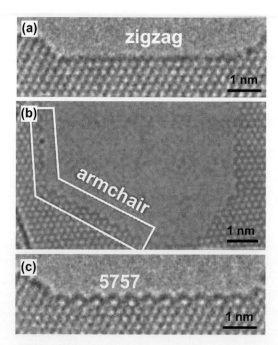

FIGURE 5.4.11 Edge configurations in graphene: (a) HRTEM micrograph of a zigzag edge. (b) HRTEM micrograph of an armchair edge. (c) HRTEM micrograph of a reconstructed zigzag edge, consisting of alternating pentagons and heptagons. *(a,c) are reproduced with permission from Chuvilin et al. (2009). Copyright (2010) IOP Publishing Ltd.; (b) is reproduced with permission from Song et al. (2011). Copyright (2011) American Chemical Society.*

the front surface (the latter is facing the electron beam). If an atom is sputtered from the back surface is does not have to be incorporated into the lattice structure, but can instead be ejected into the vacuum in the direction of momentum transfer (Warner et al., 2009a). As a result, material from the back plane is lost, which is known as electron beam sputtering. This becomes even more pronounced if unsaturated atoms are situated in the back surface, e.g. a sputtered edge, because these atoms have fewer bonds than saturated atoms. Similar to the results reported by Girit et al. (2009) from graphene monolayers, Warner et al. (2009a) find that the zigzag termination has a higher stability over the armchair configuration. Under the influence of the electron beam, Warner et al. could track the removal of carbon atoms from a zigzag edge; they found it to occur in a zipper-like manner, i.e. one zigzag row was removed at a time. To be able to detect this kind of transformation, a fast temporal resolution in conjunction with atomic spatial resolution is required; in the experiments described by Warner et al. (2009a), HRTEM images were acquired with a temporal resolution of 80 ms. This highlights the strong advantage of TEM imaging over scanning probe techniques.

Examination of the intrinsic edge structure of FLG prepared via liquid-phase exfoliation supports the observation of a preference of the zigzag edge (Warner et al., 2009b). While intrinsic edges that are aligned parallel to a zigzag lattice direction exhibit only small random variations in the edge structure, edges aligned parallel to an armchair direction display large corrugations with sawtooth structure along the zigzag directions (Warner et al., 2009b).

Driving electron beam sputtering further eventually leads to the observation of one-dimensional carbon chains, which are envisaged as building blocks in molecular devices. Jin et al. (2009) employed electron beam irradiation to induced holes in a graphene sheet in a controlled manner. Growth of two neighbouring holes results into the formation of nanoribbons in the area in between. Continuous thinning of the ribbon then leads to the production of free-standing linear chains of carbon. These carbon chains are most likely sp-hybridised cumulene (with the carbon atoms bonded by double-bonds) or polyyne chains (characterised by alternating single and triple bonds); however, the exact type of atomic chain could not be experimentally determined due to mechanical instability of the specimen. Jin et al. (2009) also frequently observe the formation of a double-chain. Similar findings have been reported by Chuvilin et al. (2009). They observe that upon thinning the graphene strip below a width of ~1 nm, the hexagonal atom arrangement reconstructs into a configuration with more pentagons and heptagons as compared to hexagons. They also point out that in 50% of the cases, where a graphene ribbon (GNR) has been thinned down, the final product is a single-carbon chain. These linear chains are found to be stable for up to two minutes under continuous 80 kV electron beam irradiation. This indicates that this reconstructed linear config-uration is energetically favoured and not a mere product of atom removal by the electron beam (Chuvilin et al., 2009).

5.4.7. IN-SITU MANIPULATION OF GRAPHENE IN A TEM

In previous sections it already became apparent that TEM is a versatile tool to study in situ transformations. Some examples for transformations that are induced by the electron beam are described above, such as the formation of defects, reconstruction of edges and the electron beam sputtering of holes in the graphene membrane. The development of special TEM holders with additional capabilities, e.g. to carry out STM measurements inside a TEM or apply a bias voltage to a suspended sample, enables researchers to manipulate graphene in situ.

One example is the observation of carbon atom sublimation from graphene and edge reconstruction through Joule heating (Huang et al., 2009; Jia et al., 2009). A piece of graphene is suspended on a TEM grid and contacted with a STM tip of the specialised TEM holder. Applying a bias voltage causes Joule heating of the graphene flake, which, in combination with electron irradiation results into the removal of contaminants and carbon atom sublimation from the graphene lattice. The formation of sharp edges terminating along zigzag and armchair directions is observed (Huang et al., 2009; Jia et al., 2009). Similar to observation of the formation of closed graphene edges after high-temperature ex situ annealing at 2000 °C in vacuum, Huang et al. report that more than 99% of the graphene edges after Joule heating are 'bilayer edges' (i.e. closed edges), rather than monolayer edges (open edges), which indicates that bilayer edges are energetically favoured as they minimise the number of dangling bonds (Huang et al., 2009; Liu et al., 2009). The temperatures reached during in situ Joule heating are estimated to be around ~2000 °C, similar to the ex situ experiment. Carbon sublimation driven by a high bias also represents an alternative tool for nanostructuring graphene (Barreiro et al., 2012). Under high bias, cracks form at the graphene edges and slowly propagate through the sample leading to the formation of ultranarrow graphene constrictions which can be employed to produce graphene nanoribbons. These constrictions are found to withstand enormous current densities before breakdown (Barreiro et al., 2012). Further, the heat-induced evolution of hydrocarbons on graphene has been studied in situ (Westenfelder et al., 2011). In these experiments the graphene sheet supporting the carbon adsorbates serves as an in situ heater. As the temperature (the current) increases, a transformation of physisorbed hydrocarbons into polycrystalline graphene is observed.

To summarise, this section of the book shows the immense variety of structural phenomena in graphene that can be studied using TEM. It becomes clear that with the development of aberration-corrected electron optics, this technique has made strong progress that now enables researchers to examine graphene with atomic resolution and allows accurate mapping of, e.g., defect structures that cannot be so easily accessed by other techniques.

REFERENCES

An, J., Voelkl, E., Suk, J.W., Li, X., Magnuson, C.W., Fu, L., Tiemeijer, P., Bischoff, M., Freitag, B., Popova, E., Ruoff, R.S., 2011. Domain (grain) boundaries and evidence of "twinlike" structures in chemically vapor deposited grown graphene. ACS Nano 5, 2433–2439.

Banhart, F., Kotakoski, J., Krasheninnikov, A.V., 2011. Structural defects in graphene. ACS Nano 5, 26–41.

Barreiro, A., Börrnert, F., Rümmeli, M.H., Büchner, B., Vandersypen, L.M.K., 2012. Graphene at high bias: cracking, layer by layer sublimation, and fusing. Nano Lett. 12, 1873–1878.

Cai, W., Moore, A.L., Zhu, Y., Li, X., Chen, S., Shi, L., Ruoff, R.S., 2010. Thermal transport in suspended and supported monolayer graphene grown by chemical vapor deposition. Nano Lett. 10, 1645–1651.

Casiraghi, C., Hartschuh, A., Qian, H., Piscanec, S., Georgi, C., Fasoli, A., Novoselov, K.S., Basko, D.M., Ferrari, A.C., 2009. Raman spectroscopy of graphene edges. Nano Lett. 9, 1433–1441.

Christenson, K.K., Eades, J.A., 1986. On "parallel" illumination in the transmission electron microscope. Ultramicroscopy 19, 191–194.

Chuvilin, A., Meyer, J.C., Algara-Siller, G., Kaiser, U., 2009. From graphene constrictions to single carbon chains. New. J. Phys. 11, 083019.

Cretu, O., Krasheninnikov, A.V., Rodríguez-Manzo, J.A., Sun, L., Nieminen, R.M., Banhart, F., 2010. Migration and localization of metal atoms on strained graphene. Phys. Rev. Lett. 105, 196102.

Enoki, T., Kobayashi, Y., Fukui, K.-I., 2007. Electronic structures of graphene edges and nano-graphene. Int. Rev. Phys. Chem. 26, 609–645.

Erni, R., Rossell, M.D., Nguyen, M.-T., Blankenburg, S., Passerone, D., Hartel, P., 2010. Stability and dynamics of small molecules trapped on graphene. Phys. Rev. B 82, 165443.

Gan, Y., Sun, L., Banhart, F., 2008. One and two-dimensional diffusion of metal atoms in graphene. Small 4, 587–591.

Gao, L., Guest, J.R., Guisinger, N.P., 2010. Epitaxial graphene on Cu(111). Nano Lett. 10, 3512–3516.

Garvie, L.A.J., Craven, A.J., Brydson, R., 1994. Use of electron-energy loss near-edge fine structure in the study of minerals. Am. Mineral 79, 411–425.

Gass, M.H., Bangert, U., Bleloch, A.L., Wang, P., Nair, R.R., Geim, A.K., 2008. Free-standing graphene at atomic resolution. Nat. Nanotechnol. 3, 676–681.

Girit, ÇÖ, Meyer, J.C., Erni, R., Rossell, M.D., Kisielowski, C., Yang, L., Park, C.-H., Crommie, M.F., Cohen, M.L., Louie, S.G., Zettl, A., 2009. Graphene at the edge: stability and dynamics. Science 323, 1705–1708.

Gómez-Navarro, C., Meyer, J.C., Sundaram, R.S., Chuvilin, A., Kurasch, S., Burghard, M., Kern, K., Kaiser, U., 2010. Atomic structure of reduced graphene oxide. Nano Lett. 10, 1144–1148.

Grantab, R., Shenoy, V.B., Ruoff, R.S., 2010. Anomalous strength characteristics of tilt grain boundaries in graphene. Science 330, 946–948.

Guisinger, N.P., Rutter, G.M., Crain, J.N., First, P.N., Stroscio, J.A., 2009. Exposure of epitaxial graphene on SiC(0001) to atomic hydrogen. Nano Lett. 9, 1462–1466.

Hartel, P., Rose, H., Dinges, C., 1996. Conditions and reasons for incoherent imaging in STEM. Ultramicroscopy 63, 93–114.

Hashimmoto, A., Suenaga, K., Gloter, A., Urita, K., Iijima, S., 2004. Direct evidence for atomic defects in graphene layers. Nature 430, 870–873.

Huang, J.Y., Ding, F., Yakobson, B.I., Lu, P., Qi, L., Li, J., 2009. In situ observation of graphene sublimation and multi-layer edge reconstructions. Proc. Natl. Acad. Sci. U.S.A. 106, 10103–10108.

Huang, P.Y., Ruiz-Vargas, C.S., van der Zande, A.M., Whitney, W.S., Levendorf, M.P., Kevek, J.W., Garg, S., Alden, J.S., Hustedt, C.J., Zhu, Y., Park, J., McEuen, P.L., Muller, D.A., 2011. Grains and grain boundaries in single-layer graphene atomic patchwork quilts. Nature 469, 389–392.

Hÿtch, M.J., Snoeck, E., Kilaas, R., 1998. Quantitative measurement of displacement and strain fields from HREM micrographs. Ultramicroscopy 74, 131–146.

Ishigami, M., Chen, J.H., Cullen, W.G., Fuhrer, M.S., Williams, E.D., 2007. Atomic structure of graphene on SiO_2. Nano Lett. 7, 1643–1648.

Jia, X., Hofmann, M., Meunier, V., Sumpter, B.G., Campos-Delgado, J., Romo-Herrera, J.M., Son, H., Hsieh, Y.-P., Reina, A., Kong, J., Terrones, M., Dresselhaus, M.S., 2009. Controlled formation of sharp zigzag and armchair edges in graphitic nanoribbons. Science 323, 1701–1705.

Jiao, L., Wang, X., Diankov, G., Wang, H., Dai, H., 2010. Facile synthesis of high-quality graphene nanoribbons. Nat. Nanotechnol. 5, 321–325.

Jin, C., Lan, H., Peng, L., Suenaga, K., Iijima, S., 2009. Deriving carbon atomic chains from graphene. Phys. Rev. Lett. 102, 205501.

Kim, K.S., Zhao, Y., Jang, H., Lee, S.Y., Kim, J.M., Kim, K.S., Ahn, J.H., Kim, P., Choi, J.Y., Hong, B.H., 2009. Large-scale pattern growth of graphene films for stretchable transparent electrodes. Nature 457, 706–710.

Kim, K., Lee, Z., Regan, W., Kisielowski, C., Crommie, M.F., Zettle, A., 2011. Grain boundary mapping in polycrystalline graphene. ACS Nano 5, 2142–2146.

Kisielowski, C., Freitag, B., Bischoff, M., van Lin, H., Lazar, S., Knippels, G., Tiemeijer, P., van der Stam, M., von Harrach, S., Stekelenburg, M., Haider, M., Uhlemann, S., Muller, H., Hartel, P., Kabius, B., Miller, D., Petrov, I., Olson, E.A., Donchev, T., Kenik, E.A., Lupini, A.R., Bentley, J., Pennycook, S.J., Anderson, I.M., Minor, A.M., Schmid, A.K., Duden, T., Radmilovic, V., Ramasse, Q.M., Watanabe, M., Erni, R., Stach, E.A., Denes, P., Dahmen, U., 2008. Detection of single atoms and buried defects in three dimensions by aberration-corrected electron microscope with 0.5-angstrom information limit. Microsc. Microanal. 14, 469–477.

Klein, D.J., 1994. Graphitic polymer strips with edge states. Chem. Phys. Lett. 217, 261–265.

Kobayashi, Y., Fukui, K., Enoki, T., Kusakabe, K., 2006. Edge state on hydrogen-terminated graphite edges investigated by scanning tunneling microscopy. Phys. Rev. B 73, 125415.

Koskinen, P., Malola, S., Häkkinen, H., 2008. Self-passivating edge reconstructions of graphene. Phys. Rev. Lett. 101, 115502.

Koskinen, P., Malola, S., Häkkinen, H., 2009. Evidence for graphene edges beyond zigzag and armchair. Phys. Rev. B 80, 073401.

Kosynkin, D.V., Higginbotham, A.L., Sinitskii, A., Lomeda, J.R., Dimiev, A., Price, B.K., Tour, J.M., 2009. Longitudinal unzipping of carbon nanotubes to form graphene nanoribbons. Nature 458, 872–876.

Kotakoski, J., Meyer, J.C., Kurasch, S., Santos-Cottin, D., Kaiser, U., Krasheninnikov, A.V., 2011. Stone–Wales-type transformations in carbon nanostructures driven by electron irradiation. Phys. Rev. B 83, 245420.

Krivanek, O.L., Chisholm, M.F., Nicolosi, V., Pennycook, T.J., Corbin, G.J., Dellby, N., Murfitt, M.F., Own, C.S., Szilagyi, Z.S., Oxley, M.P., Pantelides, S.T., Pennycook, S.J., 2010. Atom-by-atom structural and chemical analysis by annular dark-field electron microscopy. Nature 464, 571–574.

Lahiri, J., Lin, Y., Bozkurt, P., Oleynik, I.I., Batzill, M., 2010. An extended defect in graphene as a metallic wire. Nat. Nanotechnol. 5, 326–329.

Lee, G.-D., Wang, C.Z., Yoon, E., Hwang, N.-M., Ho, K.M., 2010. Reconstruction and evaporation at graphene nanoribbon edges. Phys. Rev. B 81, 195419.

Li, Z.Y., Young, N.P., Di Vece, M., Palomba, S., Palmer, R.E., Bleloch, A.L., Curley, B.C., Johnston, R.L., Jiang, J., Yuan, J., 2008. Three-dimensional atomic-scale structure of size-selected gold nanoclusters. Nature 451, 46–49.

Lin, Y.-C., Jin, C., Lee, J.-C., Jen, S.-F., Suenaga, K., Chiu, P.-W., 2011. Clean transfer of graphene for isolation and suspension. ACS Nano 5, 2362–2368.

Lin, Y.-C., Lu, C.-C., Yeh, C.-H., Jin, C., Suenaga, K., Chiu, P.-W., 2012. Graphene annealing: how clean can it be? Nano Lett. 12, 414–419.

Liu, Z., Suenaga, K., Harris, P.J.F., Iijima, S., 2009. Open and closed edges of graphene layers. Phys. Rev. Lett. 102, 015501.

Malola, S., Häkkinen, H., Koskinen, P., 2010. Structural, chemical, and dynamical trends in graphene grain boundaries. Phys. Rev. B 81, 165447.

Meyer, J.C., Geim, A.K., Katsnelson, M.I., Novoselov, K.S., Obergfell, D., Roth, S., Girit, C., Zettl, A., 2007. On the roughness of single- and bi-layer graphene membranes. Solid State Commun. 143, 101–109.

Meyer, J.C., Kisielowski, C., Erni, R., Rossell, M.D., Crommie, M.F., Zettl, A., 2008a. Direct imaging of lattice atoms and topological defects in graphene membranes. Nano Lett. 8, 3582–3586.

Meyer, J.C., Girit, C.O., Crommie, M.F., Zettl, A., 2008b. Imaging and dynamics of light atoms and molecules on graphene. Nature 454, 319–322.

Neubeck, S., You, Y.M., Ni, Z.H., Blake, P., Shen, Z.X., Geim, A.K., Novoselov, K.S., 2010. Direct determination of the crystallographic orientation of graphene edges by atomic resolution imaging. Appl. Phys. Lett. 97, 053110.

Park, H.J., Meyer, J.C., Roth, S., Skákalová, V., 2010. Growth and properties of few-layer graphene prepared by chemical vapor deposition. Carbon 48, 1088–1094.

Reina, A., Jia, X., Ho, J., Nezich, D., Son, H., Bulovic, V., Dresselhaus, M.S., Kong, J., 2009. Large area, few-layer graphene films on arbitrary substrates by chemical vapor deposition. Nano Lett. 9, 30–35.

Ritter, K.A., Lyding, J.W., 2009. The influence of edge structure on the electronic properties of graphene quantum dots and nanoribbons. Nat. Mater. 8, 235–242.

Robertson, A.W., Bachmatiuk, A., Wu, Y.A., Schäffel, F., Rellinghaus, B., Büchner, B., Rümmeli, M.H., Warner, J.H., 2011. Atomic structure of interconnected few-layer graphene domains. ACS Nano 5, 6610–6618.

Sasaki, T., Sawada, H., Hosokawa, F., Kohno, Y., Tomita, T., Kaneyama, T., Kondo, Y., Kimoto, K., Sato, Y., Suenaga, K., 2010. Performance of low-voltage STEM/TEM with delta corrector and cold field emission gun. J. Electron Microsc. 59, S7–S13.

Schäffel, F., Wilson, M., Warner, J.H., 2011. Motion of light adatoms and molecules on the surface of few-layer graphene. ACS Nano 5, 9428–9441.

Sloan, J., Liu, Z., Suenaga, K., Wilson, N.R., Pandey, P.A., Perkins, L.M., Rourke, J.P., Shannon, I.J., 2010. Imaging the structure, symmetry, and surface-inhibited rotation of poly-oxometalate ions on graphene oxide. Nano Lett. 10, 4600–4606.

Smith, B.W., Luzzi, D.E., 2001. Electron irradiation effects in single wall carbon nanotubes. J. Appl. Phys. 90, 3509–3515.

Song, B., Schneider, G.F., Xu, Q., Pandraud, G., Dekker, C., Zandbergen, H., 2011. Atomic-scale electron-beam sculpting of near-defect-free graphene nanostructures. Nano Lett. 11, 2247–2250.

Stone, A.J., Wales, D.J., 1986. Theoretical study of icosahedral C_{60} and some related species. Chem. Phys. Lett. 128, 501–503.

Suenaga, K., Koshino, M., 2010. Atom-by-atom spectroscopy at graphene edge. Nature 468, 1088–1090.

Tsen, A.W., Brown, L., Levendorf, M.P., Ghahari, F., Huang, P.Y., Havener, R.W., Ruiz-Vargas, C.S., Muller, D.A., Kim, P., Park, J., 2012. Tailoring electrical transport across grain boundaries in polycrystalline graphene. Science 336, 1143–1146.

Warner, J.H., Rümmeli, M.H., Ge, L., Gemming, T., Montanari, B., Harrison, N.M., Büchner, B., Briggs, G.A.D., 2009a. Structural transformations in graphene studied with high spatial and temporal resolution. Nat. Nanotechnol. 4, 500–504.

Warner, J.H., Schäffel, F., Rümmeli, M.H., Büchner, B., 2009b. Examining the edges of multi-layer graphene sheets. Chem. Mater. 21, 2418–2421.

Warner, J.H., Schäffel, F., Zhong, G., Rümmeli, M.H., Büchner, B., Robertson, J., Briggs, G.A.D., 2009c. Investigating the diameter-dependent stability of single-walled carbon nanotubes. ACS Nano 3, 1557–1563.

Warner, J.H., Margine, E.R., Mukai, M., Robertson, A.W., Giustino, F., Kirkland, A.I., 2012. Dislocation driven deformations in graphene. Science 337, 209–212.

Warner, J.H., 2010. The influence of the number of graphene layers on the atomic resolution images obtained from aberration-corrected high resolution transmission electron microscopy. Nanotechnology 21, 255707.

Westenfelder, B., Meyer, J.C., Biskupek, J., Kurasch, S., Scholz, F., Krill, C.E., Kaiser, U., 2011. Transformations of carbon adsorbates on graphene substrates under extreme heat. Nano Lett. 11, 5123–5127.

Williams, D.B., Carter, C.B., 2009. Transmission Electron Microscopy – A Textbook for Materials Science. Springer.

Wu, Y.A., Fan, Y., Speller, S., Creeth, G.L., Sadowski, J.T., He, K., Robertson, A.W., Allen, C.S., Warner, J.H., 2012. Large single crystals of graphene on melted copper using chemical vapor deposition. ACS Nano 6, 5010–5017.

Yazyev, O.V., Louie, S.G., 2010. Topological defects in graphene: dislocations and grain boundaries. Phys. Rev. B 81, 195420.

Zan, R., Bangert, U., Ramasse, Q., Novoselov, K.S., 2011. Metal-graphene interaction studied via atomic resolution scanning transmission electron microscopy. Nano Lett. 11, 1087–1092.

Chapter 5.5

Electron Diffraction

Franziska Schäffel

University of Oxford, Oxford, UK

5.5.1. INTRODUCTION

TEM is not only a tool for imaging; a variety of electron diffraction (ED) and spectroscopy techniques can also be carried out using a TEM and are highly appealing to study graphene. In graphene research, ED is used to obtain information on the number of graphene layers, their stacking order, stacking faults and layer roughness.

As introduced in Section 2.1 the reciprocal lattice of graphene is hexagonal. In Fig. 5.5.1a, a normal incidence selected area electron diffraction (SAED)

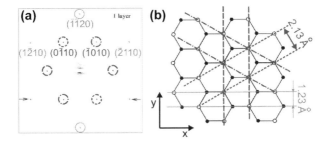

FIGURE 5.5.1 (a) Normal incidence SAED pattern of a graphene monolayer. *Adapted from Meyer et al. (2007b). Copyright (2007), with permission from Elsevier.* (b) Schematic representation of the real space lattice of graphene. The lattice planes that are highlighted with lines of different style (full, dashed, dash–dot–dash) in the schematic representation give rise to the diffraction spots highlighted with circles of respective lines style in (a).

pattern is depicted, which shows the typical sixfold symmetry expected for graphene (Meyer et al., 2007b). The topmost diffraction spots are labelled with Bravais-Miller indices (*hkil*). Each set of real space-lattice planes gives rise to two spots in the ED pattern. In order to gain a better understanding of this, Fig. 5.5.1b shows a schematic representation of the real space lattice of graphene. Circles with different line styles are drawn around pairs of spots in the ED pattern; the lattice planes that give rise to these spots are highlighted with respective lines in the schematic lattice. As an example, the lattice planes indicated by the dash–dot–dash line in the real space lattice (Fig. 5.5.1b, from top left to bottom right) generate the two spots with the respective line style in the top right and the bottom left of the inner hexagon. The inner set of six diffraction spots corresponds to the $\{10\bar{1}0\}$ planes of graphene with a lattice spacing of $d = 2.13$ Å. The larger set of diffraction spot can be assigned to the $\{11\bar{2}0\}$ planes with a shorter lattice spacing of $d = 1.23$ Å, as discussed in Section 2.1. In a similar way a 2D fast Fourier transform can be used to identify lattice distances and directions, e.g. the relative directions of armchair and zigzag orientations in a graphene sheet, from a HRTEM micrograph (cf. (Warner et al., 2009b)).

5.5.2. DETERMINING THE NUMBER OF LAYERS USING ELECTRON DIFFRACTION

As mentioned in Section 5.4, the number of graphene layers can be determined by counting lines of contrast along a backfolded edge of a graphene sheet, comparable to determining the number of walls of a carbon nanotube (Robertson and Warner, 2011). An alternative route to identify the layer number is through analysis of ED patterns (Hernandez et al., 2008; Meyer et al., 2007a,b). Figure 5.5.2a,b show the normal incidence SAED patterns of a graphene

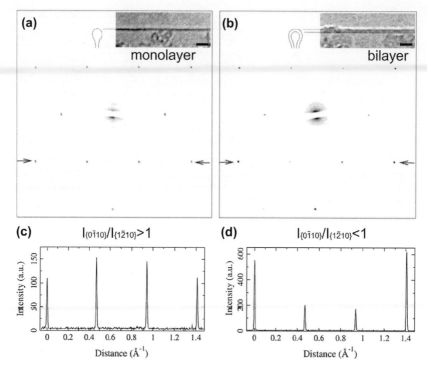

FIGURE 5.5.2 Normal incidence SAED patterns of (a) a graphene monolayer and (b) a bilayer. The insets show HRTEM micrographs of backfolded edges of the respective graphene sheets. Scale bars are 2 nm. (c, d) Intensity profiles plotted along the lines between the arrows in (a) and (b). *Adapted from Meyer et al. (2007b). Copyright (2007), with permission from Elsevier.*

monolayer and a bilayer, respectively (Meyer et al., 2007b). The insets in Fig. 5.5.2a,b show a HRTEM micrograph of a folded edge, confirming the layer number. In Fig. 5.5.2c,d, intensity profiles taken along the line between the arrows in Fig. 5.5.2a,b are plotted. While in the monolayer, the inner $\{10\bar{1}0\}$ spots are more intense than the outer $\{11\bar{2}0\}$ spots (Fig. 5.5.2c), the relative intensity is reversed in the bilayer (Fig. 5.5.2d). From computational studies it is known that in the case of AB (Bernal)-stacked graphene multilayers, the intensity ratio $I_{\{0\bar{1}10\}}/I_{\{1\bar{2}10\}}$ is below 1 while it is larger than 1 for a graphene monolayer (Horiuchi et al., 2003). Therefore, assuming the samples retain AB stacking, e.g. from a graphitic precursor, a graphene monolayer can be identified from the intensity ratios of the diffraction peaks (cf. Fig. 5.5.2c,d). However, selected area electron diffraction can also give rise to monolayer-like patterns for multi-layers (Hernandez et al., 2008). An unambiguous identification of the layer number can either be carried out by counting the number of dark lines along the foldings at the rim of a graphene sheet or by obtaining a tilt series, as will be discussed in the following (Meyer et al., 2007b).

Meyer et al. (2007b) demonstrated that a direct and unambiguous identification of single- vs. multilayer graphene membranes can be obtained from a series of SAED patterns by varying incidence angles between the electron beam and the graphene sheet. In this way, they probed the whole 3D reciprocal space. Figure 5.5.3a,b show calculated 3D Fourier transforms of single- and AB-stacked bi-layer graphene, respectively (Meyer et al., 2007b). These calculations were carried out without incorporating the atomic scattering factors so that the intensities are only qualitatively correct. In the case of single-layer graphene, the intensities in reciprocal space are a set of continuous rods arranged on the 2D reciprocal lattice with a weak, monotonic intensity change normal to the plane (Fig. 5.5.3a). The dark grey plane indicates a diffraction pattern as obtained at normal incidence. The light grey plane marks a pattern observed at a tilt angle of 20°. From this it can be derived that the intensities in

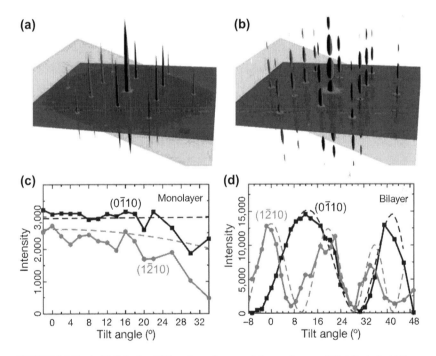

FIGURE 5.5.3 (a, b) Calculated 3D reciprocal space of a) monolayer and b) AB-stacked bilayer graphene. For monolayer graphene the intensities in reciprocal space are continuous rods with a weak, monotonic intensity change normal to the plane. For bilayer (or multilayer) graphene, an additional intensity modulation along the rods is present. The dark grey planes indicate a diffraction pattern as obtained at normal incidence, the light grey plane for a tilt angle of 20°. *Reprinted from Meyer et al. (2007b). Copyright (2007), with permission from Elsevier.* (c, d) Total reflection intensities of c) monolayer and AB stacked bilayer graphene as a function of tilt angle. The solid lines correspond to experimental data as obtained from Gaussian fits; the dashed lines are derived from respective electron diffraction simulations. *Adapted by permission from Macmillan Publishers Ltd: Nature (Meyer et al., 2007a), copyright (2007).*

the diffraction pattern only slightly change when adjusting the tilt angle between the monolayer graphene membrane and the incident electron beam. In Fig. 5.5.3c, the experimentally observed total intensities of the $(0\bar{1}10)$ and the $(1\bar{2}10)$ reflections of monolayer graphene are plotted as a function of tilt-angle (solid lines) (Meyer et al., 2007a). To obtain the total intensity values, the Bragg reflections were fitted by a Gaussian distribution revealing the peaks' intensities, positions, heights and widths (Meyer et al., 2007a). The dashed lines correspond to numerical electron diffraction simulations (Meyer et al., 2007a). This weak and monotonous change of the total intensity with tilt angle is a signature specific to a suspended graphene monolayer (Meyer et al., 2007a).

For graphene samples with two (or more) layers, the intensity strongly varies along the rods in 3D reciprocal space (Fig. 5.5.3b) (Meyer et al., 2007b). A look at the light-grey plane (corresponding to a tilt angle of 20°) reveals that diffraction peaks are suppressed at certain tilt angles. In Fig. 5.5.3d, the total intensities of the $(0\bar{1}10)$ and $(1\bar{2}10)$ diffraction peaks of an AB-stacked bilayer graphene membrane are plotted as a function of tilt angle (solid lines), which is in agreement with data from ED simulation of an AB stacked bilayer membrane (dashed lines) (Meyer et al., 2007a). Here, the peak intensities strongly vary with tilt angle and even become suppressed at some tilt angles.

To summarise, while in the case of all multilayer samples, independent of the stacking sequence, variations of only a few degrees in the tilt angle lead to strong alterations in the diffraction intensities, for a single-layer, only a weak, monotonous intensity variation occurs, which can be used for its unambiguous identification in TEM.

5.5.3. DETERMINING THE GRAPHENE TOPOGRAPHY

A tilt series of this kind can also be used to study the topography of suspended graphene membranes as already mentioned in Section 2.1 (Meyer et al., 2007a,b). Although the total intensity of the Bragg reflections agrees well with the model of a flat graphene membrane, the shape and widths of the reflections show strong and unexpected deviations from the standard diffraction behaviour of 3D crystals (Spence, 2003; Williams and Carter, 2009). Meyer et al. (2007a,b) observed a peak broadening with increasing tilt angle. In Fig. 5.5.4a–c, SAED patterns obtained at different tilt angles are depicted. The tilt axis corresponds to the horizontal dashed lines as marked in the images. The reflections in the SAED pattern generated at zero-tilt angle (Fig. 5.5.4a) are relatively sharp, whereas the reflections in the patterns with 14°- and 16°- tilt angles (Fig. 5.5.4b and c, respectively) spread isotropically to an approximately Gaussian shaped peak area and become progressively blurred with increasing tilt angle and increasing distance from the tilt axis. Figure 5.5.4d shows the full width at half maximum (FWHM) of the peak width as a function of the tilt angle for a graphene monolayer, a bilayer and a graphitic flake with approximately 50 layers. This plot reveals that the peak

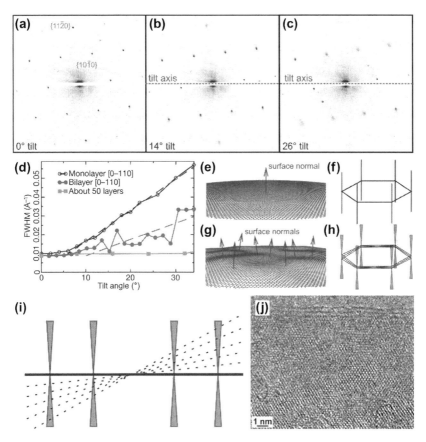

FIGURE 5.5.4 Selected area ED (SAED) patterns from a graphene monolayer under incidence angles of (a) 0°, (b) 14° and (c) 26°, respectively. The tilt axis is marked with a dashed line in (b) and (c). (d) FWHM of the peak width for monolayer, bilayer and multi-layer graphene. (e–i) Schematic explanation of peak broadening through microscopic corrugations in the membrane: while for perfectly flat graphene (e), the reciprocal space is a set of rods arranged perpendicular to the reciprocal lattice (f), in corrugated graphene (g), several locally flat parts of the graphene sheet yield diffraction peaks at slightly different positions, resulting in the rods turning into cone-shaped volumes in reciprocal space (h). The diffraction spots broaden and the effect becomes more pronounced with increasing distance from the tilt axis, as indicated by the dashed lines in (i). (j) Atomic resolution TEM micrograph of FLG with 'domains' of different brightness indicative of their different orientation with respect to the electron beam. *Reprinted by permission from Macmillan Publishers Ltd: Nature (Meyer et al., 2007a), copyright (2007), and (Meyer et al., 2007b). Copyright (2007), with permission from Elsevier.*

broadening is a very prominent feature for monolayer graphene, whereas it is significantly reduced to approximately half the width in a bilayer and does not occur in thin graphite (Meyer et al., 2007a).

From their data Meyer et al. (2007a) concluded that a suspended graphene membrane is not perfectly flat, which will be elucidated in the following.

Figure 5.5.4e depicts a flat graphene sheet in perspective view, together with a schematic 3D Fourier transform in Fig. 5.5.4f, which consists of a set of continuous rods arranged on the 2D reciprocal lattice, as discussed previously (Fig. 5.5.3a: here only, the $\{10\overline{1}0\}$ reflexes are depicted for simplicity). Meyer et al. (2007b) explained that, taking microscopic corrugations into account, the graphene membrane can be viewed as an ensemble of locally flat pieces with slightly different orientations as highlighted by the varying surface normals in Fig. 5.5.4g. Each piece causes the diffraction peak to appear at a slightly different position which results into an isotropic spread of the peak intensities. In 3D reciprocal space, a superposition of the diffraction beams of the microscopically flat pieces effectively turns the rods into cones as indicated in Fig. 5.5.4h. Figure 5.5.4i further illustrates that this corrugated graphene model also confirms the experimental observations that the peaks are sharp at normal incidence and rapidly become wider as the tilt angle increases and as the distance from the tilt axis increases (Meyer et al., 2007a). The quantitative analysis of the evolution of the FWHM of the peak width with tilt angle (Fig. 5.5.4d), thus provides a direct measure of the roughness of the graphene membrane. The linear slope (dashed lines) can be directly related to the cone angle. For the mono-layer sample the cone angle was found to be between 8° and 11°, which corresponds to a deviation of the surface normal from the mean surface by approximately 5°. In graphene bilayers, the peak broadening effect is only half as strong as compared to a monolayer; therefore, the variation of the surface normal only amounts to ca. 2° (Meyer et al., 2007a,b). The fact that the diffraction peaks broaden isotropically implies that the surface normals vary in all directions and that the microscopic rippling of the membrane is omni-directional. If the graphene membrane was curved only in one direction the diffraction peaks would be elongated into lines indicating the direction of curvature, much like it is observed in ED patterns of carbon nanotubes (Meyer et al., 2006, 2007a).

Meyer et al. (2007a) further carried out numerical simulations of ED patterns in order to determine the spatial size of the microscopic corrugations. In this way, they estimated a lateral extent of the ripples of 5 nm to 20 nm and an out-of-plane deformation of approximately 1 nm (Meyer et al., 2007a). These estimates have also been supported by atomic resolution TEM imaging. As discussed previously (cf. Fig. 5.5.3d), the visibility of the hexagonal lattice of multilayered graphene membranes strongly varies with tilt angle. Therefore, the corrugations should result into areas of different brightness in the TEM image, which are indeed experimentally observed (Meyer et al., 2007a). Figure 5.5.4j shows an HRTEM micrograph of an FLG membrane in which 'domains' of different brightness can clearly be made out. Their lateral size is on the order of a few nanometres, which expectedly is a bit smaller than the estimate obtained from the simulations of a monolayer membrane, where the corrugations are larger (Meyer et al., 2007a). However, for a mono-layer, direct imaging with atomic resolution TEM cannot be employed to visualise corrugations, since the diffraction intensities hardly vary with tilt angle (cf. Fig. 5.5.3c). The atomic resolution TEM image further reveals that the rippling is static, since, otherwise, the additional 'domain' contrast would disappear (Meyer et al., 2007a).

A different technique to visualise static corrugations by taking advantage of the multi-layer's strong variation in ED intensity with sample tilt (cf. Fig. 5.5.3d) is convergent beam electron diffraction (CBED) (Meyer et al., 2007b). When focussing the electron probe onto the graphene membrane, the diffraction pattern consists of disks with a smooth intensity distribution. However, with the sample offset from the beam focus, local variations of the orientation of the graphene membrane translate into intensity variations within the diffraction spots (Meyer et al., 2007b). Meyer et al. (2007b) examined a bi-layer graphene membrane using CBED and found the intensity variations within the diffraction disks to be in agreement with the $\pm 2°$ deviation of the mean surface normal as derived from the peak broadening in their SAED study (Meyer et al., 2007b).

SAED studies were also performed on suspended reduced GO by Wilson et al. (2009, 2010). Similar to mechanically exfoliated graphene, an intensity ratio of $I_{\{0\bar{1}10\}}/I_{\{1\bar{2}10\}} > 1$ was indicative of mono-layer GO (Wilson et al., 2010). However, in contrast to the linear dependence of the peak broadening with tilt angle that was observed for mechanically cleaved graphene, reduced GO exhibited a nonlinear broadening dependence on the tilt angle. This was related to 'short-wavelength' corrugations that undulate with a wavelength of a length scale of a few nanometres or less (Wilson et al., 2010). Wilson et al. revealed that these distortions correspond to atomic displacements on the order of 10% of the carbon–carbon distance thus, leading to large strain in the lattice that can be attributed to functional hydroxyl and epoxide group that are present on both sides of the membrane (Lerf et al., 1998; Thompsom-Flagg, 2009; Wilson et al., 2010). These short range distortions have also been observed for bi-layer GO sheets, which is again in contrast to findings from mechanically exfoliated graphene where bi-layers showed a reduced peak broadening with tilt angle (Wilson et al., 2010).

5.5.4. DETERMINATION OF STACKING ORDER AND IDENTIFICATION OF ROTATIONAL STACKING FAULTS

As mentioned above, it is widely known that, most commonly, AB-Bernal stacking is observed in graphite and FLG. Bulk natural graphite exhibits a volume fraction of AB:ABC:turbostratic of about 80:14:6 (Haering, 1958). While the total energy difference between ABC- and AB-stacked graphite is 0.11 meV/atom, the energy difference between AA- and AB-stacked graphite amounts to 17.31 meV/atom, which explains the presence of ABC-stacked and the absence of AA-stacked graphite in *natural* graphite (Charlier et al., 1994). Yet, AA stacking, where the carbon atoms of the second layer reside directly above the first layer, has been observed in nanofilms of GO (Horiuchi et al., 2003) and is also known to occur at folds of graphene single- and bilayers (Liu et al., 2009; Roy et al., 1998). ABC stacking has been reported for multilayer graphene epitaxially grown on silicon-terminated SiC (Norimatsu and Kusunoki, 2010). Recently, HRTEM studies showed that FLG grown via CVD from a copper metal catalyst can also adopt ABC rhombohedral stacking (Warner et al., 2012). Analysis of the stacking order has been carried out, using STM (Roy et al., 1998; Varchon et al., 2008) or atomic resolution TEM imaging (Liu et al., 2009; Warner et al., 2012). Also, ED can be employed to provide information on the stacking order of FLG and multilayer graphene membranes and to identify rotational stacking faults (Norimatsu and Kusunoki, 2010; Warner et al., 2009a).

Theoretical calculations of ED intensities have been carried out for graphene with different stacking order as a function of the number of graphene layers N and are summarised in Table 5.5.1 (Horiuchi et al., 2003). As previously indicated, single-layer graphene has a $I_{\{0\bar{1}10\}}/I_{\{1\bar{2}10\}}$ ratio larger than 1 (cf. Fig. 5.5.2c). From these calculations, it can further be derived that the $I_{\{0\bar{1}10\}}/I_{\{1\bar{2}10\}}$ ratio gives information on the stacking order of graphene membranes that have been identified to consist of more than one layer, e.g. by imaging along a fold or a flake edge. Few-layer membranes with an intensity ratio larger than 1 exhibit AA stacking order. Membranes with a $I_{\{0\bar{1}10\}}/I_{\{1\bar{2}10\}}$ ratio of approximately 0.3 most likely are AB stacked (Horiuchi et al., 2003; Wilson et al., 2010). In the case of ABC-stacked graphene, the intensity ratio is close to 0, i.e. the $\{10\bar{1}0\}$ diffraction spots almost vanish.

Norimatsu and Kusunoki (2010) analysed the stacking sequence of graphene grown on Si-terminated SiC by comparing simulated ED patterns with FFTs of the experimentally observed atomically resolved cross-sectional TEM images. The FFT pattern of an HRTEM image generally gives equivalent information to an ED pattern. Figure 5.5.5a shows simulated ED patterns along

TABLE 5.5.1 Calculated Diffraction Intensity Ratio $I_{\{0\bar{1}10\}}/I_{\{1\bar{2}10\}}$ for AA-, AB- and ABC-stacked FLGs as a Function of the Number of Graphene Layers N. Reproduced With Permission from Horiuchi et al. (2003). Copyright By the Japan Society of Applied Physics

	Number of graphene layers N							
Stacking order	1	2	3	4	5	10	11	20
AA	1.1	1.1	1.1	1.1	1.1	1.1	1.1	1.1
ABAB	1.1	0.28	0.37	0.28	0.31	0.28	0.29	0.28
ABCABC	1.1	0.28	0.0	0.069	0.044	0.011	0.0	0.0028

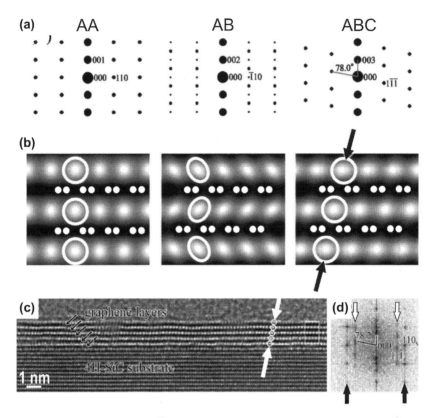

FIGURE 5.5.5 (a) Simulated [11$\bar{2}$0] ED patterns of AA-, AB- and ABC-stacked graphites (from left to right). (b) Corresponding simulated HRTEM images. The small white spots correspond to the carbon atom positions. The white circles mark specific features to clarify the stacking sequence. (c) HRTEM micrograph of the cross section of ABC-stacked graphene grown on 4H–SiC. (d) Corresponding FFT pattern. The diffraction spots along the lines marked by the black arrows belong to SiC, reflections along the lines marked by white arrows are generated by the graphene layers and correspond to those in the right panel in a) indicating ABC stacking. *Adapted with permission from Norimatsu and Kusunoki, (2010). Copyright (2010) by the American Physical Society.*

the [11$\bar{2}$0] direction of AA, AB and ABC stacked graphite (from left to right). The diffraction spots are indexed in *P6/mmm* (left), *P6₃/mmc* (middle) and *R$\bar{3}$ m* (right) notations. The diffraction patterns strongly differ from each other. In particular ABC-stacked graphite can be identified from a 78° angle between the (003) and the ($\bar{1}$11) reflections. Figure 5.5.5b shows the respective simulated HRTEM images of AA, AB and ABC stacked graphite for a 40 nm underfocus and a sample thickness of 3 nm (Norimatsu and Kusunoki, 2010). The small, filled white dots highlight the positions of the carbon atoms. Thus, the graphene layers are observed as dark lines in the TEM image acquired under these imaging conditions. Another image feature clearly marks out strong differences between samples of different stacking sequence. Between the dark graphene layers, characteristic bright features can be observed. In AA-stacked graphite, these features are circular and arranged perpendicular to the graphene layer as marked by the white circles (Norimatsu and Kusunoki, 2010). In the case of AB stacked graphene, the bright features are slightly elongated and tilted. They are aligned perpendicular to the graphene planes as well but exhibit some kind of zigzag pattern (Norimatsu and Kusunoki, 2010). In contrast to this, the bright features in ABC stacked graphite are circular and arranged along a line which is not perpendicular to the graphene plane, as marked with black arrows in the simulated TEM image (Fig. 5.5.5b, right panel). Figure 5.5.5c shows a cross-sectional HRTEM image of graphene grown on a 4H–SiC substrate acquired at approximately 40 nm underfocus (Norimatsu and Kusunoki, 2010). In such a side view, the graphene layers appear dark and are marked with black arrows (at the left side of the micrograph). As discussed before, information on the stacking sequence can be obtained from the characteristic bright features between the dark contrast lines of the graphene layers. In this image the bright circular contrast features are arranged along a line which is tilted with respect to the graphene layers as it was observed from the image simulations of ABC stacked graphite (cp. Fig. 5.5.5b, right panel). The presence of ABC stacking is further corroborated by the corresponding FFT pattern presented in Fig. 5.5.5d (Norimatsu and Kusunoki, 2010). Here, diffraction spots originating from the SiC substrate are arranged along the lines indicated with black arrows, whereas spots generated by the graphene layers are oriented along the lines marked with white arrows. The simulated ED pattern of ABC-stacked graphite matches the experimentally observed FFT pattern very well. The typical 78° angle between the (003) and the $(\bar{1}11)$ reflections has also been marked in the FFT pattern. These studies, carried out by Norimatsu and Kusunoki (2010) thus highlight the feasibility of the combined approach of simulating ED patterns and their comparison to FFT patterns obtained from HRTEM imaging in order to obtain information on the stacking sequence of graphene.

Not only can the stacking order be determined using TEM and ED, also information on stacking faults can be obtained by combined HRTEM and ED or FFT analysis. Warner et al. (2009a) studied rotational stacking faults in FLG membranes. From aberration corrected HRTEM micrographs, they determined the number of graphene layers to range between one to six layers. Figure 5.5.6a shows an atomic resolution TEM micrograph of a section of a few-layer graphene membrane which exhibits AB Bernal stacking. In the top left corner of the image (marked with the dotted line) an extra monolayer of graphene resides on top of the few-layer membrane. No change in the atomic structure of the image can be observed, which is indicative of AB-Bernal stacking (Warner et al., 2009a).

A different contrast has been observed from the edge of a graphene bilayer shown in Fig. 5.5.6b (Warner et al., 2009a). Compared to the AB Bernal

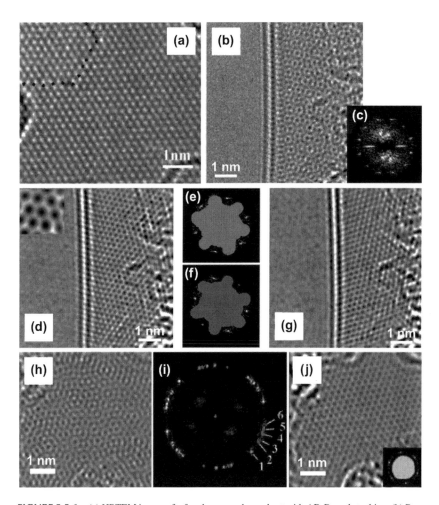

FIGURE 5.5.6 (a) HRTEM image of a few-layer graphene sheet with AB-Bernal stacking. (b) Raw HRTEM micrograph of a graphene bilayer exhibiting a Moiré pattern. (c) Corresponding FFT in which two sets of six spots corresponding to the 2.13 Å spacing are observed. (d) Reconstructed TEM image of the back graphene layer after filtering in the frequency domain using the inclusive mask in (e). Inset in d): clearly the hexagonal graphene lattice can be resolved. (f) Inclusive mask used to reconstruct the TEM image of the front graphene layer shown in g). (h) Complex raw HRTEM micrograph of a few-layer graphene sheet consisting of at least 6 layers. (i) Corresponding FFT with six sets of spots, where each can be attributed to a graphene layer with a specific orientation. (j) TEM image reconstructed from spot set 4 revealing the hexagonal graphene lattice. *Reprinted with permission from Warner et al. (2009a). Copyright (2009) American Chemical Society.*

stacked few-layer graphene in Fig. 5.5.6a, this contrast pattern is more complex consisting of spots and larger circular features. Figure 5.5.6c shows the respective FFT. Since the FFT pattern of an HRTEM image is generally equivalent to the ED pattern, a single-layer of graphene or AB stacked few-layer graphene exhibits only six spots of 2.13 Å spacing. In the FFT in Fig. 5.5.6c, a total of 12 spots was observed indicating that the graphene layers within the bilayer are rotated with respect to each other. In Fig. 5.5.6d, a reconstructed image of the back graphene layer is shown after filtering in the frequency domain. Clearly, the hexagonal graphene lattice is revealed, as can be seen in the inset. The inclusive mask used for image reconstruction is marked grey in the FFT in Fig. 5.5.6e. The same procedure has been applied to resolve the front graphene layer with the inclusive mask shown in Fig. 5.5.6f and the reconstructed image in Fig. 5.5.6g. The layers are rotated by 30° with respect to each other, which gives rise to the Moiré pattern observed in the original TEM image (Fig. 5.5.6b) (Warner et al., 2009a).

Warner et al. (2009a) also detected more complex Moiré patterns containing up to five rotational stacking faults in a single FLG sheet. Figure 5.5.6h shows a section of a graphene membrane with at least 6 layers, which exhibits a complex Moiré pattern. In the respective FFT in Fig. 5.5.6i, six different sets of spots are observed of which each can be attributed to a graphene layer with a specific orientation. Figure 5.5.6j shows the reconstructed image for the layer giving rise to the spots labelled with number 4 in the FFT. The inclusive mask used for filtering in the frequency domain is indicated in the inset. Warner et al. (2009a) demonstrated that the hexagonal carbon network of each graphene sheet for all six orientations is resolved and determined the relative angles of rotation.

Other techniques like AFM (Novoselov et al., 2004) and optical imaging of graphene (Jung et al., 2007), which are regularly used to determine the number of graphene layers, do not allow to detect the presence of rotational stacking faults within few-layer graphene. SAED and HRTEM in combination with the analysis of the frequency domain are capable of measuring rotational stacking faults, which is important due to their strong influence on graphene's electronic properties (Warner et al., 2009a).

5.5.5. LOW-ENERGY ELECTRON DIFFRACTION

Another ED technique often employed to study graphene is low-energy electron diffraction (LEED). LEED is a surface sensitive method which is applied to examine the surface structure of crystalline materials and provides information on the arrangement of atoms on surfaces and in thin films. In graphene research, LEED is often used to study graphene epitaxially grown on various substrates (de Heer et al., 2007; N'Diaye et al., 2008; Ogawa et al., 2012; Sutter et al., 2008). A low energy electron beam (20–200 eV) is incident perpendicular to the surface of the sample, and elastically scattered electrons are detected by a fluorescent viewing screen. An advantage of LEED is its ability to follow in situ processes, which is very important for process monitoring purposes and to understand the growth mechanisms involved.

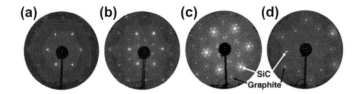

FIGURE 5.5.7 LEED patterns of graphene grown on the Si-terminated (0001) face of single-crystal 6H–SiC at different growth stages: (a) 1050 °C, 10 min, acquired at 177 eV. (b) 1100 °C, 3 min, acquired at 171 eV. (c) 1250 °C, 20 min, acquired at 109 eV. (d) 1400 °C, 8 min, acquired at 98 eV. *Reprinted with permission from Berger et al. (2004), Copyright (2004) American Chemical Society.*

As discussed in Section 4.7 graphene can be grown by sublimating silicon from SiC through heating above 1200 °C under UHV conditions (Berger et al., 2004; Forbeaux et al., 2000; Hass et al., 2006). As an example, Fig. 5.5.7 shows LEED patterns acquired at different stages during the growth of a 2.5 mono-layer graphite film on the Si-terminated (0001) face of single-crystal 6H–SiC (Berger et al., 2004). After oxide removal and heating the sample to 1050 °C for 10 min a SiC 1×1 pattern was observed at 177 eV (Fig. 5.5.7a). Further heating to 1100 °C for 3 min led to the $\sqrt{3} \times \sqrt{3}$ reconstruction of Si adatoms (Fig. 5.5.7b, 171 eV). Heating above 1150 °C resulted in the carbon-rich $6\sqrt{3} \times 6\sqrt{3}$ reconstruction, marking the beginning of graphene growth (Emtsev et al., 2008). This $6\sqrt{3} \times 6\sqrt{3}$ unit cell can be seen in the LEED pattern in Fig. 5.5.7c, which was observed after annealing at 1250 °C for 20 min at 109 eV. This pattern corresponds to a graphene monolayer, as confirmed by Auger electron spectroscopy (AES) where a C:Si intensity ratio of 2 is obtained. After heating the sample to 1400 °C for 8 min (Fig. 5.5.7d, 98 eV), the AES ratio is 7.5 which corresponds to approximately 2.5 carbon monolayers. Diffraction spots corresponding to the SiC(0001) substrate and the graphene layer(s) are marked with white or black arrows, respectively. The LEED pattern provides information on the relative rotation between substrate and the graphene layer(s). In the case of Si-terminated SiC(0001), the graphene layers are rotated by 30° with respect to the SiC substrate (Berger et al., 2004). For carbon-terminated SiC-(000$\bar{1}$), ultrathin graphene, i.e. one to three layers, grows epitaxially on the SiC-(000$\bar{1}$) surface with the graphene layers rotated by 30° with respect to the SiC, as observed for graphene growth on Si-terminated SiC (Berger et al., 2006; de Heer et al., 2007). However, four-layer graphene and thin graphitic films grown on SiC-(000$\bar{1}$) show rotational disorder with the diffraction spots becoming dashed diffraction rings (de Heer et al., 2007; Forbeaux et al., 2000), suggesting that epitaxial growth occurs at the interface, but that succeeding graphene layers do not have strong rotational order. From high-resolution LEED studies, information on the in-plane strain in the graphene layers can be extracted (Berger et al., 2004). Further, fingerprints in the spot intensity spectra plotted as a function of the electron energy allow for the determination of the number of graphene layers (Riedl et al., 2008).

LEED has also been employed to study graphene grown on other substrates such as ruthenium (Sutter et al., 2008), iridium (N'Diaye et al., 2008), MgO (Gaddam et al., 2011), cobalt (Ago et al., 2010), nickel (Iwasaki et al., 2011), palladium (Murata et al., 2010) and copper (Ogawa et al., 2012). As a recent example, Ogawa et al. (2012) examined the domain structure of monolayer graphene grown on Cu(111) and Cu(100) films using LEED. Studies on graphene grown from Cu films are extremely important, since this is one of the most frequently used synthesis methods with high potential for scale-up, as highlighted in Section 4.5 (Bae et al., 2010). Ogawa et al. (2012) could show that graphene is epitaxially formed on Cu(111) proving uniform monolayer graphene whose orientation is consistent with the underlying Cu lattice over large areas of up to 1 mm^2. Graphene grown on Cu(100), on the other hand, showed a more complex multidomain structure. From the LEED pattern Ogawa et al. (2012) conclude that graphene is oriented at two preferential orientations with angles of $0 \pm 2°$ and $30 \pm 2°$ with respect to the underlying Cu(100) lattice. They also observed weak diffraction streaks between the $0°$ and $30°$ positions which correspond to misaligned graphene domains (Ogawa et al., 2012).

REFERENCES

Ago, H., Ito, Y., Mizuta, N., Yoshida, K., Hu, B., Orofeo, C.M., Tsuji, M., Ikeda, K., Mizuno, S., 2010. Epitaxial chemical vapor deposition growth of single-layer graphene over cobalt film crystallized on sapphire. ACS Nano 4, 7407–7414.

Bae, S., Kim, H., Lee, Y., Xu, X., Park, J.-S., Zheng, Y., Balakrishnan, J., Lei, T., Kim, H.R., Song, Y.I., Kim, Y.-J., Kim, K.S., Özyilmaz, B., Ahn, J.-H., Hong, B.H., Iijima, S., 2010. Roll-to-roll production of 30-inch graphene films for transparent electrodes. Nat. Nanotechnol. 5, 574–578.

Berger, C., Song, Z., Li, T., Li, X., Ogbazghi, A.Y., Feng, R., Dai, Z., Marchenkov, A.N., Conrad, E.H., First, P.N., de Heer, W.A., 2004. Ultrathin epitaxial graphite: 2D electron gas properties and a route toward graphene-based nanoelectronics. J. Phys. Chem. B 108, 19912–19916.

Berger, C., Song, Z., Li, X., Wu, X., Brown, N., Naud, C., Mayou, D., Li, T., Hass, J., Marchenkov, A.N., Conrad, E.H., First, P.N., Walt, A., Heer, de, 2006. Electronic confinement and coherence in patterned epitaxial graphene. Science 312, 1191–1196.

Charlier, J.-C., Gonze, X., Michenaud, J.-P., 1994. First-principle study of the stacking effect on the electronic properties of graphite(s). Carbon 32, 289–299.

de Heer, W.A., Berger, C., Wua, X., First, P.N., Conrad, E.H., Li, X., Li, T., Sprinkle, M., Hass, J., Sadowski, M.L., Potemski, M., Martinez, G., 2007. Epitaxial graphene. Solid State Commun. 143, 92–100.

Emtsev, K.V., Speck, F., Seyller, Th., Ley, L., Riley, J.D., 2008. Interaction, growth, and ordering of epitaxial graphene on SiC{0001} surfaces: a comparative photoelectron spectroscopy study. Phys. Rev. B 77, 155303.

Forbeaux, I., Themlin, J.-M., Charrier, A., Thibaudau, F., Debever, J.-M., 2000. Solid-state graphitization mechanisms of silicon carbide 6H–SiC polar faces. Appl. Surf. Sci. 162–163, 406–412.

Gaddam, S., Bjelkevig, C., Ge, S., Fukutani, K., Dowben, P.A., Kelber, J.A., 2011. Direct graphene growth on MgO: origin of the band gap. J. Phys. Cond. Matter. 23, 072204.

Haering, R.R., 1958. Band structure of rhombohedral graphite. Can. J. Phys. 36, 352–362.

Hass, J., Feng, R., Li, T., Li, X., Zong, Z., de Heer, W.A., First, P.N., Conrad, E.H., Jeffrey, C.A., Berger, C., 2006. Highly ordered graphene for two dimensional electronics. Appl. Phys. Lett. 89, 143106.

Hernandez, Y., Nicolosi, V., Lotya, M., Blighe, F.M., Sun, Z., De, S., Mc-Govern, I.T., Holland, B., Byrne, M., Gun'Ko, Y.K., Boland, J.J., Niraj, P., Duesberg, G., Krishnamurthy, S., Goodhue, R., Hutchison, J., Scardaci, V., Ferrari, A.C., Coleman, J.N., 2008. High-yield production of graphene by liquid-phase exfoliation of graphite. Nat. Nanotechnol. 3, 563–568.

Horiuchi, S., Gotou, T., Fujiwara, M., Sotoaka, R., Hirata, M., Kimoto, K., Asaka, T., Yokosawa, T., Matsui, Y., Watanabe, K., Sekita, M., 2003. Carbon nanofilm with a new structure and property. Jpn. J. Appl. Phys. 42, L1073–L1076.

Iwasaki, T., Park, H.J., Konuma, M., Lee, D.S., Smet, J.H., Starke, U., 2011. Long-range ordered single-crystal graphene on high-quality heteroepitaxial Ni thin films grown on MgO(111). Nano Lett. 11, 79–84.

Jung, I., Pelton, M., Piner, R., Dikin, D.A., Stankovich, S., Watcharotone, S., Hausner, M., Ruoff, R.S., 2007. Simple approach for high-contrast optical imaging and characterization of graphene-based sheets. Nano Lett. 7, 3569–3575.

Lerf, A., He, H., Forster, M., Klinowski, J., 1998. Structure of graphite oxide revisited. J. Phys. Chem. B 102, 4477–4482.

Liu, Z., Suenaga, K., Harris, P.J.F., Iijima, S., 2009. Open and closed edges of graphene layers. Phys. Rev. Lett. 102, 015501.

Meyer, J.C., Paillet, M., Duesberg, G.S., Roth, S., 2006. Electron diffraction analysis of individual single-walled carbon nanotubes. Ultramicroscopy 106, 176–190.

Meyer, J.C., Geim, A.K., Katsnelson, M.I., Novoselov, K.S., Booth, T.J., Roth, S., 2007a. The structure of suspended graphene sheets. Nature 446, 60–63.

Meyer, J.C., Geim, A.K., Katsnelson, M.I., Novoselov, K.S., Obergfell, D., Roth, S., Girit, C., Zettl, A., 2007b. On the roughness of single- and bi-layer graphene membranes. Solid State Commun. 143, 101–109.

Murata, Y., Starodub, E., Kappes, B.B., Ciobanu, C.V., Bartelt, N.C., McCarty, K.F., Kodambaka, S., 2010. Orientation-dependent work function of graphene on Pd(111). Appl. Phys. Lett. 97, 143114.

N'Diaye, A.T., Coraux, J., Plasa, T.N., Busse, C., Michely, T., 2008. Structure of epitaxial graphene on Ir(111). New J. Phys. 10, 043033.

Norimatsu, W., Kusunoki, M., 2010. Selective formation of ABC-stacked graphene layers on SiC(0001). Phys. Rev. B 81, 161410.

Novoselov, K.S., Geim, A.K., Morozov, S.V., Jiang, D., Zhang, Y., Dubonos, S.V., Grigorieva, I.V., Firsov, A.A., 2004. Electric field effect in atomically thin carbon films. Science 306, 666–669.

Ogawa, Y., Hu, B., Orofeo, C.M., Tsuji, M., Ikeda, K., Mizuno, S., Hibino, H., Ago, H., 2012. Domain structure and boundary in single-layer graphene grown on Cu(111) and Cu(100) films. Phys. Chem. Lett. 3, 219–226.

Riedl, C., Zakharov, A.A., Starke, U., 2008. Precise in situ thickness analysis of epitaxial graphene layers on SiC(0001) using low-energy electron diffraction and angle resolved ultraviolet photoelectron spectroscopy. Appl. Phys. Lett. 93, 033106.

Robertson, A.W., Warner, J.H., 2011. Hexagonal single crystal domains of few-layer graphene on copper foils. Nano Lett. 11, 1182–1189.

Roy, H.V., Kallinger, C., Sattler, K., 1998. Study of single and multiple foldings of graphitic sheets. Surf. Sci. 407, 1–6.

Spence, J.C.H., 2003. High-resolution Electron Microscopy. Oxford University Press.

Sutter, P.W., Flege, J.-I., Sutter, E.A., 2008. Epitaxial graphene on ruthenium. Nat. Mater. 7, 406–411.

Thompson-Flagg, R.C., Moura, M.J.B., Marder, M., 2009. Rippling of graphene. EPL 85 46002.

Varchon, F., Mallet, P., Magaud, L., Veuillen, J.-Y., 2008. Rotational disorder in few-layer graphene films on 6H–SiC-(000$\overline{1}$): a scanning tunneling microscopy study. Phys. Rev. B 77, 165415.

Warner, J.H., Rümmeli, M.H., Gemming, T., Büchner, B., Briggs, G.A.D., 2009a. Direct imaging of rotational stacking faults in few layer graphene. Nano Lett. 9, 102–106.

Warner, J.H., Schäffel, F., Rümmeli, M.H., Büchner, B., 2009b. Examining the edges of multi-layer graphene sheets. Chem. Mater. 21, 2418–2421.

Warner, J.H., Mukai, M., Kirkland, A.I., 2012. Atomic structure of ABC rhombohedral stacked trilayer graphene. ACS Nano 6, 5680–5686.

Williams, D.B., Carter, C.B., 2009. Transmission Electron Microscopy – A Textbook for Materials Science. Springer.

Wilson, N.R., Pandey, P.A., Beanland, R., Young, R.J., Kinloch, I.A., Gong, L., Liu, Z., Suenaga, K., Rourke, J.P., York, S.J., Sloan, J., 2009. Graphene oxide: structural analysis and application as a highly transparent support for electron microscopy. ACS Nano 3, 2547–2556.

Wilson, N.R., Pandey, P.A., Beanland, R., Rourke, J.P., Lupo, U., Rowlands, G., Römer, R.A., 2010. On the structure and topography of free-standing chemically modified graphene. New J. Phys. 12, 125010.

Chapter 5.6

Scanning Tunnelling Microscopy

Jamie H. Warner

University of Oxford, Oxford, UK

5.6.1. INTRODUCTION TO SCANNING TUNNELLING MICROSCOPY

Scanning tunnelling microscopy (STM) is a scanning probe imaging technique that provides atomic resolution of structures typically on surfaces. It was invented by Gerd Binning and Heinrich Rohrer in the early 1980's and led to the award of a Nobel prize in 1986 (Binnig and Rohrer, 1986). It is one of the most important tools used by surface scientists to probe materials at the atomic level. STM involves positioning a metallic tip with an end point of near atomic sharpness within a nanometre of a surface. Tips are normally made from tungsten, platinum–iridium, gold or even carbon nanotubes (Dai et al., 1996). Applying a voltage between the tip and the surface leads to the quantum tunnelling of electrons that generates a current between the tip and surface. Generally, either the sample or the substrate needs to be conductive to enable a current to flow and be measured. The tunnelling current depends upon the distance between the tip and surface, the local density of states and the applied voltage. The tip acts as a probe and is raster scanned across the sample using a piezo-controller that has picometre stepping capabilities. The distance between the tip and surface can also be adjusted with picometre sensitivity, using piezo-controllers. There are two main image types that can be acquired with STM, (1) topographic (constant-current) images obtained by keeping the current constant and adjusting the z distance between the tip and surface, or (2) current (constant-height) images obtained by keeping the height constant and monitoring changes in the current. Topographic images can be taken with different tip-surface bias voltages, and current images at different heights. Figure 5.6.1 shows a schematic illustration of an STM setup.

If the tip is positioned at a specific location in a sample, the current can be recorded as a function of tip-surface bias voltage. The resulting I–V curves provide information about the local electronic structure and I–V data can be collected at every point during an X–Y raster scan in constant height mode to map out the electronic structure in 2D. Constant height current images are faster to obtain due to the tip not needing to continually move up and down, but constant-current topographical images provide higher precision of irregular surfaces. Care must be taken when interpreting so-called topographical images, since it is actually the electronic density of states at the surface that is being probed rather than physical structure.

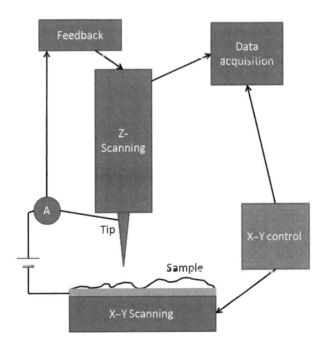

FIGURE 5.6.1   Schematic illustration of a scanning tunnelling microscopy (STM) setup.

5.6.2. STM STUDIES OF GRAPHITE

Graphite was one of the first materials examined using STM due to its abundance, ease of acquisition, conductivity and atomically smooth surfaces obtained by cleaving (Batra et al., 1987; Soler et al., 1986). Mechanically cleaving of graphite yields a smooth surface that is also suitable as a substrate for STM investigation of other materials. STM images of graphite are complicated by the density of states in the top region being influenced by underlying graphene layers. In AB Bernal stacked graphite the top graphene layer has some atoms that sit directly above an atom in the second top most layer, and some that do not (see chapter 2 for more details), which is predicted to give rise to different density of states for these two classes of atoms (Batra et al., 1987; Tomanek and Louie, 1988). Atomic resolution STM images of graphite often result in a triangular pattern, with a hexagonal symmetry modulation of 0.246 nm periodicity (Wong et al., 2009), whereas the C–C distance within a graphene layer is actually 0.142 nm. There have been reports of hexagonal lattice structure in STM images of graphite as well, sometimes coexisting with the trigonal pattern and this has led to considerable debate regarding the interpretation of STM images of graphite (Moriarty and Hughes, 1992; Ouseph et al., 1995; Paredes et al., 2001; Wang et al., 2006, Zeinalipour-Yazdi and Pullman, 2008). Figure 5.6.2 shows (a) an STM image of graphite from (Wong et al., 2009) with both trigonal and hexagonal patterns, (b) line-profile along the triangular pattern (green line in (a)) showing 0.246 nm periodicity, (c) line-profile along the hexagonal pattern (green line in (a)) showing 0.142 nm periodicity.

The hexagonal pattern in STM images of graphite has been attributed to tip-modification (Mizes et al., 1987), displacement of the top layer (Luican et al., 2009) and modified tip-sample interactions (Atamny et al., 1999). Recently,

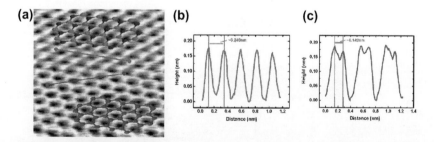

FIGURE 5.6.2 (a) Atomic resolution of bare graphite showing coexistence of the triangular and honeycomb lattice (0.1 V, 0.40 nA) with the atomic model fitted (2×2 nm^2). The bottom half of the image where the honeycomb traverses across the entire length of the linescan is termed 'full-displacement'. (b) Line profile along triangular lattice showing the distance between every other atom. (c) Line profile along honeycomb lattice revealing two pronounced maxima corresponding to two adjacent carbon atoms measured to be 0.142 nm. *Reprinted with permission from Wong (2009). Copyright (2009) American Chemical Society.*

Wong et al. (2009) showed that variation in interlayer coupling produces switches in STM images from triangular to honeycomb patterns. They confirmed that it is a displacement in the top graphene layer of its decoupling that causes the variation in image patterns and not a modification of the layer stacking from AB to AA (Wong et al., 2009). Further clarity in understanding the STM images of graphite has arisen from recent studies on monolayer and few layer graphene sheets on insulating surfaces (Stolyarova et al., 2007).

5.6.3. STM OF GRAPHENE ON METALS

STM is ideal to study synthetic graphene grown on a catalytic substrate such as Ir (Gall et al., 2004; Klusek et al., 2005; Makarenko et al., 2007; Ni'Diaye et al., 2006), Pt (Land et al., 1992; Ueta et al., 2004), Ru (Pan et al., 2009), and Cu (Cho et al., 2011; Gao et al., 2010; Rasool et al., 2011; Yu et al., 2011; Zhang et al., 2011), as well as graphene that forms from high temperature annealing of SiC (Berger et al., 2006; Charrier et al., 2002; Mallet et al., 2007; Rutter et al., 2007). STM images of graphitic lattice structure provided confirmation that synthetic graphite has been formed using novel growth techniques. Using STM, Coraux et al. (2008) showed that graphene grown on Ir(111) had continuity over Ir terraces and step edges larger than micrometres and contained very few defects, with the only defects observed being edge dislocations (shown in Fig. 5.6.3). As the graphene flows across Ir step edges, it is bent with a radius of curvature typical of small carbon nanotubes (Coraux et al., 2008). Differences in lattice constants between graphene and Ir(111) results in a Moire pattern, and this is tracked in Fig. 5.6.4 across several Ir step edges.

Similar results were reported by Pan et al. (2009) for graphene grown on Ru(0001), with the graphene extending seamlessly over millimetre scales and a Moire pattern in the STM images due to interference between the lattices of the graphene and Ru crystal. Marchini et al. (2007) conducted a detailed study of graphene on Ru(0001) using STM and were able to resolve both carbon atoms in the graphene unit cell, with contrast depending on the position above the Ru atoms. They showed that significantly stronger chemical interaction occurs between graphene and the metal surface compared to graphene residing on bulk graphite (Marchini et al., 2007).

In the report of Gao et al. (2010), graphene domains grown on Cu(111) were studied with STM and also scanning tunnelling spectroscopy (STS). They show that the differential conductance (dI/dV) image (Fig. 5.6.5(b)) provides improved contrast of graphene domains compared to a topography image of the same area (Fig. 5.6.5(a)). The graphene domains sometimes extended over step edges of the Cu, whilst most of the time terminated at Cu(111) step edges (Gao et al., 2010). Further examination of some of the graphene regions reveal that the Moire pattern is not uniform, as shown in Fig. 5.6.5(c), indicating that there are different crystallites within these graphene domains. By increasing the tunnelling current, Gao et al. (2010) were able to obtain atomic resolution

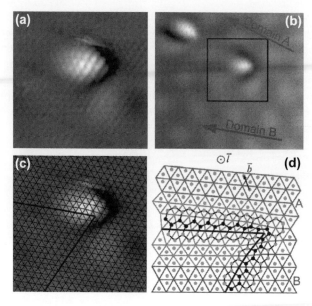

FIGURE 5.6.3 (a) Atomic resolution 5.5 nm × 5.5 nm STM topograph (Fourier filtered, 0.1 V, 9 nA) of a graphene layer grown on Ir- (111) at 1120 K. (b) Demagnified view (14 nm × 12.4 nm) showing two defects at the boundary between two Moire' domains (A, B); the area within the black frame is panel a demagnified. (c) Line network connecting the centres of the C rings in panel a. In regions where the contrast is lost, the atomic arrangement was extrapolated (dashed lines). Thick solid lines indicate the presence of extra lines. (d) Scheme of the correspondence between the network in panel c and the C atoms in the vicinity of the wedge, lying at the tilt boundary between domains A and B, shown in panel b. A and B are misoriented by a rotation along the 1 normal to the surface. Two extra atomic rows corresponding to an edge dislocation with Burger's vector b ($b = a_c = 0.2452$ nm) are outlined (black circles). Note the relative orientation of the wedge and of the C pentagon and heptagon. *Reprinted with permission from Coraux (2008). Copyright (2008) American Chemical Society.*

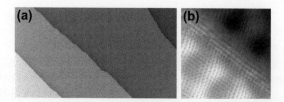

FIGURE 5.6.4 (a) 125 nm × 250 nm STM topograph (0.10 V, 30 nA) of graphene grown on Ir(111) at 1320 K, crossing several Ir steps. (b) Continuous atomic arrangement in graphene across a step edge (5 nm × 5 nm, 0.04 V, 30 nA). *Reprinted with permission from Coraux (2008). Copyright (2008) American Chemical Society.*

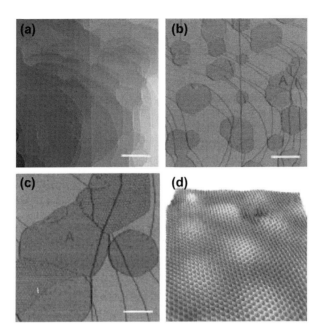

FIGURE 5.6.5 STM topography and differential conductance dI/dV images of graphene islands on Cu(111). (a) STM topography image of 0.35 monolayer of graphene on Cu(111). Scale bar: 200 nm. (b) Differential conductance dI/dV image ($U = -200$ mV) recorded simultaneously with the topography image (a). dI/dV signal can differentiate graphene from copper surface. Graphene shows dark contrast on this image. The areas highlighted with dotted lines are graphene domains showing Moire′ patterns. Scale bar: 200 nm. (c) A close-up dI/dV image of a graphene island (marked with 'A' in (b)). Scale bar: 60 nm. (d) Atomic resolution STM topography image of graphene showing the Moire′ pattern and the honeycomb structure. *Reprinted with permission from Gao (2010). Copyright (2010) American Chemical Society.*

images of the graphene, Fig. 5.6.5(d), exhibiting a Moire pattern. Zhang et al. (2011) studied graphene grown on polycrystalline Cu and found perfect continuity of the graphene extending over both crystalline and noncrystalline regions, suggestive of weak interactions between the graphene and Cu. They also extended their study to show that wrinkle and ripples occur in graphene upon cooling and found three-for-six lattices form on high-curvature surfaces due to strain effects (Zhang et al., 2011). Rasool et al. (2011) also studied graphene grown on polycrystalline Cu as well and revealed that growth models based on a stagnant catalytic Cu surface were not appropriate and growing macroscopic pristine graphene is not hindered by the underlying Cu structure.

Graphene domains grown on Cu can form well-defined hexagonal shapes before they merge together into a continuous film, shown in Fig. 5.6.6, that are obviously linked to its hexagonal lattice structure (Wu et al., 2012). STM is the

FIGURE 5.6.6 SEM image of a graphene hexagon grown on Cu surface Wu et al. (2012). *Reprinted with permission from Wu (2012). Copyright (2012) American Chemical Society.*

perfect technique for examining the edge termination of these graphene hexagons without any postgrowth transfer that may perturb the original structure. Figure 5.6.7(a) shows an STM topography image of Yu et al. (2011) taken near a corner of a graphene hexagon on Cu. Atomic resolution images from the three boxed areas in (a) are shown in Fig. 5.6.7(b)–(d) respectively, and reveal that the edge of the hexagon aligns with the graphene zig-zag lattice direction. This was further supported by SAED and transmission electron microscopy (Yu et al., 2011).

Edges of graphene ribbons (GNR) are also suitable for imaging with STM and details about the edges states can be gained from STS. Cai et al. (2010) relied upon STM to show that graphene ribbons (GNR) could be produced using bottom-up growth methods from molecular precursors on Au substrates with atomically precise structure, shown in Fig. 5.6.8. An alternative way of obtaining graphene nanoribbons is by unzipping carbon nanotubes in solution or by lithographic methods combined with etching (Jiao et al., 2009; Kosynkin et al., 2009). Tao et al. (2011) examined graphene nanoribbons using both STM and STS produced by chemical unzipping that were dispersed on Au(111) substrates and their STS measurements revealed the presence of 1D edge states. Figure 5.6.9(a) shows a STM atomically resolved topography image of the edge of a (8, 1) graphene nanoribbon and (b) an atomic model. Local information regarding the energy-resolved density of states was acquired through a series of dI/dV spectra, Fig. 5.6.9(c) and (d), obtained both perpendicular and parallel to the GNR at positions indicated by black and red dots in Fig. 5.6.9(a). Spectra taken 2.4 nm from the edge were similar to those from the middle of a large graphene sheet, and spectra taken close to the edge show the emergence of

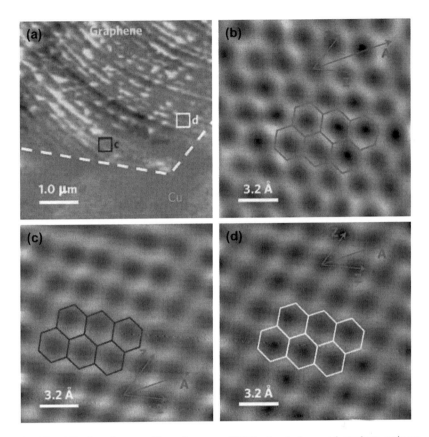

FIGURE 5.6.7 Scanning tunnelling microscopy (STM) of a single-crystal graphene grain on Cu. (a), STM topography image taken near a corner of a graphene grain on Cu. The image was acquired with a sample–tip bias $V_b = -2$ V and tunnelling current $I = 50$ pA. Dashed lines mark the edges of this grain. b–d, Atomic resolution STM topography images (filtered to improve contrast) taken from three different areas in the grain as indicated in a. The green (b), black (c) and white (d) squares (not to scale) indicate the approximate locations where the images were taken ($V_b = -0.2$ V, $I = 20$ nA). A few model hexagons are superimposed on the images to demonstrate the graphene honeycomb lattice. Select special crystal directions ('Z' for zigzag, 'A' for armchair) are indicated by arrows. A small distortion in images c and d was due to a slight hysteresis in the movement of the STM tip. *Reprinted from Yu (2011). Copyright (2011) Nature Publishing Group.*

new peaks, attributed to 1D spin-polarised edge states coupled across the width of the chiral nanoribbon (Tao et al., 2011).

In cases where the underlying metal substrate is uneven and perturbs the graphene lattice, unusual properties can emerge. Levy et al. studied graphene grown on Pt(111) surfaces and focussed on the strained regions of graphene within nanobubbles. They found that the nanobubbles had Landau

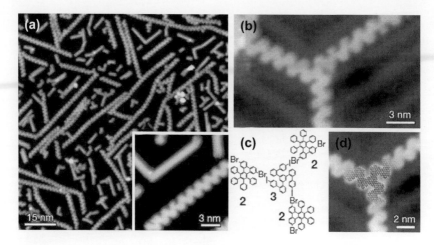

FIGURE 5.6.8   Versatility of bottom-up GNR synthesis. (a) STM image of coexisting straight $N = 7$ and chevron-type GNRs sequentially grown on Ag(111) ($T = 5$ K, $U = -2$ V, $I = 0.1$ nA). (b) Threefold GNR junction obtained from a 1,3,5-tris(4″-iodo-2′-biphenyl) benzene monomer **3** at the nodal point and monomer **2** for the ribbon arms: STM image on Au(111) ($T = 115$ K, $U = -2$ V, $I = 0.02$ nA). Monomers **2** and **3** were deposited simultaneously at 250 °C followed by annealing to 440 °C to induce cyclodehydrogenation. (c) Schematic model of the junction fabrication process with components **3** and **2**. (d) Model (blue, carbon; white, hydrogen) of the colligated and dehydrogenated molecules forming the threefold junction overlaid on the STM image from b. *Reprinted from Cai (2010). Copyright (2010) Nature Publishing Group.*

levels due to strain-induced pseudo-magnetic fields greater than 300 tesla, opening a way forward for strain engineering of graphene (Levy et al., 2010).

In many of the studies of synthetic graphene grown on metal substrates, the strong interaction between the graphene layers and the underlying substrate make it difficult to ascertain the electronic properties associated with an isolated graphene sheet. Differences in the coupling between the substrate and graphene layers may arise from different sample growth methods and yield variations in the STM images. Since graphene itself is highly conductive, it is possible to perform STM and STS using an insulating substrate.

5.6.4. STM OF GRAPHENE ON INSULATORS

The early work of mechanically exfoliating graphene often resulted in graphene residing on an insulating surface of SiO_2, as this was compatible with device fabrication. Stolyarova et al. (2007) conducted one of the first detailed STM studies of graphene on an insulating surface (SiO_2) and showed they were able to image the full hexagonal honeycomb atomic lattice for monolayer graphene. Tunnelling was found to only occur between the tip and the graphene sample.

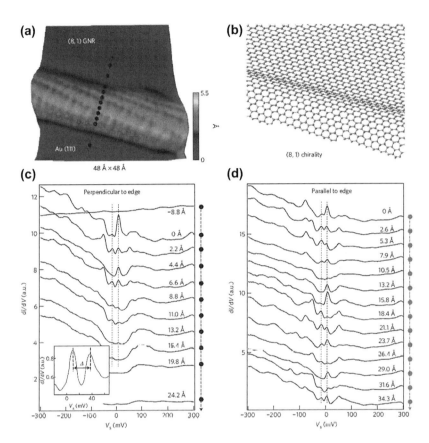

FIGURE 5.6.9   Edge states of GNRs. (a) Atomically resolved topography of the terminal edge of an (8, 1) GNR with measured width of 19.5 ± 0.4 nm ($V_s = 0.3$ V, $I = 60$ pA, $T = 7$ K). (b) Structural model of the (8, 1) GNR edge shown in a. (c) dI/dV spectra of the GNR edge shown in a, measured at different points (black dots, as shown) along a line perpendicular to the GNR edge at $T = 7$ K. Inset shows a higher resolution dI/dV spectrum for the edge of a (5, 2) GNR with width of $15.6 \pm$ 0.1 nm (initial tunnelling parameters $V_s = 0.15$ V, $I = 50$ pA; wiggle voltage $V_{r.m.s.} = 2$ mV). The dashed lines are guides to the eye. (d) dI/dV spectra measured at different points (red dots, as shown) along a line parallel to the GNR edge shown in 'a' at $T = 7$ K (initial tunnelling parameters for c and d are $V_s = 0.3$ V, $I = 50$ pA; wiggle voltage $V_{r.m.s.} = 5$ mV). *Reprinted from Tao (2011). Copyright (2011) Nature Publishing Group.*

Regions of monolayer graphene were initially identified from other few layer areas by Raman spectroscopy. STM topographic images of few layer graphene showed the characteristic three-for-six patterning associated with bulk graphite (Stolyarova et al., 2007). This work showed that STM could identify monolayer graphene from multilayered graphene, based on the topography images, provided an insulating substrate is used.

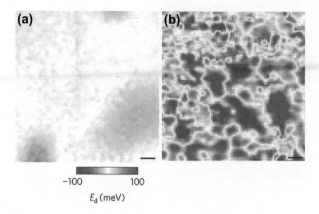

FIGURE 5.6.10 Spatial maps of the density of states of graphene on h-BN and SiO$_2$. (a) Tip voltage at the Dirac point as a function of position for graphene on h-BN. (b) Tip voltage at the Dirac point as a function of position for graphene on SiO$_2$. The colour scale is the same for a and b. The scale bar in all images is 10 nm. *Reprinted from Dean (2010). Copyright (2010) Nature Publishing Group.*

Using SiO$_2$ as an insulating surface for STM studies of graphene is not ideal due to its appreciable surface roughness. Ultraflat graphene can be obtained by transferring it to layered insulators such as mica or boron nitride (BN) (Lui et al., 2009; Xue et al., 2011). Graphene devices fabricated on BN substrates showed improved performance, indicating the importance of the underlying substrate (Dean et al., 2010). Xue et al. (2011) used STM to show that graphene conforms to the hexagonal lattice structure of BN. Local STS measurements revealed that the predicted opening of a band gap did not emerge due to the rotational misalignment of the graphene with respect to the BN. Analysis of the STS revealed that electron-hole charge fluctuations were reduced by two orders of magnitude when graphene resides on BN as compared to SiO$_2$ (Xue et al., 2011). Figure 5.6.10 shows 2D images of the tip voltage at the Dirac point as a function of position for graphene on (a) BN and (b) SiO$_2$, with the BN substrate showing less fluctuations. These images were produced by measuring dI/dV at 1 nm intervals over an area of 100 × 100 nm and then extracting the tip voltage corresponding to the Dirac point. The regions of red and blue in Fig. 5.6.11 represent puddles of electrons and holes and show the variation in the energy of the Dirac point (Xue et al., 2011).

5.6.5. SUMMARY

STM and STS are valuable surface science tools that enable atomic resolution characterisation of graphene and few layer graphene on both metal and insulating substrates. For synthetic graphene grown on metal catalysts or from SiC annealing, STM provides important conclusive evidence that graphitisation of carbon had been achieved and enabled the development and refinement of growth methods. Metal substrates interact with graphene and this can hinder the interpretation of the STM images and recent work on using insulating substrates enabled the intrinsic properties of graphene to be studied. STS provides a means for mapping out spatial variations in the electronic structure, which is important for examining edge states of graphene domains, GNRs and understanding how surface roughnesses of substrates result in puddles of holes and electrons. There is no doubt that STM will continue to be an important characterisation tool for graphene, and can even be used to lithographically pattern graphene at the nanoscale (Tapaszto et al., 2008). The challenges of STM characterisation are that it requires very clean surfaces and ultrahigh vacuum, meaning turnaround time for analysis can be long compared to other characterisation techniques such as SEM or TEM. Other complimentary characterisation techniques are needed to bridge length scales from the nano (STM) to the micro and macroscopic levels. Optical microscopy is one approach or SEM.

REFERENCES

Atamny, F., Spillecke, O., Schlogl, R., 1999. On the STM imaging contrast of graphite: towards a "true" atomic resolution. Phys. Chem. Chem. Phys. 1, 4113–4118.

Batra, I.P., Garcia, N., Rohrer, H., Salemink, II., Stoll, E., Ciraci, S., 1987. A study of graphite surface with STM and electronic structure calculations. Surf. Sci. 181, 126–138.

Berger, C., Song, Z., Li, X., Wu, X., Brown, N., Naud, C., Mayou, D., Li, T., Hass, J., Marchenkov, A.N., Conrad, E.H., First, P.N., de Heer, W.A., 2006. Electronic confinement and coherence in patterned epitaxial graphene. Science 312, 1191–1196.

Binnig, G., Rohrer, H., 1986. Scanning tunneling microscopy. IBM J. Res. Dev. 30, 4.

Cai, J., Ruffieux, P., Jaafar, R., Bieri, M., Braun, T., Blankenburg, S., Muoth, M., Seitsonen, A.P., Saleh, M., Feng, X., Mullen, K., Fasel, R., 2010. Atomically precise bottom-up fabrication of graphene nanoribbons. Nature 466, 470–473.

Charrier, A., Coati, A., Argunova, T., Thibaudau, F., Garreau, Y., Pinchaux, R., Forbeaux, I., Debever, J.-M., Sauvage-Simkin, M., Themlin, J.-M., 2002. Solid-state decomposition of silicon carbide for growing ultra-thin heteroepitaxial graphite films. J. Appl. Phys. 92, 2479–2484.

Cho, J., Gao, L., Tian, J., Cao, H., Wu, W., Yu, Q., Yitamben, E.N., Fisher, B., Guest, J.R., Chen, Y.P., Guisinger, N.P., 2011. Atomic-scale investigation of graphene grown on Cu foil and the effects of thermal annealing. ACS Nano 5, 3607–3613.

Coraux, J., N'Diaye, A.T., Busse, C., Michely, T., 2008. Structural coherency of graphene on Ir(111). Nano Lett. 8, 565–570.

Dai, H., Hafner, J.H., Rinzler, A.G., Colbert, D.T., Smalley, R.E., 1996. Nanotubes as nanoprobes in scanning probe microscopy. Nature 384, 147–150.

Dean, C.R., Young, A.F., Meric, I., Lee, C., Wang, L., Sorgenfrei, S., Watanabe, K., Taniguchi, T., Kim, P., Shepard, K.L., Hone, J., 2010. Boron nitride substrates for high-quality graphene electronics. Nat. Nanotechnol. 5, 722–726.

Gall, N.R., Rut'kov, E.V., Tontegode, A.Y., 2004. Phys. Solid State 46, 371.

Gao, L., Guest, J.R., Guisinger, N.P., 2010. Epitaxial graphene on Cu(111). Nano Lett. 10, 3512–3516.

Jiao, L.Y., Zhang, L., Wang, X.R., Diankov, G., Dai, H.J., 2009. Narrow graphene nanoribbons from carbon nanotubes. Nature 458, 877–880.

Klusek, Z., Kozlowski, W., Waqar, Z., Datta, S., Burnell-Gray, J.S., Makarenko, I.V., Gall, N.R., Rutkov, E.V., Tontegode, A.Y., Titkov, A.N., 2005. Local electronic edge states of graphene layer deposited on Ir(1 1 1) surface studied by STM/CITS. Appl. Surf. Sci. 252, 1221–1227.

Kosynkin, D.V., et al., 2009. Longitudinal unzipping of carbon nanotubes to form graphene nanoribbons. Nature 458, 872–875.

Land, T.A., Michely, T., Behm, R.J., Hemminger, J.C., Comsa, G., 1992. Surf. Sci. 264, 261–270.

Levy, N., Burke, S.A., Meaker, K.L., Panlasigui, M., Zettl, A., Guinea, F., Castro Neto, A.H., Crommie, M.F., 2010. Strain-induced pseudo-magnetic fields greater than 300 tesla in graphene nanobubbles. Science 329, 544–547.

Lui, C.-H., Liu, L., Mak, K.-F., Flynn, G.W., Heinz, T.F., 2009. Ultraflat graphene. Nature 462, 339–341.

Luican, A., Li, G., Andrei, E.Y., 2009. Scanning tunneling microscopy and spectroscopy of graphene layers on graphite. Solid State Commun. 149, 27–28.

Makarenko, I.V., Titkov, A.N., Waqar, Z., Dumas, P., Rutkov, E.V., Gall, N.R., 2007. Phys. Solid State 49, 371.

Mallet, P., Varchon, V., Naud, C., Magaud, L., Berger, C., Veuillen, J.-Y., 2007. Electron states of mono- and bilayer grapheen on sic probed by scanning-tunneling-microscopy. Phys. Rev. B 76, 041403.

Marchini, S., Gunther, S., Wintterlin, J., 2007. Scanning tunneling microscopy of graphene on Ru(0001). Phys. Rev. B 76, 075429.

Mizes, H.A., Park, S.-I., Harrison, W.A., 1987. Multiple-tip interpretation of anomalous scanning-tunnelling-microscopy images of layered materials. Phys. Rev. B 36, 4491–4494.

Moriarty, P., Hughes, G., 1992. Appl. Phys. Lett. 60, 2338.

N'Diaye, A.T., Bleikamp, S., Feibelman, P., Michely, T., 2006. Phys. Rev. Lett. 97, 215501.

Ouseph, P.J., Poothackanal, T., Mathew, G., 1995. Phys. Lett. A 205, 65.

Pan, Y., Zhang, H., Shi, D., Sun, J., Du, S., Liu, F., Gao, H-j., 2009. Highly ordered, millimeter-scale, continuous, single-crystalline graphene monolayer formed on Ru (0001). Adv. Mater. 21, 2777.

Paredes, J.I., Martinez-Alonso, A., Tascon, J.M.D., 2001. Carbon 39, 476.

Rasool, H.I., Song, E.B., Allen, M.J., Wassei, J.K., Kaner, R.B., Wang, K.L., Weiller, B.H., Gimzewski, J.K., 2011. Nano Lett. 11, 251–256.

Rutter, G.M., Crain, J.N., Guisinger, N.P., Li, T., First, P.N., Stroscio, A.A., 2007. Scattering and interference in epitaxial graphene. Science, 219–222.

Soler, J.M., Baro, A.M., Garcia, N., 1986. Interactomic forces in scanning tunneling microscopy: giant corrugations of the graphite surface. Phys. Rev. Lett. 57, 444.

Stolyarova, E., Rim, K.T., Ryu, S., Maultzsch, J., Kim, P., Brus, L.E., Heinz, T.F., Hybertsen, M.S., Flynn, G.W., 2007. High-resolution scanning tunneling microscopy imaging of mesoscopic graphene sheets on an insulating surface. Proc. Natl. Acad. Sci. U.S.A. 104, 9209.

Tao, C., Jiao, L., Yazyev, O.V., Chen, Y.-C., Feng, J., Zhang, X., Capaz, R.B., Tour, J.M., Zettl, A., Louie, S.G., Dai, H., Crommie, M.F., 2011. Spatially resolving edges states of chiral graphene nanoribbons. Nat. Phys. 7, 616–620.

Tapaszto, L., Dobrik, G., Lambin, P., Biro, L.P., 2008. Tailoring the atomic structure of graphene nanoribbons by scanning tunneling microscope lithography. Nat. Nanotechnol. 3, 397–401.

Tomanek, D., Louie, S.G., 1988. First-principles calculation of highly asymmetric structure in scanning-tunneling-microscopy images of graphite. Phys. Rev. 37, 8327.

Ueta, H., Saida, M., Nakai, C., Yamada, Y., Sasaki, M., Yamamoto, S., 2004. Surf. Sci. 560, 183–190.

Wang, Y., Ye, Y., Wu, K., 2006. Simultaneous observation of the triangular and honeycomb structures on highly oriented pyrolytic graphite at room temperature: an STM study. Surf. Sci. 600, 729.

Wong, H.S., Durkan, C., Chandrasekhar, N., 2009. Tailoring the local interaction between graphene layers in graphite at the atomic scale and above using scanning tunneling microscopy. ACS Nano 3, 3455–3462.

Wu, Y., Fan, Y., Speller, S., Creeth, G., Sadowski, J., He, K., Robertson, A.W., Allen, C., Warner, J.H., 2012. Large single crystals of graphene on melted copper using chemical vapour deposition. ACS Nano 6, 5010–5017.

Xue, J., Sanchez-Yamagishi, J., Bulmash, D., Jacquod, P., Deshpande, A., Watanabe, K., Taniguchi, T., Jarillo-Herrero, P., LeRoy, B.J., 2011. Scanning tunneling microscopy and spectroscopy of ultra-flat graphene on hexagonal boron nitride. Nat. Mater. 10, 282–285.

Yu, Q., Jauregui, L.A., Wu, W., Colby, R., Tian, J., Su, Z., Cao, H., Liu, Z., Pandey, D., Wei, D., Chung, T.-F., Peng, P., Guisinger, N.P., Stach, E.A., Bao, J., Pei, S.-S., Chen, Y.P., 2011. Control and characterization of individual grains and grain boundaries in graphene grown by chemical vapour deposition. Nat. Mater. 10, 443–449.

Zeinalipour-Yazdi, C.D., Pullman, D.P., 2008. Interpretation of the scanning tunneling microscope image of graphite. Chem. Phys. 348, 233–236.

Zhang, Y., Gao, T., Bao, Y., Xie, S., Ji, Q., Peng, H., Liu, Z., 2011. Defect-like structures of graphene on copper foils for strain relief investigation by high-resolution scanning tunneling microscopy. ACS Nano 5, 4014–4022.

⟨ Chapter 5.7 ⟩

AFM as a Tool for Graphene

Mark Rümmeli

IFW Dresden, Germany

5.7.1. INTRODUCTION

Atomic force microscopy (AFM) has emerged as an invaluable tool for the characterisation of graphene in terms of determining the number of layers residing on a support, the shape of the deposited flakes and the level of creasing in the flake or sheet. Moreover it can show (height) differences between graphene and graphene oxide, which is not only helpful in distinguishing the two but also in engineering graphene. AFM has proven as invaluable tool to help experimentally determine the mechanical properties of graphene. This section highlights the potential of AFM in the characterisation and engineering of graphene.

The roots of AFM lie with the initial development of the scanning tunnelling microscope by researchers in 1981 (Binnig et al., 1982). Binning and Quate (1986) demonstrated the potential of AFM. In 1987 Wickramsinghe demonstrated the first vibrating cantilever AFM (Martin et al., 1987). Since then, AFM has made major contributions in the investigation of DNA, proteins and cells, polymers as well as helped in understanding mechanical properties of materials.

Atomic force microscopy allows 3D profiling of surfaces at the nanoscale. This is accomplished by measuring the forces between a sharp tip (probe), usually less than 10 nm, and close to the sample surface (proximity 0.2–10 nm). The probe is supported on a cantilever and is usually made of Si_3N_4 or Si. More expensive variants are carbon nanotube tips. By reflecting a laser beam off the cantilever into a sensitive photodiode detector the cantilever deflections that occur as the tip moves over the sample can be monitored. Recording the (z) deflections as a function of the sample's X and Y positions a three dimensional image of the sample's surface can be constructed. In addition, the size of force between the tip and sample is dependant on the spring constant of the cantilever and the distance between the probe and sample surface. This force can be described by Hook's law and allows AFM to provide information on the mechanical properties of materials.

The primary interactions between the sample surface and the tip for short probe-sample separations (<0.5 nm) are van der Waals interactions. Further away from the surface long range interactions such as capillary, electrostatic and magnetic interaction are significant.

Three primary imaging modes are normally available:

Contact mode AFM: Here the tip-sample distances (<0.5 nm) lead to repulsive van der Waals forces viz. the force on the tip is repulsive. By keeping a constant cantilever deflection (using feedback loops) the force between the tip and the sample stays constant and allows an image of the surface to be obtained. Relatively speaking, fast scan rates are possible and it is a good mode for rough samples.

Tapping mode AFM: the imaging process is similar to contact mode, however, in this mode the cantilever oscillates at its resonant frequency and the tip-sample distance varies between 0.5 and 2 nm. A small piezoelectric element mounted on the tip holder drives the oscillation. To collect an image a constant oscillation amplitude is maintained, which provides a constant tip-sample interaction. This mode is good for high resolution imaging of samples that are less easily damaged or loosely held to a surface, however slower scanning rates are obtained.

Non-contact mode AFM: as the name suggests, in this mode the tip never touches the surface. The cantilever oscillates slightly above its resonance frequency with the amplitude of oscillation being a few nanometres. Between 1 and 10 nm, the van der Waals forces are strongest and this force (or any other long range force) decreases the resonance frequency of the cantilever. Using the decrease in resonant frequency along with a feedback loop keeps the oscillation amplitude or frequency constant by adjusting the average tip-sample distance. This mode does not suffer from tip or sample degradation effects.

5.7.2. GRAPHENE ON DIFFERENT SURFACES

The topology of graphene is sensitive to the surface that it is deposited on. For example, height variations are observed for monolayer graphene on insulating surfaces (Booth et al., 2008; Geringer et al., 2009; Ishigami et al., 2007; Knox et al., 2008; Stöberl et al., 2008; Stolyarova et al., 2007). Lui et al. (2009) investigated the topology of monolayer graphene on several surfaces in detail.

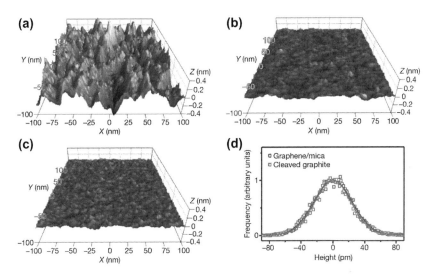

FIGURE 5.7.1   Comparison of surface roughness for graphene on SiO_2 and on mica, and for cleaved graphite. a, b, Three-dimensional representations of the AFM topographic data for graphene on SiO_2 (a) and on mica (b) substrates. The images correspond to the regions in Fig. 1a and b designated by the blue squares. (c), AFM image of the surface of a cleaved kish graphite sample. Images a, b and c correspond to 200×200 nm areas and are presented with the same height scale. (d), Height histograms of the data in b as blue squares and in c as red squares. The histograms are described by Gaussian distributions (solid lines) with standard deviations σ of 24.1 pm and 22.6 pm, respectively (Lui et al., 2009).

They explored SiO_2, mica and cleaved graphite as supports. The graphene layer deposited was obtained by mechanical exfoliation of kish graphite[1]. Using amplitude-modulation AFM in the non-contact mode they characterised the topography of the graphene samples. The lateral resolution[2] was 7 nm and the height resolution was 23 pm. An overview of the raw data (no smoothing has been applied) is provided in Fig. 5.7.1.

In the case of pristine silicon (no graphene deposited) the surface is rather corrugated and shows a height variation of 168 pm and a correlation length of 16 nm. In the case of SiO_2 with a monolayer of graphene deposited on the surface, the degree of roughness is slightly diminished with a height variation of 154 pm and a length correlation of 22 nm. This indicates that monolayer graphene tends to follow the underlying substrate morphology. The topographic images for pristine mica and monolayer graphene on mica stand in a stark

---

1. Kish graphite is obtained as a by product of the steel making process.

2. In AFM, the lateral resolution is limited due to tip-sample convolution. It can be reduced with a fine tip.

contrast to the silica substrate in that the mica exhibits a far smoother land-scape. For a bare mica surface the height variation was found to be 34.3 pm and the length correlation was 2 nm. For graphene on mica the height variation was 24.1 pm and the length correlation was 2 nm. In other words the topography for graphene on mica is at least 5 times smoother than that of graphene on SiO_2. The authors highlight that since the correlation length of 2 nm for bare mica and graphene on mica is smaller than the AFM spatial (lateral) resolution of 7 nm the observed roughness primarily arises from noise, since any true feature that can be measured can only be measured at or above the scale of the spatial resolution (7 nm). The work not only highlighted the effect of substrate topology but also that van der Waals interactions between the support and monolayer graphene suppresses the formation of intrinsic ripples that can be observed, for example, in free standing graphene. Typically, the ripples have height variations ~1 nm and span laterally in the scale of 10–25 nm. These corrugations are usually argued as being necessary to stabilise the graphene against thermal instabilities that exist in ideal two dimensional systems (Fasolino et al., 2007).

Ishigami et al. (2007) looked at graphene, as part of graphene based devices, on SiO_2. In agreement with Lui et al. (2009), they found the graphene to primarily follow the underlying morphology of the SiO_2 support. In addition they showed that the graphene sheets do have finite intrinsic stiffness. This prevents the graphene from fully conforming to the underlying substrate morphology. They also showed that residue material on the surface of litho-graphically prepared graphene devices is ubiquitous.

The fact that graphene deposited on substrates tends to follow the under-lying support can make identification of the number of layers by height measurements somewhat tricky. Huntzinger et al. (Tiberj et al., 2009) looked at mechanically cleaved highly oriented pyrolytic graphite (HOPG) deposited on polished 6H–SiC. They found that, due to SiC atomic steps and kinks at boundaries, the graphene layer determination had an uncertainty of ± 1 layer. Severin et al. (2011) exploited the behaviour of graphene to follow the underlying support to replicate single macromolecules. They mechanically exfoliated single-layer graphene and FLG onto mica which has an atomically flat surface. The mica surfaces were covered in isolated double stranded plasmid DNA rings. Using AFM in both the contact and intermittent contact modes, they could demonstrate that graphenes do replicate the topography of the underlying DNA with high precision. When using the intermittent contact mode (c.f. noncontact mode) as compared to the contact mode, they found that the double-stranded DNA appears broader and slightly higher. They attributed this to a liquidlike layer that encloses the double stranded DNA. In contact mode, the tip forces are higher and are sufficient to 'squeeze' away from the liquid layer during the scanning process. The technique may have promise to modify the electronic properties of graphene, since its electronic properties are sensitive to strain/deformation.

Kellar et al. (2010) compared STM, AFM, lateral force microscopy and conductive AFM. They showed that under ambient conditions (viz. with no ultrahigh vacuum was employed to minimise vibrations), graphene could still be characterised with nanometre spatial resolution with AFM. cAFM, on the other hand, showed the highest imaging contrast between the epitaxially grown graphene (on SiC) and the SiC substrate (6v3 surface). This is because of the strong differences between the tip-sample resistances (recall graphene is conductive and SiC is not). Hauquier et al. (2012) looked at graphene sheets bonded to gold surfaces using cAFM. The electrical-resistance images showed rather different conduction properties which they attributed to different local conduction properties, being linked to the number of underlying graphene sheets.

5.7.3. AFM STUDIES ON GRAPHENE OXIDE

Graphene oxide, put simply, has a variety of functional groups such as epoxides (bridging oxygen atoms), carbonyls (=CO), hydroxyls (–OH) and phenol groups attached to its surfaces. The attachment of these groups leads to sp^3 hybridisation which leads to a certain amount of buckling. Thus, the interlayer spacing between GO sheets is larger, 0.7–0.8 nm which is roughly twice that found in graphite. Thus we might expect the height measurements obtained by atomic force microscopy to exceed that of graphene. Shen et al. (2009) investigated graphene oxide and reduced graphene oxide nano-platelets. Cross-sectional views of the material were investigated, using AFM in the tapping mode. Exfoliated graphene oxide showed an average thickness of individual sheets of ca. 1.3 nm while the reduced graphene oxide nano-platelets showed a bumpy structure with flat areas exhibiting heights of 0.2–0.4 nm and some high regions showing an average height of 1.5 nm. They attributed the bumpy surface to regions of dead space because of extensive edge functionalisation (Dideykin et al., 2011). Shen et al. (2009) also investigated reduced GO on supports and found an average height of 1.3 nm.

Pandey et al. (Shen et al., 2009) investigated exfoliated graphene oxide deposited on HOPG. The surface morphology of the HOPG substrate had atomic steps. The steps were decorated by nanometre particulates of unknown origin. The deposited GO was shown to conform to the underlying support morphology. In addition, wrinkles and folds were observed in the GO sheets.

5.7.4. AFM AS A TOOL TO INVESTIGATE AND ENGINEER PHYSICAL PROPERTIES

AFM has turned out to be an invaluable tool with which to experimentally investigate the mechanical properties of graphene and its derivatives. Examples include experimental determinations of Young's modulus, spring constants and intrinsic strength. These aspects are discussed in detail in Section 3.4 and are not discussed further here.

The basic tribological properties of graphene have been investigated to understand its mechanical behaviour under sliding contact situations using AFM. Initial studies show that the frictional properties of graphene vary with the number of layers forming the graphene film. In essence, the friction decreases as the number of layers increases (Filleter et al., 2009; Lee et al., 2009, 2010). This variation of friction with respect to the number of graphene layers is attributed to the interplay of surface attractive forces, electron–phonon coupling and a puckering effect (Filleter et al., 2009; Lee et al., 2009, 2010). However, for more than 5 layers the friction properties are similar to that from bulk graphite. Lin et al. (2011) conducted systematic studies on the friction and wear characteristics of multilayer graphene films, with 16–15 layers deposited on silicon substrates. The multi-layer graphene films were prepared by mechanical exfoliation. Their systematic studies used both positive and negative loads to the AFM tip because the attractive force between the tip and the specimen can provide a significant contribution to the effective applied load at the nanoscale. For comparison the friction of a bare Si surface was measured under identical experimental conditions. Applied loads from 30 to −5 nN were used and initially the AFM tip came into contact with the specimen at a high load. The friction force was then measured over a distance of 20 μm for 2 cycles after which the cantilever was retracted slightly to lower the applied force, and the friction was measured again. This process was repeated until the tip detached from the specimen surface due to excessive negative loading. The process is illustrated schematically in Fig. 5.7.2.

The surface morphology and the thickness (height) of the graphene was the same before and after the friction tests confirming no basal plane slip (layers sliding relative to each other) occurred during the friction measurements. Thus, it is reasonable to assume the frictional energy dissipation arises mostly from the sliding interface between the tip and the graphene surface, viz. the results can be attributed to the intrinsic properties of graphene. The results showed that the graphene had a significantly lower friction than the Si substrate. For graphene, friction values between 0.36 and 0.62 nN where obtained, while for Si they ranged between 1.1 and 4.3 nN. The investigators also investigated the wear properties of the graphene. To achieve this they used higher applied loads from 10 to 3 μN for at least 100 cycles of reciprocating tip motion. These forces are sufficient to cause bond breakage in the graphene (usually bond breakage does not occur below 250 nN). The process leads to the formation of wear tracks with depths in the range of 1.14 to 0.47 nm. In other

FIGURE 5.7.2 Schematic of AFM tip/specimen contact under negative- and positive applied loads.

words, the wear track depths don't correspond to multiples of the thickness of single layer graphene (0.35 nm), which was attributed to measurement artefacts. Nonetheless careful evaluation of the data suggests that 5 µN is the critical load for which a single layer of graphene is removed after 100 cycles of sliding. The wear process is considered to occur by the breakage of in-plane bonds. In short, graphene exhibits high wear resistance.

Adhesion along with deformation is the main mechanism behind friction. Ding et al. (2011) measured the adhesion force for GO and various degrees of reduced GO through to graphene. They accomplished this using AFM in the tapping mode with a colloidal probe[3]. Under ambient conditions the tip or colloid probe oscillates above an adsorbed fluid layer on the surface during scanning (n.b.: all samples unless in UHV or placed in an environmental chamber have some liquid adsorbed on the surface – e.g. water). In this study, a water bridge formed between the AFM colloid tip and the surface. This nanoscale bridge exerts a substantial force on the probe tip. This force can be used for imaging or, as in this case, the tip can record the force felt by the cantilever as the tip is first brought close to the sample surface and then pulled away. The adhesion force between the tip and the sample surface is determined by capillary force. In this case, it is assumed to be the sum of the water surface tension and pressure difference on the water meniscus surface. The adhesive forces between the tip and the graphene oxide and reduced graphene oxide samples diminished as the graphene oxide was further reduced. The adhesive force between the AFM colloid probe and graphene oxide reached 170 nN, and it reduced to 66 nN for fully reduced graphene oxide.

A technical difficulty when working with graphene is the avoidance of surface contaminants. The use of solvents to clean graphene for the most part does not remove all residues. High-temperature annealing is most commonly implemented as a cleaning route. In essence the graphene is annealed in an inert atmosphere, typically Ar/H_2. This process removes residue material by desorption, however, the coupling between the graphene and the support may increase which tends to induce mechanical deformation of the graphene. If the graphene is to be used as the basis of a device the mechanical deformations can alter the graphene's electronic properties, just as the presence of residues can since charged impurities, surface contaminants and structural deformation contribute to local doping potentially degrading the devices performance. Moreover, some substrates cannot sustain elevated temperatures making annealing processes for cleaning unsuitable. An alternative cleaning route is through current annealing (Barreiro et al., 2012; Bolotin et al., 2008).

---

3. A colloid probe is a tip that has a sphere at the tip instead of a sharp tip. A sharp tip is not beneficial for adhesion measurements since, because the contact area depends strongly on the direction of indentation, it is difficult to measure the radius and they cause high pressure at the contact location. Colloid tips avoid these complications.

With this technique the graphene (under bias) is locally heated by Joule heating, however the graphene can crack or ripple under these conditions.

Lindvall et al. (2012) established a technique to clean graphene used in devices. Their work highlighted the advantages of clean graphene for efficient device performance. In their study they deposited mechanically exfoliated graphene on silicon with a 290 nm oxide layer (thermally grown) and on Nb doped strontium titanium oxide substrates with 50 nm barium-STO (BSTO) thin films (grown by pulsed laser deposition). The later substrates were selected, because it is incompatible with temperature annealing, viz. the graphene cannot be annealed for cleaning purposes. Graphene Hall bar devices were fabricated using electron beam lithography (EBL) and subsequent oxygen plasma etching. The AFM studies were conducted in the tapping mode. The tip was scanned in a back and forth manner over the sample. In this process, contaminants are mechanically pushed to the sides in a broomlike way. The procedure was repeated up to five times, however they observed that graphene was usually clean after only two scans. Thereafter, only very minor improvements were obtained. The roughness of the graphene was shown to depend on the applied force during cleaning rather than the type of oxide substrate. This information was found by exploiting two cantilevers, namely, a stiff cantilever with a large contact force of ca. 180 nN and a softer cantilever with a contact force of around 30 nN. Several of the samples were measured before and after cleaning by measuring the resistance as a function of the gate voltage Vg. The data showed consistent changes in the charge-neutrality point (gate voltage and drain voltage are the same) – see Fig. 5.7.3. The shift in the drain voltage is towards slightly negative voltages which indicates weak electron doping, perhaps due to charges trapped in the SiO_2. In nearly all cases, the mobility was shown to have improved after the cleaning process, however samples cleaned with only a small contact force showed the best improvement. Their work suggests mechanical cleaning of graphene might warrant further development.

Giesbers et al. (2008) demonstrated how one can use an atomic force microscope to locate and manipulate single and few-layer graphene sheets. In this study the potential of an AFM tip to peel graphene layers off as well as to cut graphene flakes was investigated. Initial attempts to peel-off graphene layers or cut layers with mechanically cleaved graphene placed on a silica substrate were unsuccessful. This was attributed to the low sticking force of the graphene on rough substrates, such as silica. Hence they switched support material to epi-ready GaAs and successfully showed that one can displace and even tear graphene flakes. However, the tearing process was rather crude and might be more efficient if using sharp diamond coated tips. Nevertheless the successful peeling of graphene layers paves the way of creating single layer flakes rather than randomly searching for one. The investigators also explored the use of local anodic oxidation (Bolotin et al., 2008). In essence, a bias

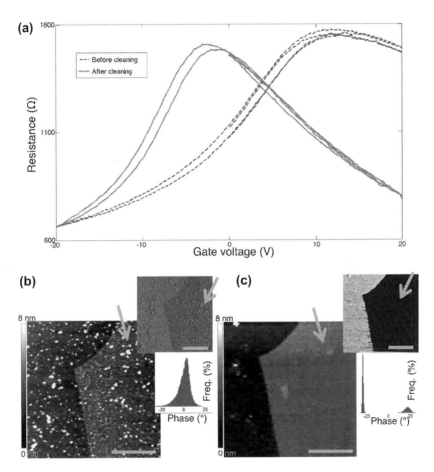

FIGURE 5.7.3 (a) Resistance as a function of gate before (red, solid line) and after (blue, dashed line). The charge neutrality point moves towards zero after cleaning, and the estimated mobility increases from 4300 cm^2 to 7700 cm^2. The Z-scale is 8 nm. (b) and (c) AFM height images before and after cleaning of graphene on BSTO, respectively. The heavy contamination is removed and the atomic steps in the BSTO are clearly seen, including through graphene. Upper insets: AFM phase images. The phase response of both substrate and graphene is almost flat after cleaning, evident from the phase histograms (lower insets). All scale bars are 500 nm. *From Lindvall et al. (2012).*

voltage is applied between the AFM tip and the substrate in a humid environment. The humid environment leads to an adsorbed fluid layer on the sample's surface. Applying a bias causes the substrate to oxidise. Thus by controlling the tip position and the applied bias voltage, one can engineer different surface morphologies and structures. In this case, graphene deposited on silica and then electrically contacted was used. Upon the application of

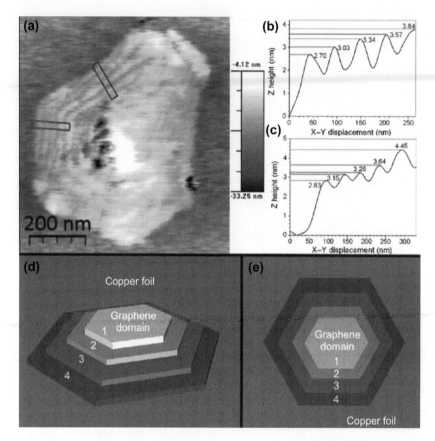

FIGURE 5.7.4 (a) A contact mode AFM image of a FLG domain, scanned from left to right. (b,c) Box-averaged z-height line profiles shown for the black (b) and blue (c) highlighted regions in (a). (d) A 3D schematic representation of a terraced few-layer graphene (FLG) domain on copper foil. (e) Overhead view of a schematic representation of a single terraced, FLG domain on copper foil. Novoselov et al. (2005, 2007); Tan et al. (2007).

a positive voltage to the AFM tip the graphene was locally oxidised below the tip (electrochemical oxidation). At the cathode the current induced oxidation of the carbon (graphene) forms a variety of carbon-based oxides and acids that escape from the surface forming a groove in the graphene sheet. The groove forming procedure does require that the initial process starts at the edge of a graphene sheet, though. The reason for this is that breakage of C–C bond on a pristine surface is inefficient while edge bonds are weaker and consequently easier to break.

Wei et al. (2010) took the concept of AFM tip based engineering further by introducing tip-based thermochemical nanolithography. In their method they used graphene oxide (on SiO_2) as their starting material. By applying a heated tip to the graphene oxide, it can be reduced (reduced graphene oxide). Indeed, they could form arbitrary reduced graphene oxide features, for example a cross, by scanning the heated tip over graphene oxide-isolated flakes. The reduction process, in essence, removes oxygen-rich functional groups from the surface and this is reflected by a reduction in sheet height of 2–5 Å. Scanning with an unheated tip did not result in any height changes indicating the reduction is caused by the local heating process. Moreover, the rate of thermal reduction was shown to depend on the tip temperature. For temperatures ~100 °C no apparent reduction was observed. For temperatures above 150 °C reduction was obtained. The reduced graphne-oxide showed lower friction values as compared to the original GO and is in line with studies by others, since the functional groups forming on GO make the surface rougher because sp^3 hybridisation is introduced. Four point and two point I-V measurements on various fabricated structures showed how one can change the transport properties from insulating to metallic in a controlled manner, depending on the degree of reduction applied to the initial GO. The technique could be applied to the manufacture of graphene nanoelectronics, by using arrays of heated tips. Alternatively heated AFM probes could read or write nanostructures on a surface and in large arrays might even address wafer scale areas at high speed.

As a characterisation tool for graphene growth, atomic force microscopy is invaluable. Usually the implementation of AFM enables scientists to evaluate the number of graphene layers formed on a substrate subsequent its synthesis on a substrate as for example found in the chemical vapour synthesis of graphene. A nice example of this is shown in Fig. 5.7.4 (Robertson and Warner, 2011). In the figure few layer graphene domains with a hexagonal structure are measured in AFM using contact mode. The technique highlights the domain form and also the layer structure. In other words important three dimensional profile information is obtained and provides useful insight into the growth of graphene, in this case CVD growth of graphene of copper.

5.7 REFERENCES

Barreiro, A., Börrnert, F., Rümmeli, M.H., Büchner, B., Vandersypen, L.M.K., 2012. Graphene at high bias: cracking, layer by layer sublimation, and fusing. Nano Lett. 12, 1873–1878.

Binnig, G., Quate, C.F., 1986. Atomic force microscope. Phys. Rev. Lett. 56 (9), 930–933.

Binnig, G., Rohrer, H., Gerber, Ch., Weibel, E., 1982. Surface studies by scanning tunneling microscopy. Phys. Rev. Lett 49 (1), 57–61.

Bolotin, K.I., Sikes, K.J., Jiang, Z., Klima, M., Fudenberg, G., Hone, J., Kim, P., Stormer, H.L., 2008. Ultrahigh electron mobility in suspended graphene. Solid State Commun. 146 (9–10), 351–355.

Booth, T.J., Blake, P., Nair, R.R., Jiang, D., Hill, E.W., Bangert, U., Bleloch, A., Gass, M., Novoselov, K.S., Katsnelson, M.I., Geim, A.K., 2008. Macroscopic graphene membranes and their extraordinary stiffness. Nano Lett. 8 (8), 2442–2446.

Dideykin, A., Aleksenskiy, A.E., Kirilenko, D., Brunkov, P., Goncharov, V., Baidakova, M., Sakseev, D., Ya. Vul, A., 2011. Monolayer graphene from graphite oxide. Diamond Relat. Mater. 20, 105–108.

Ding, Y.-H., Zhang, P., Ren, H.-M., Zhuo, Q., Yang, Z.-M., Jiang, X., Jiang, Y., 2011. Surface adhesion properties of graphene and graphene oxide studied by colloid-probe atomic force microscopy. App. Surf. Sci. 258, 1077–1081.

Fasolino, A., Los, J.H., Katsnelson, M.I., 2007. Intrinsic ripples in graphene. Nat. Mater. 6 (11), 858–861.

Filleter, T., McChesney, J.L., Bostwick, A., Rotenberg, E., Emtsev, K.V., Seyller, Th., Horn, K., Bennewitz, R., 2009. Friction and dissipation in epitaxial graphene films. Phys. Rev. Lett 102 (8) 086102(4).

Geringer, V., Liebmann, M., Echtermeyer, T., Runte, S., Schmidt, M., Rückamp, R., Lemme, M.C., Morgenstern, M., 2009. Intrinsic and extrinsic corrugation of monolayer graphene deposited on SiO_2. Phys. Rev. Lett. 102 (7) 076102(4).

Giesbers, A.J.M., Zeitler, U., Neubeck, S., Freitag, F., Nososelev, K.S., Maan, J.C., 2008. Nano-lithography and manipulation of graphene using an atomic force microscope. Solid State Commun. 147 (9–10), 366–369.

Hauquiera, F., Alamarguya, D., Vielb, P., Noëla, S., Filoramoc, A., Hucd, V., Houzéa, F., Palacinb, S., 2012. Conductive-probe AFM characterization of graphene sheets bonded to gold surfaces. App. Surf, Sci. 258 (7), 2920–2926.

Ishigami, M., Chen, J.H., Cullen, W.G., Fuhrer, M.S., Williams, E.D., 2007. Atomic structure of graphene on SiO_2. Nano Lett. 7 (6), 1643–1648.

Kellar, J.A., Alaboson, J.M.P., Wang, Q.H., Hersam, M.C., 2010. Identifying and characterizing epitaxial graphene domains on partially graphitised SiC(0001) surfaces using scanning probe miscroscopy. App. Phys. Lett. 96 143103(3).

Knox, K.R., Wang, S., Morgante, A., Cvetko, D., Locatelli, A., Mentes, T.O., Niño, M.A., Kim, P., Osgood Jr., R.M., 2008. Spectromicroscopy of single and multilayer graphene supported by a weakly interacting substrate. Phys. Rev. B 78 201408(R)(4).

Lee, C.G., Wei, X.D., Li, Q.Y., Carpick, R., Kysar, J.W., Hone, J., 2009. Elastic and frictional properties of graphene. Phys. Status Solidi B 246 (11–12), 2562–2567.

Lee, H.S., Lee, N.S., Seo, Y.H., Eom, J.H., Lee, S.W., 2009. Comparison of frictional forces on graphene and graphite. Nanotechnology 20 (32) 325701(6).

Lee, C.G., Li, Q.Y., Kalb, W., Liu, X.Z., Berger, H., Carpick, R.W., Hone, J., 2010. Frictional characteristics of atomically thin sheets. Science 328 (5974), 76–80.

Lin, L.-Y., Kim, D.-E., Kim, W.-K., Jun, S.-C., 2011. Friction and wear characteristics of multi-layer graphene films investigated by atomic force microscopy. Surf. Coat. Technol. 205 (20), 4864–4869.

Lindvall, N., Kalabukhov, A., Yurgens, A., 2012. Cleaning graphene using atomic force micro-scope. J. Appl. Phys. 111 064904(4).

Lui, C.H., Liu, L., Mak, K.F., Flynn, G.W., Heinz, T.F., 2009. Ultraflat graphene. Nature 462, 339–341.

Martin, Y., Williams, C.C., Wickramasinghe, H.K., 1987. Atomic force microscope-force mapping and profiling on a sub 100-Å scale. J. Appl. Phys. 61 (10), 4723.

Pandey, D., Reifenberger, R., Piner, R., 2008. Scanning probe microscopy study of exfoliated oxidized graphene sheets. Surf. Sci. 602, 1607–1613.

Robertson, A.W., Warner, J.H., 2011b. Hexagonal single crystal domains of few-layer graphene on copper foils. Nano Lett. 11, 1182–1189.

Severin, N., Dorn, M., Kalachev, A., Rabe, J.P., 2011. Replication of single macromolecules with graphene. Nano Lett. 11, 2436–2439.

Shen, J., Hu, Y., Shi, M., Lu, X., Qin, C., Li, C., Ye, M., 2009. Fast and facile preparation of graphene oxide and reduced graphene oxide nanoplatelets. Chem. Mater. 21, 3514–3520.

Stöberl, U., Wurstbauer, U., Wegscheider, W., Weiss, D., Eroms, J., 2008. Morphology and flexibility of graphene and few-layer graphene on various substrates. Appl. Phys. Lett. 93 (5) 051906(3).

Stolyarova, E., Taeg Rim, K., Ryu, S., Maultzsch, J., Kim, P., Brus, L.E., Heinz, T.F., Hybertsen, M.S., Flynn, G.W., 2007a. High-resolution scanning tunneling microscopy imaging of mesoscopic graphene sheets on an insulating surface. Proc. Natl Acad. Sci. U.S.A. 104 (22), 9209–9212.

Tiberj, A., Martin, M., Camara, N., Poncharal, P., Michel, T., Sauvajol, J.L., Godignon, P., Camassel, J., 2009. AFM and Raman studies of graphene exfoliated on SiC, silicon carbide and related materials 2008. Mater. Sci. Forum 615–617, 215–218.

Wei, Z., Wang, D., Kim, S., Kim, S.-Y., Hu, Y., Yakes, M.K., Laracuente, A.R., Dai, Z., Marder, S.R., Berger, C., King, W.P., de Heer, W.A., Sheehan, P.E., Riedo, E., 2010. Nanoscale tunable reduction of graphene oxide for graphene electronics. Science 328 (5984), 1373–1376.

Chapter 5.8

Hall Mobility and Field-effect Mobility

Christopher S. Allen and Jamie H. Warner

University of Oxford, Oxford, UK

5.8.1. INTRODUCTION TO THE HALL EFFECT

Any moving charged particle that is subjected to a magnetic field **B** experience a Lorentz force given by

$$F = q(E + v \times B) \tag{5.8.1}$$

where q is the charge on the particle, E the electric field and v the velocity of the particle.

When electrons of current density j_x are passed through a wire in the presence of a transverse magnetic field B_z, the Lorentz force acts on them causing a build-up of charge on one side of the wire (see Fig. 5.8.1) resulting in the presence of an electric field, E_H. The charge builds up very quickly along one side of the wire until the resultant force on the electrons eE_H due to the electric field exactly balances the Lorentz force. The magnitude of E_H depends upon the values of the current density and applied magnetic field and is given by

$$E_H = R_H B \times j \tag{5.8.2}$$

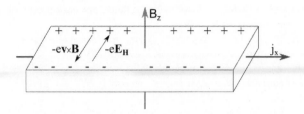

FIGURE 5.8.1 The Hall effect. In the presence of a transverse magnetic field $\boldsymbol{B}_z$, electrons with current density j_x experience a Lorentz force equal to $-e\boldsymbol{v} \times \boldsymbol{B}$. In equilibrium, this is balanced by the Hall electric field E_H.

where R_H is called the Hall coefficient and is defined as

$$R_H = \frac{E_H}{B_z j_x} = \frac{1}{ne} \tag{5.8.3}$$

The Hall coefficient therefore depends only on the density of carriers, n an intrinsic material property.

From the Drude model the conductivity of a material is given by

$$\sigma = \mu n e \tag{5.8.4}$$

where μ is the mobility of the material. Mobility takes the units of $cm^2 V^{-1} s^{-1}$ and is a measure of the speed at which charge carriers move through a material in the presence of an electric field.

5.8.2. MEASUREMENT OF THE HALL MOBILITY ON GRAPHENE SAMPLES

To perform Hall measurements on graphene samples, it is necessary to make electrical contact to the graphene and ideally cut the graphene to a suitable shape. In order to fabricate a device the graphene must first be on a suitable substrate, typically heavily doped silicon with a 300 nm SiO_2 capping layer. The standard procedure to shape graphene is to lithographically create a mask, in which the graphene to remain is covered with a polymer mask. The rest of the graphene is removed by dry O_2 plasma etching. Using standard lithographic techniques, contacts can be defined and then metallised. For simple Hall bar devices evaporated films of Ti(5 nm)/Au(60 nm) are typical. The Ti acts as an adhesion layer for the Au, Cr is also commonly used for this task.

The ideal device geometry for a Hall measurement is the Hall bar (see Fig. 5.8.2), which has two end contacts (marked 1 and 4 in Fig. 5.8.2(a)) and four- or sometimes six-side contacts (marked 2, 3, 5 and 6 in Fig. 5.8.2(a)).

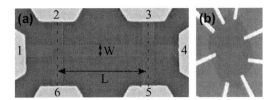

FIGURE 5.8.2 (a) Scanning electron microscope image of a six terminal graphene Hall bar device (the patterned graphene has been highlighted in green. *Reprinted from Novoselov (2005). Copyright (2005) Nature Publishing Group.* (b) Shows an optical image of the geometry for performing a Hall effect measurement on an un-patterned graphene flake. *Reprinted from Tan (2007). Copyright (2007) American Physical Society.*

Substituting $E_H = V_y W$ and $j_x = I_x/Wt$ (where W is the width and t the thickness of the sample) into Eqn (5.8.3) gives an expression for the Hall coefficient in terms of easily measurable parameters:

$$R_H = \frac{1}{ne} = \frac{V_y t}{I_x B_z} \tag{5.8.5}$$

$$V_y = \frac{B_z I_x}{net} \tag{5.8.6}$$

where V_y is the voltage drop perpendicular to the current flow, I_x, i.e. in Fig. 5.8.2(a) the current would flow between contacts 1 and 4 and the voltage drop measured between 2 and 6 (or 3 and 5). In Eqn (5.8.5), t stands for the thickness of the sample. In a two-dimensional system, the sample thickness has no meaning so we drop this term. Two-dimensional or sheet carrier density therefore has units of cm^{-2} (as opposed to carrier density in a bulk material which has units of cm^{-3}). In the two dimensional case, then

$$V_y = \frac{B_z I_x}{ne} \tag{5.8.7}$$

$$\therefore n = \frac{B_z I_x}{V_y e} = \frac{B_z}{R_{xy} e} = \frac{B_z}{\rho_{xy} e} \tag{5.8.8}$$

where $R_{xy} = V_y/I_x$ and $\rho_{xy} = V_y/I_x$ are the transverse or Hall resistance and resistivity respectively and are identical and used interchangeably. There is not a simple inverse relationship between the Hall conductivity σ_{xy} and the Hall resistivity ρ_{xy}. Instead they follow an inverse tensor relation:

$$\sigma_{xy} = \frac{\rho_{xy}}{\rho_{xy}^2 + \rho_{xx}^2} \tag{5.8.9}$$

And correspondingly

$$\sigma_{xx} = \frac{\rho_{xx}}{\rho_{xx}^2 + \rho_{xy}^2}$$ (5.8.10)

where $\rho_{xx} = \left(\frac{W}{L}\right)\left(\frac{V_x}{I_x}\right)$ is the longitudinal resistivity, which at zero magnetic field is simply the sample resistivity. Once the carrier density is known, the Hall mobility μ_H can simply be calculated from the Drude formula (Eqn (5.8.4)) combined with a four-point measurement of the conductivity of the sample σ in zero magnetic field:

$$\sigma^{-1} = \rho = \left(\frac{W}{L}\right)\left(\frac{V_{23}}{I_{14}}\right)$$ (5.8.11)

where V_{23} is the voltage drop measured between contacts 2 and 3 in Fig. 5.8.2(a) and I_{14} the current flow between contacts 1 and 4. W and L are the width and length of the sample as defined in Fig. 5.8.2(a).

For measurements on graphene, which for some reason cannot be patterned into the Hall bar geometry, the van der Pauw technique should be used (van der Pauw, 1958).

To perform a van der Pauw measurement it is necessary to have at least four and preferably eight contacts to your sample. These contacts must be small and on the sample perimeter. The sample must also be continuous and of the uniform thickness. So long as these criteria are fulfilled the van der Pauw technique is valid for measuring both the sheet resistivity and Hall effect on arbitrarily shaped samples.

To calculate the sheet resistivity, ρ the measurements depicted in Fig. 5.8.3(a) and (b) must be performed. Firstly R_A must be determined

$$R_A = \frac{1}{4}\left(\frac{V_{24}}{I_{86}} + \frac{V_{42}}{I_{68}} + \frac{V_{86}}{I_{24}} + \frac{V_{68}}{I_{24}}\right)$$ (5.8.12)

where V_{24} corresponds to the voltage measured between contacts 2 and 4, and I_{86} to the current passed between contacts 8 and 6. In Eqn (6.1) the measurement is performed four times and then averaged, with each measurement switching either the polarity or interchanging the voltage probes with the current source. This is done to improve the accuracy of the measured R_A and eliminate any errors caused by, for example thermal gradients or instrumentation offsets. Similarly R_B is given by:

$$R_B = \frac{1}{4}\left(\frac{V_{28}}{I_{64}} + \frac{V_{82}}{I_{46}} + \frac{V_{64}}{I_{28}} + \frac{V_{46}}{I_{82}}\right)$$ (5.8.13)

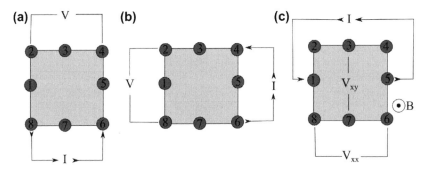

FIGURE 5.8.3   (a) and (b) Example measurement geometries for performing van der Pauw and (c) Hall effect measurements on arbitrarily shaped samples.

The sheet resistivity, ρ can be found by numerically solving the van der Pauw equation:

$$\exp(-\pi R_A/\rho) + \exp(-\pi R_B/\rho) = 1 \qquad (5.8.14)$$

By applying a magnetic field normal to the sample plane and using the measurement setup shown in Fig. 5.8.2(c), V_y and I_x can be measured and the carrier density determined using Eqn (5.8.8). Combined with ρ calculated from Eqn (5.8.14), μ_H can be found using the Drude Eqn (5.8.4).

5.8.3. MEASUREMENT OF THE FIELD EFFECT MOBILITY IN GRAPHENE

An alternate approach to extracting the mobility from measurements of a graphene sample is via the field effect. Graphene samples are, in general, fabricated on heavily doped silicon with a ~ 300 nm oxide cap. A voltage applied to this heavily doped silicon (termed a gate voltage, V_g) induces a surface charge density and correspondingly shifts the position of the Fermi level in the graphene sample. For intrinsic graphene with zero applied gate voltage the Fermi level sits at the Dirac point (see Section 3.1), at which the carrier density is zero. The carrier density induced in the graphene by the application of a gate voltage can be approximated using a simple capacitive model (Novoselov et al., 2005).

$$n = \frac{C_g V_g}{e} \qquad (5.8.15)$$

where $C_g = \varepsilon_0 \varepsilon / t$ is the oxide capacitance (per unit area), with ε_0 the permittivity of free space, ε the relative permittivity of the oxide material and t the oxide thickness. In practice, the Dirac point is often shifted by several volts or more away from $V_g = 0\,V$ due to doping of the graphene sample. To compensate for this we replace V_g in Eqn (5.8.15) with $(V_g - V_D)$, with V_D the gate voltage at the Dirac point. Induced carrier densities estimated using Eqn (5.8.15) show good agreement with those determined from Hall effect measurements (Zhang et al., 2005). A more precise model in which the quantum capacitance of the two-dimensional electrons are included extracts the induced carrier density from the expression: (Kim et al., 2009)

$$V_g - V_D = \frac{en}{C_g} + \frac{\hbar v_F \sqrt{\pi n}}{e} \qquad (5.8.16)$$

with v_F the Fermi velocity of electrons in graphene.

The field effect mobility can be measured from the variation of the sample conductivity with applied gate voltage and is defined as the derivative of the Drude formula:

$$\mu_{FE} = \frac{1}{C_g} \frac{d\sigma}{dV_g} \qquad (5.8.17)$$

For this model values of μ_{FE} vary dramatically with carrier density and, in general, μ_{FE} measured at high carrier densities (away from the Dirac point) is reported.

An alternate approach to obtaining sample mobility from field effect measurements was introduced by Kim et al. (2009) in which a constant mobility, μ_c is extracted from fitting the total resistance through the device, R_{total} as a function of V_g to the expression:

$$R_{total} = 2R_c + \frac{L/W}{\sqrt{\left(n_0^2 + n^2\right)e\mu_c}} \qquad (5.8.18)$$

Using this model, the contact resistance R_c, residual carrier density at minimum resistance, n_0 and μ_c are extracted from the fitting procedure.

For high-carrier concentration (high V_g), μ_H, μ_{FE} and μ_c measured on the same sample have been shown to be in good agreement (Venugopal et al., 2011).

5.8.4. MAXIMISING MOBILITY

The early work on exfoliated monolayer graphene reported values for mobility in the region of $\mu = 3000$–$10{,}000\ cm^2V^{-1}s^{-1}$ with typical carrier densities of $n \approx 5 \times 10^{12}\ cm^{-2}$ (Novoselov et al., 2004). For graphene on a SiO_2 surface, maximum mobility limits of between 40,000 and 70,000 $cm^2V^{-1}s^{-1}$ have been suggested (Chen et al., 2008, 2009), and predictions of mobilities up to 200,000 $cm^2V^{-1}s^{-1}$ have been made for graphene in the absence of any extrinsic disorder (Morozov et al., 2008).

Graphene grown by CVD has struggled to reach the high mobilities of exfoliated graphene. This is at least partly due to the increased amount of processing necessary to fabricate devices from CVD graphene. Until recently, mobilities reaching those of the early experiments on exfoliated graphene had not been reported (Petrone et al., 2012).

The mobility in epitaxial graphene (formed by Si sublimation from the surface of SiC, and the subsequent reconstruction of the remaining carbon) is dependent on which crystal face the graphene is grown. Graphene grown on the C face has reached mobilities of ~5000 $cm^2V^{-1}s^{-1}$ compared to ~1000 $cm^2V^{-1}s^{-1}$ for that grown on the Si face (Kedzierski et al., 2008).

In order to fabricate electronic devices from graphene, it is necessary to perform a number of processing steps. These invariably involve employing lithographic techniques in which a radiation sensitive polymer is spun onto the graphene. This polymer is then selectively exposed to the relevant form of radiation (generally UV light or electrons) and the exposed areas chemically removed. This creation of a polymer mask allows for the selective etching of graphene or the accurate placement of metallic contacts. For graphene devices it has become increasingly clear that the difficulty in completely removing the polymer from the surface of the graphene is a likely candidate for the introduction of disorder and subsequent suppression of mobility (Pirkle et al., 2011).

Other extrinsic mobility limiting factors are thought to be largely due to the substrate. Ripples in the graphene caused by the presence of the substrate, (Morozov et al., 2008) any charge inhomogeneity in the substrate dielectric (Hwang et al., 2007) and interfacial phonons (Chen et al., 2008) are all potential sources of scattering resulting in the suppression of mobility.

There are various steps which can be taken to minimise or completely eliminate these extrinsic scattering sources and increase the mobility of graphene samples. Here, we describe, in some detail a range of the state of the art in graphene device-fabrication techniques.

5.8.4.1. Ensuring Clean Graphene, the Effects of Annealing

Resist residues left on the surface of a graphene sample after processing have been shown to degrade device performance (Dan et al., 2009). Due to strong van der Waals interactions between the polymer resists and graphene, standard chemical removal (using for example acetone) fails to completely remove all resist from the graphene surface (Cheng et al., 2011). One route towards obtaining postprocessing clean graphene is to thermally decompose any residual polymer. This is generally performed either under vacuum (Cheng et al., 2011) or a flow of H_2/Ar (Dean et al., 2010).

Using a combination of AFM and transport measurements, Cheng et al (2011) found that annealing for 3 h under vacuum at temperatures of 300 °C or greater removed almost all resist residue from the graphene surface. However annealing at these high temperatures also caused the graphene to adhere more closely to the underlying Si substrate resulting in a degradation of mobility and increased doping. A peak mobility and minimum doping was found at annealing temperatures of 200 °C even though AFM inspection clearly showed remaining resist on the graphene surface.

An alternate approach to ensuring clean graphene is to Joule heat the sample under an inert atmosphere (Moser et al., 2007). Currents densities of the order of $\sim 10^8$ A/cm^2 are passed through the graphene device, generally within a cryostat under He atmosphere. Under such high current densities the graphene heats up significantly driving off absorbed species and residual resist. This results in improved device properties including dramatic decreases in doping level. Device properties do, however, tend to degrade on exposure to atmosphere.

For CVD graphene the problem of residue degradation of device performance is compounded. CVD graphene must first be transferred from the growth substrate – thus, increasing the amount of processing necessary to fabricate an electronic device. The transfer process typically involves the application of a support scaffold to the graphene (generally a standard resist polymer such as PMMA), chemical etching of the growth substrate followed by numerous rinsing steps in de-ionised water. The graphene is then transferred to the desired substrate and allowed to dry the scaffold material is removed and processing can continue as with exfoliated graphene. These extra processing steps not only expose the graphene to an additional layer of PMMA which must be effectively removed but also result in a layer of water drying between the graphene and the substrate which can cause further degradation of device properties.

The use of isopropanol instead of water in the final transfer step has been shown to improve measured mobilities for CVD graphene, presumably by minimising the amount of water trapped between the graphene and the substrate (Chan et al., 2012). In a study by Chan et al. devices transferred from water were found to have mobilities of $\mu \sim 300$ cm^{2}V^{-1}s^{-1} and those transferred from IPA to have mobilities of $\mu \sim 2000$ cm^{2}V^{-1}s^{-1}. Both samples were then annealed in UHV to 300 °C and remeasured (after a brief exposure to ambient conditions). Post anneal both IPA and water-transferred graphene showed mobilities of $\mu \sim 3000$ cm^{2}V^{-1}s^{-1}. It was concluded that the annealing process helps to remove both interfacial water molecules and PMMA residue.

5.8.4.2. Minimising Substrate Interactions

Interactions between graphene and the substrate are believed to be one of the major sources of intrinsic disorder and as such a number of techniques have emerged to minimise these interactions.

High mobilities have been achieved for exfoliated graphene when the substrate has been coated in hydrophobic hexa-methyldisilazane (HMDS). (Lafkioti et al., 2010). This is thought to prevent the absorption of molecules onto the Si surface. This not only prevents changes to the graphene morphology due to molecules trapped between the graphene and the substrate but also minimises any charge transfer which may dope the graphene sample. Mobilities of ~ 6000 cm^{2}V^{-1}s^{-1} have also been reported for CVD grown graphene utilising similar substrate preparation (Chan et al., 2012).

Hexagonal boron-nitride (h-BN) with its lack of dangling bonds, atomically smooth surface, similar lattice constant to graphene and large electronic band gap have made it an attractive substrate to use for graphene electronic devices.

Dean et al. (2010) produced the first report of graphene devices fabricated on h-BN substrates. Boron nitride flakes were deposited onto a silicon substrate using a micromechanical cleavage method similar to that first developed to isolate monolayer graphene. Graphene was then exfoliated onto a silicon substrate covered with PMMA layer/water soluble laminate (Fig. 5.8.4(a)). The underlayer was then dissolved away to leave the PMMA floating on the water (Fig. 5.8.4(b)), which was then lifted from the water by adhering to a glass slide (Fig. 5.8.4(c)). The glass slide was attached to a micromechanical manipulator and, with the aid of an optical microscope, the graphene placed on top of the previously prepared h-BN. During transfer, the silicon substrate, on which the h-BN had been prepared was heated to 110 °C to minimise trapped water. After transfer was complete the PMMA was dissolved using acetone (Fig. 5.8.4(d)) and electrical contacts patterned using standard electron beam lithography techniques. Before measurement the devices were annealed in a H_2/Ar flow at 340 °C for three and a half hours. Values for Hall mobility of $\mu_H \sim 25,000$ cm^{2}V^{-1}s^{-1} are reported for high carrier concentration (with field effect mobility μ_{FE} of 140,000 cm^{2}V^{-1}s^{-1} at low carrier density).

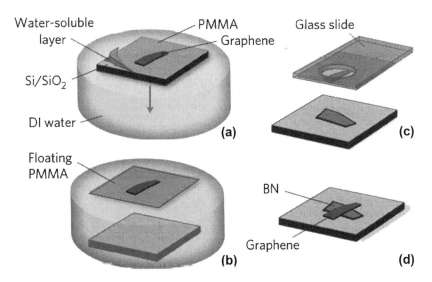

FIGURE 5.8.4   Transfer of exfoliated graphene to h-BN substrates. (a) Graphene is exfoliated onto a Si chip covered with a PMMA/water-soluble laminate. (b) The PMMA is removed from the Si chip by dissolving away the water-soluble layer, and is then lifted from the water bath using a glass slide (c). The graphene is then placed onto a predeposited h-BN crystal and the PMMA scaffold removed. *Reprinted from Dean (2010). Copyright (2010) Nature Publishing Group.*

Petrone et al. (2012) use a modified transfer technique to fabricate high-mobility devices from large grain (> 150 μm) CVD grown graphene. Their process is as follows: PMMA is spun onto the graphene surface as usual. A piece of polyimide tape with a small window cut out is then attached to the other side of the copper foil. The exposed copper foil is then chemically etched and rinsed in IPA, leaving graphene covering the window supported by the surrounding copper. This allows the graphene to be thoroughly dried with nitrogen before transfer to the substrate thus minimising any trapped residues. The graphene is then transferred onto a pre-selected h-BN flake, etched using a PMMA mask and contacts defined using standard electron beam lithography techniques. Prior to measurement devices were annealed in forming gas at 345 °C to remove PMMA residue. This process results in devices from CVD grown graphene with field-effect mobilities in excess of 50, 000 cm^2/V s at carrier densities of $n < 5 \times 10^{11}$ cm^{-2} ($\mu_{FE} \sim 30{,}000$ cm^{2}V^{-1}s $^{-1}$at higher carrier densities).

In order to reach the maximum possible mobility, it is necessary to remove all extrinsic forms of scattering. In order to achieve this the substrate must be completely removes. Fully suspended multi-terminal graphene devices have been successfully fabricated by Bolotin et al. (2008). After mechanical exfoliation, an appropriate graphene flake is selected, and metal contacts are patterned using standard electron beam lithography techniques. The authors do not employ any plasma etching to shape the graphene to avoid introducing additional defects to the graphene. The entire device is then dipped into a buffered oxide etch which removes approximately 150 nm of the SiO$_2$ from the surface of the substrate leaving the graphene suspended. The device is then rinsed in ethanol and dried using a critical-point-drying technique. The completed suspended graphene device is shown in Fig. 5.8.5. Before measurement at 5 K within a cryostat, the

FIGURE 5.8.5 A fully suspended six-terminal graphene device. Etching away, the SiO$_2$ underneath the graphene leaves it suspended approximately 150 nm above the substrate. Due to the lack of extrinsic scattering, mobilities in excess of 200,000 cm^2/V s are reported. *Reprinted from Bolotin 2008. Copyright (2008) Elsevier.*

graphene is annealed to 400 K under vacuum to remove any contaminants. Hall mobilities of $\mu_H = 200{,}000$ cm^{2}V^{-1}s^{-1} were recorded for carrier densities of $n = 2 \times 10^{11}$ cm^{-2} (Bolotin et al., 2008).

5.8.5. SUMMARY

Being able to accurately determine mobility is critical for the development of graphene as a viable material for the microelectronics industry. In the literature details of how mobility is extracted from experimental data are not always given. It is therefore not always possible to perform direct comparisons between samples (Schwierz, 2010).

Measurement of the Hall effect allows for a direct determination of the sample carrier density. This combined with a four probe measurement of the sample resistivity gives a value for mobility which is not dependant on any estimation of sample capacitance or contact resistance. It does however require specific sample design and the ability to apply a magnetic field.

Field effect mobility measurements impose less restrictions on sample geometry and do not require a magnetic field. The resultant mobility can, however be carrier density dependant and the inclusion of fitting procedure which involves many free parameters can make direct comparison between samples difficult.

Ideally μ_H, μ_{FE} (at both low and high n) and μ_C should be quoted along with details of how they were extracted from experimental data. This allows easy and transparent comparison between different works.

5.8 REFERENCES

Bolotin, K.I., Sikes, K.J., Jiang, Z., Klima, M., Fudenberg, G., Hone, J., et al., 2008. Ultrahigh electron mobility in suspended graphene. Solid State Commun. 146, 351–355.

Chan, J., Venugopal, A., Pirkle, A., Mcdonnell, S., Hinojos, D., Magnuson, C.W., et al., 2012. Reducing extrinsic performance- limiting factors in graphene grown by chemical vapor deposition. ACS Nano, 3224–3229.

Chen, J.-H., Jang, C., Xiao, S., Ishigami, M., Fuhrer, M.S., 2008. Intrinsic and extrinsic performance limits of graphene devices on SiO$_2$. Nat. Nanotechnol. 3, 206–209.

Chen, F., Xia, J., Ferry, D.K., Tao, N., 2009. Dielectric screening enhanced performance in graphene FET. Nano Lett. 9, 2571–2574.

Cheng, Z., Zhou, Q., Wang, C., Li, Q., Wang, C., Fang, Y., 2011. Toward intrinsic graphene surfaces: a systematic study on thermal annealing and wet-chemical treatment of SiO$_2$-supported graphene devices. Nano Lett. 11, 767–771.

Dan, Y., Lu, Y., Kybert, N.J., Luo, Z., Johnson, a T.C., 2009. Intrinsic response of graphene vapor sensors. Nano Lett. 9, 1472–1475.

Dean, C.R., Young, a F., Meric, I., Lee, C., Wang, L., Sorgenfrei, S., et al., 2010. Boron nitride substrates for high-quality graphene electronics. Nat. Nanotechnol. 5, 722–726.

Hwang, E., Adam, S., Sarma, S., 2007. Carrier transport in two-dimensional graphene layers. Phys. Rev. Lett. 98, 2–5.

Kedzierski, J., Hsu, P.-L., Healey, P., Wyatt, P.W., Keast, C.L., Sprinkle, M., et al., 2008. Epitaxial graphene transistors on SiC substrates. IEEE Trans. Electron Devices 55, 2078–2085.

Kim, S., Nah, J., Jo, I., Shahrjerdi, D., Colombo, L., Yao, Z., et al., 2009. Realization of a high mobility dual-gated graphene field-effect transistor with Al(2)O(3) dielectric RID B-4944-2011. Appl. Phys. Lett. 94, 62107.

Lafkioti, M., Krauss, B., Lohmann, T., Zschieschang, U., Klauk, H., Klitzing, K.V., et al., 2010. Graphene on a hydrophobic substrate: doping reduction and hysteresis suppression under ambient conditions. Nano Lett. 10, 1149–1153.

Morozov, S., Novoselov, K., Katsnelson, M., Schedin, F., Elias, D., Jaszczak, J., et al., 2008. Giant intrinsic carrier mobilities in graphene and its bilayer. Phys. Rev. Lett. 100, 11–14.

Moser, J., Barreiro, A., Bachtold, A., 2007. Current-induced cleaning of graphene. Appl. Phys. Lett. 91, 163513.

Novoselov, K.S., Geim, a K., Morozov, S.V., Jiang, D., Zhang, Y., Dubonos, S.V., et al., 2004. Electric field effect in atomically thin carbon films. Science (New York, N.Y.) 306, 666–669.

Novoselov, K.S., Geim, a K., Morozov, S.V., Jiang, D., Katsnelson, M.I., Grigorieva, I.V., et al., 2005. Two-dimensional gas of massless Dirac fermions in graphene. Nature 438, 197–200.

Novoselov, K.S., Jiang, Z., Zhang, Y., Morozov, S.V., Stormer, H.L., Zeitler, U., et al., 2007. Room-temperature quantum hall effect in graphene. Science 315, 2007.

Petrone, N., Dean, C.R., Meric, I., van der Zande, A.M., Huang, P.Y., Wang, L., et al., 2012. Chemical vapor deposition-derived graphene with electrical performance of exfoliated graphene. Nano Lett. 12, 2751–2756.

Pirkle, a., Chan, J., Venugopal, a., Hinojos, D, WMagnuson, C., McDonnell, S., et al., 2011. The effect of chemical residues on the physical and electrical properties of chemical vapor deposited graphene transferred to SiO_2. Appl. Phys. Lett. 99, 122108.

Schwierz, F., 2010. Graphene transistors. Nat. Nanotechnol. 5, 487–496.

Tan, Y.-W., Zhang, Y., Bolotin, K., Zhao, Y., Adam, S., Hwang, E.H., et al., 2007. Measurement of scattering rate and minimum conductivity in graphene. Phys. Rev. Lett. 99, 10–13.

van der Pauw, L.J., 1958. A method of measuring the resistivity and hall coefficient on lamellar of arbitrary shape. Phillips Tech. Rev.

Venugopal, A., Chan, J., Li, X., Magnuson, C.W., Kirk, W.P., Colombo, L., et al., 2011. Effective mobility of single-layer graphene transistors as a function of channel dimensions. J. Appl. Phys. 109, 104511.

Zhang, Y.B., Tan, Y.W., Stormer, H.L., Kim, P., 2005. Experimental observation of the quantum Hall effect and Berry's phase in graphene. Nature 438, 201–204.

Applications of Graphene

Electronic Devices

Alexander W. Robertson and Jamie H. Warner

University of Oxford, Oxford, UK

6.1.1. INTRODUCTION

This chapter will discuss some of the promising avenues available for the incorporation of graphene into electronic devices and the current state of the art in the field. In particular, the chapter will outline how graphene may be utilised in various future transistor and circuit designs and how its unique material properties lend itself to them. We shall also look at how graphene's sensitivity to doping may be used in gas-sensor applications and how the manifestation of the quantum Hall effect (QHE) in graphene can allow the definition of unit of electrical resistance measure, the Ohm, for use in measurements and standards.

6.1.2. METAL-OXIDE-SEMICONDUCTOR FIELD EFFECT TRANSISTORS (MOSFETs)

The bedrock of modern computer hardware technology is the concept that the semiconductor industry works to double the number of transistors per integrated circuit every two years; colloquially known as "Moore's Law" (Fig. 6.1.1) (Moore, 1965; Reprint available, 2006). Complementary metal-oxide-semiconductor (CMOS) is the incumbent technology used for integrated circuits and currently relies on using silicon as the semiconducting channel layer in the metal-oxide-semiconductor field-effect transistors (MOSFETs) that comprise the logic gates of the circuit. These transistors have been shrunk continuously over the past several decades to the stage, where the physical limit of the silicon-based devices is inexorably approaching.

Graphene. http://dx.doi.org/10.1016/B978-0-12-394593-8.00006-0

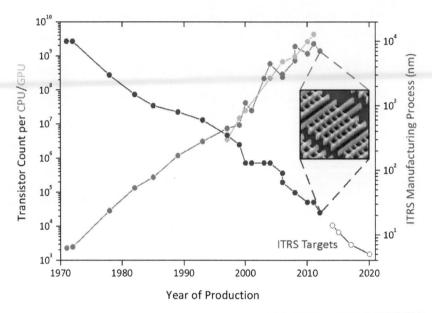

FIGURE 6.1.1 The exponential increase in transistor count per central, and graphics, processor unit (CPU/GPU) with time. On the right (blue) is shown the international technology roadmap for semiconductors (ITRS), a designated name for the characteristic scale of the manufacturing process, typically referring to the gate length. Hollow blue circles denote the future ITRS targets. The inset shows an SEM image of Intel's 2011 IvyBridge 22-nm process tri-gate transistors.

The operation of a MOSFET is illustrated briefly in Fig. 6.1.2. Broadly, a voltage between the source and drain electrodes, which are separated by a region of opposite charge, acts as the supply voltage to drive current flow through the device. A voltage across the gate and body electrodes performs the switching, by causing the inversion of the charge in the intermediate layer between the source and drain, to allow current flow. There are three main operational modes:

(1) **Subthreshold**. In insert (i), a schematic cross section of a MOSFET is shown, with the gate voltage (V_{GS} – the body terminal is typically linked to the source) set below the threshold voltage (V_{th}) required to switch the MOSFET to the ON state. Under these conditions, negligible carrier flow, limited to only those that are sufficiently thermally excited, can occur between the source and drain electrodes. This is due to a depletion region (an area where the charge carriers have diffused elsewhere or been pushed back by an electric field) which forms around, and between, the source and drain contacts.

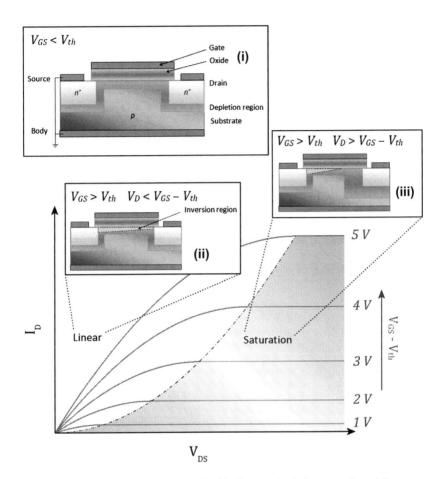

FIGURE 6.1.2 A plot illustrating the relationships between the drain current, I_D, and the source-drain voltage, V_{DS}, and the relative strength of the gate voltage, V_{GS}, to the threshold voltage, V_{th}, of the device. The inserts (i–iii) show the three principal operating modes of an n-channel MOSFET, shown in cross section, with (i) the subthreshold mode, (ii) linear mode and (iii) saturation mode.

(2) **Linear**. Increasing V_{GS} above the threshold voltage leads to the formation of a conducting channel region, also known as an inversion layer, as illustrated in panel (ii). A current, I_D, can now flow between the drain and source, which increases according to Eqn (6.1.1) (μ_n, electron mobility, C_{ox}, gate oxide capacitance, W, channel width, L, channel length, V_D, source drain voltage).

$$I_D \approx \mu_n C_{ox} \frac{W}{L} (V_{GS} - V_{th}) V_D \qquad (6.1.1)$$

(3) **Saturation**. A concomitant increase in the drain voltage V_D, beyond that of V_G, leads the channel region to be tapered towards the source and the

reemergence of the depletion region near the drain (panel (iii)). This is known as pinch-off and limits the maximum drain current that can be achieved, as described in Eqn (6.1.2).

$$I_{\text{D,Sat}} \approx \mu_\text{n} C_\text{ox} \frac{W}{2L} (V_\text{GS} - V_\text{th})^2 \qquad (6.1.2)$$

An expression often used to characterise the performance of a MOSFET is the transconductance, which quantifies the strength of the gating effect on the channel and the formation of the inversion layer.

$$g_\text{m} = \frac{\Delta I_\text{out}}{\Delta V_\text{in}} = \left. \frac{\delta I_\text{D}}{\delta V_\text{G}} \right|_{V_\text{D}=\text{constant}} \qquad (6.1.3)$$

6.1.3. THE GRAPHENE MOSFET

The promise of graphene as a straight semiconductor replacement in CMOS chips – one that could beat the silicon limit – was first muted in the same paper that announced the initial electronic characterisation of isolated monolayer graphene, written by Novoselov et al. (2004), and has remained an often-cited potential application of graphene in both the scientific literature and the wider popular press. It is not difficult to appreciate why graphene research builds on the expertise developed in the study of fullerene film transistors (Haddon et al., 1995) and high performance single-carbon nanotube transistors (Tans et al., 1998). The ultrahigh carrier mobility and monatomic thickness of graphene making it an attractive CMOS candidate, as both these properties are two of the key fundamental limits of silicon (The International Technology, 2011). Furthermore, a critical advantage that graphene possesses over other alternative channel materials, such as carbon nanotubes, is the capability to process in a planar form.

With the continued miniaturisation of the silicon-channel MOSFETs, the reduction to the channel lengths eventually falls below that of the depletion region widths around the source and drain junctions, leading to adverse 'short channel effects' (Packan, 1999; Wong, 2002). These include both reductions to the gate-oxide thickness and drain-induced barrier lowering. The atomically thin-gate oxides used allow for quantum tunnelling of electrons across the insulator into the gate, increasing the power consumption. Drain-induced barrier lowering is where the threshold voltage is decreased due to the disproportionate effect of the drain voltage over the gate voltage on the source-channel flow of charge, effectively causing the device to allow current flow under the subthreshold (OFF) condition.

Of particular interest with regard to graphene are two short-channel effects that act to impede the charge mobility: namely, the field dependence of charge carriers in the inversion layer, leading to carrier scattering at the dielectric–semiconductor interface (Ohmi et al., 1991), and also carrier velocity saturation, where the high electric fields limit the drain current before the pinch-off (Lundstrom, 1997). Both of these can be partly mitigated by using a channel made from materials exhibiting higher mobility (Lundstrom, 2001). This would also result in improved device transconductance (Eqn (6.1.3)), as increasing the mobility leads to a concomitant increase in drain current (approximately expressed earlier in Eqns (6.1.1) and (6.1.2)).

The relativistic charge carriers of graphene have been found to exhibit exceptionally high mobilities in characterised devices, with reports of ultrahigh electron mobilities of 200,000 $cm^2V^{-1}s^{-1}$ recorded in suspended devices at low temperature (Bolotin et al., 2008a; Du et al., 2008; Morozov et al., 2008). Carbon nanotubes, another high-mobility carbon nanomaterial, have already attracted a large body of research (Bachtold et al., 2001; Javey et al., 2003; Misewich et al., 2003). As such, there has been a great deal of interest in attempting to harness the high mobilities of graphene in a MOSFET-type device, with a number of key milestones and developments discussed in the below sections.

Two other attractive features of graphene are the single-atomic thickness and the planar nature of the graphene sheet. The thickness of a semiconductor is a crucial consideration in CMOS scaling, as the depletion width (the depth into the channel that charge is depleted by the action of the gating field) must also scale down with channel length (Brews et al., 1980). The threshold voltage of the device is affected only by the channel length and drain potential in a manner that is strongly dependent on the depletion width (Taur, 2002), hence, minimising the channel thickness should assist in optimising the device transconductance (Schwierz, 2010). The planar nature of graphene – its structure as a two-dimensional sheet – allows for the transfer of many standard techniques to graphene from existing CMOS fabrication processes, such as lithographic patterning, etching and layered structures from deposition.

However, in spite of the advantages outlined above, there are numerous obstacles to the implementation of graphene in this capacity. These, and the progress towards overcoming them, shall now be discussed.

The ITRS outlined in their 2011, 'emerging research materials,' document the following critical issues to the incorporation of graphene in FETs (International Technology Roadmap, 2011):

(1) Graphene deposition over large areas with controlled grain size, thickness and crystallographic orientation. This is discussed in Chapter 4.
(2) A method for manifesting a band gap in graphene.
(3) Control surface and interface effects on carrier mobility and scattering.
(4) Achieve a high mobility on a silicon compatible substrate.
(5) Deposition of a high-k dielectric with a high-quality passivated interface.

(6) Form reproducible low-resistance contacts to graphene, without damage to the graphene layer.

(7) Integration, doping and compatibility with CMOS.

A number of these challenges also apply to the wider field of graphene research, not just with regard to incorporation of graphene into CMOS, and whilst they are significant, a great deal of progress has been made in a relatively short time.

6.1.4. OPENING A BAND GAP

The lack of an intrinsic band gap in graphene is a fundamental obstacle to the use of graphene as a channel layer in a MOSFET. The graphene band structure leads to both low on–off ratios and the inability to switch the device to an off state of minimal drain current. For CMOS-type logic computing, opening a band gap in graphene is essential. Several methods for accomplishing this have been put forward, such as straining the graphene, applying a perpendicular electric field to a graphene bilayer and confining graphene into pseudo-one-dimensional nanoribbons.

6.1.5. STRAIN ENGINEERING A BAND GAP

Experimental work done on both metallic and semiconducting carbon nano-tubes demonstrated the possibility of altering the band structure, with strain applied to the semiconducting tube resulting in band gap changes of ± 100 meV per 1% stretch (Minot et al., 2003; Tombler et al., 2000). It has been shown by a Raman spectroscopy study that it is possible to alter the electronic structure of graphene, as measured by a shift of the G and 2D peaks (Ni et al., 2008), and further modelling (Pereira et al., 2009) and theoretical (Cocco et al., 2010) studies have suggested that a bang gap of up to 0.9 eV could be engineered.

However, the complete picture is uncertain with different strain along different axes, not yielding equivalent results and the effect on monolayer graphene differing from the few-layer case. Moreover, no experimental device measurement has yet demonstrated the existence of a band gap in strained graphene.

6.1.6. FIELD INDUCED BAND GAP IN BILAYER GRAPHENE

Bilayer graphene, like its monolayer, does not exhibit a band gap. However, it has been found that the system may manifest a separation between the valence and conduction bands when under heavy doping, such as that seen in an 'angle-resolved photo-emission spectroscopy' (ARPES) study (Ohto et al., 2006). A similar effect can be observed by the application of an electric field perpendicular to the bilayer, allowing for the specific control of the carrier concentration. This has been demonstrated experimentally and, in conjunction with a rigorous theoretical description (Gava et al., 2009; McCann, 2006), was found to allow for controllably adjustable band gaps of as much as 0.2 eV (Castro et al., 2007). Further, more recent experiments have confirmed this effect (Oostinga et al., 2007), with the most recent thorough experimental study using a dual-gate sandwich structure to control the band gap and the FET switching independently, achieving a band gap of 0.25 eV (Zhang et al., 2009).

6.1.7. GRAPHENE NANORIBBONS

Of the methods discussed here, the constricting of graphene into a pseudo one-dimensional structure, known as a nanoribbon, has attracted the most intensive research effort. A great deal of theoretical work, some dating back to before the formal discovery of freestanding graphene, demonstrated the influence nanoribbon edge structure, termination and width had on the band structure (Areshkin et al., 2007; Brey and Fertig, 2006; Ezawa, 2006; Miyamoto et al., 1999; Nakada et al., 1996; Peres et al., 2006; Son et al., 2006; Wakabayashi et al., 1999). The first experimental results of a graphene nanoribbon (GNR) were from ribbons patterned by electron beam lithography, with GNR width variation demonstrating an appreciable effect on the band gap (Chen et al., 2007; Han et al., 2007). The measured energy gaps were found to be inversely proportional to the GNR width (Han et al., 2010; Kim et al., 2009). However, lithographic and graphene etching definition is limited in resolution, with lithographic patterning in the sub-10 nm regime challenging (Avouris et al., 2007). The etching process also prevents the creation of GNR with the smooth, pristine zigzag edges that are desired.

In an attempt to address these issues, solution processing was employed to break the exfoliated graphite sheets down into a finer size, with centrifugation used to siphon away the unwanted larger debris (Li et al., 2008). When deposited on a silicon substrate, among the detritus of the exfoliated solution were found to be GNRs of sub-10 nm in width and also exhibited relatively smooth edge profiles. By employing molecular precursors, the chemical patterning of GNRs was accomplished, yielding GNR of a variety of shapes and with repeatable uniformity (Cai et al., 2010). A top-down solution to the same problems was to utilise scanning tunnelling microscopy (STM) to examine and modify lithographically patterned GNR, resulting in the engineering of structures with nanoscale precision and band gaps of up to 0.5 eV (Tapasztó et al., 2008). Alternatively, unzipping of carbon nanotubes by plasma etching (Jiao et al., 2009), or ultrasonication (Jiao et al., 2010), can lead to the formation of nanoribbons with suitably narrow widths.

A closely related structure to the GNR is the graphene nanomesh: a periodic array of holes in a graphene sheet, with the necks between adjacent holes narrowing to 5 nm, which provides the confinement to open a band gap (Bai et al., 2010). A copolymer block with an array of patterned cylindrical holes is used as a mask for the oxygen plasma etch, which forms the holes in the graphene. The nanomesh, when used as a channel layer in a transistor, provides broadly comparable performance to GNRs.

6.1.8. FURTHER TECHNIQUES

Hydrogenation of graphene, graphane, has been predicted to yield a semi-conductor (Sofo et al., 2007). By hydrogen doping of graphene in a regular patchwork lattice, it is possible to confine the graphene charge carriers in the remaining pristine regions, generating a band gap of 0.45 eV, as measured by ARPES (Balog et al., 2010). A substrate–graphene interaction has been shown to lead band gap formation in epitaxial graphene on silicon carbide, with ARPES demonstrating a band gap of 0.26 eV (Zhou et al., 2007). However, electron doping prevents the graphene from exhibiting semiconducting behaviour. By incorporation of boron and nitrogen into the graphitic lattice, it is possible to make hybridised boron–nitrogen–carbon films (h-BNC) by a chemical vapour deposition (CVD) method. The patchwork nature of the boron–nitrogen and graphene domains leads to quantum confinement and the opening of a small bandgap (Ci et al., 2010).

6.1.9. THE OPTIMISATION OF MOBILITY

The carriers in graphene are massless, relativistic Dirac fermions, and thus graphene exhibits exceptionally high mobilities. The highest mobilities are reported in samples where the graphene is suspended, as charged impurities and other scattering centres from substrate interactions are minimal, and where the current annealing has removed contamination. These mobilities are of the order of 10^5 $cm^2V^{-1}s^{-1}$ at temperatures up to 240 K (Bolotin et al., 2008a,b; Du et al., 2008), with near-ballistic transport at 20 K and mobilities as high as 10^6 $cm^2V^{-1}s^{-1}$ reported in strained samples (Castro et al., 2010). In the afore-mentioned works on suspended graphene, evidence suggests that flexural phonons are the limiting factor for mobility at room temperature, which require straining to suppress.

For useful electronic devices, it is more important to understand the behaviour of graphene at room temperature, also taking into account the effect of the substrate and the gate dielectric. In these devices, mobilities are typi-cally within the range of $1–2 \times 10^4$ $cm^2V^{-1}s^{-1}$ (Geim and Novoselov, 2007). Within the graphene itself, the effect of charged impurities and disorder on scattering becomes important at room temperature (Tan et al., 2007), although, if these sources of disorder are removed, intrinsic room-temperature mobilities of 2×10^5 $cm^2V^{-1}s^{-1}$ could be achievable for particular carrier concentrations, due to the weak effect of electron–phonon scattering (Morozov et al., 2008). The substrate can be expected to contribute further extrinsic scattering sources, in the form of substrate roughness (Ishigami et al., 2007; Katsnelson and Geim, 2007) and impurities (Chen et al., 2008; Das Sarma et al., 2011; Hwang et al., 2007), ultimately leading to an inhomogeneous distribution of charge localisation throughout the sample. These Dirac point charge localisations are often referred to as electron and hole puddles (Martin et al., 2008). Even after minimising these extrinsic sources of disorder, however, the effect of polar optical phonons of the most frequently used graphene substrate, the dielectric silicon dioxide, can severely limit the carrier mobility (Fratini and Guinea, 2008). This is known as remote polar phonon scattering (Hess, 1979), where the SiO_2 polar optical phonons couple with the graphene charge carriers and would also be expected to occur with any additional polar top-gate dielectric. Selection of appropriate substrates and gate dielectrics should minimise this effect (Berger et al., 2006). Hexagonal boron nitride (h-BN) may prove to be such an ideal insulating substrate, due to its atomically smooth surface, large optical phonon modes and a lattice constant comparable to that of graphene. By employing h-BN, graphene devices with mobilities of 6×10^4 $cm^2V^{-1}s^{-1}$ have been achieved (Dean et al., 2010), some three times greater than with equivalent silicon dioxide substrate samples.

Fabrication of useful FET devices will also require the deposition of a top-gate dielectric, allowing for both control of the charge concentration and the gating field, however, will also lead to further mobility impeding effects, similar to those discussed already for the substrate. Encapsulation of the graphene in a h-BN sandwich structure, thus providing both a gate and a body electrode, as well as isolating the graphene from unwanted environmental doping, leads to mobilities in the 1×10^5 $cm^2V^{-1}s^{-1}$ range (Mayorov et al., 2011). On conventional SiO_2-based substrates, an aluminium-oxide top-gate dielectric initially yielded mobilities of 8×10^3 $cm^2V^{-1}s^{-1}$, comparable to the low end of the range reported for nontop-gated devices (Kim et al., 2009). More recently, mobilities of 2.3×10^4 $cm^2V^{-1}s^{-1}$ were achieved (Liao et al., 2010b), with negligible measured changes in device mobilities before and after the gate deposition now possible (Cheng et al., 2012), although this is likely due to the mobility limiting factors being elsewhere in the fabricated devices.

6.1.10. DEPOSITION OF A HIGH-κ GATE DIELECTRIC AND LOW-RESISTANCE METAL CONTACTS

As discussed in Section 6.1.3, the deposition of a uniform high-κ gate dielectric on top of the graphene is essential for providing improved control of the channel carrier concentration. Atomic layer deposition (ALD) is the technique that is utilised for the growth of dielectric layers; however, due to the nonreactive nature of graphene, the ALD of Al_2O_3 leads to the formation of broken layers along bands of defective graphene (Xuan et al., 2008). Furthermore, the hydrophobic graphene surface prevents standard water-based ALD methods

(Lee et al., 2008). Deposition of a 1-nm layer of aluminium by standard metal evaporation, followed by its oxidation in the air, can act as a sufficient nucleation layer for the uniform ALD growth of Al_2O_3 (Kim et al., 2009), and similarly, a polymer buffer layer may also act as a seed (Farmer et al., 2009). Oxidation of evaporated aluminium or hafnium metals allows for dielectric formation without the need for a nucleation layer (Pirkle et al., 2009). Another method is to dry-transfer a prefabricated Al_2O_3 dielectric layer to the device (Cheng et al., 2012; Liao et al., 2010b), and one novel procedure utilised self-aligning $Co_2Si–Al_2O_3$ nanowires as a gate (Liao et al., 2010c).

The deposition of metal contacts can lead to induced doping in the low density of states graphene, which can extend along the graphene channel (Blake et al., 2009; Golizadeh-Mojarad and Datta, 2009; Nagashio et al., 2009), and is found to severely impair FET device characteristics (Parrish and Akinwande, 2011; Russo et al., 2010; Venugopal et al., 2010). Systematic studies on the behaviour of graphene–metal junctions have been performed (Song et al., 2012; Xia et al., 2011), which suggest that palladium or gold contacts provide the best contact resistance, with palladium marginally preferable due to superior wetting to the graphene surface. More research is needed in this area, as contact resistances need to be lowered by at least an order of magnitude to be viable in miniaturised devices (Nagashio and Toriumi, 2011; Schwierz, 2010).

6.1.11. THE VIABILITY OF GRAPHENE IN CMOS

Whilst considerable progress has been made in surmounting the shortcomings of graphene as a suitable channel material, there are still crucial issues to overcome. Arguably, the most critical problem is the lack of a band gap, with the bandgap generation methods, remaining commercially unviable or unproven. The reported band gaps in graphene have been limited to the narrow regime, short of the 1.14 eV of silicon, and more akin to the other high mobility, narrow gap semiconductors, such as indium arsenide at 0.35 eV. This would prohibit the naive direct replacement of silicon with graphene in the current inversion-based MOSFET–CMOS architecture, as the device cannot be efficiently turned off (Passlack, 2006). Issues with performance limits from scattering caused by silicon dioxide substrates have been addressed by employing h-BN, although it remains to be seen if h-BN is sufficiently compatible, and scalable, for use in CMOS.

To be a competitive consideration for the replacement of silicon, researchers will need to have demonstrated solutions for all of these issues by around 2014 (International Technology, 2011). In spite of the rapid progress, it seems unlikely that this will be achieved, and as such, graphene transistor research has increasingly shifted focus towards both radio-frequency analogue devices, and novel architectures that are specifically designed to harness graphene's unique properties.

6.1.12. RADIO-FREQUENCY (RF) ELECTRONICS

The low on–off ratios of graphene channel transistors, stemming from their lack of a band gap, would not prove to be such a fatal shortcoming in radio-frequency or analogue applications, such as amplifiers and transmitters. These devices benefit from graphene's high transconductance and carrier mobility, as the transconductance is the crucial parameter when considering an amplifiers performance. A further important parameter in the RF field is the frequency cut-off; the frequency at which the current gain for the device is reduced to one, and is thus the limiting frequency at which a signal will be carried through. Higher maximum operating frequencies (>100 GHz) would permit the use of graphene electronics in radar applications. Current commercial technology for wireless communications in the 60-GHz regime already exists; however, high-frequency graphene electronics (122 GHz and 240 GHz) could allow for the commercial exploitation of these bands (Pasanen et al., 2012), as demand for the spectrum bandwidth ever increases (Economic Benefits, 2012). Figure 6.1.3 shows

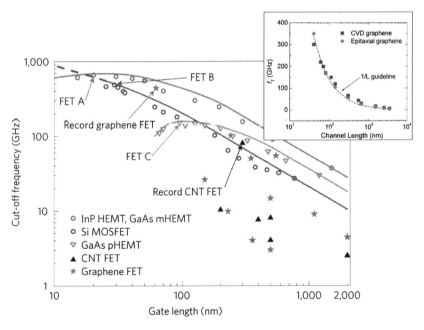

FIGURE 6.1.3 Cut-off frequencies of different FET materials against gate length. FET A refers to InP HEMT (High Electron Mobility Transistors) and GaAs metamorphic HEMT, FET B to silicon MOSFETs and FET C to GaAs pseudomorphic HEMT. Figure reproduced (and updated) from (Schwierz (2010). *Copyright (2010) Nature Publishing Group.* The record graphene transistor performance is 427 GHz at a gate length of 67 nm (Cheng et al., 2012). The insert illustrates the reciprocal relationship between channel length and the cut-off frequency (f_T). *Reprinted from Wu et al. (2012). Copyright (2012) American Chemical Society.*

a comparison of graphene high-frequency FETs to several competing technologies (Schwierz, 2010), with promising progress having been achieved in a short period; currently devices exhibit cut-off frequencies as high as 427 GHz (Cheng et al., 2012), with performances into the terahertz regime, thought possible in ideal cases (Liao et al., 2010a).

There are several recent papers which display the current state of the art in graphene radio-frequency devices. The work by IBM researchers has consistently been leading the field, with successive papers pushing device performance forward, with cut-off frequencies of 26 GHz in 2009 (Lin et al., 2009), 100 GHz in 2010 (Lin et al., 2010), 155 GHz with CVD-grown graphene in 2011 (Wu et al., 2011), and 300 GHz with both CVD and SiC epitaxial graphene in 2012 (Wu et al., 2012). Recent work from Duan et al. has demonstrated innovative ways to overcome the problem of deposition of a quality-gate dielectric (as discussed in Section 0) in order to achieve superior devices (Cheng et al., 2012; Liao et al., 2010c). Progress has also been made towards graphene-based integrated circuits, with a broadband radio-frequency mixer fabricated on silicon carbide that operates at 10 GHz (Lin et al., 2011), which shows great promise towards future integration of graphene into existing technologies.

The current limiting factors, with regard to device performance and commercial exploitation, are principally the same as those outlined in Section 0; such as graphene synthesis, dielectric deposition and mobility debilitating sources of scattering. Further problems unique to RF-FETs are the comparatively poor maximum oscillation-frequency performance, a measure of the maximum power transmission, stemming from the weak saturation behaviour (Kim et al., 2011; Palacios et al., 2010; Schwierz, 2010).

6.1.13. NOVEL FIELD EFFECT TRANSISTOR DESIGNS

The problems arising from the attempts to develop logic MOSFET-type transistors with graphene encourage the querying of whether the conventional device architectures are the best way to utilise graphene's properties. Three novel transistor designs have been put forward – the tunnel FET, the triode barristor and the bilayer pseudospin FET (BiSFET) – that exploit these properties.

It is possible to fabricate a high on/off ratio device by harnessing the quantum mechanical tunnelling effect across a potential barrier. A gating field is applied across a layered device, with two layers separated by a thin dielectric, and is used to tune the tunnelling rate, with the gate voltage in the ON state permitting tunnelling between the source and the channel and inhibited in the OFF state (Boucart and Ionescu, 2007; Fiori, 2009). The experimental realisation of such a device, using bilayer graphene intercalated with an atomically thin dielectric, manifesting switching rations of up to 10^4, demonstrates the strong potential of the concept (Britnell et al., 2012).

The graphene barristor relies on the formation of a Schottky barrier between doped/biased metallic graphene and a semiconductor, such as silicon

(Yang et al., 2012). Critically, the graphene does not interact with the hydrogen-terminated silicon employed in the barristor, and as such, few interface states are formed. The gating field from the top gate can be used to mediate the Schottky barrier height and thus control the drain current. Depending on the source-drain bias regime used, the barristor can work as a regular transistor, displaying current saturation behaviour and an on/off ratio of approximately 300 in the reverse bias, or the bias can be flipped to forward, which whilst not showing any current saturation behaviour, still exhibits a low-drain current leakage in the off-state and a high on/off ratio of 10^5.

The BiSFET employs two semiconducting sheets, separated by a thin dielectric layer, that can couple across the layer and form excitons. These excitons can condense into a Bose–Einstein condensate, causing the coherence of states across both layers and thus acting as one unified layer, ignoring the intermediate insulator. The term 'pseudospin' arises as a way of addressing which layer the carrier inhabits, so that top and bottom are analogous to spin up and down (Su and MacDonald, 2008). It may be possible to observe this effect at room temperature in a graphene BiSFET system, with one p-type and one n-type graphene layer, where the charge concentrations in each layer are roughly equal (Banerjee et al., 2009). The switching effect would be caused by controlling this charge density, likely with a gating field, with imbalances in the relative charge densities between the two layers leading to decoherence, and thus lower current flow. Such a device would exhibit different digital logic operations to a MOSFET; however, conventional logic functions can be imitated. Rough simulations suggest that such a graphene BiSFET device in an inverter circuit would consume 8×10^{-21} J per clock cycle, 2–3 orders of magnitude less than current silicon CMOS technology (Banerjee et al., 2010). As of yet, though, no graphene-based BiSFET has been experimentally tested.

6.1.14. GAS SENSORS

The two-dimensional and readily dopable nature of graphene makes it a promising potential candidate for the use in sensor applications (Hill et al., 2011). Interactions of various gas molecules with the graphene slightly modify its electronic properties by doping, which can subsequently be measured (Wehling et al., 2008). It has been shown that this technique can be used to detect individual gas molecule binding events with the graphene. This is possible due to the low-electronic noise found in graphene, the quality of the graphene crystal used and the ability to employ fourprobe measurement geometries (Fchedin et al., 2007). A scalable gas sensor fabricated from reduced graphite oxide has been demonstrated (Fowler et al., 2009). Surface acoustic wave (SAW) sensors from graphene oxide/graphene nanosheets have been employed to detect carbon monoxide and hydrogen gas, responding both to the doping and also to mass, with the sorbed molecules perturbing the acoustic wave (Arsat et al., 2009).

6.1.15. METROLOGY AND THE DEFINITION OF THE OHM

The quantum Hall effect (QHE) is a quantisation of resistance, exhibited by two-dimensional electronic systems, that is defined by the electron charge e and Planck's constant h. In metrology, the field of standards and defining of SI units, the QHE seen in the 2D electron gas (2DEG) formed in semiconductor GaAs/AlGaAs heterojunctions has been used to define the 'ohm'. Graphene also exhibits its own variety of the QHE, and as such, it has attracted interest as a potential calibration standard – one that can leverage the potential low cost of QHE-graphene devices to be widely disseminated beyond just the few international centres for measurement and unit calibration (European Association of National Metrology Institutes, 2012). The employment of graphene in the QHE metrology is particularly prescient, with SI units for mass and current to in future also be defined by h and e (Mills et al., 2011).

Epitaxially grown graphene on silicon carbide has been used to fabricate Hall devices that reported Hall resistances accurate to a few parts per billion at 300 mK, comparable to the best incumbent Si and GaAs heterostructure semiconductor devices (Tzalenchuk et al., 2010, 2011). The maturity of graphene as a QHE standard has allowed for the fine comparison of the quantisation behaviour with that of GaAs heterostructures. Experiments demonstrated no difference in the resistance values between the two device types within the experimental uncertainty of $\sim 10^{-10}$, thus both verifying the value of the QHE quantum of resistance and demonstrating the universality of the QHE in fundamentally different material systems (Janssen et al., 2012).

REFERENCES

Areshkin, D.A., Gunlycke, D., White, C.T., 2007. Ballistic transport in graphene nanostrips in the presence of disorder: importance of edge effects. Nano Lett. 7 (1), 204–210.

Arsat, R., Breedon, M., Shafiel, M., Spizziri, P.G., Gilje, S., Kaner, R.B., Kalantar-sadeh, K., Wlodarski, W., 2009. Graphene-like nano-sheets for surface acoustic wave gas sensor applications. Chem. Phys. Lett. 467, 344–347.

Avouris, P., Chen, Z., Perebeinos, V., 2007. Carbon-based electronics. Nat. Nanotechnol. 2, 605–615.

Bachtold, A., Hadley, P., Nakanishi, T., Dekker, C., 2001. Logic circuits with carbon nanotube transistors. Science 294, 1317–1320.

Bai, J., Zhong, X., Jiang, S., Huang, Y., Duan, X., 2010. Graphene nanomesh. Nat. Nanotechnol. 5, 190–194.

Balog, R., Jørgensen, B., Nilsson, L., Anderson, M., Rienks, E., Bianchi, M., Fanetti, M., Lægsgaard, E., Baraldi, A., Lizzit, S., Slhivancanin, Z., Besenbacher, F., Hammer, B., Pedersen, T.G., Hofmann, P., Hornekær, L., 2010. Bandgap opening in graphene induced by patterned hydrogen adsorption. Nat. Mater. 9, 315–319.

Banerjee, S.K., Register, L.F., Tutuc, E., Reddy, D., MacDonald, A.H., 2009. Bilayer pseudospin field-effect transistor (BiSFET): a proposed new logic device. IEEE Electron Devices Lett. 30 (2), 158–160.

Banerjee, S.K., Register, L.F., Tutuc, E., Basu, D., Kim, S., Reddy, D., MacDonald, A.H., 2010. Graphene for CMOS and beyond CMOS Applications. Proc. IEEE 98 (12), 2032–2046.

Berger, C., Song, Z., Li, X., Wu, X., Brown, N., Naud, C., Mayou, D., Li, T., Hass, J., Marchenkov, A.N., Conrad, E.H., First, P.N., de Heer, W.A., 2006. Electronic confinement and coherence in patterned epitaxial graphene. Science 312, 1191–1196.

Blake, P., Yang, R., Morozov, S.V., Schedin, F., Ponomarenko, L.A., Zhukov, A.A., Nair, R.R., Grigorieva, I.V., Novoselov, K.S., Geim, A.K., 2009. Influence of metal contacts and charge inhomogeneity on transport properties of graphene near the neutrality point. Solid Sate Commun. 149, 1068–1071.

Bolotin, K.I., Sikes, K.J., Jiang, Z., Klima, M., Fudenberg, G., Hone, J., Kim, P., Stormer, H.L., 2008a. Ultrahigh electron mobilty in suspended graphene. Solid State Commun. 146, 351–355.

Bolotin, K.I., Sikes, K.J., Hone, J., Stormer, H.L., Kim, P., 2008b. Temperature-dependent transport in suspended graphene. Phys. Rev. Lett. 101, 096802.

Boucart, K., Ionescu, A.M., 2007. Double-gate tunnel FET with high-κ gate dielectric. IEEE Trans. Electron Devices 54 (7), 1725–1733.

Brews, J.R., Fichtner, W., Nicollian, E.H., Sze, S.M., 1980. Generalized Guide for MOSFET Miniaturization. IEEE Electron Device Lett. 1 (1), 2–4.

Brey, L., Fertig, H.A., 2006. Electronic strates of graphene nanoribbons studied with the Dirac equation. Phys. Rev. B 73, 235411.

Britnell, L., Gorbachev, R.V., Jalil, R., Belle, B.D., Schedin, F., Mishchenko, A., Georgiou, T., Katsnelson, M.I., Eaves, L., Morozov, S.V., Peres, N.M.R., Leist, J., Geim, A.K., Novoselov, K.S., Ponomarenko, L.A., 2012. Field-effect tunneling transistor based on vertical graphene heterostructures. Science 335, 947–950.

Cai, J., Ruffieux, P., Jaafar, R., Bieri, M., Braun, T., Blankenburg, S., Muoth, M., Seitsonen, A.P., Saleh, M., Feng, X., Müllen, K., Fasel, R., 2010. Atomically Precise bottom-up fabrication of graphene nanoribbons. Nature 466, 470–473.

Castro, E.V., Novoselov, K.S., Morozov, S.V., Peres, N.M.R., Santos, J. M. B. L. d., Nilsson, J., Guinea, F., Geim, A.K., Castro Neto, A.H., 2007. Biased bilayer graphene: semiconductor with a gap tunable by the electric field effect. Phys. Rev. Lett. 99, 216802.

Castro, E.V., Ochoa, H., Katsnelson, M.I., Gorbachev, R.V., Elias, D.C., Novoselov, K.S., Geim, A.K., Guinea, F., 2010. Limits on charge carrier mobility in suspended graphene due to flexural phonons. Phys. Rev. Lett. 105, 266601.

Chen, Z., Lin, Y. –M., Rooks, M.J., Avouris, P., 2007. Graphene nano-ribbon electronics. Physica E: Low-Dimensional Syst. Nanostruct. 40 (2), 228–232.

Chen, J. –H., Jang, C., Ziao, S., Ishigami, M., Fuhrer, M.S., 2008. Intrinsic and extrinsic perfor-mance limits of graphene devices on SiO_2. Nat. Nanotechnol. 3, 206–209.

Cheng, R., Bai, J., Liao, L., Zhou, H., Chen, Y., Liu, L., Lin, Y. -C., Jiang, S., Huang, Y., Duan, X., 2012. High-frequency self-aligned graphene transistors with transferred gate stacks. Proc. Natl. Acad. Sci. U.S.A. 109 (29), 11588–11592.

Ci, L., Song, L., Jin, C., Jariwala, D., Wu, D., Li, Y., Srivastava, A., Wang, Z.F., Storr, K., Balicas, L., Liu, F., Ajayan, P.M., 2010. Atomic layers of Hybridized boron nitride and graphene domains. Nat. Mater. 9, 430–435.

Cocco, G., Cadelano, E., Colombo, L., 2010. Gap opening in graphene by shear strain. Phys. Rev. B 81, 241412 (R).

Dean, C.R., Young, A.F., Meric, I., Lee, C., Wang, L., Sorgenfrei, S., Watanabe, K., Taniguchi, T., Kim, P., Shepard, K.L., Hone, J., 2010. Boron nitride substrates for high-quality graphene electronics. Nat. Nanotechnol. 5, 722–726.

Du, X., Skachko, I., Barker, A., Andrei, E.Y., 2008. Approaching ballistic transport in suspended graphene. Nat. Nanotechnol. 3, 491–495.

The Economic Benefits of New Spectrum for Wireless Broadband, Executive Office of the President Council of Economic Advisers (2012). http://www.whitehouse.gov/sites/default/files/cea_spectrum_report_2-21-2012.pdf.

European Association of National Metrology Institutes, 2012. Quantum Resistance Metrology Based on Graphene. EMRP Call.

Ezawa, M., 2006. Peculiar width dependence of the electronic properties of carbon nanoribbons. Phys. Rev. B 73, 045432.

Farmer, D.B., Chiu, H.Y., Lin, Y.M., Jenkins, K.A., Xia, F., Avouris, P., 2009. Utilization of a buffered dielectric to achieve high field-effect carrier mobility in graphene transistors. Nano Lett. 9, 4474–4478.

Fchedin, F., Geim, A.K., Morozov, S.V., Hill, E.W., Blake, P., Katsnelson, M.I., Novoselov, K.S., 2007. Detection of individual gas molecules absorbed on graphene. Nat. Mater. 6, 652–655.

Fiori, G., 2009. Ultralow-voltage bilayer graphene tunnel FET. IEEE Electron Device Lett. 30 (10), 1096–1098.

Fowler, J.D., Allen, M.J., Tung, V.C., Yang, Y., Kaner, R.B., Weiller, B.H., 2009. Practical chemical sensors from chemically derived graphene. ACS Nano 3 (2), 301–306.

Fratini, S., Guinea, F., 2008. Substrate-limited electron dynamics in graphene. Phys. Rev. B 77, 195415.

Gava, P., Lazzeri, M., Saitta, A.M., Mauri, F., 2009. Ab initio study of gap opening and screening effects in gated bilayer graphene. Phys. Rev. B 79, 165431.

Geim, A., Novoselov, K., 2007. The rise of graphene. Nat. Mater. 6, 183–191.

Golizadeh-Mojarad, R., Datta, S., 2009. Effect of contact induced states on minimum conductivity in graphene. Phys. Rev. B 79, 085410.

Haddon, R.C., Perel, A.S., Morris, R.C., Palstra, T.T.M., Hebard, A.F., 1995. C60 thin film transistors. Appl. Phys. Lett. 67, 121.

Han, M.Y., Özyilmaz, B., Zhang, Y., Kim, P., 2007. Energy band-gap engineering of graphene nanoribbons. Phys. Rev. Lett. 98, 206805.

Han, M.Y., Brant, J.C., Kim, P., 2010. Electron transport in disordered graphene nanoribbons. Phys. Rev. Lett. 104, 056801.

Hess, K., 1979. Remote polar phonon scattering in silicon inversion layers. Solid State Commun. 30 (12), 797–799.

Hill, E.W., Vijayaragahvan, A., Novoselov, K., 2011. Graphene sensors. IEEE Sensors J. 11 (12), 3161–3170.

Hwang, E.H., Adam, S., Das Sarma, S., 2007. Carrier transport in two-dimensional graphene layers. Phys. Rev. Lett. 98, 186806.

The International Technology Roadmap for Semiconductors 2011 Edition – Emerging Research Materials. http://www.itrs.net/Links/2011ITRS/Home2011.htm

Janssen, T.J.B.M., Williams, J.M., Fletcher, N.E., Goebel, R., Tzalenchuk, A., Yakimova, R., Lara-Avila, S., Kubatkin, S., Fal'ko, V.I., 2012. Precision comparison of the quantum hall effect in graphene and gallium arsenide. Metrologia 49, 294–306.

Javey, A., Guo, J., Wang, Q., Lundstrom, M., Dai, H., 2003. Ballistic carbon nanotube field-effect transistors. Nature 424, 654–657.

Jiao, L., Zhang, L., Wang, X., Diankov, G., Dai, H., 2009. Narrow graphene nanoribbons from carbon nanotubes. Nature 458, 877–880.

Jiao, L., Wang, X., Diankov, G., Wang, H., Dai, H., 2010. Facile synthesis of high-quality graphene nanoribbons. Nat. Nanotechnol. 5, 321–325.

Kim, P., Han, M.Y., Young, A.F., Meric, I., Shepard, K.L., 2009. Graphene nanoribbon devices and quantum heterojunction devices. Electron Devices Meeting (IEDM). http://dx.doi.org/10.1109/IEDM.2009.5424379.

Kim, S., Nah, J., Jo, I., Shahrjerdi, D., Colombo, L., Yao, Z., Tutuc, E., Banerjee, S.K., 2009. Realization of a high mobility dial-gated graphene field-effect transistor with Al_2O_2 dielectric. Appl. Phys. Lett. 94, 062107.

Kim, K., Choi, J. –Y., Kim, T., Cho, S. –H., Chung, H. –J., 2011. A role for graphene in silicon-based semiconductor devices. Nature 479, 338–344.

Lee, B., Park, S.Y., Kim, H.C., Cho, K.J., Vogel, E.M., Kim, M.J., Wallace, R.M., Kim, J., 2008. Conformal Al2O3 dielectric layer deposited by atomic layer deposition for graphene-based nanoelectronics. Appl. Phys. Lett. 92, 203102.

Li, X., Wang, X., Zhang, L., Lee, S., Dai, H., 2008. Chemically derived, Ultrasmooth graphene nanoribbon semiconductors. Science 319, 1229–1232.

Liao, L., Bai, J., Cheng, R., Lin, Y. -C., Jiang, S., Qu, Y., Huang, Y., Duan, X., 2010a. Sub-100 nm channel length graphene transistors. Nano Lett. 10 (10), 3952–3956.

Liao, L., Bai, J., Qu, Y., Lin, Y. -C., Li, Y., Huang, Y., Duan, X., 2010b. High-κ oxide nanoribbons as gate dielectrics for high mobility top-gated graphene transistors. Proc. Natl. Acad. Sci. U.S.A. 107 (15), 6711–6715.

Liao, L., Lin, Y. –C., Bao, M., Cheng, R., Bai, J., Liu, Y., Qu, Y., Wang, K.L., Huang, Y., Duan, X., 2010c. High-speed graphene transistors with a self-aligned nanowire gate. Nature 467, 305–308.

Lin, Y. –M., Jenkins, K.A., Valdes-Garcia, A., Small, J.P., Farmer, D.B., Avouris, P., 2009. Operations of graphene transistors at gigahertz frequencies. Nano Lett. 9 (1), 422–426.

Lin, Y. –M., Dimitrakopoulos, C., Jenkins, K.A., Farmer, D.B., Chiu, H. –Y., Grill, A., Avouris, Ph, 2010. 100-GHz transistors from wafer-scale epitaxial graphene. Science 327, 662.

Lin, Y. -M., Valdes-Garcia, A., Han, S. –J., Farmer, D.B., Meric, I., Sun, Y., Wu, Y., Dimitrakopoulos, C., Grill, A., Avouris, P., Jenkins, K.A., 2011. Wafer-scale graphene integrated circuit. Science 332, 1294–1297.

Lundstrom, M., 1997. Elementary scattering theory of the Si MOSFET. IEEE Electorn Device Lett. 18 (7), 361–363.

Lundstrom, M.S., 2001. On the mobility versus drain current relation for nanoscale MOSFET. IEEE Electron Device Lett. 22 (6), 293–295.

Martin, J., Akerman, N., Ulbricht, G., Lohmann, T., Smet, J.H., von Klitzing, K., Yacoby, A., 2008. Observation of electron-hole puddles in graphene using a scanning single-electron transistor. Nat. Phys. 4, 144–148.

Mayorov, A.S., Gorbachev, R.V., Morozov, S.V., Britnell, L., Jalil, R., Ponomarenko, L.A., Blake, P., Novoselov, K.S., Watanabe, K., Taniguchi, T., Geim, A.K., 2011. Micrometer-scale ballistic transport in encapsulated graphene at room temperature. Nano Lett. 11, 2396–2399.

McCann, E., 2006. Asymmetry gap in the electronic band structure of bilayer graphene. Phys. Rev. B 74, 161403.

Mills, I.M., Mohr, P.J., Quinn, T.J., Taylor, B.N., Williams, E.R., 2011. Adapting the international system of units to the twenty-first century. Philos. Trans. R. Soc. A 369, 3907–3924.

Minot, E.D., Yaish, Y., Sazanova, V., Park, J. –Y., Brink, M., McEuen, P.L., 2003. Tuning carbon nanotube band gaps with strain. Phys. Rev. Lett. 90 (15), 156401.

Misewich, J.A., Martel, R., Avouris, Ph., Tsang, J.C., Heinze, S., Tersoff, J., 2003. Electically induced optical emission from a carbon nanotube FET. Science 300, 783–786.

Miyamoto, Y., Nakada, K., Fujita, M., 1999. First-principles study of edge states of H-terminated graphitic ribbons. Phys. Rev. B 59, 9858–9861.

Moore, G.E., 1965. Cramming more components onto integrated circuits. Electronics 38 (8), 114.

Nagashio, K., Toriumi, A., 2011. Density-of-states limited contact resistance in graphene field-effect transistors. Jpn. J. Appl. Phys. 50, 070108.

Nagashio, K., Nishimura, T., Kita, K., Toriumi, A., 2009. Metal/graphene contact as a performance killer of ultra-high mobility graphene – analysis of intrinsic mobility and contact resistance. Electron Devices Meet. (IEDM), 1–4.

Nakada, K., Fujita, M., Dresselhaus, G., Dresselhaus, M.S., 1996. Edge state in graphene ribbons: nanometer size effect and edge shape dependence. Phys. Rev. B 54 (24), 17954.

Ni, Z.H., Yu, T., Lu, Y.H., Lu, Y.H., Wang, Y.Y., Feng, Y.P., Shen, Z.X., 2008. Uniaxial strain on graphene: Raman spectroscopy study and band-gap opening. ACS Nano 2 (11), 2301–2305. Erratum: 3(2), 483 – 483(2009).

Novoselov, K.S., Geim, A.K., Morozov, S.V., Jiang, D., Zhang, Y., Dubonos, S.V., Grigorieva, I.V., Firsov, A.A., 2004. Electric field effect in atomically thin carbon films. Science 306, 666–669.

Ohmi, T., Kotani, K., Teramoto, A., Miyashita, M., 1991. Dependence of electron channel mobility on Si–SiO_2 interface microroughness. IEEE Electron Device Lett. 12 (12), 652–654.

Ohto, T., Bostwick, A., Seyller, T., Horn, K., Rotenberg, E., 2006. Controlling the electronic structure of bilayer graphene. Science 313, 951–954.

Oostinga, J.B., Heersche, H.B., Li, X., Morpurgo, A.F., Vandersypen, L.M.K., 2007. Gate-induced insulating state in bilayer graphene devices. Nat. Mater. 7, 151–157.

Packan, P.A., 1999. Pushing the limits. Science 285 (5436), 2079–2081.

Palacios, T., Hsu, A., Wang, H., 2010. Applications of graphene devices in RF communications. IEEE Commun. Mag. 48 (6), 122–128.

Parrish, K.N., Akinwande, D., 2011. Impact of contact resistance on the transconductance and linearity of graphene transistors. Appl. Phys. Lett. 98, 183505.

Pasanen, P., Voutilainen, M., Helle, M., Song, X., Hakonen, P.J., 2012. Graphene for future electronics. Phys. Scr. T146, 014025.

Passlack, M., 2006. Off-state current limits of narrow bandgap MOSFETs. IEEE Trans. Electron Devices 53 (11), 2773–2778.

Pereira, V.M., Neto, A.H.C., Peres, N.M.R., 2009. Tight-binding approach to uniaxial strain in graphene. Phys. Rev. B 80, 045401.

Peres, N.M.R., Castro Neto, A.H., Guinea, F., 2006. Conductance quantization in mesoscopic graphene. Phys. Rev. B 73, 195411.

Pirkle, A., Wallace, R.M., Colombo, L., 2009. In situ studies of Al_2O_3 and HfO_2 dielectrics on graphite. Appl. Phys. Lett. 95, 133106.

Reprint available at IEEE Solid-State Circuits Newsletter, 11(5), 33–35, 2006.

Russo, S., Craciun, M.F., Yamamoto, M., Morpurgo, A.F., Tarucha, S., 2010. Contact resistance in graphene-based devices. Physica E 42, 677–679.

Das Sarma, S., Adam, S., Hwang, E.H., Rossi, E., 2011. Electronic transport in two-dimensional graphene. Rev. Mod. Phys. 83, 407–470.

Schwierz, F., 2010. Graphene transistors. Nat. Nanotechnol. 5, 487–496.

Sofo, J.O., Chaudhari, A.S., Barber, G.D., 2007. Graphane: a two-dimensional hydrocarbon. Phys. Rev. B 75, 153401.

Son, Y. –W., Cohen, M.L., Louie, S.G., 2006. Half-metallic graphene nanoribbons. Nature 444, 347–349.

Song, S.M., Park, J.K., Sul, O.J., Cho, B.J., 2012. Determination of work function of graphene under a metal electrode and its role in contact resistance. Nano Lett. ASAP, http://dx.doi.org/10.1021/nl300266p.

Su, J. –J., MacDonald, A.H., 2008. How to make a bilayer exciton condensate flow. Nat. Phys. 4 (10), 799–802.

Tan, Y. –W., Zhang, Y., Bolotin, K., Zhao, Y., Adam, S., Hwang, E.H., Sarma, S.D., Stormer, H.L., Kim, P., 2007. Measurement of scattering rate and minimum conductivity in graphene. Phys. Rev. Lett. 99, 246803.

Tans, S.J., Verschueren, A.R.M., Dekker, C., 1998. Room-temperature transistor based on a single carbon nanotube. Nature 393, 49–52.

Tapasztó, L., Dobrik, G., Lambin, P., Biró, L.P., 2008. Tailoring the atomic structure of graphene nanoribbons by scanning tunneling microscope lithography. Nat. Nanotechnol. 3, 397–401.

Taur, Y., 2002. CMOS design near the limit of scaling. IBM J. Res. Dev. 46, 213–222.

The International Technology Roadmap for Semiconductors: 2011 Edition.

Tombler, T.W., Zhou, C., Alexseyev, L., Kong, J., Dai, H., Liu, L., Jayanthi, C.S., Tang, M., Wu, S.-Y., 2000. Reversible electromechanical characteristics of carbon nanotubes under local-probe manipulation. Nature 405, 769–772.

Tzalenchuk, A., Lara-Avila, S., Kalaboukhov, A., Paolillo, S., Syväjärvi, M., Yakimova, R., Kazakova, O., Janssen, T.J.B.M., Fal'ko, V., Kubatkin, S., 2010. Towards a quantum resistance standard based on epitaxial graphene. Nat. Nanotechnol. 5, 186–189.

Tzalenchuk, A., Lara-Avila, S., Cedergren, K., Syväjärvi, M., Yakimova, R., Kazakova, O., Janssen, T.J.B.M., Moth-Poulsen, K., Bjørnholm, T., Kopylov, S., Fal'ko, V., Kubatkin, S., 2011. Engineering and metrology of epitaxial graphene. Solid State Commun. 151, 1094–1099.

Venugopal, A., Colombo, L., Vogel, E.M., 2010. Contact resistance in few and multilayer graphene devices. Appl. Phys. Lett. 96, 013512.

Wakabayashi, K., Fujita, M., Ajiki, H., Sigrist, M., 1999. Electronic and magnetic properties of nanographite ribbons. Phys. Rev. B 59, 8271–8282.

Wehling, T.O., Novoselov, K.S., Morozov, S.V., Vdovin, E.E., Katsnelson, M.I., Geim, A.K., Lichtenstein, A.I., 2008. Molecular doping of graphene. Nano Lett. 8 (1), 173–177.

Wong, H. -S.P., 2002. Beyond the conventional transistor. IBM J. Res. Dev. 46, 133–168.

Wu, Y., Lin, Y., Bol, A.A., Jenkins, K.A., Xia, F., Farmer, D.B., Zhu, Y., Avouris, P., 2011. High-frequency, scaled graphene transistors on diamond-like carbon. Nature 472, 74–78.

Wu, Y., Jenkins, K.A., Valdes-Garcia, A., Farmer, D.B., Zhu, Y., Bol, A.A., Dimitrakopoulos, C., Zhu, W., Xia, F., Avouris, P., Lin, Y. -M., 2012. State-of-the-art graphene high-frequency electronics. Nano Lett. 12 (6), 3062–3067.

Xia, F., Perebeinos, V., Lin, Y., Wu, Y., 2011. Avouris. The origins and limits of metal-graphene junction resistance. Nat. Nanotechnol. 6, 179–184.

Xuan, Y., Wu, Y.Q., Shen, T., Qi, M., Capano, A., Cooper, J.A., Ye, P.D., 2008. Atomic-layer-deposited nanostructures for graphene-based nanoelectronics. Appl. Phys. Lett. 92, 013101.

Yang, H., Heo, J., Park, S., Song, H.J., Seo, D.H., Byun, K. –E., Kim, P., Yoo, I., Chung, H. –J., Kim, K., 2012. Graphene barristor, a triode device with a gate-controlled Schottky barrier. Science 336, 1140–1143.

Zhang, Y., Tang, T. –I., Girit, C., Hao, Z., Martin, M.C., Zettl, A., Crommie, M.F., Shen, Y.R., Wang, F., 2009. Direct observation of a widely tunable bandgap in bilayer graphene. Nature 459, 820–823.

Zhou, S.Y., Gweon, G. –H., Fedorov, A.V., First, P.N., de Heer, W.A., Lee, D. -H., Guinea, F., Castro Neto, A.H., Lanzara, A., 2007. Susbstrate-induced bandgap opening in epitaxial graphene. Nat. Mater. 6, 770–775.

Spintronics

Jamie H. Warner

University of Oxford, Oxford, UK

6.2.1. INTRODUCTION

Spintronics is the name associated with technology that utilises both the intrinsic spin of an electron as well as its charge in transport devices. It is primarily concerned with solid-state systems and how manipulation of the electron spin state can result in appreciable changes in conductance. Its most famous incarnation is in the form of giant magnetoresistance (GMR), which earned Nobel prizes for its discoverers; Albert Fert and Peter Grünberg. Most modern memory storage devices utilise the GMR effect, and this in part has led to the massive increase in the high-density data storage that is integral to laptops, ipods, and video recorders.

Graphene seems to have potential in future spintronic applications due to carbon having a low natural abundance of the spin-isotope C^{13} and graphene having small spin–orbit interactions. In Chapter 3.3, we discussed the electron spin properties of graphene and discussed spins associated with defects, adatoms, edge states, as well as conduction electrons. In this chapter, we will focus on electronic devices that aim to measure the spin lifetime as well as device architectures used to transit and process spin information.

The simplest spintronic device geometry is the spin valve, which is comprised of two ferromagnetic electrodes with a nonmagnetic material in-between. The direction of the injected charge carrier's spin state relative to the electrode's magnetic polarisation determines the current flow. If both electrodes have parallel magnetisation, then current is able to flow; however, if they are antiparallel, current is restricted. The magnetoresistance is determined by the difference between these two electrode magnetisation configurations. Figure 6.2.1 shows a schematic illustration of the two-terminal spin valves in both the vertical and lateral geometries. The majority of work on the spin-valve research on nanotubes and graphene has been with lateral geometries.

6.2.2. MAGNETORESISTANCE USING CARBON NANOTUBES

Spin-dependent transport experiments have been undertaken using carbon nanotubes since 1999, when Tsukagoshi et al. reported coherent transport of electron spin in ferromagnetically contacted carbon nanotubes (Tsukagoshi et al., 1999). Figure 6.2.2 shows two-terminal differential resistance as

High resistance   Low resistance

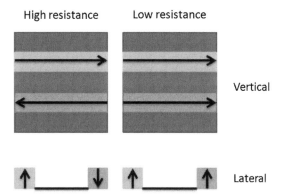

Vertical

Lateral

FIGURE 6.2.1 Schematic illustration of two-terminal spin-valve devices, both vertical and lateral.

a function of magnetic field for three different multiwalled carbon nanotube devices (Tsukagoshi et al., 1999). When the magnetisations of the two contacts were parallel, the resistance was lower than when the magnetisation of the two contacts were antiparallel. The magnetoresistance reported was at most 9% (Tsukagoshi et al., 1999). The authors explain that this requires the spin scattering length in the nanotube to be of the order of the contact separation, and the scattering at the contact interface must not completely randomise the spin. The ferromagnetic electrodes used in these devices were polycrystalline cobalt (Co) deposited by thermal evaporation, the length of the conducting channel (i.e. nanotube) was 250 nm and the multiwalled carbon nanotubes (MWNTs) varied between 10 and 40 nm in diameter (Tsukagoshi et al., 1999). A spin-flip scattering length of at least 130 nm was estimated from the results. The hysteresis in the magnetoresistance is attributed to spin-polarised electron tunnelling and arises from a magnetic tunnel junction (Julliere, 1975; Meservey and Tedrow, 1994; Miyazaki and Tezuka, 1995; Moodera et al., 1995; Tsukagoshi et al., 1999).

Jensen et al. (2005) investigated the two-terminal magnetoresistance in single-walled carbon nanotubes using ferromagnetic electrodes that were either semiconducting ((Ga,Mn)As) or metallic (Fe) and found both led to strong hysteretic magnetoresistance below 30 K. Their devices exhibited a wide range of sign and magnitude in the hysteretic magnetoresistance. Zhao et al. (2002) reported a maximum magnetoresistance ratio of 30% at a junction bias current of 1 nA using Co electrodes and MWNTs in a two-terminal device geometry. Further refinement in the device geometry and electrode fabrication was reported by Sahoo et al. (2005) where gate-field-controlled magnetoresistance response of carbon nanotubes (both MWNT and SWNT) was reported. They used ferromagnetic Pd1-xNix with $x = 0.7$ as electrodes and claim that they

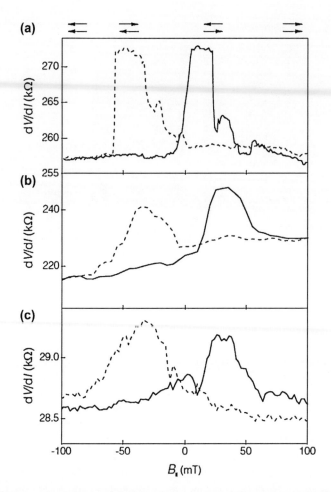

FIGURE 6.2.2 Two-terminal differential resistance as a function of magnetic field for three different MWNT devices. The magnetic field $B_{\parallel}$ is directed parallel to the substrate, and the temperature is 4.2 K. The solid (dashed) trace corresponds to the positive (negative) sweep direction. The differential resistance shows a large variation among devices – the device shown in c has a resistance an order of magnitude lower than the devices shown in a and b. This is probably due to variations in the contact resistance. The two-terminal resistance could sometimes be lowered by thermal annealing. Each device shows a large hysteretic magnetoresistance peak. The magnetisation direction of the left and right contacts is represented by the direction of the arrows at the top of the figure. The percent difference DR/Ra between the tunnel resistance in the parallel and the antiparallel states is approximately 6% in a, 9% in b and 2% in c. *Reproduced from Tsukagoshi et al. (1999), Fig. 2. Copyright (1999) Nature Publishing Group.*

took advantage of the good contacting properties of Pd to nanotubes and its paramagnetism (Sahoo et al., 2005). Figure 6.2.3 shows the changes in the tunnelling magnetoresistance (TMR) for different back-gate voltages, with the inset showing an SEM image of the device geometry.

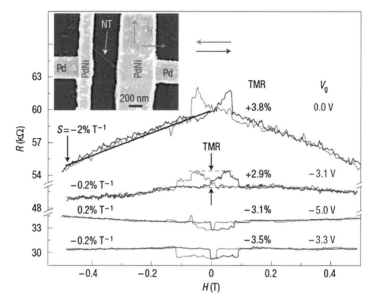

FIGURE 6.2.3 The TMR changes sign with V_g. Inset: SEM picture of a carbon nanotube contacted to ferromagnetic PdNi strips. The separation between the contacts along the nanotube amounts to $L = 400$ nm. The magnetic field H was applied in plane. No qualitative difference has been seen for the field direction parallel and perpendicular to the long axis of the ferromagnetic electrodes. Main panel: linear response resistance R as a function of H at temperature $T = 1.85$ K for different V_g. The blue (red) arrow indicates the up (down) magnetic field sweep direction, respectively. The observed amplitude and the sign of the TMR depend on V_g, but not on the high-field magnetoresistance, which is expressed by S, denoting the percentage change of the magnetoresistance with magnetic field. *Reproduced from Sahoo et al. (2005), Fig. 1. Copyright (2005) Nature Publishing Group.*

Hueso et al. (2007) showed that careful choice of the material used for the ferromagnetic electrode could lead to further increases in the magnetoresistance effect. They showed that epitaxial electrodes of the highly spin-polarised manganite $La_{0.7}Sr_{0.3}MnO_3$ (LSMO) combined with MWNTs enables magnetoresistance values up to 61%. In LSMO electodes, high spin polarisation approaching 100% can be achieved, compared to <40% for elemental ferromagnets (Hueso et al., 2007). The oxide nature of the electrode is also compatible with environmental stability. Figure 6.2.4 shows (a) an optical image of the LSMO electrode layout, (b) an SEM image of the CNT across the electrodes and (c) a schematic side view (Hueso et al., 2007). Figure 6.2.5 shows the large magnetoresistance effect observed for the LSMO contacts with the MWNT (Hueso et al., 2007).

Tombros et al. (2006) showed that the two-terminal geometry used by Tsukagoshi et al. and others is not ideal, as it is difficult to separate spin transport from other effects such as Hall effects, anisotropic magnetoresistance,

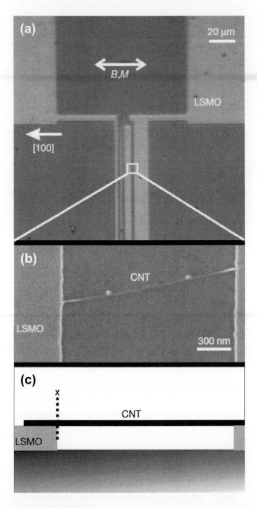

FIGURE 6.2.4 LSMO–CNT–LSMO device. (a) optical micrograph of four variable-width LSMO electrodes and two of the four associated contact pads. In electrically conducting devices, two adjacent electrodes were connected by an overlying CNT, in regions such as that in the white square. Magnetic fields B were applied along the orthorhombic [100] direction in which the magnetisation M is expected to lie due to uniaxial magnetocrystalline anisotropy. (b) Scanning electron microscope image of a CNT running between LSMO electrodes; magnified view corresponding to the boxed area in a. (c) Schematic side view of b with the plane through the CNT at the edge of the LSMO electrode denoted X. *Reproduced from Hueso et al. (2007), Fig 1. Copyright (2007) Nature Publishing Group.*

interference effects, tunnelling anisotropic magnetoresistance-like effects and magneto-Coulomb effects. They instead use a four-terminal nonlocal spin-valve geometry to separate the spin-current path from the charge-current path and demonstrated spin accumulation in SWNTs (Tombros et al., 2006). One

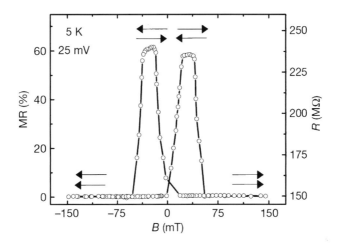

FIGURE 6.2.5   MR for an LSMO–CNT–LSMO device. Data recorded at 5 K with a bias voltage of 25 mV show two distinct states of resistance R, as the magnetic configuration of the two LSMO electrodes is switched by an applied magnetic field B. The arrows indicate the relative magnetic orientation of the electrodes, which possess different switching fields because of their different widths. The data points and interconnecting lines were generated by averaging over 25 cycles; MR (%) was calculated as MR (%) = $100 \times [R(B) - R(0)]/R(B)$. *Reproduced from Hueso et al. (2007), Fig. 3. Copyright (2007) Nature Publishing Group.*

major drawback of using carbon nanotubes compared to graphene for spin-tronics is the existence of spin–orbit coupling in nanotubes. Kuemmeth et al. (2008) showed that in clean carbon nanotubes, the spin and orbital motions of electrons are coupled, revealed through splitting of the fourfold degeneracy of a single electron in a quantum dot.

6.2.3. MAGNETORESISTANCE USING GRAPHENE

Hill et al. (2006) reported graphene spin-valve devices in 2006, using mechanically exfoliated graphene and NiFe ferromagnetic electrodes in a two-electrode configuration, shown schematically in Fig. 6.2.6. Magnetoresistive response of 10% was observed from the device without any optimisation of the device geometry. In order to gain true insights into the nature of the observed magnetoresistance of graphene, the four-terminal nonlocal method was employed by Tombros et al. (2007). Figure 6.2.7 shows (a) an SEM image of the four-terminal device with Co electrodes, (b) a schematic side view, (c) and (d) schematic illustrations, showing the spin injection and diffusion for electrodes for parallel and antiparallel magnetisations (Tombros et al., 2007). They found that the spin coherence extended underneath all contacts, with no changes in the spin signals, as the temperature varied from 4.2 – room temperature (Tombros et al., 2007). A spin relaxation length of between 1.5 and

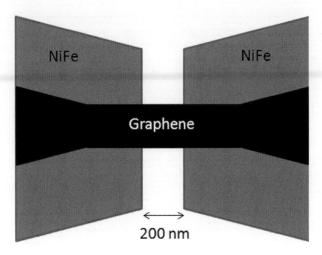

FIGURE 6.2.6 Schematic illustration of the two electrode configuration of graphene based spin valve used in Hill et al. (2006).

2 μm was determined (Tombros et al., 2007). A key part of their experiment was the use of tunnel barriers between the graphene and the Co electrodes. A thin (0.8 nm) Al_2O_3 layer was used to create a spin-dependent tunnel barrier that was transparent enough to enable measureable resistances (<1 MΩ) and also allow carriers to pass underneath with conservation of spin direction (Tombros et al., 2007). The Co electrodes achieved a spin polarisation of ~10% and shows there is room for improvement by using high spin-polarised electrodes such as LSMO. The group further improved their device performances by suspending the graphene, which is known to reduce the detrimental substrate–graphene interactions (Guimaraes et al., 2012). This led to mobilities above 100,000 cm^2/V s and increase in the spin relaxation length to 4.7 μm and an order of magnitude increase in the spin diffusion coefficient (Guimaraes et al., 2012). They report a higher bound for the spin–orbit coupling in their devices as 50 μeV, which is lower than those fabricated on SiO_2 substrates (Guimaraes et al., 2012). Cho et al. (2007) reported gate-tunable graphene spin valves using the nonlocal four-probe geometry. The spin-valve signal changed magnitude and sign with application of the back-gate voltage (Cho et al., 2007).

The graphene-based spin valves, reported in Cho et al. (2007); Guimaraes et al. (2012); Hill et al. (2006); Tombros et al. (2007), used graphene obtained by mechanical exfoliation. Whilst this provides a high-quality source of graphene, it is only suitable for making one-off devices, and alternative sources of graphene are required for arrays of devices. CVD growth of synthetic graphene is an excellent method for producing large areas of monolayer graphene and can be transferred onto silicon wafers with an oxide layer to fabricate nonlocal

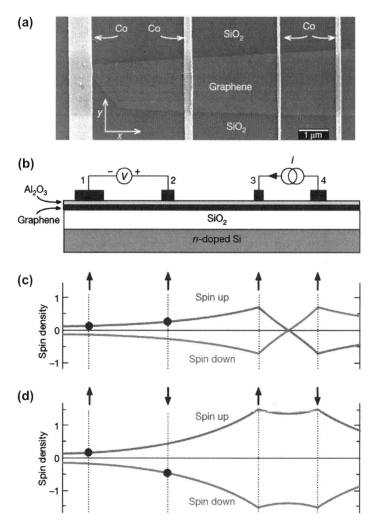

FIGURE 6.2.7 Spin transport in a four-terminal spin-valve device. (a) Scanning electron micrograph of a four-terminal single-layer graphene spin valve. Cobalt (Co) electrodes are evaporated across a single-layer graphene strip prepared on a SiO₂ surface. (b) The nonlocal spin-valve geometry. A current I is injected from electrode 3 through the Al_2O_3 barrier into graphene and is extracted at contact 4. The voltage difference is measured between contacts 2 and 1. The nonlocal resistance is $R_{nonlocal} = (V_+ - V_-)/I$. (c) illustration of spin injection and spin diffusion for electrodes having parallel magnetiations. Injection of up spins by contact 3 results in an accumulation of spin-up electrons underneath contact 3, with a corresponding deficit of spin-down electrons. Owing to spin relaxation, the spin density decays on a scale given by the spin relaxation length. The dots show the electric voltage measured by contacts 1 and 2 in the ideal case of 100% spin selectivity. A positive nonlocal resistance is measured. (We note that a larger positive signal can be obtained by reversing the magnetisation direction of contact 1). (d) Spin injection and spin diffusion for antiparallel magnetisations. The voltage contacts probe opposite spin directions, resulting in a negative nonlocal resistance. *Reproduced from Tombros et al. (2007), Fig. 1. Copyright (2007) Nature Publishing Group.*

FIGURE 6.2.8 Bottom: optical image of a 5 × 5 device array. CVD graphene allows the fabrication of large arrays of identical lateral spin valves. Top: scanning electron micrograph of CVD single-layer graphene spin sample with multiple nonlocal spin-valve devices. Electrode widths range from 0.3 to 1.2 μm. *Reprinted with permission from Avsar (2011), Fig. 1(c). Copyright (2011) American Chemical Society.*

spin valves along the same lines as reported in Cho et al. (2007); Guimaraes et al. (2012); Hill et al. (2006); Tombros et al. (2007). Chapter 4 describes the methods to obtain such graphene. Avsar et al. (2011) fabricated arrays of graphene-based spin valves, using this approach with an MgO barrier and 35-nm Co electrodes, shown in Fig. 6.2.8. The typical length and width of the graphene in the channel was between 1–2 μm, and mobility varied between 1000 to 2000 $cm^2V^{-1}s^{-1}$. A bipolar nonlocal spin signal was observed, shown in Fig. 6.2.9, with a transverse spin relation time of 180 ps, a spin diffusion constant of 0.007 $m^2 s^{-1}$ and a spin relaxation length of 1.1 μm (Avsar et al., 2011). There was no major difference in these numbers for monolayer graphene compared to bilayer graphene (Avsar et al., 2011). At room temperature, the difference between the spin relaxation length of CVD graphene and mechanically exfoliated graphene was minimal, and the limiting mechanisms that increase spin dephasing are the Elliot–Yafet (EY) mechanism for monolayer CVD graphene and the D'yakonov–Perel' (DP) mechanism for CVD bilayer graphene (Avsar et al., 2011). The EY spin-dephasing mechanism relates to momentum scattering, and DP spin-dephasing occurs between momentum scattering events and is suggested to result from Bychkov–Rashba-like spin–orbit fields (Avsar et al., 2011; D'yakonov & Perel', 1972; Elliott, 1954; Ertler et al., 2009).

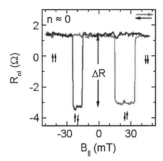

FIGURE 6.2.9   Bipolar spin signal obtained near the charge neutrality point. *Reprinted with permission from Avsar (2011), Fig. 2(b). Copyright (2011) American Chemical Society.*

The theoretical predictions of Munoz-Rojas et al. (2009) reveal that ultrasmall zigzag ribbons should enable 100% magnetoresistance to be achieved in a two-terminal device made entirely from carbon. The experimental realisation of very large magnetoresistance in graphene ribbons was reported in Bai et al. (2010). They used a novel approach of a nanowire mask to pattern a ribbon in graphene, shown in Fig. 6.2.10 (Bai et al., 2010). Figure 6.2.11 shows the differential conductance as a function of gate voltage for 0 T and 8 T fields, normal to the device plane, with suppression of conduction in a the gate region of 0.4–6.6 V (Bai et al., 2010). Figure 6.2.12 shows the current ratio I(8T)/I (0T) as a function of source-drain bias for three gate voltages of 0, 1 and 3 V, revealing a large increase in the current ratio at the edge of the blockade (Bai et al., 2010). These results indicate that GNRs are of high interest in spintronic applications and should stimulate further investigation.

The spin-valve devices presented so far have used pretty standard geometry with either the 2 or 4 terminal lateral architectures. However, there

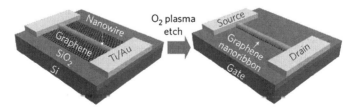

FIGURE 6.2.10   The device was fabricated on a heavily doped silicon substrate with a 300-nm-thick layer of SiO_2 as the gate dielectric. Electron-beam lithography was used to define and electron-beam evaporation to deposit a titanium/gold (7 nm/90 nm) film on a graphene block to be used as the source and drain electrodes. *Reproduced from Bai et al. (2010), Fig. 1. Copyright (2010) Nature Publishing Group.*

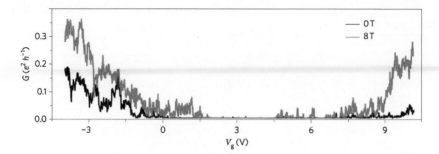

FIGURE 6.2.11 Differential conductance versus gate voltage with a magnetic field of 0 T (black) and 8 T (red) normal to the device plane. Measurements were carried out at 1.6 K on a graphene ribbon FET with channel width of ~15 nm and length of 800 nm. *Reproduced from Bai et al. (2010), Fig. 2(a). Copyright (2010) Nature Publishing Group.*

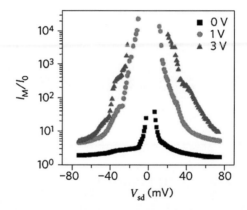

FIGURE 6.2.12 Current ratio I(8 T)/I(0 T) versus source-drain bias at $V_g = 3$ V. The middle interval for each plot is in the range of suppressed conductance and is beyond the equipment measurement limits. *Reproduced from Bai et al. (2010), Fig. 3d. Copyright (2010) Nature Publishing Group.*

is scope for more complex structures in spintronic applications. Cobas et al. (2012) showed that graphene acts as an insulator for transport perpendicular to the plane when sandwiched between two ferromagnetic layers. Vertical tunnel junctions were fabricated using single-layer graphene between Co and NiFe electrodes using scalable photolithography, shown in Fig. 6.2.13 (Cobas et al., 2012). TMR values were typically between 1–2% (Cobas et al., 2012). Lu et al. (2011) produced an extraordinary magnetoresistance device, based on monolayer graphene with an embedded metal disk, shown in Fig. 6.2.14.

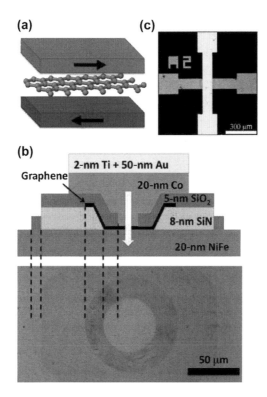

FIGURE 6.2.13 Graphene tunnel junction devices. (a) Conceptual diagram of the FM/graphene/FM junction, (b) cross-sectional diagram and optical image of the junction area prior to top contact deposition, and (c) photo of a completed four-probe device. *Reprinted with permission from Cobas (2012), Fig. 1. Copyright (2012) American Chemical Society.*

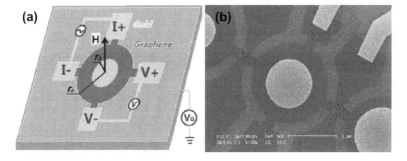

FIGURE 6.2.14 (a) Schematic illustration of the structure of a graphene EMR device. (b) A scanning electron microscope image (false colour) of an actual EMR device. The two adjacent electrodes are for current injection ($I+$ and $I-$), whereas the other two are for voltage detection ($V+$ and $V-$). For two-terminal measurements, only $I+$ and $I-$ are used, both for current injection and for voltage detection. *Reprinted with permission from Lu (2011), Fig. 1. Copyright (2011) American Chemical Society.*

The ratio between the radii of the two circles r_a and r_b shown in Fig. 6.2.14 are important factors influencing the device performance. Figure 6.2.15(a) shows the changes in resistance of a device with $r_a/r_b = 3/4$ with the back gate and for different magnetic fields, −9 T to 9 T (Lu et al., 2011). Figure 6.2.15(b) shows magnetoresistance values up to 55,000% at 9 T (Lu et al., 2011). These results show the potential of this device geometry in spin-based applications.

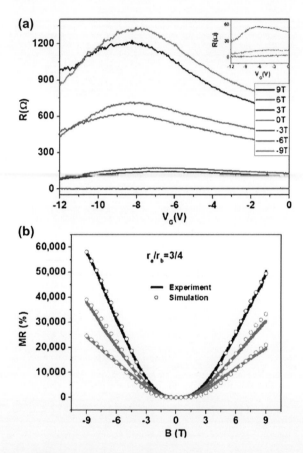

FIGURE 6.2.15 EMR device with a diameter ratio ra/rb = 3/4 and Pd as the metal for electrodes and the central metallic disk. (a) Resistance versus back gate under various magnetic fields. Inset: enlarged resistance at magnetic fields 2, 1, 0 T (from the top to down). Gate-voltage dependence is noted to almost disappears at zero magnetic field. (b) Magnetoresistance with different back-gate voltages: VG = −8, −6, −3 V (from the top to down), solid curves are experimental data and empty circles denote the simulation results with the central disk misalignment error (determined via a SEM image of the actual device) taken into account. *Reprinted with permission from Lu (2011), Fig. 2. Copyright (2011) American Chemical Society.*

6.2.4. SUMMARY

Graphene seems highly promising as a material for spin-based technologies, and with the advances in synthetic graphene, we are now seeing arrays of spin valves being produced. New architectures open up a variety of possibilities, and it is likely that the field will mature rapidly. The major challenge seems to be high spin-polarised electrodes and their interface with graphene. Whilst LSMO is shown to be a great electrode for spintronics, it is challenging to grow high-quality electrodes directly on top of graphene. The oxidizing elements that are essential to forming LSMO are destructive to graphene. Transfer of spin between electrodes forms the basis of spintronics, but it also has important impact on quantum information processing. Transmitting spin information from one location to another without decohering is a challenge likely to arise in quantum computing. Graphene may play an important role in this field in the years ahead. Through progressive advances in the development of synthetic graphene and tailored spin polarised electrodes, it is possible that graphene may reach its potential in spintronics.

REFERENCES

Avsar, A., Yang, T.-Y., Bae, S., Balakrishnan, J., Volmer, F., Jaiswal, M., Yi, Z., Ali, S.R., Guntherodt, G., Hong, B.H., Beschoten, B., Ozyilmaz, B., 2011. Toward wafer scale fabrication of graphene based spin valve devices. Nano Lett. 11, 2363–2368.

Bai, J., Cheng, R., Xiu, F., Liao, L., Wang, M., Shailos, A., Wang, K.L., Huang, Y., Duan, X., 2010. Very large magnetoresistance in graphene nanoribbons. Nat. Nanotechnol. 5, 655–659.

Cho, S., Y-Chen, F., Fuhrer, M.S., 2007. Gate-tunable graphene spin valve. Appl. Phys. Lett. 91, 123105.

Cobas, E., Friedman, A.L., van't Erve, O.M.J., Robinson, J.T., Jonker, B.T., 2012. , graphene as a tunnel barrier: graphene-based magnetic tunnel junctions. Nano Lett. 12, 3000–3005.

D'yakonov, M.I., Perel', V.I., 1972. Sov. Phys. Solid State 13, 3023–3026.

Elliott, R.J., 1954. Phys. Rev. 96, 266–279.

Ertler, C., Konschuh, S., Gmitra, M., Fabian, J., 2009. Phys. Rev. B 80, 041405.

Guimaraes, M.H.D., Veligura, A., Zomer, P.J., Maasen, T., Vera-Marun, I.J., Tombros, N., van Wees, B.J., 2012. Nano Lett. 12, 3512–3517.

Hill, E.W., Geim, A.K., Novoselov, K., Schedin, F., Blake, P., 2006. Graphene spin valve devices. IEEE Trans. Magnetics 42, 2694.

Hueso, L.E., Pruneda, J.M., Ferrari, V., Burnell, G., Valdes-Herrera, J.P., Simons, B.D., Littlewood, P.B., Artacho, E., Fert, A., Mathur, N.D., 2007. Transformation of spin information into large electrical signals using carbon nanotubes. Nature 445, 410.

Jensen, A., Hauptmann, J.R., Nygard, J., Lindelof, P.E., 2005. Magnetoresistance in ferromagnetically contacted single-walled carbon nanotubes. Phys. Rev. B. 72, 035419.

Julliere, M., 1975. Tunnelling between ferromagnetic films. Phys. Lett. A 54, 225–226.

Kuemmeth, F., Illani, S., Ralph, D.C., McEuen, P.L., 2008. Coupling of spin and orbital motion of electrons in carbon nanotubes. Nature 452, 448.

Lu, J., Zhang, H., Shi, W., Wang, Z., Zheng, Y., Zhang, T., Wang, N., Tang, Z., Sheng, P., 2011. Graphene Magnetoresistance Device in van der Pauw Geometry. Nano Lett. 11, 2973–2977.

Graphene

Meservey, R., Tedrow, P.M., 1994. Spin-polarised electron tunnelling. Phys. Rep. 238, 173–243.

Miyazaki, T., Tezuka, N., 1995. Giant magnetic tunnelling effect in Fe/Al2O3/Fe junction. J. Magn. Magn. Mater. 139, L231–L234.

Moodera, J.S., et al., 1995. Large magnetoresistance at room temperature in ferromagnetic thin film tunnel junctions. Phys. Rev. Lett. 74, 3273–3276.

Munoz-Rojas, F., Fernandez-Rossier, J., Palacios, J.J., 2009. Giant magnetoresistance in ultrasmall graphene based devices. Phys. Rev. Lett. 102, 136810.

Sahoo, S., Kontos, T., Furer, J., Hoffmann, C., Graber, M., Cottet, A., Schonenberger, C., 2005. Electric field control of spin transport. Nat. Phys. 1, 99.

Tombros, N., van der Molen, S.J., van Wees, B.J., 2006. Separating spin and charge transport in single wall carbon nanotubes. Phys. Rev. 73, 233403.

Tombros, N., Jozsa, C., Popinciuc, M., Jonkman, H.T., van Wees, B.J., 2007. Nature 448, 571.

Tsukagoshi, K., Alphenaar, B.W., Ago, H., 1999. Coherent transport of electron spin in a ferro-magnetically contacted carbon nanotube. Nature 401, 572.

Zhao, B., Monch, I., Vinzelberg, H., Muhl, T., Schneider, C.M., 2002. Spin-Coherent transport in ferromagnetically contacted carbon nanotubes. Appl. Phys. Lett. 80, 3144.

Chapter 6.3

Transparent Conducting Electrodes

Franziska Schäffel

University of Oxford, Oxford, UK

Transparent conductors (TCs) are used in a wide variety of applications ranging from low-emissivity windows, transparent electromagnetic shields, electrochromic mirrors and windows for display and touch screen technologies (Gordon, 2000). Transparent conducting electrodes are essential components in the device stack of modern electronic products. Here, liquid crystal display (LCD) technologies so far require the largest amount of transparent conducting material (Hecht et al., 2011). However, with the markets for touch screens, thin film solar cells, printable electronics and solid-state lighting devices such as (organic) light emitting diodes ((O)LEDs) growing, the need for TCs will not ebb away (Burroughes et al., 1990; Peumans et al., 2003; Wassei and Kaner, 2010). In a variety of TC applications, e.g. high resolution displays, it is critical to obtain a smooth topography and therefore reduce height variations of the devices, and thus, the TC has to be fabricated as thin as possible. Here, the conductivity becomes the important material parameter (Gordon, 2000).

It should first be noted that the minimum standards required for a transparent conducting material to be industrially useful are a sheet resistance $R_S < 100$ Ω/sq together with a transmittance of >90% in the visible range (De and Coleman, 2010). For touch screen applications, where sheet resistances of only $R_S < 500$ Ω/sq are required, these specifications are sufficient (De and Coleman, 2010; Geng et al., 2007). However, transparent conducting electrodes in flat panel displays require much lower sheet resistances down to $R_S \approx 10$ Ω/sq with tin doped indium oxide (aka: indium tin oxide, ITO) matching these requirements (De and Coleman, 2010). Apart from a high optical transparency in the visible region and high electrical conductivity (i.e. low sheet resistance), further factors for choosing a TC are its mechanical and chemical stability, the etchability, a favourable work function (WF) and, of course, the cost for the raw material and the processing (Gordon, 2000; Lee et al., 2004). In most of the TC applications ITO is the dominant transparent electrode material in use, as it possesses high electrical conductivity, high transparency and has been studied and optimised over several decades. Commercially available ITO thin films exhibit a sheet resistance as low as $R_S = 6$ Ω/sq and an absorption coefficient in the visible region of 0.04 (Gordon, 2000). ITO sheets with a transmittance larger than 90% typically exhibit a sheet resistance of 30–80 Ω/sq. However, ITO comes with some flaws. The high cost and scarcity of indium, which is only a by-product of the mining of zinc or lead ores, is a major drawback (Gordon, 2000). In 2011 the annual average price of indium increased by approximately 25% to 720 \$/kg with peak prices up to \$ 875 per kilogram (US, 2012). With the increasing demand in TCs in recent years this has led to extreme fluctuation of the indium price on the market (Hecht et al., 2011).

Alternative materials include other transparent conducting oxides (TCOs) (Minami 2008; Nomoto et al., 2011), metallic thin films (O'Connor et al., 2008) or metal grids (Tvingstedt and Inganäs, 2007). Impurity-doped zinc oxides, e.g. Al- and Ga-doped ZnO, are lower in material cost and easier to etch and thus have been proposed to replace ITO in various applications (Minami 2008; Nomoto et al., 2011). They have attracted significant attention in the growing field of thin film solar cells, however, degradation of the electronic properties during processing is a major issue (Lee et al., 2007; Vinnichenko et al., 2010). Further, to achieve 10 Ω/sq with ITO a film thickness of ~180 nm is required. When using Al- or Ga-doped ZnO much thicker films are necessary therefore somewhat overriding the advantage in material cost. In addition, the trend towards flexible touch screen, display and solar technologies requires bendable TCs. Here, the ceramic nature and brittleness of ITO and other TCOs are a significant downside, since this is the reason for a significant reduction in conductivity when bending the thin film (Lewis et al., 2004). Further, flexible substrates, such as polymers, usually cannot withstand high temperatures during thin film deposition, which has made the development of novel deposition and transfer processes necessary.

Among other novel transparent conductors, such as metallic nanostructures and conducting polymers, carbon nanomaterials, i.e. carbon nanotubes, graphene or hybrids thereof, have been proposed as the new generation TCs for electrode application (Tung et al., 2009; Wang et al., 2008; Wu et al., 2004).

Especially graphene offers several potential advantages over commonly used TCOs due to the combination of a high charge carrier mobility, high transparency in the visible region and the additional flexibility and high stretchability. As discussed in Section 3.1, single-layer graphene is a zero gap semiconductor with massless charge carriers, which leads to extremely high charge carrier mobilities. At 20 K, suspended graphene has been shown to exhibit carrier mobilities up to $\mu = 185,000$ $cm^2V^{-1}s^{-1}$ (Du et al., 2008). Graphene deposited on an insulating substrate reaches carrier mobilities up to $\mu = 15,000$ $cm^2V^{-1}s^{-1}$ at 100 K (Novoselov et al., 2005). It has been demonstrated that graphene can be chemically doped at doping levels up to $n = 10^{12}$ cm^{-2} and still maintain charge carrier mobilities on the order of $\mu = 10^5$ $cm^2V^{-1}s^{-1}$ (Schedin et al., 2007). The sheet resistance of graphene can be derived from $R_S = (\sigma_{2D} \cdot N)^{-1}$, where N is the number of graphene layers and σ_{2D} is the conductivity of the two-dimensional sheet (Wang et al., 2011). The intrinsic sheet resistance of a graphene monolayer of 6.45 kΩ is inferior to that of ITO (Novoselov et al., 2005). However, it may be tuned by increasing the number of graphene layers (N) through layer by layer stacking or increasing the conductivity through doping (Bae et al., 2010).

Nair et al. (2008) determined the optical absorbance of graphene. A single layer of mechanically exfoliated graphene absorbs only 2.3% of incident white light (Nair et al., 2008). Therefore the transmittance T varies as $T \sim 100 - 2.3N$ (in: %). Graphene synthesised via CVD shows slightly decreased optical transmittance due to contaminations produced during the transfer process (Kim et al., 2009). To achieve the opto-electronic properties required for application in transparent electrodes, graphene has to be produced defect free, and the flake size has to be large to reduce charge carrier scattering at the grain boundaries.

Graphene is readily available as scalable synthesis routes have now been developed including various solution-based (cf. Sections 4.2 and 4.3) and gas-based methods (cf. Sections 4.5 and 4.6), therefore, promising lower material costs. On top of that graphene offers a high chemical stability and reduced weight. The highlights of graphene-related research towards transparent electrode application will be presented next.

While mechanical exfoliation (Section 4.1) and thermal decomposition of silicon carbide (Section 4.7) produce high quality graphene, these processes are cost intensive and cannot easily be scaled up. The challenge of producing graphene covering large surface areas for flexible electrode applications has recently been overcome by introducing scalable synthesis methods, i.e. liquid phase exfoliation (De et al., 2010; Hernandez et al., 2008), reduction of graphene oxide (Becerril et al., 2008; Eda et al., 2008a) and CVD (Bae et al., 2010; Kim et al., 2009; Li et al., 2009a). In all these approaches a trade-off between transparency and conductivity is obvious (Bae et al., 2010; Becerril et al., 2008; De et al., 2010). In the following, these methods will be briefly reviewed with their potential for the fabrication of TCs in mind.

Both solution based techniques, i.e. the production of exfoliated graphite and the reduction of graphene oxide, make use of cheap and abundant graphitic precursors and relatively simple procedures therefore promising lower material and production costs as compared to the current TCOs. These low-material costs can further be combined with inexpensive and established thin film deposition techniques. Large area transparent conducting films from exfoliated graphene suspensions have been prepared by spray-coating (Blake et al., 2008), Langmuir–Blodgett (LB) assembly (Li et al., 2008) or vacuum filtration (De et al., 2010). Blake et al. (2008) derived graphene flakes via sonication of natural graphite in dimethylformamide (DMF) and obtained suspensions with a monolayer content of up to 50%. Figure 6.3.1a shows a light transmission image of glass (left) and of a spray-coated graphene film derived from the DMF approach (right). While these films exhibited a high transparency of 90%, their room-temperature sheet resistance was only on the order of 5 kΩ, which is too high for most applications as transparent electrodes.

Further, De et al. (2010), Li et al. (2008) demonstrated the feasibility of using surfactants to stabilise the graphene suspensions. Li et al. report on depositing varying numbers of LB films of graphene sheets onto quartz substrates. A double-layer LB film is shown in Fig. 6.3.1b. Figure 6.3.1c summarises the sheet resistances and transparencies of the thus obtained LB films with varying layer number N. The single-, double- and triplelayer LB films exhibit sheet resistances of ~150, 20 and 8 kΩ at room temperature and transparencies of ~ 93, 88 and 83% (measured at a wavelength of 1000 nm), respectively (Li et al., 2008). De et al. (2010) applied vacuum filtration to prepare large-area graphene films with thicknesses between 6 to 88 nm from their suspension. After an additional annealing at 500 °C in Ar/H$_2$, they obtain

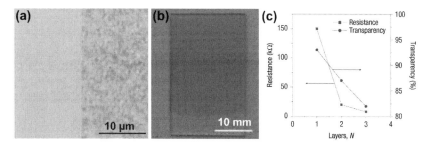

FIGURE 6.3.1 (a) Light-transmission image through glass (left) and through a thin film, consisting of overlapping graphene and few-layer graphene flakes with a thickness of ~1.5 nm. *Reprinted with permission from Blake et al. (2008), Copyright 2008 American Chemical Society.* (b) Large-scale two-layer Langmuir–Blodgett film of graphene sheets on quartz and (c) resistances and transparencies of Langmuir–Blodgett graphene films as a function of the number of graphene layers, highlighting the trade off between conductivity and transmittance. *Reprinted by permission from Macmillan Publishers Ltd: Nat. Nanotechnol. (Li et al., 2008), copyright (2008).*

thin films that exhibit transmittances ranging from 90 to 35%. Over the same range, the sheet resistance falls between 10^6 Ω/sq and 10^3 Ω/sq, again highlighting the trade-off between conductivity and transmittance.

Graphene oxide (GO) solutions are prepared by oxidation of graphite which involves strong acids and subsequent intercalation and exfoliation in water as discussed in Section 4.3. GO is electrically insulating, but conductivity may be recovered by chemical reduction using hydrazine, hydrogen or thermal annealing (Becerril et al., 2008; Stankovich et al., 2006a, 2007). However, reduced graphene oxide (rGO) inevitably contains lattice defects that degrade its electrical properties. Conductivity values often are orders of magnitude below that of pristine mechanically cleaved graphene. Currently, the GO route is being pursued further in many research labs and could lead to niche applications, where high conductivity is not a critical factor (Wassei and Kaner, 2010).

In order to prepare GO thin films, techniques such as dip coating, drop casting, spin coating, spray deposition and vacuum filtration have been utilised (Eda et al., 2008a; Gilje et al., 2007; Wang et al., 2008; Wilson et al., 2010). Reduced GO films then typically showed sheet resistances of more than 10 kΩ/sq (Cote et al., 2009; Eda et al., 2008b; Zhu et al., 2009). Becerril et al. (2008) analysed the efficiency of different chemical and thermal reduction treatments on the electronic properties of rGO. They find that controlled graphitisation treatments at a temperature of 1100 °C in vacuum ($<10^{-5}$ Torr) are most effective in deoxygenating the spin-coated GO films and restoring conductivity. The produced rGO films display sheet resistances on the order of 10^2–10^3 Ω/sq at a light transmittance of 80%. Similar results were reported by Wang et al. (2008). It should however be noted that for practical use of such film, the temperature for graphitisation needs to be reduced in order to make the process compatible with transparent substrates such as glass or plastics (Becerril et al., 2008).

A further improvement of sheet resistance can be achieved via chemical doping (Blake et al., 2008; Zheng et al., 2011). Here, Zheng et al. (2011) present remarkable results from reduced graphene oxide samples. They show a method to produce gram quantities of ultralarge graphene oxide sheets with a lateral size of 50–200 μm. Employing ultralarge graphene flakes should already result into improvement of the electronic properties, since effects that limit charge carrier transport, such as scattering from grain boundaries, are reduced. Zheng et al. (2011) prepared Langmuir–Blodgett (LB) films using these ultra-large GO sheets and thermally reduced them at 1100 °C. After the thermal treatment the sheet resistance of a LB film with a thickness of 3.7 nm was on the order of 600 Ω/sq at a transmittance of 85 %. Films with a thickness of 18.5 nm reached much lower sheet resistances of ~280 Ω/sq, however, as expected, with a significantly reduced transmittance of ~55%. As indicated above, chemical doping should result into enhancement of the electronic properties. Zheng et al. (2011) showed that the sheet resistance of their ultralarge reduced GO films could significantly be reduced by about 30–50% through additional chemical treatments with HNO$_3$ and SOCl$_2$ (Zheng et al., 2011). The 3.7 nm LB film now exhibited a sheet resistance of ~490 Ω/sq at a transmittance of 90%.

Given the fact that graphite costs only about $5/kg, solution processed films with 90% transmittance cost approximately 0.02 ¢/m^2 (thickness ~10 nm, equivalent to ~25 mg/m^2) (De and Coleman, 2010). Further, the ease with which graphene and graphene oxide solutions can be handled makes these solution based approaches highly appealing for large scale preparation of TCs. Solution based graphene and graphene oxide films will certainly play a role in the electronics industry and intensive research efforts in this area will continue.

The preparation of graphene films via chemical vapour deposition (CVD) appears to be one of the most promising routes towards their application as transparent electrodes and make the viability of graphene considerably stronger. Recent reports demonstrate inch-scale growth of graphene with very good optoelectrical properties (Bae et al., 2010; Kim et al., 2009). As discussed in Section 4.5, graphene CVD primarily requires a catalyst and a carbon source. While a vast variety of carbon sources is applied ranging from more commonly used gaseous precursors, like CH$_4$ or C$_2$H$_4$, (Addou et al., 2012; Li et al., 2009a), polymers (Peng et al., 2011; Sun et al., 2010) and highly oriented pyrolytic graphite (HOPG) (Xu et al., 2011) to more inventive carbon sources as food, insects or waste (Ruan et al., 2011), the catalyst variety is more modest. Most commonly and most successfully, nickel thin films or copper foils are used as catalysts (Kim et al., 2009; Li et al., 2009a; Reina et al., 2009). In the case of nickel catalysts carbon is dissolved in the nickel layer and precipitates upon cooling. In order to suppress precipitation of multiple graphene layers a rapid cooling rate is necessary (Kim et al., 2009; Reina et al., 2009). To be able to use these graphenes as transparent electrodes, they have to be transferred onto transparent substrates such as glass or polymers, as already discussed in Section 4.8. Kim et al. (2009) demonstrated a method to transfer the as-grown graphene films using polydimethylsiloxane (PDMS) as a scaffold for the graphene film while etching away the nickel catalyst layer. They further prepared bendable and stretchable transparent electrodes, which exhibited a sheet resistance of R_S ~ 280 Ω/sq at about 80% optical transparency (Kim et al., 2009). Bearing in mind a single-layer absorbs ~2.3%, this corresponds to an average number of six to ten graphene layers (Nair et al., 2008). Figure 6.3.2a shows a graphene film as transferred onto a bendable PDMS stamp. Kim et al. further evaluated the foldability of the graphene films. They transferred graphene onto a polyethylene terephthalate (PET) substrate of 0.1 mm thickness coated with a thin PDMS layer of 0.2 mm thickness. In Fig. 6.3.2b, the resistance variation of the graphene film is plotted with respect to the bending radius. The resistance varies slightly upon bending and is perfectly restored after unbending.

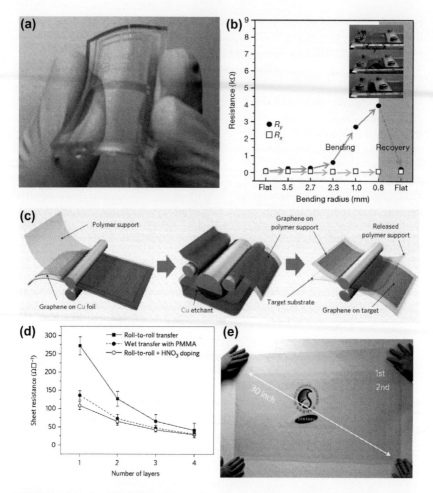

FIGURE 6.3.2 (a) Bendable graphene film on a flexible transparent PDMS substrate. (b) Dependence of the resistance of a 0.3-mm-thick PDMS/PET substrate on the bending radius, parallel (R_y) and perpendicular (R_x) to the bending direction. *Reprinted by permission from Macmillan Publishers Ltd: Nature (Kim et al., 2009), copyright (2009).* (c) Schematic illustration of the roll-to-roll transfer process of Cu-CVD-grown graphene onto a PET substrate. (d) Dependence of the sheet resistance on the number of graphene layers transferred onto PET for different transfer processes (i.e. roll-to-roll and wet transfer) and doped graphene. (e) 30-inch four-layer graphene film as transferred onto flexible PET. *Reprinted by permission from Macmillan Publishers Ltd: Nat. Nanotechnol. (Bae et al., 2010), copyright (2010).*

While nickel-CVD yields few-layer graphene films with relatively small domain sizes and varying number of layers, low pressure CVD growth of graphene on copper is surface related and self-limiting to a mono-layer and domains with sizes on the order of millimetres can be obtained thereby reducing the probability of charge carrier scattering (Li et al., 2011). Li et al.

(2009a,b) were the first to prepare graphene via CVD using copper foil as the catalyst and identified the growth mechanism, which allowed quick access to high quality material. The use of copper foil as catalyst is highly appealing for industrial scale-up, since it is readily available, inexpensive, flexible and can be etched with common solvents (e.g. aqueous $FeCl_3$) in order to transfer the graphene layer onto a substrate of choice. Bae et al. (2010) reported on large scale transfer of Cu-CVD graphene via the roll-to-roll approach. Figure 6.3.2c shows a schematic of this transfer process. In the first rolling step, the graphene film, grown on a copper foil, is attached to a thermal release tape, i.e. a polymer film, coated with an adhesive layer (left panel) which is used as polymer scaffold in this case. The copper foil is then etched by electrochemical reaction with aqueous ammonium persulphate solution $(NH_4)_2S_2O_8$ leaving the graphene layer attached to the adhesive film (middle panel). Finally, the graphene film on the thermal release tape is inserted between the rollers together with a target substrate and released onto the target substrate via exposure to 90–120 °C. Figure 6.3.2d summarises the results from electrical characterisation. Monolayer graphene films with a sheet resistance of $R_S = 272$ Ω/sq at 97.4% transmittance were prepared, using the thermal release tape (filled square in the top left corner of Fig. 6.3.2d) (Bae et al., 2010).

Other, more commonly used methods to transfer graphene onto insulating substrate involve a polymeric scaffold, e.g. PDMS or poly(methylmethacrylate) (PMMA), to adhere to the graphene film while etching the catalyst layer (Kim et al., 2009; Li et al., 2009c). Bae et al. (2010) compared their results from electrical characterisation to single-layer graphene films that were transferred using a PMMA scaffold (filled circles in Fig. 6.3.2d). This wet transfer method yields graphene films with a lower sheet resistance ($R_S = 125$ Ω/sq) as compared to transfer with thermal tape or PDMS stamps. This can be attributed to the formation of cracks during transfer, which is inevitable and will impair the optoelectronic properties (Kim et al., 2009; Srivastava et al., 2010). Using PMMA as the transfer scaffold appears to result in less defects and cracks, ultimately leading to better optoelectronic properties (Mattevi et al., 2011). Bae et al. (2010) further demonstrate that, by repeating the transfer steps on the same substrate, multilayered graphene films with even further reduced sheet resistance can be prepared. Placing four graphene layers on top of each other results into a transparent conducting film with a sheet resistance of $R_S \sim 50$ Ω/sq at ~90% transmittance. Figure 6.3.2e shows a 30-inch four-layer graphene film as transferred onto a flexible PET substrate via the roll-to-roll process (Bae et al., 2010). A further reduction of sheet resistance can be achieved by wet-chemical doping, as demonstrated by other groups via solution-based graphene preparation (Blake et al., 2008; Zheng et al., 2011). In the case of the roll-to-roll approach, doping with HNO_3 resulted in a reduction of the sheet resistance from 272 to 108 Ω/sq (~59%) for a graphene mono-layer (97.4%

transmittance) (Bae et al., 2010). A HNO_3-doped four-layer graphene films exhibited a sheet resistance of only $R_S \sim 30$ Ω/sq at ~90% transmittance, which is comparable with values obtained for ITO thin films (Bae et al., 2010). This large-scale graphene fabrication and transfer method is the first to surpass the minimum industry standard for transparent electrodes (De and Coleman, 2010). However, the stability of the graphene-dopant system is unknown. From investigations into doped carbon nanotube (CNT) electrodes, it is known that exposure to air and moderate temperature results into an increased sheet resistance and limits the performance of these electrodes over time (Jackson et al., 2008). Therefore, with long lifetime applications, i.e. solar cells or LEDs, in mind, doping stability and doping effects on the graphene properties have to be investigated in detail. Figure 6.3.3 compares the optoelectronic properties of carbon nanotubes, ITO and various CVD-grown graphene films derived from different preparation and transfer methods (Bae et al., 2010).

However, the layer-by-layer transfer realised with this roll-to-roll approach is time-consuming and cost intensive. For technological implementation other routes, e.g. the direct synthesis of multi-layer graphene, may be economically more viable and may also lead to superior opto-electronic properties of large area electronics (Mattevi et al., 2011). Direct growth of graphene on inexpensive transparent substrates such as glass or plastics is also an issue with synthesis temperatures as high as ~1000 °C. Advances in reducing the synthesis

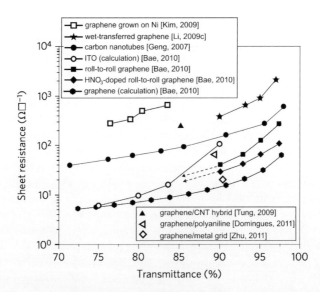

FIGURE 6.3.3 Sheet resistance vs. transmittance plots comparing results of CVD-grown graphene, CNTs, hybrid materials and ITO. *Adapted by permission from Macmillan Publishers Ltd: Nat. Nanotechnol. (Bae et al., 2010), copyright (2010).*

temperature have been made using plasma enhanced CVD techniques (Zhang et al., 2011) or employing more reactive carbon sources, like ethylene or acetylene, as compared to commonly used methane (Addou et al., 2012; Lee and Choi, 2011). However, the optoelectronic properties are not yet at parity with those of Cu-CVD grown graphene or ITO.

A further notable example has been demonstrated by Tung et al. (2009). They showed that a nanocomposite comprised of reduced graphene oxide and carbon nanotubes (CNTs) is a viable alternative to transparent conductors made from graphene or CNTs alone in terms of relatively low material costs and opto-electronic properties. Tung et al. deposited thin nanocomposite films onto glass substrates via spin coating. The thinnest films exhibited a sheet resistance of 636 Ω/sq at 88% transmittance. Further exposure to $SOCl_2$ vapour resulted in chemical doping and yielded films with an improved sheet resistance of 240 Ω/sq at only a slightly reduced transmittance of 86% (Tung et al., 2009). Another suitable transparent conducting nanocomposite was introduced by Domingues et al. (2011). They developed a route to prepare a graphene/polyaniline nanocomposite dissolving reduced graphene oxide and aniline in a mixture of aq-HCl and ammonium persulfate via magnetic stirring at 1500 rpm for 22 h. Upon interrupting the magnetic stirring a continuous transparent graphene/polyaniline nanocomposite film spontaneously formed via interfacial polymerisation. In this way, transparent nanocomposite films with a sheet resistance of 60.6 Ω/sq at a transmittance of 89% were achieved, which is comparable to ITO-based electrodes (Domingues et al., 2011). Further, the combination of metal grids with graphene promises to outperform ITO electrodes (Zhu et al., 2011). Zhu et al. demonstrated that photolithographically prepared Au or Cu grids covered by CVD-grown graphene exhibit sheet resistances as low as 20 Ω/sq at a transmittance of 91%. At lower transmittance, sheet resistances down to 3 Ω/sq have been reported (Zhu et al., 2011).

Table 6.3.1 summarises the sheet resistance and transmittance values for different graphene synthesis and graphene film preparation methods. The results selected for Table 6.3.1 are chosen with focus on highest combined opto-electronic performance achieved in the different research areas of solution based and CVD-based graphene synthesis.

Examples of prototype systems, where graphene transparent electrodes have been demonstrated, are manifold. Blake et al. (2008) demonstrated a liquid crystal device based on mechanically exfoliated graphene (Blake et al., 2008). The bottom transparent electrode was made of graphene while the top electrode consisted of ITO. Optical micrographs of the liquid crystal device with different voltages applied across the cell are shown in Fig. 6.3.4a, demonstrating the general feasibility of graphene electrodes in LCD technology. Mechanically exfoliated graphene derived from cleaving HOPG provides the best quality in terms of structural integrity and, therefore offers low sheet resistance as well as high transmittance. Since the electrodes in

TABLE 6.3.1 Summary of the Sheet Resistance and Transmittance Values of Graphene Films Derived via Different Synthesis and Film Preparation techniques. For Each Technique, the Reference With Best Optoelectronic Performance has Been Selected.

Graphene synthesis	Film deposition/transfer	Sheet resistance	Transmittance	Reference
Exfoliated graphite (DMF)	Spray coating	5000 Ω/sq	90%	(Blake et al., 2008)
Exfoliated graphite (sodium cholate)	Vacuum filtration	4000 Ω/sq	75%	(De et al., 2010)
rGO + high temperature reduction (1100 °C)	Spin coating	800 Ω/sq	82%	(Wu et al., 2010)
rGO + high temperature reduction + doping in HNO$_3$	Langmuir–Blodgett film deposition	459 Ω/sq	~90%	(Zheng et al., 2011)
CVD with Ni catalyst	Dry transfer using PDMS stamp	280 Ω/sq	~80%	(Kim et al., 2009)
rGO-CNT hybrid + anion doping in SOCl$_2$ vapour	Spin coating	240 Ω/sq	86%	(Tung et al., 2009)
CVD with Cu catalyst	Dry transfer with thermal release tape + p-doping with HNO$_3$	125 Ω/sq (monolayer) 30 Ω/sq (four layers)	97.4% ~90%	(Bae et al., 2010)
rGO-polyaniline nanocomposite	Spontaneous formation through interfacial polymerisation	60.5 Ω/sq	89%	(Domingues et al., 2011)
Cu-CVD grown graphene + Cu or Au metal grids	Wet transfer of graphene onto metal grid using PMMA sacrificial layer	Au 20 Ω/sq Cu 22 Ω/sq	91% 91%	(Zhu et al., 2011)

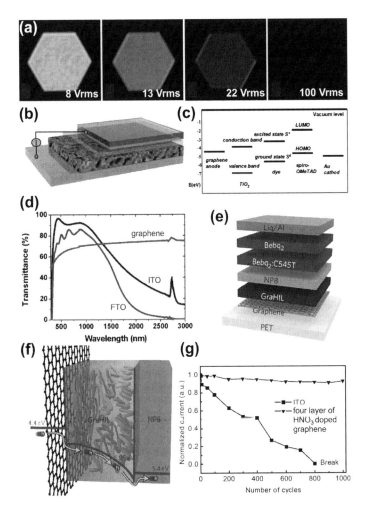

FIGURE 6.3.4 (a) Optical micrographs of a LCD using green light (505 nm, full-width half-maximum 23 nm) with different voltages applied across the cell as marked in the images; the hexagonal window is covered by graphene and surrounded by the opaque Cr/Au electrode; image width: 30 μm. *Adapted with permission from Blake et al. (2008), Copyright (2008) American Chemical Society.* (b) Schematic of a dye-sensitised solar cell with a graphene electrode: the four layers from bottom to top are Au, dye-sensitised heterojunction, compact TiO_2 and graphene electrode. (c) Energy level diagram of the graphene/TiO_2/dye/spiro-OMeTAD/Au device. (d) Comparison of the transmittance of a ~10-nm thick graphene film with that of ITO and FTO. *Reprinted with permission from Wang et al. (2008), Copyright (2008) American Chemical Society.* (e) Device structure of a flexible fluorescent OLED device. The graphene anode is modified with a hole-injection layer (HIL). (f) Schematic of the hole-injection process from the graphene anode via a self-organised HIL with work-function gradient (GraHIL) to the NPB hole-transport layer. *Reprinted by permission from Macmillan Publishers Ltd: Nat. Photonics (Han et al., 2012), copyright (2012).* (g) Comparison of the bending stability of OLEDs with graphene and ITO anode. The current at 3 V was measured as a function of bending cycles. *Reprinted with kind permission of the authors of Han et al. (2012).*

a liquid crystal device are generally in contact with an alignment layer (in this case: polyvinyl alcohol), doping effects may occur (Blake et al., 2008). In the experiment by Blake et al. (2008) the layer of polyvinyl alcohol produced n-type doping on the order of $3 \cdot 10^{12}$ cm^{-2} in the graphene electrode resulting in a relatively low sheet resistance of 400 Ω/sq at an optical transmittance of about 98%, which corresponds to the value absorbed by a single layer of graphene as discussed above.

Wang et al. (2008) demonstrated the application of graphene as window electrodes in solid-state dye-sensitised solar cells. Figure 6.3.4b shows an illustration of the solar cell structure including a graphene anode as window electrode together with the energy level diagram of the graphene/porous TiO$_2$/dye/spiro-OMeTAD/Au device (Fig. 6.3.4c). Graphene films were prepared from an aqueous graphene oxide solution through dip-coating and high-temperature vacuum reduction at 1100 °C Wang et al., 2008. A graphene film with a thickness of 10 nm exhibited a sheet resistance of ~1.8 kΩ/sq at a transmittance of ~70%. Although the transmittance of these particular graphene films was lower as compared to ITO or fluorine-doped tin oxide (FTO) in the visible region ITO and FTO showed strong absorption in the near-infrared region, while the graphene films maintained their transmittance as shown in Fig. 6.3.4d, which highlights the suitability of graphene films as window electrodes for optoelectronic devices. Further, graphene electrodes are advantageous over metal oxide electrodes in that they not only provide the necessary optoelectronic properties but also offer higher chemical and thermal stability. Wang et al. (2008) showed that after heating the graphene film at 400 °C in air, the film remained intact, and the conductivity was comparable to that of the original graphene film.

Wu et al. (2008) prepared graphene electrodes for application as transparent anodes in solid-state thin-film organic photovoltaic (OPV) cells. They spin-coated graphene oxide onto quartz. After high temperature vacuum annealing they obtained sheet resistances of 100–500 kΩ/sq at an optical transmittance of >80%. The OPV cell was then directly deposited onto the graphene anode exhibiting a layer structure consisting of graphene anode, copper phthalocyanine, fullerene, bathocuproine and a 100 nm Ag cathode. Wu et al. (2008) could show that the performance of these OPV cells approaches that of simultaneously prepared cells, based on ITO anodes. Later, the same group demonstrated the use of graphene anodes in green fluorescent organic light-emitting diodes (OLED) (Wu et al., 2010). A luminous efficiency of 0.35 lm W^{-1} and a sheet resistance of 800 kΩ/sq at an optical transmittance of 82% are reported. The sheet resistance is much higher than theoretically estimated which can be ascribed to the presence of defects typically related to the solution based approach. The graphene films contain only small graphene flakes and lattice defects from the oxidation/reduction process, and therefore scattering at defects and grain boundaries are major factors in reducing the electrical conductivity

(Wu et al., 2010). The device performance may however be improved by adopting new approaches such as particle sizing and chemical doping of the graphene films to increase the conductivity (Zheng et al., 2011). Further, the low work function (WF) of graphene (WF ~ 4.4 eV) limits the practical application of graphene electrodes in optoelectronic devices due to a high hole injection barrier at the interface between the graphene electrode and the hole transport layer.

Recently, Han et al. (2012) have tackled both problems, i.e. the low work function and the high sheet resistance, and demonstrated the fabrication of OLEDs with transparent electrodes made from graphene, that are not only flexible but also outperform OLEDs with ITO-based electrodes. Figure 6.3.4e shows the OLED structure. Two, three or four layers of graphene grown via copper CVD are doped using HNO_3 (sheet resistance down to ~30 Ω/sq) and transferred onto a PET substrate, as previously reported by Bae et al. (2010). A hole injection layer (HIL) with a work function gradient (GraHIL) composed of poly(3,4-ethylenedioxythiophene) and doped with poly (styrenesulphonate) (PEDOT:PSS) and a tetrafluoroethylene-perfluoro-3,6-dioxa-4-methyl-7-octenesulphonic acid copolymer (WF = 5.95 eV) enables holes to be injected efficiently to the overlying organic hole transport layer (N,N'-bis(naphthalen-1-yl)-N,N'-bis(phenyl)benzidine, NPB, WF ~ 5.4 eV), as schematically illustrated in Fig. 6.3.4f. Han et al. showed that in this way OLEDs with HNO_3-doped four-layer graphene electrodes and the gradient HIL showed current efficiencies of up to 98.1 cd A^{-1} and luminous efficiencies up to 102.7 lm W^{-1}, which is almost two orders of magnitude higher than previously reported OLEDs with graphene or CNT electrodes (Han et al., 2012; Hu et al., 2010; Wu et al., 2010). Further, the luminous and current efficiencies were also better than the said efficiencies of simultaneously prepared OLEDs based on ITO electrodes. Additional bending tests impressively demonstrated the bending stability of OLEDs with graphene anode over OLEDs with ITO anode, as depicted in Fig. 6.3.4g. While the graphene device maintained almost the same current density over 1000 times bending with a bending radius of 0.75 cm, the ITO device failed after 800 bending cycles.

The use of hybrid materials with graphene components has also been demonstrated to be suitable for optoelectronic applications. As previously mentioned, Tung et al. (2009) fabricated transparent nanocomposite electrodes that comprised of reduced graphene oxide and carbon nanotubes and achieved sheet resistances down to 240 Ω/sq at 86% transmittance. They further demonstrate an organic solar cell device, comprising a graphene/CNT based hybrid anode with a power conversion efficiency of 0.85% (Tung et al., 2009).

A further notable example of a prototype device has been presented by Bae et al. (2010). Using their roll-to-roll transfer technology (cf. Fig. 6.3.2c), they incorporated graphene electrodes into a fully functional four-wire touch-screen panel and demonstrated its outstanding flexibility. From a strain analysis they reported that the graphene based touch panels can resist up to 6% strain while ITO-based touch panels survive only up to 2–3% (Bae et al., 2010; Cairns et al., 2000). In ITO-based devices the increase in resistance is due to cracking of the ITO layer with increasing strain (Cairns et al., 2000). Bae et al. (2010) could further show that the performance of the graphene based touch panel was not limited by the graphene electrode but by silver electrodes printed onto the graphene/PET film to fabricate the touch screen panel.

To conclude, graphene and graphene oxide hold great potential for application as transparent electrode in optoelectronic devices. Respective research efforts are not only driven by the emerging necessity to replace existing transparent conducting oxide materials due to their high prize, scarcity and manufacturing costs but also by the growing market development towards bendable electronics. Here, flexible graphene based electrodes provide crucial performance advantages over state-of-the-art TCOs, which are brittle due to their ceramic nature. In the early stages of research into graphene films for application as transparent conducting electrodes, substantial approaches towards scalability with solution-based methods and CVD as well as deposition and transfer technologies have already been developed. Electrode performance reaches values sufficient to replace TCOs in many existing applications. Especially, in applications where high transparencies are required graphene may outperform existing TCs (cp. Fig. 6.3.3). Many prototype applications have been demonstrated, but there is room for improvement towards more profitable production and transfer technologies. In addition, long-time performance tests have so far scarcely been carried out but will be required for implementation of graphene based electrodes.

REFERENCES

Addou, R., Dahal, A., Sutter, P., Batzill, M., 2012. Monolayer graphene growth on Ni(111) by low temperature chemical vapor deposition. Appl. Phys. Lett. 100, 021601.

Bae, S., Kim, H., Lee, Y., Xu, X., Park, J.-S., Zheng, Y., Balakrishnan, J., Lei, T., Kim, H.R., Song, Y.I., Kim, Y.-J., Kim, K.S., Özyilmaz, B., Ahn, J.-H., Hong, B.H., Iijima, S., 2010. Roll-to-roll production of 30-inch graphene films for transparent electrodes. Nat. Nanotechnol. 5, 574–578.

Becerril, H.A., Mao, J., Liu, Z., Stoltenberg, R.M., Bao, Z., Chen, Y., 2008. Evaluation of solution-processed reduced graphene oxide films as transparent conductors. ACS Nano 2, 463–470.

Blake, P., Brimicombe, P.D., Nair, R.R., Booth, T.J., Jiang, D., Schedin, F., Ponomarenko, L.A., Morozov, S.V., Gleeson, H.F., Hill, E.W., Geim, A.K., Novoselov, K.S., 2008. Graphene-based liquid crystal device. Nano Lett. 8, 1704–1708.

Burroughes, J.H., Bradley, D.D.C., Brown, A.R., Marks, R.N., Mackay, K., Friend, R.H., Burns, P.L., Holmes, A.B., 1990. Light-emitting diodes based on conjugated polymers. Nature 347, 539–541.

Cairns, D.R., Witte II, R.P., Sparacin, D.K., Sachsman, S.M., Paine, D.C., Crawford, G.P., 2000. Strain-dependent electrical resistance of tin-doped indium oxide on polymer substrates. Appl. Phys. Lett. 76, 1425–1427.

Cote, L.J., Kim, F., Huang, J., 2009. Langmuir-Blodgett assembly of graphite oxide single layers. J. Am. Chem. Soc. 131, 1043–1049.

De, S., King, P.J., Lotya, M., O'Neill, A., Doherty, E.M., Hernandez, Y., Duesberg, G.S., Coleman, J.N., 2010. Flexible, transparent, conducting films of randomly stacked graphene from surfactant-stabilized, oxide-free graphene dispersions. Small 6, 458–464.

De, S., Coleman, J.N., 2010. Are there fundamental limitations on the sheet resistance and transmittance of thin graphene films? ACS Nano 4, 2713–2720.

Domingues, S.H., Salvatierra, R.V., Oliviera, M.M., Zarbin, A.J.G., 2011. Transparent and conducting thin films of graphene/polyaniline nanocomposites prepared through interfacial polymerization. Chem. Comm. 47, 592–2594.

Du, Xu, Skachko, Ivan, Barker, Anthony, Andrei, Eva Y., 2008. Approaching ballistic transport in suspended graphene. Nat. Nanotechnol. 3, 491–495.

Eda, G., et al., 2008a. Large-area ultrathin films of reduced graphene oxide as a transparent and flexible electronic material. Nat. Nanotechnol. 3, 270–274.

Eda, G., et al., 2008b. Transparent and conducting electrodes for organic electronics from reduced graphene oxide. Appl. Phys. Lett. 92, 233305.

Geng, H.-Z., Kim, K.K., P So, K., Lee, Y.S., Chang, Y., Lee, Y.H., 2007. Effect of acid treatment on carbon nanotube-based flexible transparent conducting films. J. Am. Chem. Soc. 129, 7758–7759.

Gilje, S., Han, S., Wang, M., Wang, K.L., Kaner, R.B., 2007. A chemical route to graphene for device applications,. Nano Lett. 7, 3394–3398.

Gordon, R., 2000. Criteria for choosing transparent conductors. MRS Bull. 25 (8), 52–57.

Han, T.-H., Lee, Y., Choi, M.-R., Woo, S.-H., Bae, S.-H., Hong, B.H., Ahn, J.-H., Lee, T.-W., 2012. Extremely efficient flexible organic light-emitting diodes with modified graphene anode. Nat. Photonics 6, 105–110.

Hecht, D.S., Hu, L., Irvin, G., 2011. Emerging transparent electrodes based on thin films of carbon nanotubes, graphene, and metallic nanostructures. Adv. Mater. 23, 1482–1513.

Hernandez, Y., Nicolosi, V., Lotya, M., Blighe, F.M., Sun, Z., De, S., Mc-Govern, I.T., HollandByrne, B., Byrne, M., Gun'Ko, Y.K., Boland, J.J., Niraj, P., Duesberg, G., Krishnamurthy, S., Goodhue, R., Hutchison, J., Scardaci, V., Ferrari, A.C., Coleman, J.N., 2008. High-yield production of graphene by liquid-phase exfoliation of graphite. Nat. Nanotechnol. 3, 563–568.

Hu, L., Li, J., Liu, J., Grüner, G., Marks, T., 2010. Flexible organic light-emitting diodes with transparent carbon nanotube electrodes: problems and solutions. Nanotechnology 21, 155202.

Jackson, R., Domercq, B., Jain, R., Kippelen, B., Graham, S., 2008. Stability of doped transparent carbon nanotube electrodes. Adv. Func. Matter. 18, 2548–2554.

Kim, Keun S., Zhao, Y., Jang, H., Lee, S.Y., Kim, J.M., Kim, Kwang S., Ahn, J.H., Kim, P., Choi, J.Y., Hong, B.H., 2009. Large-scale pattern growth of graphene films for stretchable transparent electrodes. Nature 457, 706–710.

Lee, K.H., Jang, H.W., Kim, K.-B., Tak, Y.-H., Lee, J.-L., 2004. Mechanism for the increase of indium-tin-oxide work function by O_2 inductively coupled plasma treatment. J. Appl. Phys. 95, 586–590.

Lee, K.Y., Becker, C., Muske, M., Ruske, F., Gall, S., Rech, B., Berginski, M., Hüpkes, J., 2007. Temperature stability of ZnO:AL film properties for poly-Si thin-film devices. Appl. Phys. Lett. 91, 241911.

Lee, C.M., Choi, J., 2011. Direct growth of nanographene on glass and postdeposition size control. Appl. Phys. Lett. 98, 183106.

Lewis, J., Grego, S., Vick, E., Chalamala, B., Temple, D., 2004. Mechanical performance of thin films in flexible displays. Mat. Res. Soc. Symp. Proc. 814, 189–198.

Li, X., Zhang, G., Bai, X., Sun, X., Wang, X., Wang, E., Dai, H., 2008. Highly conducting graphene sheets and Langmuir–Blodgett films. Nat. Nanotechnol. 3, 538–542.

Li, X., Cai, W., An, J., Kim, S., Nah, J., Yang, D., Piner, R., Velamakanni, A., Jung, I., Tutuc, E., Banerjee, S.K., Colombo, L., Ruoff, R.S., 2009a. Large area synthesis of high-quality and uniform graphene films on copper foils. Science 324, 1312–1314.

Li, X., Cai, W., Colombo, L., Ruoff, R.S., 2009b. Evolution of graphene growth on Ni and Cu by carbon isotope labeling. Nano Lett. 9, 4268–4272.

Li, X., Zhu, Y., Cai, W., Borysiak, M., Han, B., Chen, D., Piner, R.D., Colombo, L., Ruoff, R.S., 2009c. Transfer of large-area graphene films for high-performance transparent conductive electrodes. Nano Lett. 9, 4359–4363.

Li, X., Magnuson, C.W., Venugopal, A., Tromp, R.M., Hannon, J.B., Vogel, E.M., Colombo, L., Ruoff, R.S., 2011. Large-Area graphene single crystals grown by low-pressure chemical vapor deposition of methane on copper. J. Am. Chem. Soc. 133, 2816–2819.

Mattevi, C., Kima, H., Chhowalla, M., 2011. A review of chemical vapour deposition of graphene on copper,. J. Mater. Chem. 21, 3324–3334.

Minami, T., 2008. Present status of transparent conducting oxide thin-film development for Indium-Tin-Oxide (ITO) substitutes. Thin Solid Films 516, 5822–5828.

Nair, R.R., Blake, P., Grigorenko, A.N., Novoselov, K.S., Booth, T.J., Stauber, T., Peres, N.M.R., Geim, A.K., 2008. Fine structure constant defines visual transparency of graphene. Science 320, 1308.

Nomoto, J., Hirano, T., Miyata, T., Minami, T., 2011. Preparation of Al-doped ZnO transparent electrodes suitable for thin-film solar cell applications by various types of magnetron sputtering depositions. Thin Solid Films 520, 1400–1406.

Novoselov, K.S., Geim, A.K., Morozov, S.V., Jiang, D., Katsnelson, M.I., Grigorieva1, I.V., Dubonos, S.V., Firsov, A.A., 2005. Two-dimensional gas of massless Dirac fermions in graphene. Nature 438, 197–200.

O'Connor, B., Haughn, C., An, K.-H., Pipe, K.P., Shtein, M., 2008. Transparent and conductive electrodes based on unpatterned thin metal films. Appl. Phys. Lett. 93, 223304.

Peng, Z., Yan, Z., Sun, Z., Tour, J.M., 2011. Direct growth of bilayer graphene on SiO_2 substrates by carbon diffusion through nickel. ACS Nano 5, 8241–8247.

Peumans, P., Yakimov, A., Forrest, S.R., 2003. Small molecular weight organic thin-film photodetectors and solar cells. J. Appl. Phys. 93, 3693–3723.

Reina, A., Jia, X., Ho, J., Nezich, D., Son, H., Bulovic, V., Dresselhaus, M.S., Kong, J., 2009. Large area, few-layer graphene films on arbitrary substrates by chemical vapor deposition. Nano Lett. 9, 30–35.

Ruan, G., Sun, Z., Peng, Z., Tour, J.M., 2011. Growth of graphene from food, insects, and waste. ACS Nano 5, 7601–7607.

Schedin, F., Geim, A.K., Morozov, S.V., Hill, E.W., Blake, P., Katsnelson, M.I., Novoselov, K.S., 2007. Detection of individual gas molecules adsorbed on graphene. Nat. Mater. 6, 652–655.

Srivastava, A., Galande, C., Ci, L., Song, L., Rai, C., Jariwala, D., Kelly, K.F., Ajayan, P.M., 2010. Novel liquid precursor-based facile synthesis of large-area continuous, single, and few-layer graphene films. Chem. Mater. 22, 3457–3461.

Stankovich, S., Piner, R.D., Chen, X., Wu, N., Nguyen, S.T., Ruoff, R.S., 2006. Stable aqueous dispersions of graphitic nanoplatelets via the reduction of exfoliated graphite oxide in the presence of poly(sodium 4-styrenesulfonate). J. Mater. Chem. 16, 155–158.

Stankovich, S., Dikin, D.A., Piner, R.D., Kohlhaas, K.A., Kleinhammes, A., Jia, Y., Wu, Y., Nguyen, S.T., Ruoff, R.S., 2007. Synthesis of graphene-based nanosheets via chemical reduction of exfoliated graphite oxide. Carbon 45, 1558–1565.

Sun, Z., Yan, Z., Yao, J., Beitler, E., Zhu, Y., Tour, J.M., 2010. Growth of graphene from solid carbon sources. Nature 468, 549–552.

Tvingstedt, K., Inganäs, O., 2007. Electrode grids for ITO-free organic photovoltaic devices, Adv. Mater 19, 2893–2897.

Tung, V.C., Chen, L.-M., Allen, M.J., Wassei, J.K., Nelson, K., Kaner, R.B., Yang, Y., 2009. Low-temperature solution processing of graphene-carbon nanotube hybrid materials for high-performance transparent conductors. Nano Lett. 9, 1949–1955.

US, 2012. http://minerals.usgs.gov/minerals/pubs/commodity/indium/mcs-2012-indiu.pdf.

Vinnichenko, M., Gago, R., Cornelius, S., Shevchenko, N., Rogozin, A., Kolitsch, A., Munnik, F., Möller, W., 2010. Establishing the mechanism of thermally induced degradation of ZnO:Al electrical properties using synchrotron radiation. Appl. Phys. Lett. 96, 141907.

Wang, X., Zhi, L., Müllen, K., 2008. Transparent, conductive graphene electrodes for dye-sensitized solar cells. Nano Lett. 8, 323–327.

Wang, Y., Tong, S.W., Xu, X.F., Özyilmaz, B., Loh, K.P., 2011. Interface engineering of layer-by-layer stacked graphene anodes for high performance organic solar cells. Adv. Mater. 23, 1514–1518.

Wassei, J.K., Kaner, R.B., 2010. Graphene, a promising transparent conductor. Mater. Today 13, 52–59.

Wilson, N.R., Pandey, P.A., Beanland, R., Rourke, J.P., Lupo, U., Rowlands, G., Römer, R.A., 2010. On the structure and topography of free-standing chemically modified graphene. New J. Phys. 12, 125010.

Wu, Z., Chen, Z., Du, X., Logan, J.M., Sippel, J., Nikolou, M., Kamaras, K., Reynolds, J.R., Tanner, D.B., Hebard, A.F., Rinzler, A.G., 2004. Transparent, conductive carbon nanotube films. Science 305, 1273–1276.

Wu, J., Becerril, H.A., Bao, Z., Liu, Z., Chen, Y., Peumans, P., 2008. Organic solar cells with solution-processed graphene transparent electrodes. Appl. Phys. Lett. 92, 263302.

Wu, J., Agrawal, M., Becerril, H.A., Bao, Z., Liu, Z., Chen, Y., Peumans, P., 2010. Organic light-emitting diodes on solution-processed graphene transparent electrodes. ACS Nano 4, 43–48.

Xu, M., Fujita, D., Sagisaka, K., Watanabe, E., Hanagata, N., 2011. Production of extended single-layer graphene. ACS Nano 5, 1522–1528.

Zhang, L., Shi, Z., Wang, Y., Yang, R., Shi, D., Zhang, G., 2011. Catalyst-free growth of nano-graphene films on various substrates. Nano Res. 4, 315–321.

Zheng, Q., Ip, W.H., Lin, X., Yousefi, N., Yeung, K.K., Li, Z., Kim, J.-K., 2011. Transparent conductive films consisting of ultralarge graphene sheets produced by Langmuir-Blodgett assembly. ACS Nano 5, 6039–6051.

Zhu, Y., Cai, W., Piner, R.D., Velamakanni, A., Ruoff, R.S., 2009. Transparent self-assembled films of reduced graphene oxide platelets. Appl. Phys. Lett. 95, 103104.

Zhu, Y., Sun, Z., Yan, Z., Jin, Z., Tour, J.M., 2011. Rational design of hybrid graphene films for high-performance transparent electrodes. ACS Nano 5, 6472–6479.

Nanoelectromechanical Systems (NEMS) using Graphene

Mark H. Rümmeli

IFW Dresden, Germany

Nanoelectromechanical systems (NEMS) are devices at the nanoscale that integrate both electrical and mechanical functionality. Typically NEMS assimilate transistor-like nanoelectronics with mechanical actuators, pumps, or motors to form physical, biological and chemical sensors. The fact that these devices exist at the nanometre scale implies low mass, high mechanical resonance frequencies and potentially large mechanical effects and a high surface-to-volume ratio. Usually the gain in sensitivity by decreasing size implies a decrease in mechanical stability and thickness. Graphene though, despite being only one atom thick is remarkably strong and stiff (see Section 3.4). Moreover, it is electrically conductive lending itself to integrated electrical transconductance and, in addition its planar geometry makes it suitable for lithographic processing (Barton et al., 2011). Hence there is an increasing interest in graphene as a material for nanoelectromechanical devices such as accelerometers, pressure sensors, resonators and much more.

6.4.1. ACTUATION, DETECTION AND QUALITY FACTOR OF NEMS

6.4.1.1. Actuation and Detection

The core element of mechanical devices is, for the most part, a section of material that is in some way suspended so that it is free to resonate. In the case of graphene various fabrication routes for suspended sheets have been developed, some of which are presented during discussion of the developed NEMS devices to date. It is often argued that the performance of a nanomechanical resonator is limited by the read-out technology (Ekinci et al., 2004). Thus the transduction (transfer of mechanical motion to some other form, e.g. an electrical signal) of graphene is crucial. Various methods have been developed for graphene and no doubt new ones will emerge. The first motion detection for suspended graphene involved an optical interferometric effect. In short, a laser beam is fired on to the device perpendicular to the graphene membrane and reflects off both the graphene sheet and the support backplane beneath the device. The double reflection surfaces for a Fabry-Perot interferometer so that mechanical variations in the graphene sheet translate into variations in the reflected light intensity which are easily detected by a fast photodiode. An advantage of this detection route is that electrodes are not required. On the down side the optics involved are bulky. The first electrical readout of mechanical motion in graphene took advantage of a technique initially developed for carbon nanotubes. In this method, the resonator itself is used to mix down the signal to a lower frequency and in so doing overcomes impedance difficulties associated with monitoring electrical signals from NEMS. The process involves the mixing of two high frequency signals, f, (applied to the gate to drive graphene sheet oscillation) and a second rf signal, $f + \Delta f$ applied to the source. This results in a mixed down source drain current at Δf. A draw back of the technique is the obtained operating frequency Δf (typically ca. 1 kHz), limits the operational bandwidth. The use of local back gates minimises stray capacitance can improve the operational bandwidth.

In terms of actuating or driving the motion of graphene this is usually achieved by electrical or optical routes. In the electrical approach, an alternating electrical signal relative to a gate to drive motion by capacitive attraction (electrostatic tension) is used. The optical route implements a pulsed photon signal from a laser. Some of the laser beam is absorbed, leading to heating, and hence the oscillating photon signal leads to thermal expansion and contraction, viz. temperature changes lead to device motion. Greater details on these processes can be found in reference Barton et al. (2011) and references within.

6.4.1.2. Quality Factor

The performance of any nanoelecromechanical resonator is usually quantified in terms of its quality factor. It is defined as:

$$Q = 2\pi(E_{\text{total}}/\Delta E_{\text{cycle}}) = f/\Delta f$$

where f is the resonant frequency of the resonator, Δf is the full width half power maximum of the Lorentzian amplitude response peak, E_{total} is the total energy stored in the resonator and ΔE_{cycle} is the energy lost per cycle. High quality factors then have a narrow resonance peak and are desirable for the efficient performance of nanomechanical resonators serving as a sensor or signal processor. At room temperature Q factors between 80 and 300 have been observed for single layer graphene while thicker films can reach values of 4000 (Barton et al., 2011). At lower temperature (90 mK) a Q factor of ca. 100,000 was observed (Eichler et al., 2011). The dissipation processes in graphene resonantors is still not fully understood and are related to temperature and size. Recent evidence hints that non linear damping determines the Q factor (Eichler et al., 2011). Theoretical studies suggest the influence of mechanical forces on ribbon conductance is strongly dependant on the ribbon edge symmetry. Zigzag edge nanoribbons are robust against high strain deformations, while armchair edge configurations are sensitive to a stretching-induced metal–semiconductor transition, which would make them better suited for electromechanical applications (Poetschke et al., 2010). A computational study by Lam et al. (2009) suggests further device performance optimisation could be achieved by increasing channel length, adjustment of the Fermi level via electrostatic doping and reducing the effective gap thickness.

6.4.2. GRAPHENE ELECTROMECHANICAL RESONATORS

An early experimental study investigated NEMS fabricated from single- and multilayer graphene sheets obtained by mechanical exfoliation (Bunch et al., 2007). The exfoliated sheets were then deposited over predefined trenches etched into a SiO_2 surface. This yielded a micron scale beam clamped at opposing ends. The clamping force is provided by the van der Waals interaction between the SiO_2 and the overlying graphene. Some of the devices had gold electrodes placed for electrical measurements. The actual resonator measurements were performed at room temperature in high vacuum ($<10^{-6}$ Torr). Both electrical (rf frequency) and optical modulation were explored as activation mechanisms. The amplitude versus frequency measurements showed multiple resonances with a single most prominent peak at the lowest frequency. The dominant frequency was ascribed as the fundamental vibrational mode, since its detected intensity was greatest when the motion is in-phase over the whole suspended region. Results from 33 resonators with thicknesses ranging from monolayer graphene to multilayer graphene up to 75 nm thick were investigated. The fundamental vibration frequencies varied from 1 MHz to 170 MHz, and the Q factors varied from 20 to 850. Their computational and experimental data showed the thicker resonators operate in the bonding-dominated limit with a characteristic Young's modulus, E, characteristic of bulk graphite. This places the resonators amongst the highest modulus systems (0.5–2 TPa). Si cantilevers and carbon nanotubes exhibit modulus values in the GPa regime. The high Young's modulus, low mass and large areas of these type of resonators make them ideal for use as mass, force and charge sensors.

Chen et al. (2009) conducted in depth studies on monolayer graphene nanomechanical resonators with electrical readouts to test their response to changes in mass and temperature. Their devices were fabricated by placing mechanically exfoliated graphene on Si/SiO_2 substrates. Monolayer flakes were identified by Raman spectroscopy after which electrodes were patterned, and then the flakes were shaped by oxygen plasma etching. The final step involved etching the underlying SiO_2 to leave suspended monolayer graphene across the electrodes. The mechanical activation was by the electrical frequency mixing technique. The researchers investigated the response of the devices to mass through two routes, namely mass removal and mass addition. Mass removal was accomplished by ohmic heating which is a technique known to improve the mobility of suspended graphene. Usually it is argued the improvement in mobility is due to the desorption of surface residue on the graphene membrane.

The devices in this study were exciting because they provided means to study this process as well as to correlate the devices performance with the amount of residue. Four ohmic heating steps were applied and the conductance of the graphene resonator measured as a function of the gate voltage. The process led to decreased contuctance at the negative gate voltage, an increased conductance at the positive gate voltage and a conductance minima (charge neutrality point) closer to zero gate voltage. After successive ohmic heating steps the resonant frequency as a function of gate voltage is seen to shift upward after each step. These effects are illustrated in Fig. 6.4.1.

The upward shift in frequency is concomitant with a loss in mass. In addition the strain in the sheets was also seen to alter slightly after each cleaning step. The room temperature Q factor did not change much throughout the steps. In the next series of experiments mass was then intentionally added to the membranes by evaporating pentacene on to the device in a vacuum chamber while simultaneously measuring the device response in situ. The mass deposited mass measured using a quartz crystal microbalance. At high gate voltages, a large gate induced tension exists and the added mass results in a decrease in the resonant frequency, as expected. However, at low gate voltages where little to no gate induced stress exists the addition of mass leads to a frequency upshift. This is attributed to the added pentacene increasing the built-in strain and shows that adsorbates can impart a tension to suspended graphene sheets. In addition the charge neutrality point was measurable in the gate voltage region. The reason for this behaviour was not determined.

The devices were also cooled and their temperature characteristics evaluated. Upon cooling the devices their resonant frequencies shifted upwards and the tenability of it by the gate voltage was diminished. This behaviour was described as being due to the geometry of the devices such that as the temperature decreases the suspended metal electrodes contract in an isotropic manner providing tension to the graphene membrane. The resonant frequency shifts were fully reversible upon heating and this indicates no slipping between the graphene sheet and the electrodes occurs during the cooling/heating process. In addition the large temperature response (5–300 K) highlights the potential to frequency tune such devices by temperature. The Q factors improved with temperature reduction.

Singh et al. (2010) also explored the temperature dependence of freestanding graphene electromechanical resonators with similar device constructions to that described above. They also found that the resonant frequency increases as the device temperature is reduced, again, because of the increase in tension due to the expansion/contraction of the substrate, the gold electrodes and the graphene itself (at low temperature, graphene has a negative coefficient of thermal expansion i.e. it shrinks). The degree of frequency shift was seen to vary between the different devices studied and was attributed to variations in the device geometry. They also investigated different modes in various devices with multiple modes at low temperature (7 K). The dispersions

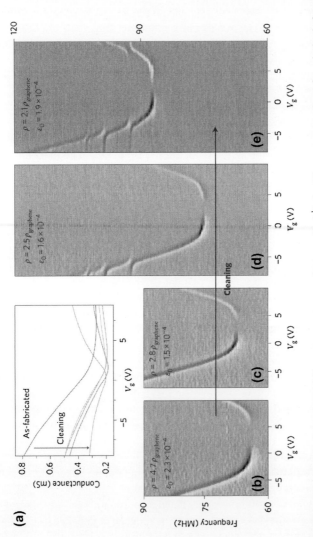

FIGURE 6.4.1 Removal of mass by ohmic heating. (a) Room-temperature conductance (V_{sd} ¼ 1 mV) versus gate voltage for a device before (black curve) and after ohmic-heating cycles to V_{sd} ¼ 1.5 V (red), 1.6 V (green), 1.8 V (blue) and 3 V (pink). Successive curves show decreased conductance at negative-gate voltage (black arrow), increased conductance at positive-gate voltage, and conductance minimal closer to zero-gate voltage. b–e, resonant response (frequency versus gate voltage plotted as a derivative with a colour scale), for the as-fabricated device (b), and after annealing at 1.5 V (c), 1.6 V (d) and 1.8 V (e). The derived strain and density for each step are shown, and confirm that residue is removed. Because of the damage to the device, it was not possible to obtain a resonance curve after the 3-V annealing step [from f4 in Chen et al. (2009)].

between modes is different and models to explain this showed different effective masses for different modes which is expected if the modes are not higher order harmonics of the fundamental mode. Their work indicates that a simple description of a rectangular membrane under tension within the continuum description is insufficient to describe such behaviour. Possible reasons for the observed behaviour were suggested. These are the following:

(1) Different mass loadings over the surface of the membranes (due to adsorbents) might alter the effective mass for different modes.

(2) The presence of edge dependent modes (Poetschke et al., 2010).

(3) The effective stiffness of the membrane might be a rather complex quantity such that the tensor nature differs significantly from the ideal value because of curvature or rippling of the graphene membrane.

(4) The capacitance of the gate for the membrane is probably mode dependent.

Eichler et al. (2011) also investigated similar devices driven by the frequency mixing route and investigated the nature of the damping process. Usually a linear damping force is used to describe oscillator systems at the macro-scale as discussed in the theory of damping in Newton's principia (Newton, 1687). However in the case of the graphene based electromechanical resonators a nonlinear damping that depended on the amplitude of motion was observed. The reason for the nonlinearity was not clear. They suggested the nonlinear damping force might be related to surface contamination, clamping configuration and suspended length. At the low temperature of 90 mK, they managed to obtain a Q factor of 100,000 for one of their devices, which at the time of writing is the highest reported Q factor for a graphene resonator.

Another team explored the use of large area CVD grown single layer graphene for the fabrication of large-scale arrays of resonators (van der Zande et al., 2010). Three types of configurations were developed. In the first the membranes were clamped at two ends over trenches, in the second square graphene membranes were clamped on all sides, and in the third electrically contacted sheets suspended between the 2 gold electrodes were fabricated. With all the approaches they were able to produce hundreds to hundreds of thousands of single-layer graphene membranes in each fabrication run. The actuation and detection of the membranes was accomplished using a resonance modulated optical route and (separately) using the electrical frequency mixing technique. The availability of multiple arrays of membranes allowed for a systematic study of single layer graphene resonators in terms of their mechanical resonance as a function of size, clamping geometry, temperature and electrostatic tuning. They were able to show that CVD produced graphene produces tensioned, electrically conducting and highly tunable resonators that compare equally well as resonator systems produced using exfoliated graphene. The study also showed that devices in which the graphene was clamped on all sides leads to a reduction in variation in resonance frequency. This has the advantage that the devices are more predictable. As with electromechanical resonance systems studied by other groups, they found the resonance frequency to be tunable by both the electrostatic gate voltage and temperature. The resonance frequency improved with temperature, and for temperatures of 10 K, they were able to obtain Q factors up to 9000. Their work showed that wafer scale processing of graphene electromechanical resonators using CVD-grown graphene comes at little or no cost in resonance performance.

6.4.2.1. Graphene-Based Cantilever Nanoelectromechanical Devices

Thus far the nanoelectromechanical systems discussed in this section have revolved around a graphene membrane clamped at two ends or on all sides. There is another class of clamping in which only a single side of a graphene membrane or nanoribbon is clamped forming a cantilever.

Conley et al. (2011) explored bi-metallic-like cantilevers as a function of temperature. The cantilevers consist of a gold or silicon nitride substrate layer and a layer of graphene. Single-, bi- and tri layer graphene films were used. The graphene used was CVD grown with Cu serving as the catalyst material. The cantilevers were fabricated using a number of transfers, lithography and etching steps (for more details see (Conley et al., 2011)). All cantilevers showed a common trait immediately after fabrication, namely they were not flat, but bent toward the graphene layer. The bending is attributed to strain mismatch between the substrate and the graphene film, viz. the unequally strained graphene and substrate layers minimise energy by bending into an arc. Analytical calculations using their curvature data and known material parameters of the substrate materials the investigators could determine the strain level for graphene. They could show all their measured cantilevers were under tensile strain and that this tensile strain was significantly larger for devices using a SiN_x substrate. When the graphene/SiN_x bi-layer cantilevers are heated for the first time and then cooled their curvature shows hysteretic behaviour and does not return to its original curvature. This hysteretic behaviour disappears after the first heating/cooling cycle so that they can then be heated and cooled multiple times reproducibly. The graphene/Au cantilever did not exhibit any hysteretic characteristics.

The non-hysteretic behaviour (that is observed after the first heat/cool cycle) where the curvature changes with temperature arises because the coefficient of thermal expansion is different for the two layers forming the bi-layer cantilever. This difference means the materials change their size with temperature (thermal expansion) differently, forming a strain between the layers, resulting in a change in curvature for the bilayer cantilever. From this data, the researchers were able to extract the thermal coefficients and compare them with theoretical values. The data suggests the coupling between graphene and the SiN_x substrate is relatively weak. This weak interaction might help explain the initial hysteretic behaviour and change in final curvature value after heating/cooling the graphene/SiN$_x$ cantilevers for the first time. In short, the weak interaction between the layers allows it to relieve strain by slipping.

Reserbat-Plantey et al. (2012) developed the use of a graphene cantilever on its own as a local optical probe to measure motion and stress in a nano-electromechanical system. This is of importance, because as we have already highlighted the various dissipation mechanisms that limit the mechanical quality factors of NEMS as well as ageing due to material degradation are poorly understood. Thus there is a need for techniques that enable one to probe the motion of NEMS and the stresses within them, viz. they need to function at the nanoscale. To achieve this they developed a non-invasive local optical probe for the quantitative measurement of motion and stresses within a NEMS based on a multilayer graphene cantilever serving as a movable mirror in an optical cavity (see Fig. 6.4.2). Under monochromatic illumination, an interference

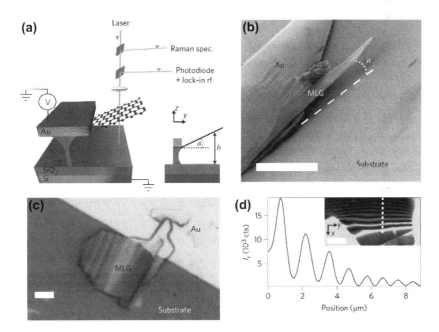

FIGURE 6.4.2 Fizeau fringes in a multilayer graphene cantilever. (a) Schematic view of the device. The cantilever can be actuated with an external voltage while its optical properties are analysed with a high-sensitivity Raman spectrometer and a fast photodiode. (b) Scanning electron micrograph showing a typical multilayer graphene (MLG) cantilever clamped to a gold film on an oxidised silicon substrate. The cantilever and the surface act as an optical cavity because they are both reflecting; light enters and leaves the cavity via the cantilever, which is semitransparent. (c) White-light optical image of a device showing iridescence. (d) Reflectance profile measured along the dashed line in the inset. The reduction in signal strength observed at large distances from the hinge is due to reduced spatial mode matching. However, the fringe contrast is preserved. Inset: reflectance confocal (x, y) scan at 633 nm. Scale bars, 5 mm (Reserbat-Plantey et al., 2012).

pattern with contrast regions of equal thickness is observed. These fringes are known as Fizeau fringes. Moreover, unlike graphene based optical cavities with fixed geometries (Ling and Zhang, 2011), in this case, because the graphene forms a cantilever, the optical length of the cavity varies linearly along the length of the cantilever (assuming the graphene remains flat). This leads to the observation of multiple fringes as can be seen in panels c and d of Fig. 6.4.2. Interference patterns are observed for both the pump laser reflection as well as the Raman scattered light. The Raman scattered light is of value as it carries local information related to the material, namely, stress, doping, defects and temperature. The investigators claim that because of the system's stiffness, semi-transparency and extremely low mass, multilayer graphene can be used as a platform for the simultaneous exploration of the spatial, temporal and spectral properties of NEMS.

Studies of graphene-based NEMS using suspended sheets as resonators revealed oscillations that were not externally driven but that are natural thermal oscillations intrinsic to the graphene sheets (Bunch et al., 2007). Intrinsic vibrational modes can be used to transport information through molecules, potentially as carriers for THz signals. Thus it may be that the natural oscillations of small graphene ribbons are appropriate as sensors or THz generators that could form a nanodevice to encode, transfer and process information. To this end, Bellido and Seminario used molecular dynamics simulations to model cantilever-like GNR's vibrational bending modes (Bellido and Seminario, 2012). Graphene ribbons of several difference sizes were used in the model to calculate the frequencies of the ribbons and determine any relationship between ribbon size and resonant frequency. Numerous ribbons with widths ranging between 7.3 and 26.7 Å and lengths ranging from 14.6 Å to 133.4 Å were investigated. The highest calculated frequency was 0.2 THz, which is within the operating frequencies used in THz technology (0.1–10 THz). The sensitivity of these cantilever-like graphene ribbons to the presence of molecules was also investigated. Specifically the change in their vibration frequency due to the adsorbance of water or isopropyl alcohol (IPA) molecules on their surface. The different molecules showed a completely different trend. With water, the frequency increases as the number of water molecules increases while for IPA the frequency decreases. In the case of water molecules in contact with graphene they stick to graphenes surface, however water is a polar sovent that has a strong interaction between other water molecules (Coulomb and van der Waals). Thus the water molecules prefer to interact with other water molecules rather than the graphene surface and this is reflected by an increase in the resonance frequency as the number of water molecules exist. The van der Waals interaction between water molecules increase the sensitivity of the vibronic device and affect the frequency of the device more than that due to the change in mass. IPA, on the other hand is a nonpolar solvent, so that the attraction between IPA molecules is solely due to van der Waals interactions. Thus they only interact for short periods of time with the graphene, and the frequency changes are affected not only by van der Waals interactions but also by mass. This suggests this type of cantilever-type graphene ribbons could be suitable as molecular sensors, that not only detect single molecules but that may also distinguish the type of molecule it is sensing.

6.4.2.2. Graphene-Based Sensors, Switches and Actuators

Yao et al. (2012) investigated the potential of graphene oxide as a material to detect water molecules to form a humidity sensor. To do this they fabricated a graphene oxide film on silicon forming a bilayer flexible structure that functions as a stress-based humidity sensor. They did this by spin coating a graphene oxide film onto a silicon microbridge. A piezoresistive Wheatstone bridge is embedded in the silicon microbridge so that any deformation of the graphene oxide/silicon bi-layer leads to a measurable output voltage. Figure 6.4.3 shows a schematic of the setup. The basic functionality of the device is as follows: graphene oxide has many functional oxygen groups such as hydroxyl and epoxy-type groups attached to its basal surfaces and carboxylic groups attached to its edges. These groups make graphene oxide highly hydrophilic which means water molecules are easily captured. This leads to the graphene oxide film swelling, with the amount of swelling, depending on the humidity in the environment. The swelling, process

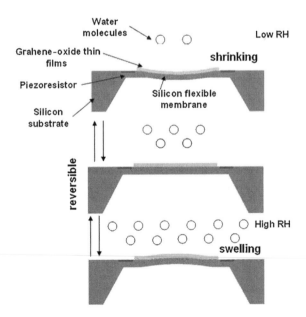

FIGURE 6.4.3 Schematic illustration of humidity sensing mechanism of graphene oxide thin film coated piezoresistive silicon membrane (Yao et al., 2012).

bends the silicon microbridge. The full piezoresistive Wheatstone bridge transforms the mechanical deformation to an output voltage. The process is reversible. The sensitivity of the swelling process with respect to humidity level can be improved with thicker graphene oxide films. The sensor is simple, easy to integrate with other systems and can be manufactured at low cost.

The swelling and shrinking effect of exposing graphene oxide films to varying degrees of humidity was also explored by Park et al. (2010). In this study they explored the potential of bi-layer films consisting of functionalised carbon nanotubes and graphene oxide. The bi-layer samples were fabricated by the sequential filtration of COOH-functionalised multiwalled carbon nanotubes and then graphene oxide platelets from an aqueous suspension. The final product consists of circular bi-layer paper that is flat at the macroscopic scale and does not delaminate. Scanning electron microscopy suggests the layers do not cross contaminate each other. The estimated thickness for each layer is ~10 μm. Upon exposure to humid conditions the structure (paper) curls up in different ways, depending on the relative humidity. At room temperature, for low relative humidity values around 12% the bilayer paper curls with the carbon nanotube side facing outward. However as the relative humidity increases, the paper then unrolled and is pretty much flat by the time humidity values of 55–60% are reached. With higher humidities, the paper starts curl up (not with the carbon nanotube film facing in) and fully closes by the time a humidity of 85% is reached. These various curling steps can be seen in Fig. 6.4.4. Upon reducing the humidity the bilayer paper gradually unrolls. A steady state is obtained after about 30 s. Similar curling actuation was obtained as a function of temperature. More in depth studies showed that the carbon nanotube layer does not absorb water molecules, while graphene oxide films do. While the full actuation or curling phases are not fully understood, the experimentalists propose the bilayer paper might be induced by the different amount of interlamellar water of both layers, depending on the relative temperature or humidity.

Shi et al. (2012) explored the use of suspended graphene membranes over SiO_2 trenches as a switch. In essence the graphene membrane serves as a movable element. The motion is triggered by an attractive electrostatic force between the graphene and a nearby static electrode. Their early experiments used a gold wire running the length of the trench base as the electrode. Upon applying a gate voltage the membrane is attracted to the electrode and eventually touches when sufficient bias is applied. However, when removing the bias after touching the electrode, the graphene would not return to its position it remained stuck. This was due to the van der Waals forces being strong enough to keep the graphene in contact with the electrode. The researchers then explored the potential of a tip electrode (STM electrode) to overcome this technical difficulty. The tip would carefully be brought into contact with the graphene and then removed. Remarkably single layer graphene still adhered to the probe and in fact torn upon sufficient extraction of the probe. This indicates

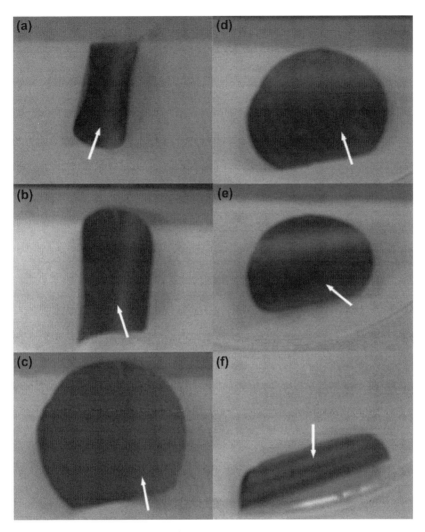

FIGURE 6.4.4   Actuation of the bilayer paper sample as a function of relative humidity (%), (a) 12, (b) 25, (c) 49, (d) 61, (e) 70, and (f) 90. White-arrowed side: surface of graphene oxide layer *Park et al. (2010).*

monolayer graphene is too fragile to serve in graphene-based NEMS switches. Multilayer graphene should be stronger and so this then investigated for its potential as a switch with a point contact. In this configuration the tip contact is by default in contact with the multi-layer graphene membrane and serves as the "on" state. By applying an electrostatic force to release the graphene (gate voltage), the "off" state can be reached. With this setup, repeatable switching behaviour could be obtained. The on–off ratios were around 10^4. Lifetime measurements showed switch failure after around 500 cycles of switching on or off. Moreover the on-state current was seen to decay with increased cycling and is an indication of contact area damage. The investigators suggest improvements could be made by using lower source-drain currents and softer tips.

REFERENCES

Barton, R.A., Parpia, J., Craighead, H.G., 2011. Fabrication and performance of graphene nano-electromechanical systems. J. Vac. Sci. Technol. B 29 (5), 050801 (10).

Bellido, E.P., Seminario, J.M., 2012. Graphene-Based vibronic devices. J. Phys. Chem. C 116, 8409–8416.

Bunch, J.S., van der Zande, A.M., Verbridge, S.S., Frank, I.W., Tanenbaum, D.M., Parpia, J.M., Craighead, H.G., McEuen, P.L., 2007. Electromechanical resonators from graphene sheets. Science 315, 490–493.

Chen, C., Rosenblatt, S., Bolotin, K.I., Kalb, W., Kim, P., Kymissis, I., Stormer, H.L., Heinz, T.F., Hone, J., 2009. Performance of monolayer graphene nanomechanical resonators woth elecrical readout. Nat. Nanotechnol. 4, 861–867.

Conley, H., Lavrik, N.V., Prasai, D., Bolotin, K.I., 2011. Graphene Bimetallic-like cantilevers: Probing Graphene/Substrate interactions. Nano Lett. 11, 4748–4752.

Eichler, A., Moser, J., Chaste, J., Zdrojek, M., Wilson-Rae, I., Bachthold, A., 2011. Nonlinear damping in mechanical resonators made from carbon nanotubes and graphene. Nat. Nano technol. 6 (6), 339–342.

Ekinci, K.L., Huang, X.M.H., Roukes, M.L., 2004. Ultrasensitive nanoelectromechanical mass detection. Appl. Phys. Lett. 84, 4469–4471.

Lam, K.-T., Lee, C., Liang, G., 2009. Bilayer graphene nanoribbon nanoelectromechanical system device: a computational study. App. Phys. Lett. 95 (3), 143107.

Ling, X., Zhang, J., 2011. Interference Phenomenon in graphene-enhanced Raman scattering. J. Phys. Chem. C 115 (6), 2835–2840.

Newton, I., 1687. Principia, Book II.

Park, S., An, J., Suk, J.W., Ruoff, R.S., 2010. Graphene-Based actuators. Small 6 (2), 210–212.

Poetschke, M., Rocha, C.G., Foa Torres, L.E.F., Roche, S., Cuniberti, G., 2010. Modeling graphene-based nanomechanical devices. Phys. Rev. B 81 (4), 193404.

Reserbat-Plantey, A., Marty, L., Arcizet, O., Bendiab, N., Bouchiat, V., 2012. A local optical probe for measuring motion and stress in a nanoelectromechanical system,. Nat. Nanotechnol. 7 (3), 151–155.

Shi, Z., Lu, H., Zhang, L., Yang, R., Wang, Y., Liu, D., Guo, H., Shi, D., Gao, H., Wang, E., Zhang, G., 2012. Studies of graphene-based nanoelectromechanical switches. Nano Res. 5 (2), 82–87.

Singh, V., Sengupta, S., Solanki, H.S., Dhall, R., Allain, A., Dhara, S., Pant, P., Deshmukh, M.M., 2010. Probing thermal expansion of graphene and modal dispersion at low-temperature using graphene nanoelectromechanical systems resonators. Nanotechnology 21 (8), 165204.

van der Zande, A.M., Barton, R.A., Alden, J.S., Ruiz-Vargas, C.S., Whitney, W.S., Pham, P.H.Q., Park, J., Parpia, J.M., Craighead, H.G., McEuen, P.L., 2010. Large-scale arrays of single-layer graphene resonators. Nano Lett. 10, 4669–4873.

Yao, Y., Chen, X., Guo, H., Wu, Z., Li, Y., 2012. Humidity sensing behaviors of graphene oxide-silicon bi-layer flexible structure. Sens. Actuators B 161, 1053–1058.

Chapter 6.5

Freestanding Graphene Membranes

Mark H. Rümmeli

IFW Dresden, Germany

One might be forgiven for thinking the potential applications for freestanding graphene are limited due to the fragility of the structure. However, in practice, a wealth of exciting uses for freestanding graphene is emerging. Thus far the greatest developments for freestanding graphene are in some way connected with transmission electron microscopy (TEM), in particular low voltage TEM, is ideally suited to investigate graphene in a variety of modes. TEM based applications include freestanding graphene as the ultimate microscope slide, cell window for liquid cells, in-situ investigations of graphene at high bias, electron structuring and engineering of graphene and even as a template for the catalyst-free fabrication of graphene with electrons. Freestanding graphene membranes have also provided the foundation for mechanical studies on graphene (see Section 3.4). In terms of bio-medical application, porous graphene has been shown to be an excellent material for transelectrode membranes, for uses such as single DNA molecule characterisation. Pristine graphene membranes are also impermeable, making them an exciting option for food packaging. On the other hand, graphene oxide films have been shown to be selectively porous to water enabling the material to serve as a nanodistillery material. The exciting world of graphene membranes is now discussed in greater detail.

6.5.1. FREE-STANDING GRAPHENE AS THE ULTIMATE MICROSCOPE SLIDE

Contemporary high-energy transmission electron microscopes use electrons with energies in the range 60–300 kV (N.B. instruments with lower acceleration voltages are in development) and are capable of delivering images with sub Angstrom resolution which makes them well suited to directly image molecules and atoms. However, to achieve this, one has to first 'suspend' the atoms or molecules of interest. Traditionally, objects to be examined in TEM were placed on amorphous carbon membranes. Typically, their thickness is in the range 3–20 nm. However, the amorphous nature of the film prohibits visualising atoms or molecules, since one cannot really discern them from the randomly placed carbon atoms in the film. Moreover the incoming electron wave upon scattering at the amorphous carbon layer leads to a rather strong background noise, reducing the quality of the object under investigation. In some cases this technical limitation can be circumvented by using a lacey or holey carbon grid so that the sample is suspended over regions free of amorphous carbon or the sample can attach to free edges. Nonetheless, lacey and holey carbon TEM grids are not suited to all samples, including molecules and atoms. To overcome these shortcomings one would ideally require a continuous structure-less electron transparent film (Schneider et al., 2010). sp^2 carbon can approximate this ideal case. The potential for sp^2 carbon networks in TEM as a tool with which to observe molecules and atoms was first pioneered using carbon nanotubes, in particular single-walled carbon nanotubes since they afford greater transparency. For example, fullerene molecules trapped within carbon nanotubes are easily imaged in TEM (Sato et al., 2007; Warner et al., 2008) and few atom carbon chains (3–8 atoms) can be imaged within a tube (Börnert et al., 2010a) or projecting from the wall of a carbon nanotube (Börnert et al., 2010b). However, the isolation of graphene and the development of various transfer techniques to handle the material, as well as its structural integrity and crystalline form make it ideal as a support film in transmission electron microscopes. The mechanical strength of graphene allows it to support samples in a stable manner and the highly ordered atoms forming the honeycomb lattice are easily filtered out by masking out graphene's signature peaks in the Fourier domain (Lee et al., 2009; Meyer et al., 2008). Moreover, because graphene is conductive, it has an equipotential surface which minimises any phase distortion to a traversing electron wave. This latter point is particularly important for holography experiments. In short free-standing graphene films are attractive as supports, and as we will shortly show, it is well suited to study molecules and atoms. Moreover, investigating molecules and atoms in TEM has certain advantages of scanning tunnelling microscopy because acquisition times are massively shorter in TEM. In addition, TEM is less demanding in terms of sample conductivity and cleanliness.

6.5.1.1. Examining Atoms and Molecules on Graphene

An early work investigating graphene membranes as TEM membranes was that by Meyer et al. (2008). In this study they used a conventional TEM without any aberration correction. Nonetheless they were able to obtain some impressive results. In essence the researchers prepared a graphene membrane by mechanical cleavage with adhesive tape followed by transfer on to commercially available TEM grids. Electron diffraction analysis was used to confirm single layer graphene. The technique yields graphene membranes with up to 50% of the membrane having very clean regions. However, closer examinations of these regions showed that they have individual adatoms and carbon chains that are readily observable by TEM. Since no aberration correction was used, multiple subsequent frames were collected and summed up to improve the signal to noise ratio. Their work also highlighted that graphene is more stable under the electron beam than single-walled carbon nanotubes. This is because the curvature of nanotubes makes them more reactive chemically as well as more sensitive to electron beam irradiation.

Graphene sheets as TEM sample supports have also been used to study the diffusion of metal atoms. For example, Pt and Au atom diffusion was investigated in situ with HRTEM by Gan et al. (2008). The large difference in atomic mass of the metal atoms with respect to the carbon support results in a strong contrast between metal atoms and the support which enables the metal atoms to be tracked and study their dynamic behaviour. Sloan et al. (2010) employed low-voltage HRTEM to study the dynamics of polyoxometalate anions on graphene oxide with high precision. The molecules contain several tungsten atoms within their structure, providing clear contrast due to their strong scattering properties. This allowed the molecules' orientation with respect to the graphene support to be determined. Schäffel et al. conducted an in depth study on the motion of light atoms and molecules on the surface of electron cleaned few-layer graphene. Adatoms trapped on the surface were found to stay localised for sufficiently long to be imaged with 1s acquisition times. In addition, they showed electron beam irradiation could induce small atomic clusters to fuse, forming larger molecular chains in the process. They even observed a molecular chain, most probably a hydrocarbon chain anchor at a site end then swing around its tether point.

Warner et al. were able to watch the removal of carbon atoms of the edge of graphene on a support graphene layer. The edge atom removal is due to sputtering by the electron beam since the unsaturated edge atoms have a reduced knock-on threshold. They showed the atom removal to occur in a zipper-like fashion (see Fig. 6.5.1) (Warner et al., 2009a).

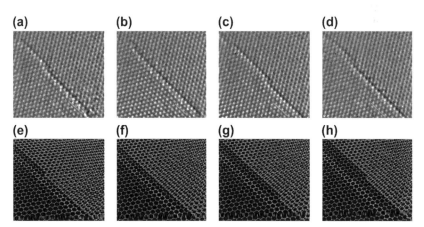

FIGURE 6.5.1 Zipper-like removal of carbon atoms. Time series of HRTEM images showing the zipper-like removal of carbon atoms along the zigzag edge (Fig. 2e–h). a–d, HRTEM images of the straight edge of a large hole in the monolayer at time 0 s (a), 2 s (b), 4 s (c) and 6 s (d). (e–h), Structural representations for each of the HRTEM images presented above. *Fig. 3 from Warner et al. (2009a).*

6.5.1.2. Examining Nanoparticles on Graphene

The attraction of graphene as a TEM sample support is not restricted to single layer graphene. The use of few-layer graphene supports may also be used without leading to a high background signal as highlighted by various works examining nanoparticles. A study by McBride et al. specifically investigated few-layer graphene supports using conventional high resolution TEM and Cs aberration corrected TEM to assess its potential as a film support. They obtained their graphitic flakes by gently pressing masking tape on the surface of a freshly cleaved piece of highly ordered pyrolytic graphene. Selected flakes that had adhered to the tape were then removed with tweezers and then sonicated in toluene. The material was then centrifuged to drive the large (heavier) pieces down. Material from the supernatant was then drop coated on commercial lacey carbon grids which was then left to dry. The process leaves a variety of few layer and single layer flakes on the grid. Similar results can obtained with other solvents, for example dichloromethane as shown by Warner et al. (2009b). To obtain CdSe nanocrystals on the graphene flakes, CdSe nanoparticles were mixed into the graphene flake/toluene solution during sonication. Initial experiments compared the quality of the images for CdSe nanoparticles on amorphous carbon with those resting on few layer graphene. The most striking difference between the two is the near featureless background of the graphene support and the quality of the edges of the nanoparticles which are sharper on a graphene membrane. Indeed, the edge contrast shows surface material on the nanoparticles, which the authors attribute to either artifacts on the material surface clustering on the nanoparticles surface or residual pre-cursor material from the CdSe nanoparticle preparation. The latter argument is more likely as residual pre-cursor material is often encountered when preparing nanoparticles using wet chemical routes as was used in this study. With very small nanoparticles (<2 nm) they can hardly be discerned when residing on an amouphous carbon support; however, with a few-layer graphene support they are readily observed.

The investigators also examined the use of few-layer graphene in aberration-corrected scanning transmission electron microscopy (STEM) with high angular dark field (HAADF) imaging. In this mode the image contrast is dependent on the atomic mass number, Z, and is referred to as Z contrast. This is because the image formation uses only highly scattered electrons. Carbon scattering is relatively low and so is essentially removed from the image. Thus when supporting nanoparticles the particles appear to sit in free space and can be imaged with atomic columns clearly visible in the crystalline CdSe nano-particles. However, the images were acquired using an acceleration voltage of 300 kV. At this energy graphene is not stable and so (depending on the electron dose) one is limited to only a few scans before the substrate starts to disinte-grate. This is still enough for imaging as well as spectroscopic measurements as was demonstrated for electron energy loss spectroscopy (EELS) studies on the nanoparticles. Usually EELS spectroscopy of nanoparticles is difficult due to the high background signal from amorphous carbon supports (Kadavanich et al., 2001; Yu et al., 2005). The use of monolayer graphene minimises this problem as it is only atomic layer thick. Z-STEM and EELS experiments using $CuInSe^2$ nanoparticles underscored the advantages of graphene supports in that elemental mappings successfully highlighted Cu and In in the core of the particle. Carbon was found around the particles surfaces. A study by Lee et al. (2009) conducted a more in depth analysis of coatings on nanoparticles residing on graphene membranes. Using a Cs corrected TEM operating at 80 kV with a monochromated beam, they conducted an atomic resolution study of the capping layers and interfaces of citrate-stabilised gold nano-particles. The atomic spacings in the gold nanoparticle and in the citrate coating were determined through intensity profiles from the collected images of the atomically resolved nanoparticles and from diffraction data from the Fourier domain. The Au particles had an average atomic spacing of 2.5 Å. The citrate coating was estimated as 2–3 layers thick and exhibited a spacing of 3.0–3.5 Å between layers. It is worth the reader noting that the coating closely resembles graphitic layers, which have similar interlayer spacings. Moreover, it is not clear whether organic molecules remain stable under 80-kV electron radiation. Organic material that does decompose under electron irradiation often forms graphitic material. Warner et al. (2010) investigated the stability of nanoparticles ($CoCl_2$ and Co) on graphene and few-layer graphene, using low-voltage aberration corrected TEM (80 kV). The graphene supports were prepared in a similar manner to that described above by the McBride team. However, in this case, the graphene flakes were sonicated in 1,2-dichloroethane. To form the nano-particles, some $CoCl_2$ was added to the sonication solution. After drop coating and drying, this results in $CoCl_2$, decorating the graphene support membrane. To obtain Co nanocrystals, the samples are annealed at 320 °C for 2 h, under vacuum. Again, they demonstrated that the think nature of monolayer and few-layer graphene enable their use as near-transparent grids for TEM anaylsis. Moreover, they showed the in-situ potential of such grids. Their in situ real-time TEM studies with $CoCl_2$ nanoparticles showed them to exhibit translational and rotational movement across the surface of the graphene membrane. They would eventually coalesce together to form a single crystalline species. The Co nanoparticles displayed unique interactions with the surface of the graphene sheets, namely, they etched the graphene (see Fig. 6.5.2).

Yuk et al. (2012) developed a liquid cell for in situ TEM based on the entrapment of a liquid film sandwiched between layers of graphene. They demonstrated the effectiveness of the technique by preparing a platinum growth solution between two laminated graphene layers over holes in a conventional TEM grid. The blister in which the solution resides is sealed by van der Waals forces between the graphene sheets around the liquid droplet. Liquid drops of various thicknesses ranging from 6 to 200 nm can be securely trapped within the blister. They employed an aberration-corrected TEM operating at 80 kV and were able to watch colloidal Pt growth in situ with remarkable resolution and

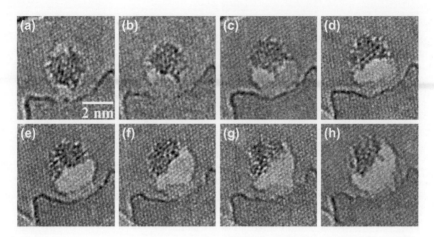

FIGURE 6.5.2 Time-series of HRTEM images showing the catalytic etching behaviour of Co nanocrystals on the surface of a few layer graphene sheet. Time between frames is 15 s. Region coloured in yellow indicates area of surface that has been etched away. *From Warner et al. (2010).*

minimal sample perturbation. They were able to directly observe nanocrystal coalescence and oriented attachment.

6.5.1.3. Examining Biological Molecules on Graphene

Nair et al. (2010) demonstrated the potential of freestanding graphene membranes for the study of biological molecules in TEM. They used a PMMA based transfer route to place commercial graphene on Quantifoil grids. Their choice of biological molecule was the Tabacco mosaic virus (TMV) in part because it was the first virus to be directly visualised in TEM in 1939. Moreover its geometrical form makes it easy to identify (it has a rod-like structure with a length of around 300 nm and a width of around 18 nm). To prepare the sample the TMV was prepared in to a suspension in which the graphene membranes were then dipped and finally dried in ambient conditions. No further processing, such as the use of dyes was implemented. The rod-like TMV structures were easily observable and have a contrast around 0.3. Given that biological molecules show little contrast in TEM (as compared to metal nanoparticles) this value is pretty good and certainly better than amorphous carbon supports.

6.5.2. GRAPHENE AS A TEMPLATE FOR CATALYST-FREE GRAPHENE FABRICATION BY ELECTRONS

Freestanding graphene has also been shown to be a remarkable support to template the growth of new graphene from amorphous carbon. This catalyst free graphene fabrication route is achieved using electrons, either as an electron beam or as an electrical current. This has the added advantage that the fabrication can be accomplished with in situ TEM.

6.5.2.1. Templated Graphene Growth with an Electron Beam

Exposing amorphous carbon to an electron beam of over 100 kV is well known to crystallise freestanding amorphous carbon into carbon onions (concentric graphitic spheres). Carbon is sensitive to a variety of irradiation effects such as knock-on displacements, electronic excitations and radiolysis, and radiation induced diffusion. These effects break chemical bonds in the amorphous carbon, enabling it to rearrange to a more stable form, namely, sp^2 carbon. In the case of freestanding amorphous carbon it graphitises in spheres as this results in a structure with no dangling bonds. Moreover, a spherical structure allows for a uniform distribution of the strain because of the out of plane geometry (Ugarte, 1992).

Börrnert (2012) first showed one can also obtain carbon onions from freestanding amorphous carbons with electrons acceleration voltages of 80 kV in low voltage HRTEM which had not previously been demonstrated. They then deposited amorphous carbon on electron cleaned mono-layer graphene, after which they exposed the supported amorphous carbon to the electron beam (in LV HRTEM) for 40 min. The electron dose was 2×10^3 C m^{-2}. After exposure the surface of the supported amorphous carbon is clearly observed to have changed, appearing crystalline with facetted steps indicating various graphene layers have formed. Closer examination shows Moiré patterns typical for graphene layers with rotational stacking faults. In addition, the hexagonal reflexes in the Fourier domain clearly showed the presence of new reflexes indicating not only the formation of new graphene but that this new graphene formed with rotated stacking relative to the underlying monolayer graphene support (see Fig. 6.5.3). They also showed graphene could be manufactured on the same manner on free-standing hexagonal boron nitride membranes. The reason why the amorphous carbon rearranges to planar graphene rather than carbon onions is attributed to the presence of van der Waals forces from the underlying support membrane.

6.5.2.2. Templated Graphene Growth with an Electron Current

In this case graphene or few layer graphene is suspended across two electrodes so that a current can be applied. Barreiro et al. (2012) established this setup in a TEM so that the process could be monitored. Initially the graphene is taken to the high bias regime for current-induced annealing (Joule heating). The procedure provides atomically clean graphene surfaces. The samples are then cooled and then amorphous carbon is deposited on the surface by electron beam cracking of hydrocarbons in the TEM. Once the amorphous carbon has been

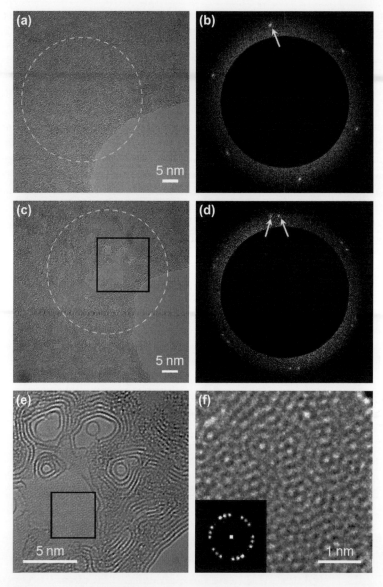

FIGURE 6.5.3 Catalyst-free fabrication of graphene from amorphous carbon supported on graphene. (a) Pristine sample of amorphous carbon residing on a single graphene layer. The dashed ring indicates the area to be irradiated. (b) The Fourier transform from micrograph shown in panel (a) – six spots from the underlying single layer graphene support are visible. (c) The sample shown in panel (a) after 12 mm irradiation. (d) The Fourier transform from micrograph (c) – An additional set of spots as compared to before irradiation (see panel (b)) have now appeared confirming new graphene has formed on the graphene support. (e) magnified section from (b), showing terrace steps of the grown planar few-layer graphene. (f) further magnified section from (e), showing Moiré patterns of the newly formed few-layer graphene. Inset: corresponding Fourier transform, indicating three different rotations from rotational sticking faults between the graphene layers (Börrnert, 2012).

deposited the graphene substrate are brought back to the high bias regime (2–3 V). Temperatures of 2000–3000 °C have been estimated to be reached with current annealing (Huang et al., 2006; Westenfelder et al., 2011). Surprisingly the amorphous carbon does not sublime, but instead it gradually transforms into graphene patches, usually consisting of a few layers. Lattice parameters determined from the Fourier domain of the obtained micrographs confirmed crystalline graphene had formed. To better comprehend the process they conducted molecular dynamics simulations, which showed that upon increasing the temperature the amorphous carbon goes through a glass-like phase at temperatures ranging from 600 to 1200 K and finally forms a graphene structure upon reaching 1800 K. The model also showed the importance of van der Waals interactions with the underlying graphene support that help encourage the formation of planar graphene parallel to the support membrane.

6.5.3. FREE-STANDING GRAPHENE AS A SUBNANOMETER TRANS-ELECTRODE MEMBRANE

In recent years interest in so-called nanopores has risen due to their potential as tools with which to interrogate single molecules. They can rapidly characterise biopolymers like DNA, RNA and DNA-ligand complexes as well as local protein structures along DNA at the single molecule level (Schneider et al., 2010 and (references therein)). A nanopore basically consists of a nanometre size hole that connects two volumes that contain electrolyte solution. The application of a voltage across the nanopore drives ions through the pore, which is recorded from the voltage source using a high gain amplifier (Merchant et al., 2010). Long biopolymers like DNA can then be driven single file through the pore by the electric field. In the process of passing through the pore the molecule reduces the ion flow through the pore which leads to a drop in the detected current. The technique potentially offers a low cost high throughput solution for molecular interrogation. This potential is obviously attractive for efficient DNA sequencing and is the primary driving force to develop the technique. Both solid-state and biological nanopores are available. Solid state pores tend to be relatively thick (30 nm), which limit the number of DNA base pairs (~100) that drop or block the detection current. Biological pores can be thinner and can even reach a DNA base pair of four (Astier et al., 2006; Clarke et al., 2009). The fact that graphene membranes can have single atom thickness and are conductive makes them attractive as pore material with improved spatial resolution. In 2010, three teams more or less at the same time successfully demonstrated the use of pores in graphene for translocation of molecules for DNA sequencing (Garaj et al., 2010; Merchant et al., 2010; Schneider et al., 2010). In all of the studies the pores were introduced to the graphene membranes by drilling through with a focused electron beam. Membranes separating the electrolyte solution with no pores exhibit very high resistance (>10 GΩ) indicating a good seal between the electrolyte volumes. Membranes with pores sow resistances on the order of MΩ as derived from linear I–V curves which indicates ion transport through the pore. Studies from numerous pores showed a limited decency of the resistance with number of graphene layers. The system is though, far more sensitive to pore size. Schneider et al. (2010) found a resistance proportional to $1/d^2$, however a $1/d$ dependence would have been expected, why this was so is still not clear. In terms of the translocation process, each molecule translocation event can be characterised in terms of average current block (reduction) and the duration of the blockade. Garaj et al. (2010) showed current drops of a few nA and event durations of a few hundred μS. From their experimental and modelling data they anticipate that, using graphene membranes with pores, it should be possible to probe molecules with subnanometre resolution. Merchant et al. (2010) found that although higher currents were obtained when using graphene with similar sized pores to SiN nanopores, graphene based pores suffered from large ion current noise and from low yield. They showed this could be improved with high-quality graphene and smaller holes in the supporting SiN membrane holding the graphene. Coating the graphene membrane with TiO_2 further improved the current noise and provided a more hydrophilic surface better suited to the translocation of DNA through the pores.

6.5.4. PERMEABILITY OF FREE-STANDING GRAPHENE

The question whether a single layer of graphene, a chemically stable and electrically conducting membrane one atom thick, is impermeable to atoms and molecules in the form of a gas was addressed by Bunch et al. (2008). In essence they suspended graphene sheets of varying layer numbers (1–75) over wells formed in silicon dioxide. The graphene membranes were clamped on all sides by van der Waals forces between the graphene sheet and the SiO_2 trapping gas in the sealed well. The internal pressure of the trapped gas was set to atmospheric pressure (1 bar). By reducing the pressure outside they could observe the internal and external pressures to equilibrate over time (from minutes to days). To measure the pressure difference, they took advantage of the fact that the membrane balloons out when the internal pressure exceeds the external pressure. The ballooning out was measured in AFM, thus they could characterise the equilibration process by measuring the pressure change and using the ideal gas law covert this to a leak rate. The study was conducted for helium, argon and air. The helium leak rates were 2 orders of magnitude faster than for argon and air. However, no leak-rate dependence on the membrane thickness was observed. This indicates that the leak was not through the graphene membranes. The leak was attributed to diffusion through the glass walls of the well or through the graphene SiO_2 interface i.e. the seal joints. Using Fick's law of diffusion, their diffusion estimate for helium (since He diffusion properties for glass are known) was found to be in excellent agreement with their measured values. This suggests diffusion through the SiO_2 dominates the leak rate. In short, freestanding graphene membranes (without defects) are impermeable to gases. This makes graphene interesting as a material for air tight packaging, for example, food packaging.

Graphene oxide films on the other hand do exhibit a degree of permeability. Nair et al. (2012) prepared graphene oxide membranes by either spray coating or spin coating a suspension of graphene oxide crystallites. This forms a layered structure sometimes referred to as graphene-oxide paper or graphene-based paper. For the permeation experiments the films were formed on Cu foils several centimetres in diameter. Apertures in the Cu were then formed by chemical etching and finally the Cu disks (with aperture) were used to seal a metal container. Membrane thicknesses of 0.1–10 μm were investigated. Initially, the metal vessels were filled with various gases including He, H_2, N_2 and Ar. They found similar leak rates to those obtained for graphene membranes (discussed above) (Bunch et al., 2008). They also investigated the permeability for various solvents such as ethanol, hexane and acetone. No permeation was observed, however, unexpectedly, for water they observed a huge weight loss. Indeed, the weight loss was nearly the same as that found without a membrane sealing the vessel. Their studies suggest the water can permeate the films through nano-capillaries inherent in the film. Such graphene oxide films may be useful in the design of filtration and separation materials and for the selective removal of water. In addition, such films have been shown to exhibit antibacterial properties (Hu et al., 2010).

REFERENCES

Astier, Y., Braha, O., Bayley, H., 2006. Toward single molecule DNA sequencing: direct identification of ribonucleoside and deoxyribonucleoside 5′-monophosphates by using an engineered protein nanopore equipped with a molecular adapter. J. Am. Chem. Soc. 128 (5), 1705–1710.

Barreiro et al., 2012. *arXiv:1201.3131*.

Branton, D., Deamer, D.W., Marziali, A., Bayley, H., Benner, S.A., Butler, T., Di Ventra, M., Garaj, S., Hibbs, A., Huang, X.H., Jovanovich, S.B., Krstic, P.S., Lindsay, S., Ling, X.S.S., Mastrangelo, C.H., Meller, A., Oliver, J.S., Pershin, Y.V., Ramsey, J.M., Riehn, R., Soni, G.V., Tabard-Cossa, V., Wanunu, M., Wiggin, M., Schloss, J.A., 2008. Nat. Biotechnol. 26, 1146–1153.

Börrnert, F., Börrnert, C., Gorantla, S., Liu, X., Bachmatiuk, A., Joswig, J.-O., Wagner, F.R., Schäffel, F., Warner, J.H., Schönfelder, R., Rellinghaus, B., Gemming, T., Thomas, J., Knupfer, M., Büchner, B., Rümmeli, M.H., 2010a. Single-wall-carbon-nanotube/single-carbon-chain molecular junctions. Phys. Rev. B 81 (5), 085439.

Börrnert, F., Gorantla, S., Bachmatiuk, A., Warner, J.H., Ibrahim, I., Thomas, J., Gemming, T., Eckert, J., Cuniberti, G., Büchner, B., Rümmeli, M.H., 2010b. In situ observations of self-repairing single-walled carbon nanotubes. Phys. Rev. B 81 (29), 201401 (R)(4)[felix prb].

Börrnert, F., Avdoshenko, S.M., Bachmatiuk, A., Ibrahim, I., Büchner, B., Cuniberti, G., Rümmeli, M. H., 2012. Amorphous carbon under 800 kV electron irradiation: a means to make or break graphene. Adv. Mater. DOI:10.1002/adma.201202173

Bunch, J.S., Verbridge, S.t S., Alden, J.S., van der Zande, A.M., Parpia, J.M., Craighead, H.G., McEuen, P.L., 2008. Impermeable atomic membranes from graphene sheets. Nano Lett. 8 (8), 2458–2462.

Clarke, J., Wu, H.-C., Jayasinghe, L., Patel, A., Reid, S., Bayley, H., 2009. Continuous base identification for single-molecule nanopore DNA sequencing. Nat. Nanotechnol. 4 (4), 265–270.

Gan, Y., Sun, L., Banhart, F., 2008. One- and two-dimensional diffusion of metal atoms in graphene. Small 4 (5), 587–591.

Garaj, S., Hubbard, W., Reina, A., Kong, J., Branton, D., Golovchenko, J.A., 2010. Graphene as a subnanometre trans-electrode membrane. Nature 467 (7312), 190–193.

Hu, W., Peng, C., Luo, W., Lv, M., Li, X., Li, D., Huang, Q., Fan, C., 2010. Graphene-based antibacterial paper. ACS Nano 4 (7), 4317–4323.

Huang, J.Y., Chen, S., Ren, Z.F., Chen, G., Dresselhaus, M.S., 2006. Real-time observation of tubule formation from amorphous carbon nanowires under high-bias Joule heating. Nano Lett. 6 (8), 1699–1705.

Kadavanich, A.V., Kippeny, T.C., Erwin, M.M., Pennycook, S.J., Rosenthal, S.J., 2001. Sublattice resolution structural and chemical analysis of individual CdSe nanocrystals using atomic number contrast scanning transmission electron microscopy and electron energy loss spectroscopy. J. Phys. Chem. B 105 (2), 361–369.

Karan, S., Samitsu, S., Peng, X., Kurashima, K., Ichinose, I., 2012. Ultrafast viscous permeation of organic solvents through diamond-like carbon nanosheets. Science 335 (6067), 444–447.

Lee, Z., Jeon, K.-J., Dato, A., Erni, R., Richardson, T.J., Frenklach, M., Radmilovic, V., 2009. Direct imaging of soft–hard interfaces enabled by graphene. Nano Lett. 9 (9) 3365–3339.

Merchant, C.A., Healy, K., Wanunu, M., Ray, V., Peterman, N., Bartel, J., Fischbein, M.D., Venta, K., Luo, Z., Johnson, A.T.C., Drndic, M., 2010. DNA translocation through graphene nanopores. Nano Lett. 10 (8), 2915–2921.

Meyer, J.C., Girit, C.O., Crommie, M.F., Zettl, A., 2008. Imaging and dynamics of light atoms and molecules on graphene. Nat. London 454 (7207), 319–322.

Nair, R.R., Blake, P., Blake, J.R., Zan, R., Anissimova, S., Bangert, U., Golovanov, A.P., Morozov, S.V., Geim, A.K., Novoselov, K.S., Latychevskaia, T., 2010. Graphene as a transparent conductive support for studying biological molecules by transmission electron microscopy. Appl. Phys. Lett. 97 (15), 153102 (3).

Nair, R.R., Wu, H.A., Jayaram, P.N., Grigorieva, I.V., Geim, A.K., 2012. Unimpeded permeation of water through helium-leak-tight graphene based membranes. Science 335, 442.

Sato, Y., Suenaga, K., Okubo, S., Okazaki, T., Iijima, S., 2007. Structures of D5d-C80 and Ih-Er3N@C80 fullerenes and their rotation inside carbon nanotubes demonstrated by aberration-corrected electron microscopy. Nano Lett. 7 (12), 3704–3708.

Schneider, G.F., Kowalczyk, S.W., Calado, V.E., Pandraud, G., Zandbergen, H.W., Vandersypen, L.M.K., Dekker, C., 2010. DNA translocation through graphene nanopores. Nano Lett. 10 (8), 3163–3167.

Sloan, J., Liu, Z., Suenaga, K., Wilson, N.R., Pandey, P.A., Perkins, L.M., Rourke, J.P., Shannon, I.J., 2010. Imaging the structure, symmetry, and surface-inhibited rotation of polyoxometalate ions on graphene oxide. Nano Lett. 10 (11), 4600–4606.

Ugarte, D., 1992. Curling and closure of graphitic networks under electron-beam irradiation. Nature 359 (6397), 707–709.

Warner, J.H., Watt, A.A.R., Ge, L., Porfyrakis, K., Akachi, T., Okimoto, H., Ito, Y., Ardavan, A., Montanari, B., Jefferson, J.H., Harrison, N.M., Shinohara, H., Briggs, G.A.D., 2008. Dynamics of paramagnetic metallofullerenes in carbon nanotube peapods. Nano Lett. 8 (4), 1005–1010.

Warner, J.H., Rümmeli, M.H., Ge, L., Gemming, T., Montanari, B., Harrison, N.M., Büchner, Bernd, Briggs, G.A.D., 2009a. Structural transformations in graphene studied with high spatial and temporal resolution. Nat. Nanotechnol. 4, 500–504.

Warner, J.H., Schäffel, F., Rümmeli, M.H., Büchner, B., 2009b. Examining the edges of multi-layer graphene sheets. Chem. Mater. 21 (12), 2418–2421.

Warner, J.H., Rümmeli, M.H., Bachmatiuk, A., Wilson, M., Büchner, Bernd, 2010. Examining Co-based nanocrystals on graphene using low-voltage aberration-corrected transmission electron microscopy. Nano 4 (1), 440–476.

Westenfelder, B., Meyer, J.C., Biskupek, J., Kurasch, S., Scholz, F., Krill III, C.E., Kaiser, U., 2011. Transformations of carbon adsorbates on graphene substrates under extreme heat. Nano Lett. 11 (12), 5123–5127.

Yu, Z., Guo, L., DuKrauss, H., Silcox, J., 2005. Shell distribution on colloidal CdSe/ZnS quantum dots. Nano Lett. 5 (4), 565–570.

Yuk, J.M., Park, J., Ercius, P., Kim, K., Hellebusch, D.J., Crommie, M.F., Lee, J.Y., Zettl, A., Alivisatos, A.P., 2012. High-resolution EM of colloidal nanocrystal growth using graphene liquid cells. Science 336 (6077), 61–64.

Chapter 6.6

Graphene-Based Energy Applications

Mark H. Rümmeli

IFW Dresden, Germany

The global energy demand is steadfastly on the rise and numerous efforts are being developed to meet these demands with minimal damage to the environment. Towards this end, one strategy is the improvement of technologies for the production and storage of electrical energy. More precisely this requires improved technology for fuel cells, advanced batteries and supercapacitors for electrical storage. These technologies are particularly relevant to the automotive industry and mobile electrical device industry. Carbon has long been used within such technology because of its abundance, processibility and electrochemical potential. In addition, it is pretty environmentally friendly. The emergence of graphene and its vast array of exciting properties have triggered widespread interest in the material, including electrochemical energy systems. In this section we briefly review the work currently developed using graphene as a material in supercapacitors, Li-ion batteries, fuel cells and solar cells. For greater detail, the reader is directed to references Hou et al., 2011; Luo et al., 2012.

6.6.1. GRAPHENE BASED MATERIALS IN SUPERCAPACITORS

Supercapacitors store and deliver energy electrochemically with high discharge rates. Their energy densities are vastly superior to conventional dielectric capacitors (by several orders of magnitude). Hence there is significant interest in supercapacitors as a future energy storage system. Their long cycle life and high power density enable them to be used either as the primary power source or in combination with fuel cells or batteries. They are used in a variety of applications including portable electronics, hybrid electric vehicles and backup power systems (Hou et al., 2011).

The make up of a supercapacitor entails two electrodes immersed in an electrolyte. A porous separator is placed in between the two electrodes (see Fig. 6.6.1). However, variations in the storage mechanism and choice of electrode material lead to three different classifications for supercapacitors (Hou et al., 2011; Huang et al., 2012).

(1) Electrochemical double-layer capacitors (EDLCs), which for the most part use high surface area carbon-based electrodes and store their energy by fast ion adsorption at the interface of the electrode/electrolyte.
(2) Psuedo-capacitors which are based on a Faradaic charge transfer process on or close to the electrode surface. In this case conducting polymers and transition metal oxides serve as the electrochemically active materials.
(3) A combination of i. and ii.

The performance of supercapacitors (also known as ultracapacitors or electrochemical capacitors) is determined in terms of its electrochemical activity and chemical kinetic properties, viz. the electron and ion kinetics (transport) both within the electrodes and the charge transfer efficiency (speed) at the

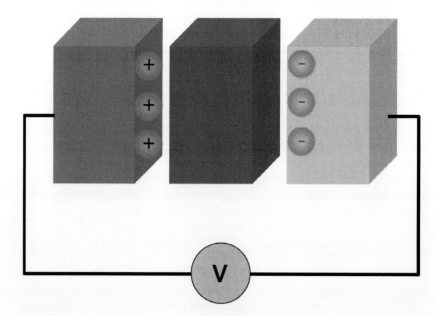

FIGURE 6.6.1 Simple schematic of a supercapacitor, showing two electrodes with a separator in between.

electrode/electrolyte interface. For high performance when using carbon-based materials with EDLCs the specific surface area, electrical conductivity and pore size and disctribution is crucial. Graphene with its high electrical conductivity, large surface area and interlayer structure is attractive for use in EDLCs. In the case of pseudo capacitors, despite delivering superior capacitances as compared to EDLCs they are, in general, limited by low power densities. This is attributed to poor electrical conductivity limiting fast electron transport. In addition, the redox process that drives the charging/discharge process can damage the electro-active materials. The high electrical conductivity of graphene and its excellent mechanical strength make it well suited as a material in pseudocapacitors.

Lv et al. (2009) conducted a variety of adsorption studies (e.g. adsorption isotherms) on graphene and showed adsorption occurs mostly on the surface of graphene sheets with access to large pores (i.e. the interlayer structure is porous providing easy access to electrolyte ions). Thus, for best performance graphene agglomeration without pores should be avoided. Performance can be further improved by modifying the surface by the attachment of functional groups, hybridisation with electrically conductive polymers and by forming graphene/metal oxide composites. Thus, the preparation routes for graphene are important and various have been developed.

6.6.2. GRAPHENE IN ELECTROCHEMICAL DOUBLE-LAYER CAPACITORS (EDLCs)

6.6.2.1. Chemically Reduced Graphene Oxide

Stoller et al. (2008) pioneered the use of graphene-based EDLCs. In their work, they used chemically modified graphene derived by suspending graphene oxide in water and then reducing them with a chemical agent, for example, hydrazine hydrate. Ruoff and his team found the individual graphene sheets agglomerated into structures 15–25 μm in diameter during the reduction process. Despite this the material exhibited a relatively high specific surface area (705 m^2/g) which was sufficient for a good electrochemical performance. Using aqueous and organic electrolytes, they achieved specific capacitances of 135 and 99 F/g respectively. In addition they showed a low variation in the specific capacitance with increasing voltage-scan rates, which was attributed to the high electrical conductivity of the material (ca. 200 S/m). These results were extremely encouraging and stimulated a significant amount of subsequent studies.

Wang et al. (2009) investigated the use of a gas solid reduction process to minimise agglomeration. The resultant material contained aggregated sheets forming a crumpled structure that formed a continuously conducting network. The overall agglomeration was reduced and the crumpled structure provides improved accessibility for electron ions which can now penetrate both the outer and inner regions leading to improved capacitance. With this material they obtained a specific capacitance of 205 F/g and a power density of 10 kW/kg and an energy density of 28.5 W h/kg, using an aqueous electrolyte. They also showed the material had very good cycle lifetimes. A 90% specific capacitance was maintained after 1200 cycle tests.

Chen et al. (2011) explored the use of a weak reducing agent (hydrobromic acid) so as to keep some oxygen functional groups on the graphene surface. The idea being the functional groups would not only promote wettability but also make penetration of the aqueous electrolyte easier. They showed a specific capacitance of 348 F/g. In addition they found the capacitance increased continuously upto 2000 cycles and hardly falls between 2000 and 3000 cycles (from 125 to 120%). This might be due to residual oxygen reduction, occurring during the first few thousand cycling steps.

6.6.2.2. Thermally Reduced Graphene

Studies show that above 550 °C graphene oxide exfoliates (Huang et al., 2012 and refs within). Vivekchand et al. (2008) exploited this by exfoliating graphene oxide at 1050 °C. They showed specific surface areas upto 925 m^2/g for their material and specific capacitance values, reaching up to 117 F/g in aqueous H_2SO_4 electrolyte. Lv et al. (2009) showed the exfoliation process can occur at temperatures as low as 200 °C when annealing in a high vaccum environment. Apparently the resultant material is an aggregated structure with large pores. They obtained capacitances of 264 and 122 F/g for aqueous and organic electrolytes, respectively. Cao et al. explored the thermal exfoliation of graphene in air at low temperature. They found specific capacitance values up to 232 F/g. Ruoff et al. explored the use of thermal treatments in solution to reduce graphene oxide. They demonstrated graphene oxide can be exfoliated and well dispersed in propylene carbonate (PC) solvent under sonication. Moreover, thermally heating the suspension to 150 °C removes a significant fraction of the oxygen functional groups. They obtained a capacitance of 112 F/g which-compares well with other electrode materials in PC-based electrolytes (80–120 F/g). Microwave irradiation can exfoliate graphene rather efficiently, and Ruoff and his team also explored this technique (Zhu et al., 2010) by simply employing a commercial microwave oven. The as-produced material consists of crumpled few-layer thick graphitic sheets with a specific surface area of 463 m^2/g. They obtained a specific capacitance of 191 F/g in a KOH electrolyte. The simplicity of the microwave route could provide a scalable and cost-effective production means of graphene-based electrode materials.

6.6.2.3. Graphene-Based Hydrogel

Despite the success of the above discussed materials less agglomerated, self-supported and binder-free graphene-based electrodes with good pore structure are still desirable (Huang et al., 2012). Xu et al. (2010) introduced a 3D self-assembled graphene hydrogel prepared by the chemical reduction of aqueous graphene oxide dispersions with sodium ascorbate. The material has a well-defined and crosslinked 3D porous structure. The pores range from submicrometre to a few micrometres. The synergetic effects between the hydrophobic reactions and π–π interactions between graphene sheets which increase upon chemical reduction, are argued as the underlying cause of the 3D assembly of the flexible assembly of graphene sheets. Using this graphene hydrogel they showed specific capacitances of up to 240 F/g with aqueous H_2SO_4 as the electrolyte. Zhang and Shi (2011) developed another graphene hydrogel via the hydrothermal reduction of graphene oxide dispersions in which a subsequent reduction with hydrazine or hydroiodic acid was implemented. The best obtained graphene-based hydrogel was obtained by a final reduction with 50% hydrazine at 100 °C for 8 h. This material yielded a specific capacitance of 220 F/g.

6.6.2.4. Activated Graphene and Intercalated Graphene

In general carbon based materials applied to supercapacitors are argued to provide improved performance upon activation, usually electrochemical activations and so it is quite natural to explore this option with graphene-based electrodes. Hantel et al. (2011) explored the electrochemical activation of partially reduced graphene oxide. After activation they obtained a whopping surface area of 2687 m^2/g which is close to the theoretical limit for graphene. They obtained a specific capacitance up to 220 F/g. Y. Li et al. (2011) chemically treated graphene in KOH solution and obtained a specific capacitance of 136 F/g.

One of the difficulties with reduced graphene oxide obtained from the chemical reduction of graphene oxide in aqueous solution is the significant loss of oxygen containing groups which lowers the electrostatic repulsion between sheets which encourages agglomeration. This agglomeration has two negative effects, first it reduces surface area and secondly it reduces access of electrolyte ions both of which limit supercapacitor performance. To overcome this, various techniques have been developed to introduce stabilisers or spacers. Si and Samulski (2008) developed a graphene-Pt nanoparticle composite. The Pt nanoparticles act as spacers to prevent face-to-face aggregation of sheets. The exfoliated graphene/Pt composite showed a specific area of 862 m^2/g and a specific capacitance of 269 F/g. An et al. (2010) developed a rather different technique in which they introduced non covalent functionalisation of graphene with 1-pyrenecarboxylic acid (PCA). The PCA directly exfoliates few layer and multilayer graphene into stable aqueous dispersions and forms stable polar functional groups on the graphene surface via noncovalent π–π stacking which does not destroy its sp^2 hybridisation. Using this material, specific capacitance values up to 120 F/g were obtained. Zhang et al. (2011) investigated the potential of various surfactants to intercalate and stabilise graphene oxide. The best surfactant was found to be tetrabutylammonium hydroxide and the resultant material yielded a capacitance of 194 F/g. Yang et al. (2011) developed an interesting strategy in which they used water as the spacer in pre-synthesised graphene sheets. They obtained a specific capacitance of 215 F/g in aqueous electrolyte. Greater details in this and other intercalation routes can be found in reference (Huang et al., 2012).

6.6.3. GRAPHENE-BASED PSEUDO CAPACITORS

6.6.3.1. Graphene-Conducting Polymer Composite Electrodes

A variety of conducting polymers, including polypyrrole (PPy), polyaniline (PANI), poly-3,4-ethylendioxythiophen (PEDOT), polythiophene (PT) and poly(p-phenylene vinylene) PPV have been implemented as electrode materials or components in supercapacitors. However, despite their relatively easy production and high capacitance, sometimes their conductivities are relatively poor limiting their performance in supercapacitors. This can be improved by doping with carbon materials including graphene. PANI is usually used since it has good environmental stability and hence has almost exclusively been used in the case of graphene studies.

H. Wang et al. (2009) developed electrode material based on fibrillar PANI doped with graphene oxide sheets. They obtained a specific capacitance of 531 F/g which is much higher than pure PANI. This highlights a positive syntergistic effect between graphene oxide and PANI. They also investigated the effect of the raw graphite size and the feeding ratios on the electrochemical properties of graphene oxide-PANI composites (Wang et al., 2010a). Their data showed the morphology of the produced composites is influenced dramatically by the mass ratio. They propose the composites combine through electrostatic interaction (doping), hydrogen bonding and π–π interactions. For graphene oxide/aniline ratios of 1:200 and 1:50, they obtained specific capacities of 746 and 627 F/g respectively.

The superior conductivity and electrochemical stability of graphene makes reduced graphene more favourable as compared to graphene oxide, and this was exploited by Zhang et al. (2010). They fabricated graphene-PANI composites by first polymerising aniline monomers in the presence of graphene oxide in acidic conditions and then reducing the graphene oxide using hydrazine. Finally the reoxidation and reprotonation of the reduce PANI followed. Composites with 80% graphene material showed the highest capacitance of 480 F/g. 70% of the capacitance was still retained after 1000 cycles which suggests the material has good cycling stability which is attributed to the PANI. Wang et al. (2010b) developed a facile three step in situ polymerisation/reduction-dedoping/doping route to produce a graphene-PANI composite suitable for as a supercapacitor electrode. The process leads to reduced graphene sheets fully covered by PANI granules. The composite exhibited better electrochemical performance than either of the pure individual components and yielded a specific capacitance of 1126 F/g and retained 80% of this value after 1000 cycles.

One of the advantages of graphene-based polymer composites as electrodes for supercapacitors is as that they can be flexible. Flexible supercapacitors can be useful in applications such as portable electronics. Freestanding and paper-like graphene-based materials have excellent flexibility, great mechanical strength and good electrical conductivity, making them attractive to be used in flexible electrodes. Wang et al. (2009b) synthesised a flexible graphene-PANI composite paper. It showed a tensile strength of 12.6 MPa and a stable capacitance of 233 F/g and 135 F/g for gravimetric and volume capacitances, respectively. Yan et al. (2010a) also fabricated flexible graphene-PANI hybrid papers. They measured a specific capacitance of 489 F/g.

A few studies have explored graphene-based polymer electrodes with other polymers outside of PANI. Bose et al. (2011) used PPy in graphene-PPy composite and found a specific capacitance of 267 F/g. After 500 cycles, a 10% reduction in capacitance was observed. Mini et al. (2011) and Zhang et al. (2010) also explored the use of PPy with graphene as electrode hybrids with good success.

6.6.3.2. Graphene-Metal Oxide and Hydroxide Composite Electrodes

A variety of metal oxides, including oxides from Ru, Fe, Ti, V, Co, Ni, Zn, Mo, Sn, and Ir, have been explored for use in supercapacitors (Hou et al., 2011). This is because metal oxides generally have much larger capacitance than carbon materials since multielectron transfer during fast Faradaic reactions is better with metal oxides. The efficiency of metal oxide-based supercapacitors can be improved by nanostructuring the oxides since decreasing the particle size increases the specific surface area which leads to an enhanced interface between the electrode and the electrolyte. However, it also reduces the path length for electronic transport (reduced conductivity). This unwanted effect can be reduced by including carbonaceous materials like graphene. F. Li et al. (2009) developed a one step approach to synthesise graphene-SnO_2 nanocomposites in acidic solution where graphite oxide was reduced by $SnCl_2$ to graphene in the presence of urea and HCl. Although the electrical conductivity improved the specific capacitance was only 43.4 F/g. The use of ZnO-graphene composites also showed low specific capacitance, namely, 61.7 F/g (Zhang et al., 2009). MnO_2 has attracted a fair bit of interest as a material for supercapacitors since it has a high specific capacitance and is cost effective. A graphene oxide-MnO_2 nanocomposite showed a specific capacity of 192.7 F/g which is close to the theoretical limit for such a composite (197.2 F/g). However, the specific capacitance for nano-MnO_2

was 211.2 F/g which suggests the interaction between MnO_2 and graphene oxide does not enhance capacitance. Some hydroxides have been explored for use in hydroxide-graphene composites. Wang et al. (2010c) grew single crystalline $Ni(OH)_2$ hexagonal nanoplates directly on graphene sheets. The material yielded a high specific capacitance of 1335 F/g. This value is strong contrast to the other metal oxide systems discussed above and suggests the interaction between nanoparticles and graphene may be important, perhaps affecting the charge transport from nanoparticles to the graphene network. Graphene-$Co(OH)_2$ nanocomposites also showed a high specific capacitance of 972.5 F/g and again suggests a synergistic effect between nanoparticles and graphene forming composites (Chen et al., 2010).

6.6.4. GRAPHENE BASED MATERIALS IN LITHIUM ION BATTERIES

The Sony company first introduced the lithium ion battery (LIB) in the early 1990s, and it has since become the battery of choice for many portable electronic device applications. Lithium ion batteries consist of an anode and a cathode and an electrolyte (separator). The energy storage is accomplished electrochemically and occurs by the intercalation and deintercalation of lithium ions such that upon charging lithium ions are taken from the cathode and introduced to the anode. In the discharge mode the reverse happens. In general the cathode is made from a lithium containing metal oxide, for example lithium cobalt oxide or lithium iron phosphate. The anode material/s is/are chosen so as to reversibly store lithium ions. Examples include carbon, transition metal oxides and other alloying materials like Si (Hou et al., 2011). There is a constant drive to improve the energy densities and performances of LIBs and to this end numerous nanomaterials are being investigated, among them graphene and its derivatives. Graphene is being explored for its potential as an anode material, as a component in hybrid anodes and as an additive in electrodes.

6.6.4.1. Graphene Anodes

Whilst graphitic carbon is commonly used as an anode material in LiBs it has a low storage capacity (<372 mA h g^{-1}). This is argued to occur because only one lithium attaches to each hexahedron in the chickenwire lattice yielding LiC_6. It is argued that for a single graphene layer this can occur on both sides of the sheet yielding Li_2C_6 which corresponds to a storage capacity of around 780 mA h g^{-1} (Dahn et al., 1995). It is also argued a covalent attachment at a covalent site on the benzene ring (LiC_2) is possible which would provide a storage capacity around 1116 mA h g^{-1} (Sato et al., 1994). The latter two configurations are obviously more attractive. One can also increase the number of Li anchor sites by introducing defects, such as edge defects and vacancies

(Abouimrane et al., 2010). There has also been significant interest to find routes which splay or separate the individual graphene sheets in graphite so as to maximise the available surface for Li insertion by the incorporation of spacers between sheets. Carbon spacers such as carbon nanotubes and fullerenes (C_{60}) have been exploited (Yoo et al., 2008). In the case of nanotube spacers a specific capacity of up to 730 mA h g^{-1} was achieved and 748 mA h g^{-1} for the case of fullerene spacers. However the cyclic performance was not good. Metallic and metallic oxide nanoparticles as spacer also been explored including SnO_2 (Wang et al., 2010), TiO_2 (Wang et al., 2009), Co_3O_4 (Yang et al., 2010a), Mn_3O_4 (Wang et al., 2010d), CuO (Mai et al., 2011), Fe_3O_4 (B. Li et al., 2011), Sn (G. Wang et al., 2009), Si (Lee et al., 2010) and $LiFePO_4$ (Wang et al., 2010). Apart from separating the sheets these nanoparticles are themselves suitable for reversible interactions with lithium ions. An alternative to the spacer route is the encapsulation of the nanoparticles with graphitic material (Cui et al., 2007a, 2008; Zhi et al., 2008) yet another approach is the use of graphene oxide paper. Here the presence of the oxygen-containing functional groups increase the interlayer spacing between the graphene sheets (Abouimrane et al., 2010; Wang et al., 2009).

6.6.4.2. Graphene in Hybrid Anodes

In hybrid systems graphene is exploited for a variety of roles. For example, while transition metal oxides can capture and release Li ions they are susceptible to crumpling and cracking due to the large volume change they incur during charging and discharging. SnO_2 is an attractive oxide, as it has a high Li storage capacity (782 mA h g^{-1}) which is much higher than graphite. To reduce cracking and crumling, Paek et al. (2009) explored the use of graphene nanosheets in ethylene glycol solution reassembled in the presence of rutile SnO_2 nanoparticles, rendering a material in which the nanosheets are homogeneously distributed between the nanoparticles with plenty of voids which in essence serve as buffering spaces. The expansion of the SnO_2 nanoparticles due to Li insertion is limited, leading to improved cyclic performance. Co_3O_4 anchored in graphene sheets has also been explored (Wu et al., 2010; Yang et al., 2010b). Again the graphene sheets help provide buffering spaces and also reduce particle agglomeration. The graphene sheets also provide a good conducting network that helps improve the power density of the networks. The nanoparticles help prevent graphene sheet agglomeration, which helps maximise Li insertion between graphene sheets.

6.6.4.3. Graphene as an Electrode Additive

The goal of graphene here is to enhance the conductivity of the electrode. For example, nanostructured TiO_2-graphene shows enhanced lithium ion insertion and extraction as compared to pure TiO_2, and this is attributed to the enhanced electrical conductivity provided by the percolated graphene network (Wang et al., 2009). The combination of TiO_2-graphene with a $LiFePO_4$ cathode showed almost no fade after more than 700 cycles (Choi et al., 2010). The formation of a $LiFePO_4$-graphene composite helped improve both the capacity and cycle performance (Zhu et al., 2010).

6.6.5. GRAPHENE-BASED MATERIALS IN FUEL CELLS

Fuel cells convert chemical energy from a fuel into electricity through a chemical reaction with oxygen or an oxidising agent. Typical fuels are hydrogen, hydrocarbons and alcohols. What differentiates fuel cells from batteries is that fuel cells require a continuous source of oxygen and fuel to run. The general makeup of a fuel cell is to place three segments together, namely, an anode, the electrolyte and the cathode. At the interfaces of these three segments, two chemical reactions occur creating a current. The process consumes the fuel and creates water or carbon dioxide. The anode is loaded with a catalyst that oxidises the fuel breaking it down into electrons and ions in the process. The released electrons travel though a wire to the load and provide the current while the ions move through the electrolyte to the cathode where they are reunited with electrons in a reaction that provides oxygen to create water or carbon dioxide. A variety of carbon materials such as carbon black, carbon nanotubes and carbon fibres have been explored as electrocatalyst supports in fuel cell anodes (Luo et al., 2012). Unsurprisingly, the fact that graphene is amenable to covalent and non-covalent functionalisation of the surface and if done correctly, does not significantly affect its conductivity as well as its high specific surface area makes it an attractive material to support the dispersion of catalysts and hence as an exciting anode material in fuel cells. To this end a variety of routes have been developed to prepare disperse Pt catalysts on graphene-based material. For example, by reducing H_2PtCl_6 and graphene oxide with $NaBH_4$ under basic conditions Y. Li et al. (2009) were able to prepare disperse Pt catalyst particles on graphene with 45% loading. Yoo et al. (2009) loaded Pt catalysts on graphene nanosheets my mixing (Pt $(NO_2)_2(NH_3)_2$) and the nano-graphene in ethanol. The Pt/nano-graphene electrocatalyst showed superior activity for the methanol oxidation reaction as compared to Pt/carbon black electrocatalyst systems. However the Pt showed little ability to resist poisonous intermediates. Binary metal catalyst systems are more stable in this respect as demonstrated for a variety of combinations including Pt–Co (Yue et al., 2010), Pt–Au (Hu et al., 2011), Pt–Ru (Dong et al., 2010) and Pt–Pd (Guo et al., 2010).

For all the examples discussed above the implemented fuel was methanol. However, graphene-based catalyst systems with other fuels have also been explored including hydrogen (Yoo et al., 2011), hydrazine (Wang et al., 2010) and formic acid (Zhang et al., 2011).

More complicated structures have also been demonstrated. Mesorpourous carbon nanotubes caonatining nanographene can be prepared through the pyrolysis of hyperbranched polyphenylene in an Al_2O_3 template as shown by Cui et al. (2007b). The material is suitable as a support for Pt catalyst nanoparticles deposited using an electrodeposition route. The oxygen–reduction reaction using the graphene composite support and depositied platinum was found to yield a current density some 1000 times better than that of a conventional Pt/C catalyst. Organo-metallic molecules can also be used as precursors to form metallic nanoparticles integrated within nanoporous carbons using a template route (Liang et al., 2010). Graphene has also found application in the development of microbial fuel cells. Wang et al. (2011) showed one can reduce graphene for direct electron transfer using the bacterium Shewanella sp.

6.6.6. GRAPHENE-BASED MATERIALS IN SOLAR CELLS

The outstanding electron transport and high carrier mobility afforded by graphene as well as its transparency make it an exciting material for photovoltaic devices. Indeed, solar cells using graphene in different parts of the cell have been reported (see for example Fig. 6.6.2). The greatest interest lies in its potential for transparent and conductive electrodes. The preparation of graphene films as transparent conductive electrodes can be achieved in different ways. Early works explored the reduction of graphene oxide to form thin transparent films. For a 10 nm thick film a transparency of 70% and conductivity of 550 S cm^{-1} was obtained. The film allowed a power-conversion efficiency of 0.26% (Wang et al., 2008a). Another work also based on reduced graphene oxide films with a sheet resistance of 40 kΩ/sq. and a transparency of 64% showed a conversion efficiency of 0.1% (Eda et al., 2008). Yang et al. (2010c) managed to form a film with a sheet resistance of a 5 kΩ/sq and ca. 80% transparency, which led to a conversion efficiency of 0.4%. Nonetheless, this is insuffient for practical purposes and efforts to improve this have been/are being conducted.

One idea being explored is the incorporation of conductive filler material to diminish the loss in conductivity that is found in processed films (e.g. solution processed graphene oxide) since the presence of many incurred defects reduces conductivity. Tung et al. (2009) explored the use of carbon nanotubes as filler material in which they demonstrated the film had a resistance of 240 Ω/sq and a transmittance of ca. 86%. The use of carbon-rich precursors to heal out defects using chemical or thermal steps has also been investigated and can improve the conductivity and power efficiency (Liang et al., 2009; López et al., 2009; Su et al., 2009).

Routes which avoid solution processing have also been explored, for example, large area CVD grown graphene. CVD-grown graphene over Ni and transferred using a dry transfer technique that was 6–30 nm thick showed

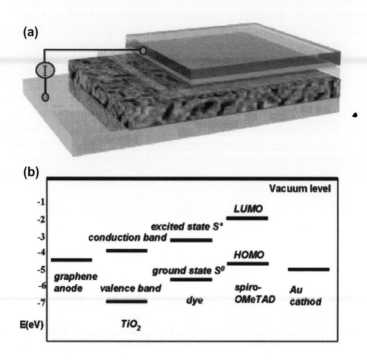

FIGURE 6.6.2 Illustration and performance of solar cell based on graphene electrodes. (a) Illustration of dye-sensitised solar cell using graphene film as electrode, the four layers from bottom to top are Au, dye-sensitised heterojunction, compact TiO_2, and graphene film. (b) The energy level diagram of graphene/TiO_2/dye/spiro-OMeTAD/Au device (Wang et al., 2008a).

film resistances between 1350 and 210 Ω/sq and optical transparencies between 72 and 91%. The material was used in a solar cell and a power conversion efficiency of 1.71% was demonstrated (Yao et al., 2009). Another group using CVD-grown graphene obtained a power conversion efficiency of 1.18% (De Arco et al., 2010). Wang et al. (2008b) explored the use of thermally reduced synthetic nanographene molecules of giant polycyclic aromatic hydrocarbons as a window for organic solar cells. The 4 nm thick film showed a transparency of 90%. Indeed the output voltage of the graphene based cell was comparable to that of indium-tin-oxide (ITO) based cells. This suggests graphene has an appropriate work function for this type of organic cell.

Aside from graphene's potential to serve as transparent conductive electrodes it has other attractive properties for photovoltaic devices such as its potential as a photoactive material (Nair et al., 2008) or as a sensitiser for dye-sensitised solar cells (Yan et al., 2010b).

REFERENCES

Abouimrane, A., Compton, O.C., Amine, K., Nguyen, S.B.T., 2010. Non-annealed graphene paper as a binder-free anode for lithium-ion batteries. J. Phys. Chem. C 114 (29), 12800–12804.

An, X., Simmons, T.J., Shah, R., Wolfe, C., Lewis, K.M., Washington, M., Nayak, S.K., Talapatra, S., Kar, S., 2010. Stable aqueous dispersions of noncovalently functionalized graphene from graphite and their multifunctional high-performance applications. Nano Lett. 10 (11), 4295–4301.

Bose, S., Kim, N.H., Kuila, T., Lau, K.T., Lee, J.H., 2011. Electrochemical performance of a graphene–polypyrrole nanocomposite as a supercapacitor electrode. Nanotechnology 22 (1), 369502.

Chen, S., Zhu, J., Wang, X., 2010. One-step synthesis of graphene–Cobalt hydroxide nanocomposites and their electrochemical properties. J. Phys. Chem. C 114 (27), 11829–11834.

Chen, Y., Zhang, X., Zhang, D., Yu, P., Ma, Y., 2011. High performance supercapacitors based on reduced graphene oxide in aqueous and ionic liquid electrolytes. Carbon 49 (2), 573–580.

Choi, D., Wang, D., Viswanathan, V.V., Bae, I.-T., Wang, W., Nie, Z., Zhang, J.-G., Graff, G.L., Liu, J., Yang, Z., Duong, T., 2010. Li-ion batteries from LiFePO$_4$ cathode and anatase/graphene composite anode for stationary energy storage. Electrochem. Commun. 12 (3), 378–381.

Cui, G., Hu, Y.-S., Zhi, L., Wu, D., Lieberwirth, I., Maier, J., Müllen, K., 2007a. A one-step approach towards carbon-encapsulated hollow tin nanoparticles and their application in lithium batteries. Small 3 (12), 2066–2069.

Cui, G., Zhi, L., Thomas, A., Kolb, U., Lieberwirth, I., Müllen, K., 2007b. One-dimensional porous carbon/platinum composites for nanoscale electrodes. Angew. Chem. Int. Ed. 46, 3464–3467. 2007.

Cui, G., Gu, L., Zhi, L., Kaskhedikar, N., van Aken, P.A., Müllen, K., Maier, J., 2008. A germanium–carbon nanocomposite material for lithium batteries. Adv. Mater. 20, 3079–3083.

Dahn, J.R., Zheng, Tao, Liu, Yinghu, Xue, J.S., 1995. Mechanisms for lithium insertion in carbonaceous materials. Science 270 (5236), 590–593.

De Arco, L.G., Zhang, Y., Schlenker, C.W., Ryu, K., Thompson, M.E., Zhou, C., 2010. Continuous, highly flexible, and transparent graphene films by chemical vapor deposition for organic photovoltaics. ACS Nano 4 (5), 2865–2873.

Dong, L., Gari, R.R.S., Li, Z., Craig, M.M., Hou, S., 2010. Graphene-supported platinum and platinum–ruthenium nanoparticles with high electrocatalytic activity for methanol and ethanol oxidation. Carbon 48 (3), 781–787.

Eda, G., Lin, Y.-Y., Miller, S., Chen, C.-W., Su, W.-F., Chhowalla, M., 2008. Transparent and conducting electrodes for organic electronics from reduced graphene oxide. Appl. Phys. Lett. 92 (3), 233305.

Guo, S., Dong, S., Wang, E., 2010. Three-dimensional Pt-on-Pd bimetallic nanodendrites supported on graphene nanosheet: facile synthesis and used as an advanced nanoelectrocatalyst for methanol oxidation. ACS Nano 4 (1), 547–555.

Hantel, M.M., Kaspar, T., Nesper, R., Wokaun, A., Kotz, R., 2011. Partially reduced graphite oxide for supercapacitor electrodes: effect of graphene layer spacing and huge specific capacitance. Electrochem. Commun. 13 (1), 90–92.

Hou, J., Shao, Y., Ellis, M.W., Moore, R.B., Yi, B., 2011. Graphene-based electrochemical energy conversion and storage: fuel cells, supercapacitors and lithium ion batteries. Phys. Chem. Chem. Phys. 13, 15384–15402.

Hu, Y., Zhang, H., Wu, P., Zhang, H., Zhou, B., Cai, C., 2011. Bimetallic Pt–Au nanocatalysts electrochemically deposited on graphene and their electrocatalytic characteristics towards oxygen reduction and methanol oxidation. Phys. Chem. Chem. Phys. 13 (9), 4083–4094.

Huang, Y., Liang, J., Chen, Y., 2012. An overview of the applications of graphene-based materials in supercapacitors. Small 8 (12), 1805–1834.

Lee, J.K., Smith, K.B., Hayner, C.M., Kung, H.H., 2010. Silicon nanoparticles–graphene paper composites for Li ion battery anodes. Chem. Commun. 46, 2025–2027.

Li, B., Cao, H., Shao, J., Qu, M., Warner, J.H., 2011. Superparamagnetic Fe_3O_4 nanocrystals@graphene composites for energy storage devices. J. Mater. Chem. 21, 5069–5075.

Li, F., Song, J., Yang, H., Gan, S., Zhang, Q., Han, D., Ivaska, A., Niu, L., 2009. One-step synthesis of graphene/SnO_2 nanocomposites and its application in electrochemical supercapacitors. Nanotechnology 20 (6), 455602.

Li, Y., Tang, L., Li, J., 2009. Preparation and electrochemical performance for methanol oxidation of pt/graphene nanocomposites. Electrochem. Commun. 11 (4), 846–849.

Li, Y., van Zijll, M., Chiang, S., Pan, N., 2011. KOH modified graphene nanosheets for supercapacitor electrodes. J. Power Sources 196 (14), 6003–6006.

Liang, Y., Frisch, J., Zhi, L., Norouzi-Arasi, H., Feng, X., P Rabe, J., Koch, N., Müllen, K., 2009. Transparent, highly conductive graphene electrodes from acetylene-assisted thermolysis of graphite oxide sheets and nanographene molecules. Nanotechnology 20 (43), 434007 (6).

Liang, Y., Schwab, M.G., Zhi, L., Mugnaioli, E., Kolb, U., Feng, X., Müllen, K., 2010. Direct access to metal or metal oxide nanocrystals integrated with one-dimensional nanoporous carbons for electrochemical energy storage. J. Am. Chem. Soc. 132 (42), 15030–15037.

López, V., Sundaram, R.S., Gómez-Navarro, C., Olea, D., Burghard, M., Gómez-Herrero, J., Zamora, F., Kern, K., 2009. Graphene monolayers: chemical vapor deposition repair of graphene oxide: a route to highly-conductive graphene monolayers, Adv. Mater 21 (46).

Luo, B., Liu, S., Zhi, L., 2012. Chemical approaches toward graphene-based nanomaterials and their applications in energy-related areas. Small 8 (5), 630–646.

Lv, W., Tang, D.-M., He, Y.-B., You, C.-H., Shi, Z.-Q., Chen, X.-C., Chen, C.-M., Hou, P.-X., Liu, C., Yang, Q.-H., 2009. Low-temperature exfoliated graphenes: vacuum-promoted exfoliation and electrochemical energy storage. ACS Nano 3 (11), 3730–3736.

Mai, Y.J., Wang, X.L., Xiang, J.Y., Qiao, Y.Q., Zhang, D., Gu, C.D., Tu, J.P., 2011. CuO/graphene composite as anode materials for lithium-ion batteries. Electrochim. Acta 56 (5), 2306–2311.

Mini, P.A., Balakrishnan, A., Nair, S.V., Subramanian, K.R.V., 2011. Highly super capacitive electrodes made of graphene/poly(pyrrole). Chem. Commun. 47, 5753–5755.

Nair, R.R., Blake, P., Grigorenko, A.N., Novoselov, K.S., Booth, T.J., Stauber, T., Peres, N.M.R., Geim, A.K., 2008. Fine structure constant defines visual transparency of graphene. Science 320 (5881), 1308.

Paek, S.-M., Yoo, E.J., Honma, I., 2009. Enhanced cyclic performance and lithium storage capacity of SnO_2/graphene nanoporous electrodes with three-dimensionally delaminated flexible structure. Nano Lett. 9 (1), 72–75.

Sato, K., Noguchi, M., Demachi, A., Oki, N., Endo, M., 1994. A mechanism of lithium storage in disordered carbons. Science 264 (5158), 556–558.

Si, Y., Samulski, E.T., 2008. Exfoliated graphene separated by platinum nanoparticles. Chem. Mater. 20 (21), 6792–6797.

Stoller, M.D., Park, S., Zhu, Y., An, J., Ruoff, R.S., 2008. Graphene-based ultracapacitors. Nano Lett. 8 (10), 3498–3502.

Su, Q., Pang, S., Alijani, V., Li, C., Feng, X., Müllen, K., 2009. Composites of graphene with large aromatic, molecules. Adv. Mater. 21 (31), 3191–3195.

Tung, V.C., Chen, L.M., Allen, M.J., Wassei, J.K., Nelson, K., Kaner, R.B., Yang, Y., 2009. Low-temperature solution processing of graphene–carbon nanotube hybrid materials for high-performance transparent conductors. Nano Lett. 9 (5), 1949–1955.

Vivekchand, S.R.C., Chandra Sekhar Rout, K.S., Subrahmanyam, A., Govindaraj, C.N., Rao, R., 2008. Graphene-based electrochemical supercapacitors. J. Chem. Sci. 120 (1), 9–13.

Wang, D., Choi, D., Li, J., Yang, Z., Nie, Z., Kou, R., Hu, D., Wang, C., Saraf, L.V., Zhang, J., Aksay, I.,A., Liu, J., 2009. Self-assembled TiO_2–graphene hybrid nanostructures for enhanced Li-ion insertion. ACS Nano 3 (4), 907–914.

Wang, D., Kou, R., Choi, D., Yang, Z., Nie, Z., Li, J., Saraf, L.V., Hu, D., Zhang, J., Graff, G.L., Liu, J., Pope, M.A., Aksay, I.A., 2010. Ternary self-assembly of ordered metal oxide– graphene nanocomposites for electrochemical energy storage. ACS Nano 4 (3), 1587–1595. 2010.

Wang, D.W., Li, F., Zhao, J., Ren, W., Chen, Z.G., Tan, J., Wu, Z.S., Gentle, I., Lu, G.Q., Cheng, H.M., 2009. Fabrication of Graphene/Polyaniline composite Paper via in situ Anodic Electropolymerization for high-performance flexible electrode. ACS Nano 3 (7), 1745–1752.

Wang, G., Wang, B., Wang, X., Park, J., Dou, S., Ahn, H., Kim, K., 2009. Sn/graphene nanocomposite with 3D architecture for enhanced reversible lithium storage in lithium ion batteries. J. Mater. Chem. 19, 8378–8384.

Wang, G., Qian, F., Saltikov, C.W., Jiao, Y., Li, Y., 2011. Microbial reduction of graphene oxide by Shewanella. Nano Res. 4 (6), 563–570.

Wang, H., Hao, Q., Yang, X., Lu, L., Wang, X., 2009. Graphene oxide doped polyaniline for supercapacitors. Electrochem. Commun. 11 (6), 1158–1161.

Wang, H., Hao, Q., Yang, X., Lu, L., Wang, X., 2010a. Effect of graphene oxide on the properties of its composite with polyaniline. ACS Appl. Mater. Interfaces 2 (3), 821–828.

Wang, H., Hao, Q., Yang, X., Lu, L., Wang, X., 2010b. A nanostructured graphene/polyaniline hybrid material for supercapacitors. Nanoscale 2, 2164–2170.

Wang, H., Casalongue, H.S., Liang, Y., Dai, H., 2010c. $Ni(OH)_2$ nanoplates grown on graphene as advanced electrochemical pseudocapacitor materials. J. Am. Chem. Soc. 132 (21), 7472–7477.

Wang, H., Cui, L.-F., Yang, Y., Casalongue, H.S., Robinson, J.T., Liang, Y., Cui, Y., Dai, H., 2010d. Mn_3O_4–Graphene hybrid as a high-capacity anode material for lithium ion batteries. J. Am. Chem. Soc. 132 (40), 13978–13980.

Wang, L., Wang, H., Liu, Z., Xiao, C., Dong, S., Han, P., Zhang, Z., Zhang, X., Bi, C., Cui, G., 2010. A facile method of preparing mixed conducting $LiFePO_4$/graphene composites for lithium-ion batteries. Solid State Ionics 181 (37–38), 1685–1689.

Wang, X., Zhi, L., Müllen, Klaus, 2008a. Transparent, conductive graphene electrodes for dye-sensitized solar cells. Nano Lett. 8 (1), 323–327.

Wang, X., Zhi, L., Tsao, N., Tomović, Ž, Li, J., Müllen, K., 2008b. Transparent carbon films as electrodes in organic solar cells. Angew. Chem. Int. Ed. 47 (16), 2990–2992.

Wang, Y., Shi, Z., Huang, Y., Ma, Y., Wang, C., Chen, M., Chen, Y., 2009. Supercapacitor devices based on graphene materials. J. Phys. Chem. 113 (30), 13103–13107.

Wang, Y., Wan, Y., Zhang, D., 2010. Reduced graphene sheets modified glassy carbon electrode for electrocatalytic oxidation of hydrazine in alkaline media. Electrochem. Commun. 12 (2), 187–190.

Wu, Z.-S., Ren, W., Wen, L., Gao, L., Zhao, J., Chen, Z., Zhou, G., Li, F., Cheng, H.-M., 2010. Graphene anchored with Co_3O_4 nanoparticles as anode of lithium ion batteries with enhanced reversible capacity and cyclic performance. ACS Nano 4 (6), 3187–3194.

Xu, Y., Sheng, K., Li, C., Shi, G., 2010. Self-assembled graphene hydrogel via a one-step hydrothermal process. ACS Nano 4 (7), 4324–4330.

Yan, X., Chen, J., Yang, J., Xue, Q., Miele, P., 2010a. Fabrication of free-standing, electro-chemically active, and biocompatible graphene oxide−polyaniline and graphene−polyaniline hybrid papers. ACS Appl. Mater. Interfaces 2 (9), 2521–2529.

Yan, X., Cui, X., Li, B., Li, L.-s., 2010b. Large, solution-Processable graphene quantum dots as light Absorbers for Photovoltaics. Nano Lett. 10 (5), 1869–1873.

Yang, S., Feng, X., Ivanovici, S., Müllen, K., 2010a. Fabrication of graphene-encapsulated oxide nanoparticles: towards high-performance anode materials for lithium storage. Angew. Chem. Int. Ed. 49, 8408–8411. 2010.

Yang, S., Cui, G., Pang, S., Cao, Q., Kolb, U., Feng, X., Maier, J., Müllen, K., 2010b. Fabrication of cobalt and cobalt oxide/graphene composites: towards high-performance anode materials for lithium ion batteries. ChemSusChem 3 (2), 236–239.

Yang, S., Feng, X., Zhi, L., Cao, Q., Maier, J., Müllen, K., 2010c. Nanographene-constructed hollow carbon spheres and their favorable electroactivity with respect to lithium storage. Adv. Mater. 22, 838–842.

Yang, X., Zhu, J., Qiu, L., Li, D., 2011. Bioinspired effective prevention of restacking in multi-layered graphene films: towards the next generation of high-performance supercapacitors. Adv. Mater. 23 (25), 2833–2838.

Yao, J., Shen, X., Wang, B., Liu, H., Wang, G., 2009. In situ chemical synthesis of SnO_2–graphene nanocomposite as anode materials for lithium-ion batteries. Electrochem. Commun. 11 (10), 1849–1852.

Yoo, E.J., Kim, J., Hosono, E., Zhou, H.-s., Kudo, T., Honma, I., 2008. Large reversible Li storage of graphene nanosheet families for use in rechargeable lithium ion batteries. Nano Lett. 8 (8), 2277–2282.

Yoo, E.J., Okata, T., Akita, T., Kohyama, M., Nakamura, J., Honma, I., 2009. Enhanced electro-catalytic activity of Pt subnanoclusters on graphene nanosheet surface. Nano Lett. 9 (6), 2255–2259.

Yoo, E.J., Okada, T., Akita, T., Kohyama, M., Honma, I., Nakamura, J., 2011. Sub-nano-Pt cluster supported on graphene nanosheets for CO tolerant catalysts in polymer electrolyte fuel cells. J. Power Sources 196 (1), 110–115.

Yue, Q., Zhang, K., Chen, X., Wang, L., Zhao, J., Liu, J., Jia, J., 2010. Generation of OH radicals in oxygen reduction reaction at Pt–Co nanoparticles supported on graphene in alkaline solutions. Chem. Commun. 46 (19), 3369–3371.

Zhang, K., Zhang, L.L., Zhao, X.S., Wu, J., 2010. Graphene/polyaniline nanofiber composites as supercapacitor electrodes. Chem. Mater. 22 (4), 1392–1401.

Zhang, K., Mao, L., Zhang, L.L., Chan, H.S.O., Zhao, X.S., Wu, J., 2011. Surfactant-intercalated, chemically reduced graphene oxide for high performance supercapacitor electrodes. J. Mater. Chem. 21, 7302–7307.

Zhang, L., Shi, G., 2011. Preparation of highly conductive graphene hydrogels for fabricating supercapacitors with high rate capability. J. Phys. Chem. C 115 (34), 17206–17212.

Zhang, L.L., Zhao, S., Tian, X.N., Zhao, X.S., 2010. Layered graphene oxide nanostructures with sandwiched conducting polymers as supercapacitor electrodes. Langmuir 26 (22), 17624–17628.

Zhang, S., Shao, Y., Liao, H.-g., Liu, J., Aksay, I.A., Yin, G., Lin, Y., 2011. Graphene decorated with PtAu alloy nanoparticles: facile synthesis and promising application for formic acid oxidation. Chem. Mater. 23 (5), 1079–1081.

Zhang, Y., Li, H., Pan, L., Lu, T., Sun, Z., 2009. Capacitive behavior of graphene–ZnO composite film for supercapacitors. J. Electroanal. Chem. 634 (1), 68–71.

Zhi, L., Hu, Y.-S., El Hamaoui, B., Wang, X., Lieberwirth, I., Kolb, U., Maier, J., Müllen, K., 2008. Precursor-controlled formation of novel carbon/metal and carbon/metal oxide nano-composites. Adv. Mater. 20, 1727–1731.

Zhu, N., Liu, W., Xue, M., Xie, Z., Zhao, D., Zhang, M., Chen, J., Cao, T., 2010. Graphene as a conductive additive to enhance the high-rate capabilities of electrospun $Li_4Ti_5O_{12}$ for lithium-ion batteries. Electrochim. Acta 55 (20), 5813–5818.

Zhu, Y., Murali, S., Stoller, M.D., Velamakanni, A., Piner, R.D., Ruoff, R.S., 2010. Microwave assisted exfoliation and reduction of graphite oxide for ultracapacitors. Carbon 48 (7), 2118–2122.

Chapter 6.7

Superstrong Graphene Composites

Imad Ibrahim and Mark H. Rümmeli

IFW Dresden, Germany

The large specific surface area, two-dimensional high aspect ratio sheet geometry, and outstanding mechanical properties of graphene make it highly attractive as a nanofiller in composite materials (Stankovich et al., 2006a). The possibility to significantly enhance the mechanical properties of graphene-based composites even with relatively low nanofiller (graphene) loading is advantageous (Ramanathan et al., 2008; Walker et al., 2011). The remarkable improvement obtained with graphene based composites performance lies in the inherent properties of the nanofiller (graphene and its derivatives) and more importantly, due to its potential for efficient dispersion and interface chemistry as well as its nanoscale morphology. This optimisation allows one to take advantage of the enormous surface area per unit volume that graphene based nanofillers have (Ramanathan et al., 2008).

6.7.1. GRAPHENE-BASED COMPOSITES

Graphene-based composites vary between graphene added to inorganic nanostructures, polymer nanocomposites, biomaterials and metal organic frameworks. The first group of graphene-based composites are the graphene-inorganic nanostructure nanocomposites. The addition of graphene allows the enhancement of the composite properties to be used in different applications, such as, electronics, optics, electrochemical energy conversion and storage. Graphene derivatives can be added to metals (e.g. Au, Ag) or oxides (e.g. TiO_2,

ZnO) either as ex situ hybridisation, or in situ crystallisation. In ex situ hybridisation, modified-surface presynthesised nanocrystals and/or graphene-based sheets are mixed in solution. The surface modifications allow non-covalent interactions or chemical bonding to take place between the nanocrystals and/or graphene sheets. Improvement of the uniform surface coverage of nanocrystals is achieved with in situ crystallisation by surface functionalisation of graphene-based sheets (graphene oxide, reduced graphene oxide). This goal can be achieved through a variety of techniques, for example, through chemical reduction in which precursors of noble metals are reduced by reduction agents in a solution containing GO or RGO in situ. The sol–gel process is another approach for preparing graphene-based nanostructures using metal precursors that undergo a series of hydrolysis and polycondensation reactions.

The second main family of graphene-based nanocomposites is the graphene-polymer composite, where graphene derivatives are either randomly distributed in polymer matrices or layer-structured on the polymer. In the latter, graphene derivatives have been composited with polymers in layered structures to be used as directional load-bearing membranes or thin films for photovoltaic applications. To prepare a layer-structured polymer, the Langmuir–Blodgett (LB) technique can be used to deposit layer-by-layer graphene oxide sheets onto films of polyelectrolyte poly(allylamine hydrochloride) (PAH) and poly(sodium 4-styrene sulfonate) (PSS) multilayer. This type of composite results in a directional elastic modulus improvement of over an order of magnitude for an 8 vol % loading of the graphene (Kulkarni et al., 2010).

On the other hand, graphene-filled polymer composites where the graphene sheets are randomly distributed in the polymer matrix are more commonly used for applications. The mixing of polymer and graphene-containing solutions, which should be compatible, is the most straightforward method for the preparation of graphene-filled polymer composites. In order to ensure a high dispersion of the graphene in the composite, these graphene structures should be oxidised or functionalised (Cai et al., 2009; Lee et al., 2009) or prepared from pristine graphene or exfoliated in liquids (Hernandez et al., 2008). However, heavily oxygenated graphene oxide sheets are strongly hydrophilic, which allow graphite oxide to readily swell and disperse in aqueous media (water soluble polymers) (Hirata et al., 2004; Szabo et al., 2005). Unfortunately, the resultant product is incompatible with most organic polymers. The exfoliation behaviour, and hence the dispersibility, of graphene oxide can be altered with thermal reduction or chemical functionalisation which changes the surface properties of graphene oxide sheets, as well as enhances its solubility in organic polymers (Rafiee et al., 2009; Stankovich et al., 2006a). The treatment of graphene oxide with organic isocyanates (Stankovich et al., 2006b) forms amide and carbamate ester bonds to the functional groups attached to the graphene oxide sheets, resulting in a reduction in its hydrophilic character (Stankovich et al., 2006a). The resultant modified graphene oxide sheets no longer exfoliate in water. However, the material readily forms stable dispersions in polar aprotic solvents with completely exfoliated, functionalised individual sheets (~1 nm thick) enabling intimate mixing with different organic polymers (Stankovich et al., 2006a).

Sonication is often carried out when the graphene derivative is to be used with noncompatible solvents. Sonication allows production of metastable dispersions of graphene derivatives, which can later be mixed with polymer solutions. High-speed shearing combined with ice-cooling has also been demonstrated for graphene-based fillers and polymer matrix preparation. Functionalisation of graphene-based fillers prior to the mixing strongly improves the composite quality, and hence its properties. This is because the technique avoids the restacking of the graphene sheets, aggregation and folding. In summary, fabrication of high-quality graphene-filled nanocomposites require three steps: firstly, preparation of hydrophilic graphene oxide sheets, secondly, the surface properties of the graphene oxide are modified by functionalisation, ensuring good dispersibility and avoiding aggregation (Wang et al., 2008; Xu et al., 2008). Finally, the functionalised graphene sheets are homogeneously mixed together with a polymer in a solvent (Liu et al., 2008).

Graphene-filled polymer composites can also be prepared by melt compounding, in this technique both high-shear forces and high-temperature are applied in order to blend the filler and matrix materials avoiding the use of a common solvent. In another approach, the graphene oxide is either thermally or chemically reduced after mixing with the polymer in a process called in situ polymerisation. Graphene epoxy composite have been fabricated by mixing both graphene-based filler with epoxy resins, followed by the addition of a curing agent to initiate the polymerisation process.

Improvement in the electrical conductivity and thermal stability and conductivity of the fabricated composites, and most importantly a significant enhancement of the composites' mechanical properties is achieved by the addition of graphene to nanocomposites. The use of graphene as nanofillers in composites allows the fabrication of superstrong composites due to the excellent mechanical properties (Young's modulus of 1.0 TPa and a breaking strength of 42 N m^{-1}) graphene has (Rafiee et al., 2009), in addition to the high aspect ratio and surface-to-volume ratio of the nanofillers (Stankovich et al., 2006a).

6.7.2. EX SITU POLYMERISATION

Uniform, densified microstructure nanocomposites consisting of silicon nitride and graphene platelets (GPL) can be prepared using colloidal processing methods in order to create a homogenously dispersed particle system in aqueous suspension. Later, GPL/Si$_3$N$_4$ nanocomposite slurries with different GPL concentrations can be prepared from such an aqueous suspension. Bulk

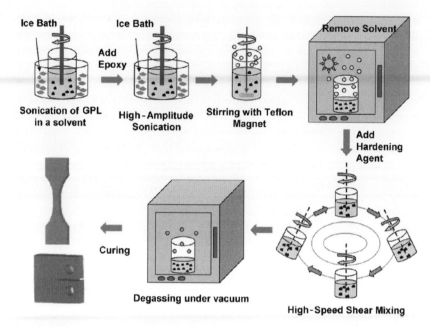

FIGURE 6.7.1 Simple schematic showing the dispersion of graphene sheets in polymer matrix by solution mixing with high-amplitude ultrasonic agitation and high-speed shear mixing (Rafiee et al., 2009). Different reports use either the same or modified route.

graphene platelets with ~3–4 sheets and thicknesses less than 2 nm have been produced by the rapid thermal expansion ($>2000\ ^\circ$C min^{-1}) of graphite oxide (Schniepp et al., 2006). The effect of the graphene concentration on the toughness of the ceramic was studied using a micro-hardness testing technique which requires relatively small amounts of graphene. In essence, the micro-hardness testing technique is used to create cracks by applying an external force. The crack areas, length as well as various other parameters, were then used to calculate the material toughness. It was found that a systematic increase in toughness from 2.8 to 6.6 MPa can be achieved by increasing the graphene platelet concentration up to 1.5%. Scanning electron microscopy characterisation of the samples exposed to the microhardness test showed the graphene sheets that are pulled-out and bridge the cracks, as shown in Fig. 6.7.2a. In addition the use of graphene platelet can block in-plane crack propagation, forcing it to climb over the wall of graphene sheets. These two phenomena become more pronounced as one increases the graphene sheet concentration and may explain the improvement of nanocomposite toughness (Walker et al., 2011). In contrast, Rafiee et al. suggested that crack bridging is less likely for graphene-based nanocomposites due to strong interfacial adhesion (Ramanathan et al., 2008; Stankovich et al., 2006a) of the graphene platelets

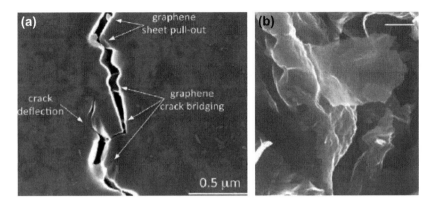

FIGURE 6.7.2 Graphene sheet pull-out over cracks (Walker et al., 2011) (a) versus graphene wrinkled surface (Ramanathan et al., 2008) (b) for enhancing mechanical interlocking and load transfer with the matrix results in enhancement in nanocomposite toughness.

with the polymer matrix. They suggest the improvement in nanocomposite toughness is due to the tilting and twisting of a crack when it encounters a rigid inclusion. This results in greater energy absorption due to the increase in the total fracture surface area. In addition, the local growth of the crack in different directions in-plane and anti-plane because of twisting (due to existence of graphene) requires a higher driving force which in turn results in a higher fracture material toughness (Rafiee et al., 2009). The wrinkled surface texture of the graphene platelets (demonstrated in Fig. 6.7.2b) is also speculated to play an important role in enhancing mechanical interlocking and load transfer with the matrix, an agreement with other studies (Ramanathan et al., 2008). In another study, graphene nanoplatelets with ~6–8 nm thickness, and average particle diameters of 5 nm were ultrasonicated for 10 min in acetone forming a uniform dispersion. This was followed by the addition of an ultrahigh molecular weight polyethylene polymer and sonication for 20 min, and results in a stable uniform suspension. Later, after settlement, the dispersant was drained, and the powder was dried in an oven at 60 °C. The elastic modulus of the nanocomposite was directly proportional to the graphene platelet concentration, reaching up to 1.19 GPa (a 124% improvement over the basic polymer). The yield strength also follows the improvement trend reaching up to 90%. However, tensile strength and toughness have their maximum improvement for moderate graphene platelet concentrations of around 80 and 54%, respectively (Lahiri et al., 2012). As often demonstrated, small amounts of graphene are preferable for polymer reinforcement. This is because low concentrations of graphene platelets in the nanocomposite material ensure better dispersion and a uniform distribution of the platelets within the polymer. In essence, this leads to superior wrapping of the inherent ripples of nicely distributed graphene platelets by the polymer, leading to the creation of wrinkles

(Rafiee et al., 2009). The ripples/wrinkles in the graphene structure, along with the strong polymer-platelet interference due to the uniform wrapping results in the absence of localised weak regions in the structure and guarantees an effective transfer of shear force from the matrix to the reinforcer (Fang et al., 2009; Gong et al., 2010). Moreover, it avoids the graphene layers' sliding against each other. The existence of a graphene nano-platelet tip at the fracture position of a nanocomposite remaining as the last destroyed part highlights the strength of polymer-graphene nano-platelet interface, which absorbs a significant amount of energy before final fracture.

To summarise, the graphene structure ripples and the uniform wrapping by the polymer induce significant improvement to the composites mechanical properties, because of the mechanical interlocking processes. On the other hand, a high concentration of graphene nanoplatelets leads to the formation of clusters due to platelet agglomeration, which may be sufficiently large to prevent full wrapping by the polymer. Clustering may also lead to nonuniform distribution of the reinforcement in the nanocomposite. Nonuniform distributions and nonfully wrapped clusters act as weak regions where the structure rapidly fails at low strain and reduces the overall toughness of the composite (Lahiri et al., 2012).

In the contrast to low graphene mass ratio studies, successful reinforcement of elastomeric materials (e.g. polyurethane) with liquid exfoliated graphene with mass fractions of up to 90% have been demonstrated. The modulus as well as the stress (at 3% strain) increase exponentially with graphene content before saturating at ~ 1.5 GPa, and at 25 MPa for mass fractions above 50 wt%, and 60 wt%, respectively. The graphene flakes size was also found to affect the nanocomposite mechanical properties. It was shown that as the flake size decreases, the Young's modulus dramatically falls, while the tensile strength and the strain at break rise slightly (Khan et al., 2010). Nawaz et al. (2012) prepared a composite using polyurethane-graphene oxide covalently functionalised with octadecylamine (ODA/GO) and characterised its mechanical properties. Various mass fractions of ODA-/GO in the range 1–50 wt% were prepared. The researchers used octadecyl-amine in order to investigate the reinforcement mechanism. They suggested a mechanism based on the formation of a graphene network which mechanically stiffens the material. They assumed that once the network forms, the GO-ODA flakes form a jammed system, which can be described for high concentration carbon black dispersions by a shear modulus (Trappe et al., 2001). They support their speculation with the behaviour of the mechanical properties of the nanocomposite as a function of the mass fractions of ODA-/GO. They showed that both the modulus and stress at low strain hardly increase for graphene contents below ~2.5 vol%, while both properties increase with graphene content as a power law above this threshold. They also found that the Young's modulus, and stress at 3% strain increase almost linearly with GO-ODA mass fractions up to 335 MPa, and ~10 MPa for the 50 wt% composite, respectively.

The possibility of high dispersibility of graphene oxide in water, even at the individual sheet level was used to achieve molecular-level dispersion of graphene in a polymer matrix (poly(vinyl alcohol) (PVA)) in which water served as the common solvent (Liang et al., 2009). This enables one to take advantage of hydrogen bonding (H-bonding), which is argued to be the strongest van der Waals interaction. As a result, maximum interfacial adhesion and effective load transfer can be achieved. Significant reinforcement was obtained for such nanocomposites, with a 76% increase in tensile strength and a 62% improvement of Young's modulus by the addition of 0.7 wt% dispersed graphene oxide to PVA in water. Another potential of graphene oxide in the reinforcement of polymers is demonstrated with the use of carbon fibres. Graphene oxide sheets were used in order to prepare stable graphene oxide-modified carbon fibre sizing, which later were used for the preparation of carbon fibre composites. The interfacial shear strength of the graphene oxide-modified carbon fibre was improved by up to 70.9%. The tensile strength and tensile modulus of the carbon fibre-reinforced composites were found to be higher as compared to normal composites (Zhang et al., 2012).

Luong et al. (2011b) explored a variety of reduced graphene oxide dispersions controllably added to Amine-functionalised nanofibrillated cellulose (A-NFC) resulting in mixtures with graphene contents ranging from 0.1 to 10 wt%. The modulus of the reduced graphene oxide/A-NFC composite paper was enhanced to 6.4 GPa for a 0.3-wt% of graphene content, while it was 7.1 GPa for a content of 1 wt%. The tensile strength of the composite with 0.3 wt% incorporated graphene was enhanced to 232 MPa. However, while increasing the graphene content up to 1 wt% the tensile strength remained almost unchanged. The excellent reinforcement is attributed to the good dispersion of both reduced graphene oxide sheets and A-NFC fibrils in the composites, as well as the strong chemical/physical interactions between them. The unchanged tensile properties at higher graphene contents may be due to the agglomeration of graphene sheets in the composite. The rough fracture surfaces reveals a strong adhesion between the cellulose fibrils and graphene sheets, which may attributed to strong hydrogen bonding and chemical bonding. This favours stress transfer to both components, as also suggested in other studies (Luong et al., 2011a). Unlike graphene oxide, reduced graphene sheets are highly hydrophobic, which makes dispersion in polar solvents challenging, due to lack of functional groups (Min et al., 2008). Polymer functionalisation of the graphene surface is often proposed in order to optimise the dispersion of reduced graphene oxide (Balazs et al., 2006). Covalently-bonded, polymer-functionalised graphene hybrid material can be prepared by combining a diazonium addition reaction with atom transfer radical polymerisation (Fang et al., 2009). Such functionalised graphene has been exploited for the reinforcement of polystyrene. It was found that the functionalisation enhanced the polymer grafting efficiency by 82%, due to reduced aggregation. The functionalised graphene sheets comprised single-layer and multi-layer with lateral sizes in the range 20–40 nm. It was possible to improve the stress–strain curve of the nanocomposite with 0.9 wt% of functionalised graphene addition from 1.45 GPa for the pristine polymer up to 2.228 GPa (an improvement of 69.5%). The Young's modulus was enhanced by 57.2%. The improvements were attributed to the efficient load transfer between graphene sheets and the polymer matrix. A similar approach was used to prepare isocyanate-treated graphite oxide. The as-produced material in a dispersion was added to polyimides resulting in a graphene-filled polymer (Luong, 2011a). While graphite oxide is insoluble in dipolar aprotic DMF solvent, isocyanate-treated graphite oxide is nicely dispersed and avoids possible agglomeration that can reduce the reinforcement ability of the graphene nanomaterial. Indeed, the produced polymer exhibited a 30% improvement in its Young's modulus for an incorporation of 0.38 wt% isocyanate-treated graphite oxide as compared to a pure polymer film. In addition, the tensile strength jumped to 131 MPa from 122 MPa. However, the incorporation of higher isocyanate-treated graphite oxide contents led to little improvement in the Young's modulus and a reduction in the tensile strength, although these values remain higher than that for the pure PI film. The rough fractured surfaces for the graphene-filled polymer compared to the pristine polymer suggest good interfacial adhesion and compatibility between the polymer matrix and functionalised graphitic sheets, which favours stress transfer from the polymer matrix to the graphene sheets, leading to an improvement in the tensile strength and modulus.

The epoxy groups that exist in graphene oxide sheets make them an ideal filler of epoxy resin, which also contains epoxy groups. Based on the principle of dissolution in a similar material structure, homogeneous graphene-filled epoxy composites were explored by Yang et al. (2009). In this study, a graphene oxide dispersion (in water) was mixed with epoxy resin, and the mixtures were vigorously shaken or stirred for a period of time in an oil bath. After standing for another period of time, the water frees itself rising to the top while the epoxy filled with graphene oxide forms at bottom of the container. Shaking and stirring of the remaining mixture after removing the water allows one to obtain a homogeneous graphene-oxide filled epoxy resin composite. The obtained composite demonstrated a remarkable increase in compressive failure strength and toughness by 48.3 and 1185.2%, respectively. This improvement is attributed to distortions introduced by the oxygen functionalisation generating defects due to the thermal treatment of the precursor graphene oxide/epoxy resin which creates a wrinkled topology owing to the small thickness of the chemically converted graphene oxide sheets. Moreover, the pendant oxygen-containing groups across the graphene oxide surfaces may form covalent bonds with the epoxy resin component. Again, the resultant surface roughness and chemistry enhance mechanical interlocking with the polymer chains and better adhesion (Ramanathan et al., 2008). In another work, graphene-oxide sheets (mono and few-layer) were introduced in small amounts (<1 wt%) to a commercially available thermo-setting epoxy system (bisphenol A/F diglycidyl ether blend) (Bortz et al., 2012). The addition of the graphene oxide enhanced the tensile modulus by 12% for a 0.1 wt% loading. However, higher graphene-oxide loading lead to a lower increase in tensile modulus than that obtained for a 0.1 wt% loading. The improvement due to the introduction of the graphene oxide lies in the composites fracture toughness. The critical stress intensity factor and the critical strain energy release rate were enhanced by approximately 28–63%, and 29–111%, respectively. It was also shown that graphene oxides deflect propagating crack fronts which introduces off-plane loading. Thus, new fractures surfaces increasing the required strain energy for the continuation of fracture were generated (Faber and Evans 1983).

Multiscale reinforcement using fibres with GPLs (0.2 wt%) in the matrix or on the surface of the fibres, was shown to enhance the fatigue life of fibre-reinforced epoxy composites. GPLs were introduced into the glass-fiber/epoxy composites, either by infiltrating into the epoxy resin matrix, or spraying graphene platelets directly onto the glass fibres, prior to curing the composites. The flexural bending fatigue life was enhanced by three orders of magnitude as compared to conventional microfibre-reinforced polymer composites. The investigators argued that the micrometre-size dimensions, high aspect ratio, and two-dimensional sheet geometry of the GPLs make them effective in deflecting cracks in bending/shear. In addition, hydrogen-bonding interactions (Ramanathan et al., 2008) or an enhanced nanofiller-polymer mechanical interlocking due to the wrinkled morphology of graphene (Rafiee et al., 2009) are additional factors that can contribute to composite reinforcement.

A remarkable improvement in mechanical properties of epoxy composites was achieved using GNRs as a nanofiller. GNRs were prepared by unzipping multiwall carbon nanotubes through a solution-based oxidative mechanism, followed by chemical reduction and thermal reduction. The as-produced ribbons had width of 50–100 nm and lengths of 1–10 μm. The prepared GNRs were uniformly dispersed in an epoxy resin by ultrasonication and high-speed shear mixing, and then cured to generate dogbone-shaped coupons for uniaxial tensile investigations. The mechanical properties for the produced nanocomposite were enhanced relative to the pristine epoxy. The tensile strength was increased by up to ~22%, while the Young's modulus showed an increase of over 30%, while its toughness was only marginally improved. The ductility of the nanocomposite was reduced by some 10–15% which was argued to arise from stress concentration in the vicinity of the nanofiller and/or agglomeration of ribbons might cause defects that act as seed points for crack initiation and premature fracture. The high defect percentage of the GNRs was suggested to contribute to the enhancement of the mechanical properties, as it provides more sites for chemical interaction with the host epoxy (Rafiee et al., 2010).

Reduced graphene oxide can also be used as a nanofiller for polymer composites. Thermally reduced graphene oxide consisting of stacked individual graphene sheets with dimensions of approximately 2.5 × 1.5 µm were also exploited by Rafiee et al. (2009) as a reinforcement material in an epoxy polymer. The weight fraction of the thermally reduced graphene oxide was limited to 0.1% in order to ensure a uniform dispersion. The resultant composite exhibited an improved tensile strength of about 40%, while the Young's modulus was enhanced by around 31% as compared to the baseline epoxy. Steurer et al. (2009) also investigated thermally reduced graphene oxides but with content additions to the base polymer of up to 12 wt%. Various polymers were investigated, including isotactic poly(propylene) (iPP), poly(styrene-co-acrylonitrile) (SAN), polyamide 6 (PA6) and polycarbonate (PC). The Young's modulus was improved strongly for all the nanocomposites. For example a Young's modulus of 4120 MPa (around 80% improvement) was achieved for a 12.5 wt% reduced graphene oxide loading in SAN. The work suggested functional groups on reduced graphene oxide are essential for promoting graphene dispersion and interfacial adhesion of material to the polymer matrix, as well as improve the grafting efficacy.

6.7.3. IN SITU POLYMERIZATION

In contrast to the above discussed works, where the graphene oxide is reduced prior the mixing with the host polymer, in the case of in situ polymerisation the graphene reduction takes place after mixing with the host polymer either by exposing the fabricated polymer to a high temperature (Shen et al., 2011; Tang et al., 2012) or chemically by introducing a chemical agent initiating the polymerisation.

The in situ thermal reduction of graphene oxide in a polyvinylidene fluoride (PVDF) polymer matrix was demonstrated by Tang et al. (2012). First a graphene oxide – N,N-dimethylformamide colloidal suspension was prepared. Polyvinylidene fluoride polymer was then added to the colloidal suspension leading to the formation of highly stable suspensions that avoid the need for aggressive sonication or high shear mixing prior to polymerisation. The graphene-oxide sheets become immobilised in the rigid polymer matrix upon evaporating the solvent leaving a PVDF/graphene oxide film. The technique prevents aggregation of the graphene oxide flakes. Hot pressing of the produced film at 200 °C for a period of time leads to the in situ reduction of the graphene oxide yielding nanocomposites with isolated single layers of reduced-graphene oxide. Likewise, polystyrene was dissolved by vigorous stirring in a colloid solution of graphene in THF to form a nanocomposite (Shen et al., 2011). THF was used as the solvent because of its high graphene solubility and low toxicity. The product was precipitated in excess ethanol and then dried at 80 °C under vacuum for 48 h. Finally, the polymer-graphene composite was melt blended at 210 °C allowing the polymer chains to enter into the gaps between graphene sheets. The invetigators speculate π–π stacking between the aromatic system of π-electrons of both the polymer and graphene in the composite occurs during the melt blending process.

In a variation, a route using graphene oxide treated under high-temperature in situ reduction and an imidisation treatment step for the fabricating graphene-polyimide nanocomposite with different graphene loadings was developed by Chen et al. (2010). The procedures were shown to partially reduce the graphene oxide into functionalised graphene sheets. The majority of the functionalised graphene sheets were successfully exfoliated in the matrix and were uniformly dispersed throughout the polymeric matrix with high alignment along the nanocomposite surface direction, resulting in a high-performance composite. Strong interactions between the graphene functional groups and PAA (poly-imide precursor) are also believed to aid homogeneous dispersion, in addition the functionalisation and sonication steps (Zhu et al., 2006). The mechanical properties of the fabricated nanocomposite were improved significantly by the incorporation of only 0.5 wt% of functionalised graphene sheets. A similar approach was used in order to fabricate chemically modified graphene poly-imide nanocomposites (Huang et al., 2012). The in situ thermal addition step of modified graphene contains interchain and interfacial cross-linking with pol-yimide, providing interfacial covalent cross-links ($\leq$1-wt% loading). The tensile strength of the produced composite increased by 30%, while the Young's modulus increased by 46%.

REFERENCES

Balazs, A.C., Emrick, T., Russell, T.P., 2006. Nanoparticle polymer composites: where two small worlds meet. Science 314, 1107.

Bortz, D.R., Heras, E.G., Martin-Gullon, I., 2012. Impressive fatigue life and fracture toughness improvements in graphene oxide/epoxy composites. Macromolecules 45, 238–245.

Cai, D.Y., Yusoh, K., Song, M., 2009. The mechanical properties and morphology of a graphite oxide nanoplatelet/polyurethane composite. Nanotechnology 20, 1–5. 085712.

Chen, D., Zhu, H., Liu, T., 2010. In situ thermal preparation of polyimide nanocomposite films containing functionalized graphene sheets. ACS Appl. Mater. Interfaces 2, 3702–3708.

Faber, K., Evans, A., 1983. Crack deflection processes-I. Theory. Acta Metall. 31, 565–576.

Fang, M., Wang, K., Lu, H., Yang, Y., Nutt, S., 2009. Covalent polymer functionalization of graphene nanosheets and mechanical properties of composites. J. Mater. Chem. 19, 7098–7105.

Gong, L., Kinloch, I.A., Young, R.J., Riaz, I., Jalil, R., Novoselov, K.S., 2010. Interfacial stress transfer in a graphene monolayer nanocomposite. Adv. Mater. 22, 2694–2697.

Hernandez, Y., Nicolosi, V., Lotya, M., Blighe, F.M., Sun, Z., De, S., Mc-Govern, I.T., HollandByrne, B., Byrne, M., Gun'Ko, Y.K., Boland, J.J., Niraj, P., Duesberg, G., Krishnamurthy, S., Goodhue, R., Hutchison, J., Scardaci, V., Ferrari, A.C., Coleman, J.N., 2008. High-yield production of graphene by liquid-phase exfoliation of graphite. Nat. Nanotechnol. 3, 563–568.

Hirata, M., Gotou, T., Horiuchi, S., Fujiwara, M., Ohba, M., 2004. Thin-film particles of graphite oxide 1: high-yield synthesis and flexibility of the particles. Carbon 42, 2929–2937.

Huang, T., Lu, R., Su, C., Wang, H., Guo, Z., Liu, P., Huang, Z., Chen, H., Li, T., 2012. Chemically modified graphene/polyimide composite films based on utilization of covalent bonding and oriented distribution. ACS Appl. Mater. Interfaces 4, 2699–2708.

Khan, U., May, P., O'Neill, A., Coleman, J.N., 2010. Development of stiff, strong, yet tough composites by the addition of solvent exfoliated graphene to polyurethane. Carbon 48, 4035–4041.

Kulkarni, D.D., Choi, I., Singamaneni, S.S., Tsukruk, V.r V., 2010. Graphene Oxide–Polyelectrolyte Nanomembranes. ACS Nano 4, 4667–4676.

Lahiri, D., Dua, R., Zhang, C., Socarraz-Novoa, I. d., Bhat, A., Ramaswamy, S., Agarwal, A., 2012. Graphene nanoplatelet-induced strengthening of UltraHigh molecular weight polyethylene and biocompatibility in vitro. ACS Appl. Mater. Interfaces 4, 2234–2241.

Lee, Y.R., Raghu, A.V., Jeong, H.M., Kim, B.K., 2009. Properties of waterborne polyurethane/functionalized graphene sheet nanocomposites prepared by an in situ method. Macromol. Chem. Phys. 210, 1247–1254.

Liang, J., Huang, Y., Zhang, L., Wang, Y., Ma, Y., Guo, T., Chen, Y., 2009. Molecular-level dispersion of graphene into poly(vinyl alcohol) and effective reinforcement of their nanocomposites. Adv. Funct. Mater. 19, 2297–2302.

Liu, Z.F., Liu, Q., Huang, Y., Ma, Y.F., Yin, S.G., Zhang, X.Y., Sun, W., Chen, Y.S., 2008. Organic photovoltaic devices based on a novel acceptor material: graphene. Adv. Mater. 20, 3924–3930.

Luong, N.D., Hippi, U., Korhonen, J.T., Soininen, A.J., Ruokolainen, J., Johansson, L.-S., Nam, J.-D., Sinh, L.H., Seppälä, J., 2011a. Enhanced mechanical and electrical properties of polyimide film by graphene sheets via in situ polymerization. Polymer 52, 5237–5242.

Luong, N.D., Pahimanolis, N., Hippi, U., Korhonen, J.T., Ruokolainen, J., Johansson, L.-S., Nam, J.-D., Seppälä, J., 2011b. Graphene/cellulose nanocomposite paper with high electrical and mechanical performances. J. Mater. Chem. 21, 13991–13998.

Min, Y.J., Akbulut, M., Kristiansen, K., Golan, Y., Israelachvili, J., 2008. The role of interparticle and external forces in nanoparticle assembly. Nat. Mater. 7, 527–538.

Nawaz, K., Khan, U., Ul-Haq, N., May, P., O'Neill, A., Coleman, J.N., 2012. Observation of mechanical percolation in functionalized graphene oxide/elastomer composites. Carbon 05, 029. http://dx.doi.org/.org/10.1016/j.carbon.

Rafiee, M.A., Lu, W., Thomas, A.V., Zandiatashbar, A., Rafiee, J., Tour, J.M., Koratkar, N.A., 2010. Graphene nanoribbon composites. ASC Nano 4, 7415–7420.

Rafiee, M.A., Rafiee, J., Wang, Z., Song, H., Yu, Z.-Z., Koratkar, N., 2009. Enhanced mechanical properties of nanocomposites at low graphene content. ACS Nano 3, 3884–3890.

Ramanathan, T., Abdala, A.A., Stankovich, S., Dikin, D.A., Herrera-Alonso, M., Piner, R.D., Adamson, D.H., Schniepp, H.C., Chen, X., Ruoff, R.S., Nguyen, S.T., Aksay, I.A., Prud'Homme, R.K., Brinson, L.C., 2008. Functionalized graphene sheets for polymer nanocomposites. Nat. Nanotechnol. 3, 327–331.

Schniepp, H.C., Li, J.L., McAllister, M.J., Sai, H., Herrera-Alonso, M., Adamson, D.H., Prud'homme, R.K., Car, R., Saville, D.A., Aksay, I.A., 2006. Functionalized single graphene sheets derived from splitting graphite oxide. J. Phys. Chem. B 110, 8535–8539.

Shen, B., Zhai, W., Chen, C., Lu, D., Wang, J., Zheng, W., 2011. Melt blending in situ enhances the interaction between polystyrene and graphene through π–π stacking. ACS Appl. Mater. Interfaces 3, 3103–3109.

Stankovich, S., Dikin, D.A., Dommett, G.H.B., Kohlhaas, K.M., Zimney, E.J., Stach, E.A., Piner, R.D., Nguyen, S.T., Ruoff, R.S., 2006a. Graphene-based composite materials. Nature 442, 282–286.

Stankovich, S., Piner, R.D., Nguyen, S.T., Ruoff, R.S., 2006b. Synthesis and exfoliation of isocyanate-treated graphene oxide nanoplatelets. Carbon 44, 3342–3347.

Steurer, P., Wissert, R., Thomann, R., Muelhaupt, R., 2009. Functionalized graphenes and thermoplastic nanocomposites based upon expanded graphite oxide. Macromol. Rapid Commun. 30, 316–327.

Szabo, T., Szeri, A., Dekany, I., 2005. Composite graphitic nanolayers prepared by self-assembly between finely dispersed graphite oxide and a cationic polymer. Carbon 43, 87–94.

Tang, H., Ehlert, G.J., Lin, Y., Sodano, H.A., 2012. Highly efficient synthesis of graphene nanocomposites. Nano Lett. 12, 84–90.

Trappe, V., Prasad, V., Cipelletti, L., Segre, P.N., Weitz, D.A., 2001. Jamming phase diagram for attractive particles. Nature 411, 772–775.

Walker, L.S., Marotto, V.R., Rafiee, M.A., Koratkar, N., Corral, E.L., 2011. Toughening in graphene ceramic composites. ACS Nano 5 (4), 3182–3190.

Wang, S., Chia, P.-J., Chua, L.-L., Zhao, L.-H., Png, R.-Q., Sivaramakrishnan, S., Zhou, M., Goh, R.G.-S., Friend, R.H., Wee, A.T.-S., Ho, P.K.-H., 2008. Band-like transport in surface-functionalized highly solution-processable graphene nanosheets. Adv. Mater. 20, 3440–3446.

Xu, Y.X., Bai, H., Lu, G.W., Li, C., Shi, G.Q., 2008. Flexible graphene films via the filtration of water-soluble noncovalent functionalized graphene sheets. J. Am. Chem. Soc. 130, 5856–5857.

Yang, H., Shan, C., Li, F., Zhang, Q., Han, D., Niu, L., 2009. Convenient preparation of tunably loaded chemically convertedgraphene oxide/epoxy resin nanocomposites from graphene oxide sheets through two-phase extraction. J. Mater. Chem. 19, 8856–8860.

Yavari, F., Rafiee, M.A., Rafiee, J., Yu, Z. Z., Koratkar, N., 2010. Dramatic increase in fatigue life in hierarchical graphene composites. ACS Appl. Mater. Interfaces 2, 2738–2743.

Zhang, X., Fan, X., Yan, C., Li, H., Zhu, Y., Li, X., Yu, L., 2012. Interfacial microstructure and properties of carbon fiber composites modified with graphene oxide. ACS Appl. Mater. Interfaces 4, 1543–1552.

Zhu, B.K., Xie, S.H., Xu, Z.K., Xu, Y.Y., 2006. Preparation and properties of the polyimide/multiwalled carbon nanotubes (MWNTs) nanocomposites. Compos. Sci. Technol. 66, 548–554.

Index

Note: Page numbers followed by "f" and "t" indicate figures and tables respectively

Printed and bound by CPI Group (UK) Ltd, Croydon, CR0 4YY

13/05/2025

01869555-0001